Introduction to Topological Quantum Matter & Quantum Computation

What is "topological" about topological quantum states? How many types of topological quantum phases are there? What is a zero-energy Majorana mode, how can it be realized in a solid-state system, and how can it be used as a platform for topological quantum computation? What is quantum computation and what makes it different from classical computation?

Addressing these and other related questions, Introduction to Topological Quantum Matter & Quantum Computation provides an introduction to and a synthesis of a fascinating and rapidly expanding research field emerging at the crossroads of condensed matter physics, mathematics, and computer science. Providing the big picture and emphasizing two major new paradigms in condensed matter physics – quantum topology and quantum information – this book is ideal for graduate students and researchers entering this field, as it allows for the fruitful transfer of ideas amongst different areas, and includes many specific examples to help the reader understand abstract and sometimes challenging concepts. It explores the topological quantum world beyond the well-known topological insulators and superconductors and unveils the deep connections with quantum computation. It addresses key principles behind the classification of topological quantum phases and relevant mathematical concepts and discusses models of interacting and noninteracting topological systems, such as the toric code and the p-wave superconductor. The book also covers the basic properties of anyons, and aspects concerning the realization of topological states in solid state structures and cold atom systems.

Topological quantum computation is also presented using a broad perspective, which includes elements of classical and quantum information theory, basic concepts in the theory of computation, such as computational models and computational complexity, examples of quantum algorithms, and key ideas underlying quantum computation with anyons. This new edition has been updated throughout, with exciting new discussions on crystalline topological phases, including higher-order topological insulators; gapless topological phases, including Weyl semimetals; periodically-driven topological insulators; and a discussion of axion electrodynamics in topological materials.

Key Features:

- Provides an accessible introduction to this exciting, cross-disciplinary area of research.
- Fully updated throughout with new content on the latest result from the field.
- Authored by an authority on the subject.

Tudor Stanescu is a professor of Condensed Matter Theory at West Virginia University, USA. He received a B.S. in Physics from the University of Bucharest, Romania, in 1994 and a Ph.D. in Theoretical Physics from the University of Illinois at Urbana Champaign in 2002. He was a Postdoctoral Fellow at Rutgers University and at the University of Maryland from 2003 to 2009. He joined the Department of Physics and Astronomy at West Virginia University in Fall 2009. Prof. Stanescu's research interests encompass a variety of topics in theoretical condensed matter physics including topological insulators and superconductors, topological quantum computation, ultra-cold atom systems in optical lattices, and strongly correlated materials, such as, for example, cuprate high-temperature superconductors. His research uses a combination of analytical and numerical tools and focuses on understanding the emergence of exotic states of matter in solid state and cold atom structures, for example, topological superconducting phases that host Majorana zero modes, and on investigating the possibilities of exploiting these states as physical platforms for quantum computation.

Introduction to Topological Quantum Matter & Quantum Computation

Second Edition

Tudor D. Stanescu

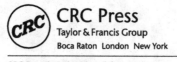

CRC Press
Taylor & Francis Group
Boca Raton London New York

CRC Press is an imprint of the
Taylor & Francis Group, an **informa** business

Designed cover image: Irina Stanescu

Second edition published 2024
by CRC Press
2385 NW Executive Center Drive, Suite 320, Boca Raton FL 33431

and by CRC Press
4 Park Square, Milton Park, Abingdon, Oxon, OX14 4RN

CRC Press is an imprint of Taylor & Francis Group, LLC

© 2025 Tudor D. Stanescu

First edition published by CRC Press 2020

Library of Congress Cataloging-in-Publication Data
Names: Stanescu, Tudor D., 1967- author.
Title: Introduction to topological quantum matter and quantum computation / Tudor D. Stanescu.
Other titles: Introduction to topological quantum matter & quantum computation
Description: Second edition. | Boca Raton, FL : CRC Press, 2024. | Identifiers: LCCN 2023057455 | ISBN 9781032126524 (hardback) | ISBN 9781032127446 (paperback) | ISBN 9781003226048 (ebook)
Subjects: LCSH: Quantum theory--Data processing. | Topology. | Quantum computing.
Classification: LCC QC174.17.D37 S73 2024 | DDC 530.1201/514--dc23/eng/20240324
LC record available at https://lccn.loc.gov/2023057455

ISBN: 978-1-032-12652-4 (hbk)
ISBN: 978-1-032-12744-6 (pbk)
ISBN: 978-1-003-22604-8 (ebk)

DOI: 10.1201/9781003226048

Typeset in CMR10
by KnowledgeWorks Global Ltd.

Publisher's note: This book has been prepared from camera-ready copy provided by the authors

To Elvira

Contents

Preface to the second edition

In the seven years since the first edition of this book, quantum topology and quantum computation have evolved into mainstream research fields and continue to attract massive interest and a lot of research effort. The recent developments, including theoretical advances, new materials and enhanced characterization, and improved device engineering, are tremendous, yet no book can do justice to such a wide range of new results. However, these advancements underscore a basic idea that should be clearly conveyed to everyone interested in this field: quantum topology and quantum information are powerful, interconnected paradigms that provide a new perspective regarding the organization of matter and, possibly, the fundamental principles of physics. The second edition of this book was prepared with this idea in mind; hopefully, the additions and the restructuring of the material will help communicate it better. It is my belief that, particularly when considered within the context of condensed matter physics, these paradigms have the potential to unveil new fundamental physics. It is also my belief that the impact of the major results in this area will be profound and intellectually rewarding, but probably slow and rather subtle; I do not expect something comparable to the brutal efficiency of the emerging AI technologies.

Thus, the new edition contains a discussion of topological crystalline phases, including so-called higher order topological phases, and clarifies related concepts, such as fragile topology and weak topology. To further emphasize the wide range of systems in which topology plays a key role, we also discuss gapless topological phases, including the so-called Weyl semimetals, as well as out-of-equilibrium, periodically-driven systems, particularly the so-called Floquet topological insulators. A chapter on axion electrodynamics in topological quantum systems was also added, to better illustrate the connection between topology and non-trivial emergent physics. Several necessary clarifications — e.g., the role of disorder in destabilizing topological superconducting phases — were added throughout the book. The old material was partially restructured, so that the first nine chapters are concerned with the basic theory, classification, and modeling of topological quantum phases, while all discussions related to quantum information and computation are now included in the last three chapters. Nonetheless, there are multiple connections between the two parts, starting with the first section, which introduces the idea of topology in connection with knots representing the world lines of anyonic quasiparticles — a theme that is expanded in the last chapter in the context of topological quantum computation. Finally, the list of references was updated and expanded. And, of course, many — but definitely not all — errors and typos were corrected.

Tudor D. Stanescu
Morgantown

Preface to the first edition

Quantum topology and quantum computation are relatively new, exciting, and rapidly developing fields that have attracted a lot of interest in recent years. Perhaps more importantly, these areas host some of the most significant research efforts to identify new paradigms for understanding physical reality. From this perspective, they should be regarded as intimately related aspects of an interdisciplinary effort that integrates ideas from physics, mathematics, and computer science and aims to understand how quantum matter is organized and how the Universe is processing information. Eventually, this effort could lead to a refinement, if not a revolution, of the fundamental physical theories.

The main goal of this book is to provide the newcomer to the field with a broad perspective of this vast and diverse landscape. Having such a perspective could facilitate a productive transfer of ideas among different related areas, could help in avoiding misconception and false targets, and, ultimately, could lead to a better understanding of the underlying physics. The book is intended as a sort of "travel guide" for students who plan to do research in this field and students who have worked for a while on a specific research problem but lack a broader perspective. No one is able to "visit" all the interesting places discussed (i.e., to become an expert in all those related areas). However, it is critical, I believe, to have an idea about their existence, their significance, and their relationship with the object of one's immediate interest. After all, climbing a summit without being able to see the panoramic view reduces it to a purely technical exercise that is less likely to lead to future success.

Being mainly about the big picture, this book is anything but comprehensive. Each chapter addresses a topic that deserves its own specialized book. In fact, these books exist and some of them are listed in the reference section. The list of references is relatively long, but again, far from comprehensive; my intention was to point the reader in the right direction. Actually reaching specific places (i.e., mastering various concepts and techniques) requires a lot more effort. After all, just reading the travel guide will not get you to the actual site.

Many of the topics discussed in this book have generated a lot of excitement in recent years. Unfortunately, with excitement oftentimes come exaggeration and false hope. I tried to avoid these pitfalls, as much as possible, and provide warnings when appropriate. I also tried to avoid oversimplifications, which are extremely tempting especially when dealing with technically complex problems. All the material covered in this book is introductory, but not necessarily elementary. Some concepts and techniques become (or maybe appear) simple only after establishing a certain familiarity with that particular topic. Therefore, readers with different backgrounds may find certain (unfamiliar) sections as particularly difficult. If it is unbearable, you can probably skip them; however, spending some time and effort with them will definitely bring a reward.

I assume that writing a book always involves an intense personal experience. For me, it was, in some sense, similar to self-taught skiing. The weather is nice, the view is outstanding, other skiers graciously pass by; it cannot be too bad. You overcome your fear and let go. What follows is a rapid succession of emotions which mixes genuinely fearing for your life and cursing the moment you decided on such a foolish adventure with feeling some hard-to-define excitement.

Ultimately, however, a book is first and foremost a collective endeavor: there are the people whose knowledge and ideas have shaped the author, making the writing of the book possible; there are those who shared some of the effort and frustration involved in this process; and, finally, there are the readers who will taste the fruits of this labor. I can only hope that someone will benefit from this work, making it meaningful and worthwhile. What is clear is that, over the years, I have benefited from the support and example of many people who have shaped my career and my views; they deserve my unconditional gratitude. In particular, I am deeply indebted to Gheorghe Ciobanu, Radu Lungu, Philip Phillips, Eduardo Fradkin, Gabriel Kotliar, Victor Galitski, and Sankar Das Sarma. I also thank my collaborators, Sumanta Tewari, Jay Deep Sau, and Roman Lutchyn, with whom I had many stimulating discussions on research problems related to some of the topics covered in this book, and my colleagues, Alan Bristow and Mikel Holcomb, for reading some of the chapters and providing valuable suggestions. This book would not have been written without my editor, Francesca McGowan, luring me into this project, and then kindly and patiently following its progress.

This work would not have been completed without the love and support of my family- my wife Elvira, and my children Alexandru and Irina. My apologies for depriving them from shared time that rightfully should have been theirs. I also thank my mom for encouraging me to finish this work and for her love of books.

Tudor D. Stanescu
College Park

I

Topological Quantum Phases: Basic Theory, Classification, and Modeling

CHAPTER 1

Topology and Quantum Theory

TOPOLOGY IS WOVEN INTO THE SUBTLE FABRIC OF quantum theory. Topological concepts and ideas developed starting in the 1980s by the high energy physics and quantum gravity communities, first viewed by many as nothing but aesthetically appealing abstract possibilities, are central components of modern condensed matter physics and play a key role in the field of quantum information and quantum computation. The discovery of integer and fractional quantum Hall effects and, more recently, of topological insulators and superconductors has brought topological quantum matter in the laboratory, with the perspective of realizing functional platforms for topological quantum computation. Rooted in a solid mathematical formalism and spanning its relevance from fundamental questions in quantum gravity and quantum information to practical aspects regarding new materials and platforms for quantum computation, quantum topology is positioned at a fertile confluence of mathematics, physics, and computer and information science. This book focuses on the aspects of quantum topology that are most relevant from the perspective of condensed matter physics and its relation to quantum computation. We start our journey by sketching the relation between topology and quantum physics and summarizing a few basic mathematical concepts that will be used throughout this work. Beyond dry formal definitions, our main goal is to grasp the delicate root that makes the topological quantum world utterly strange and beautiful and follow it into the realm of condensed matter many-body systems.

1.1 QUANTUM AMPLITUDES AND KNOT INVARIANTS

Consider a two-dimensional system initially in its ground state, with excited states corresponding to the presence of quasiparticles localized throughout the system. Note that the ground state can be viewed as the "vacuum," i.e., the

DOI: 10.1201/9781003226048-1

state with no quasiparticles. Assume now that two successive pair-creation events lead to the emergence of two particle–antiparticle pairs from the vacuum. What is the probability that the quasiparticles annihilate back to the vacuum? The corresponding probability amplitude can be determined using the path integral formalism [167, 350] as a sum over all "paths" associated with space–time processes like those shown in Figure 1.1. In general, the contribution associated with a specific path depends on details regarding the geometry of the world lines and the Hamiltonian that describes the dynamics of the system. Remarkably, there exist so-called "topological quantum systems" for which the amplitudes associated with specific space–time processes depend only on the topology of the world lines. Such systems are two-dimensional, examples of *topological quantum matter*.

But what do we mean by "topology" in relation to the world lines representing different space–time processes, e.g., those shown in Figure 1.1? In essence, topology refers to the properties of an object that remain invariant under continuous transformations. For example, the mathematical objects corresponding to the world lines in Figure 1.1 are *knots*, each representing an embedding (without intersections) of two circles into a three-dimensional manifold.[1] Two knots are *topologically equivalent* if they can be deformed smoothly into each other without cutting any strand. For example, the knot in Figure 1.1(a) is topologically equivalent to every other knot consisting of two separate rings (regardless of their detailed geometry), but it is topologically inequivalent to the knot in panel (b), which consists of two linked rings. The topological equivalence relation[2] induces equivalence classes on the set of knots, the elements of each class being mutually equivalent, but inequivalent to elements from a different class. The remarkable property of topological quantum matter mentioned above is that the quantum amplitudes associated with space–time processes involving its emergent quasiparticles are the same for all paths (i.e., world lines) that belong to a given topological equivalence class. Note that this is strikingly different from the behavior of nontopological systems, where the quantum amplitudes associated with space–time processes involving the creation and annihilation of particle-hole pairs depend on a multitude of system- and path-dependent details.

Based on the examples in Figure 1.1, it would appear that establishing the topological equivalence (or inequivalence) of two knots is an easy task. However, in general this is not the case, since the complexity of the problem increases exponentially with the number of strand crossings (which are no more than two in Figure 1.1). In this context, a useful concept is that of a *topological invariant* – a property $P(x)$ that takes the same value for all elements x within a given equivalence class. In this language, the quantum

---

[1] Technically, a *knot* is the embedding of a single circle (i.e., it consists of a single strand), while objects made of multiple strands are called *links*. Here, we ignore this distinction.

[2] An *equivalence relation* ($\sim$) is a binary relation that is: (i) reflexive, $a \sim a$; (ii) symmetric, $a \sim b$ if and only if $b \sim a$; (iii) transitive, if $a \sim b$ and $b \sim c$ then $a \sim c$. The *equivalence classes* of a set X are the subsets of X that contain all mutually equivalent elements.

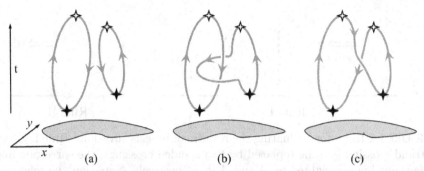

(a)   (b)   (c)

FIGURE 1.1 Topologically distinct space–time processes showing the world lines of two particle–antiparticle pairs that emerge from the vacuum of a two-dimensional system (shaded region at the bottom) and reannihilate to the vacuum. The filled (empty) stars mark the pair creation (annihilation) events. The space–time paths (world lines) with arrows pointing backward in time correspond to antiparticles.

amplitude associated with space–time processes involving the quasiparticles of a topological quantum system is a topological invariant.

As a simple example of a knot (topological) invariant, let us consider the so-called *Kauffman bracket invariant*, which is a quantity expressed in terms of integer powers of a scalar variable A. Examples involving specific values of A are provided in Chapter 2 (page 61). To construct the invariant, we first "decompose" the knot into sums of diagrams containing (weighted) simple loops by eliminating all strand crossings using rule I given in Figure 1.2, then we ascribe a factor $d = -A^2 - A^{-2}$ to each simple loop (rule II). Thus, the knot in Figure 1.1(a), which consists of two simple loops, has a Kauffman bracket invariant equal to d^2. For the knot in Figure 1.1(c), we first eliminate the crossing using rule I and we obtain a sum of two diagrams, one consisting of two simple loops (weighted by A) and the other consisting of a single loop (weighted by A^{-1}). The corresponding Kauffman bracket invariant is $Ad^2 + A^{-1}d = -A^3d$. Finally, after eliminating the two crossings in Figure 1.1(b), we obtain four weighted diagrams and a Kauffman bracket invariant equal to $2d - d^3$. Note that for a knot containing N crossings the number of weighted diagrams obtained after applying rule I is 2^N, which makes the evaluation of the Kauffman bracket invariant a hard problem on a classical computer.[3]

This introductory discussion highlights one manifestation of topological quantum matter – the quantum amplitudes associated with space–time processes involving the creation and annihilation of quasiparticles are (topological) knot invariants, the "knots" corresponding to the world lines of the quasiparticles. However, a topological quantum system is "topological" even in its ground state, i.e., in the absence of quasiparticles. Our task will be to

---

[3]See Chapter 11 (page 339) for a formal introduction to computational complexity. Also, see the discussion in Chapter 2 (page 65) on the evaluation of knot invariants using topological quantum computation.

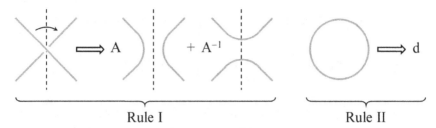

FIGURE 1.2 Rules for evaluating the Kauffman bracket invariant. Rule I: Every strand crossing is to be replaced by two avoided crossings, the corresponding diagrams being weighted by A and A^{-1}, respectively. Note that the reference dashed line (which is not part of the knot) can form an arbitrary angle with the vertical axis, the (avoided) crossings being rotated correspondingly. Rule II: Ascribe a factor $d = -A^2 - A^{-2}$ to each simple loop.

identify the corresponding mathematical structures having nontrivial topological properties and to characterize them using appropriate topological invariants. For gapped systems, for example, all ground states that can be continuously connected to one another (by varying the system parameters) without closing the bulk spectral gap are topologically equivalent and will be characterized by the same value of the topological invariant. The corresponding equivalence class constitutes a *topological phase*. In the remainder of this chapter, we briefly summarize a few basic mathematical concepts (from the areas of topology and differential geometry) that we will later use to characterize topological quantum systems. We also discuss ideas related to the concept of *Berry phase*, which will play an important role in defining certain types of topological invariants.

We should caution the reader that the following sections, particularly Sections 1.2 and 1.5, are technical and rather dry. Our goal here is twofold. On the one hand, the area of quantum topology and, more generally, the key areas involved in the ongoing "quantum transformation" (including, e.g., quantum communication, sensing, and computation) are often advertised as comprising of synergistic efforts involving several (typically distinct) fields, including physics and mathematics. These sections offer a taste of the type of mathematics relevant to these areas and provide some direction for those interested in delving deeper into the formal aspects of topological quantum theory. On the other hand, these sections should serve as a warning that topological quantum matter cannot be reduced to simple phenomenological aspects, such as the presence of gapless surface or edge states in an otherwise gapped system. The physics is rather subtle, oftentimes counterintuitive, and cannot be properly grasped without a minimal understanding of its formal aspects, which are rooted in quantum theory. Of course, a counter warning is also warranted. This abstract mathematical language is profoundly relevant precisely because it provides an understanding of actual physical systems that can be realized in the laboratory (e.g., quantum Hall systems and topological insulators). The abstract theory reflects key aspects of the physical reality and provides tools to control and exploit it. In particular, the emerging quantum technologies are likely to have profound medium- and long-term impacts, with implications for science, economy, and global security. While these implications are not yet fully explored and assessed, they include both major promises and considerable risks. What happens in the Hilbert space does not stay in the Hilbert space.

1.2 TOPOLOGY AND DIFFERENTIAL GEOMETRY: MATHEMATICAL HIGHLIGHTS

To paraphrase a well-known mathematical joke, a topologist is someone capable of realizing that a doughnut and a coffee cup are, in some sense, identical. Indeed, topology can be described as the mathematical study of properties (called *topological properties*) that remain invariant under continuous deformations (i.e., deformations that do not involve any "cutting" and "tearing") of certain objects (called *topological spaces*). Topology is not restricted to the study of geometrical objects, such as spheres, tori, Klein bottles, and knots, but also deals with structures that have no geometrical correspondent. *Point-set* (or *general*) topology deals with the basic aspects related to the construction of topological spaces and with fundamental concepts, such as *continuity, compactness, connectedness, countability*, and *separability*. Specialized areas such as differential topology, low-dimensional topology, and algebraic topology focus on certain types of topological spaces, e.g., smooth (differentiable) manifolds or low-dimensional manifolds (with up to four dimensions), or use specific tools (e.g., algebraic objects such as groups and rings) to characterize the topological properties of these spaces. Here, we sketch a few basic mathematical ideas and give some simple examples with the goal of shedding light on a key question: what is "topological" about topological quantum matter?

A *topological space* $(X, \mathcal{T})$ is a set X together with a collection $\mathcal{T}$ of subsets of X having the following properties: (1) $\emptyset$ (the empty set) and X are in $\mathcal{T}$. (2) Any union of elements from $\mathcal{T}$ is in $\mathcal{T}$. (3) Any finite intersection of elements from $\mathcal{T}$ is in $\mathcal{T}$. The collection $\mathcal{T}$ is called a *topology* on the set X and the elements of $\mathcal{T}$ are called *open sets*. We emphasize that for any given set $X \neq \emptyset$ there are many possible topologies. For example, if $X = \{a, b, c\}$, the collections of sets $\mathcal{T}_1 = \{\emptyset, X\}$, $\mathcal{T}_2 = \{\emptyset, \{a\}, \{a, b\}, \{a, c\}, X\}$, and $\mathcal{T}_3 = \{\emptyset, \{a\}, \{b\}, \{a, b\}, X\}$ satisfy the definition and hence are topologies on the three-element set X. The collection $\{\emptyset, \{a, b\}, \{b, c\}, X\}$, on the other hand, does not satisfy property (3) (since $\{a, b\} \cap \{b, c\} = \{b\}$ is not an element of the collection), hence it is not a topology. The collection of all subsets of X is called the *discrete topology*, while $\mathcal{T}_1$ (the topology consisting of X and $\emptyset$ only) is called the *trivial* or the *indiscrete topology*.

Often times, rather than describing the entire collection, a topology is specified by a smaller collection of subsets called a *basis*. If $(X, \mathcal{T})$ is a topological space, $\mathcal{B}$ is a basis for $\mathcal{T}$ if and only if $\mathcal{T}$ contains the empty set and all the subsets of X that are unions of elements of $\mathcal{B}$. For example, the so-called *standard topology* on the real line $\mathbb{R}$ is generated by the collection of all open intervals (a, b), with $a < b$. On the other hand, the basis consisting of all half-open intervals $[a, b)$ generates the *lower-limit topology* on the set $\mathbb{R}$ and the corresponding topological space (sometimes denoted by $\mathbb{R}_\ell$) has properties different from those of $\mathbb{R}$ with the standard topology (e.g., $\mathbb{R}_\ell$ is not locally compact). An interval $[a, b]$ is *closed* in the standard topology (i.e., its complement $\mathbb{R} - [a, b]$ belongs to the topology), but is not closed or open in $\mathbb{R}_\ell$; by contrast, $[a, b)$ is not closed or open in the standard topology but is both open and closed in $\mathbb{R}_\ell$. Finally, we note that another important type of subfamilies of $\mathcal{T}$ are the so–called *open coverings* of X. The family of open subsets $\mathcal{C} = \{O_\alpha\} \subset \mathcal{T}$ is called an *open covering* of X if $\cup_\alpha O_\alpha = X$.

Continuous functions represent key tools for comparing topological spaces and for identifying their topological properties. A function $f : (X, \mathcal{T}_X) \to (Y, \mathcal{T}_Y)$ is said to be *continuous* if for each open subset $V \subset Y$ the inverse image $f^{-1}(V)$ is an open set of X. We emphasize that continuity depends on the specific topologies $\mathcal{T}_X$ and $\mathcal{T}_Y$, although sometimes they are not explicitly present in the definition of the function. For example, let $\mathbb{R}$ and $\mathbb{R}_\ell$ denote the set of real numbers with the standard topology and with the lower-limit topology, respectively, and $f : \mathbb{R}_\ell \to \mathbb{R}$ be the identity function, $f(x) = x$ for every real number x. Since the inverse image of an open set (a, b), which is equal to itself, is open[4] in $\mathbb{R}_\ell$, f is continuous. By contrast, the identity function $f' : \mathbb{R} \to \mathbb{R}_\ell$ is not continuous, because the inverse image of $[a, b)$ is not open in $\mathbb{R}$. This is a consequence of $\mathbb{R}$ and $\mathbb{R}_\ell$ being different topological spaces. But when are two spaces topologically the same?

We define a *homeomorphism* as a bijective function $f : X \to Y$ with the property that both f and its inverse $f^{-1} : Y \to X$ are continuous. A homeomorphism provides a bijective correspondence between the collections of open sets of the two topological spaces and, as a result, any property of X that is entirely expressed in terms of its topology (i.e., collection of open sets) will be preserved under f (i.e., Y will be characterized by the same property). Such properties that are invariant under homeomorphisms are called *topological properties*. A simple example of homeomorphism is the function $g : (-1, 1) \to \mathbb{R}$ (both sets having the standard topology), with $g(x) = x/(1 - x^2)$. This homeomorphism stretches the segment $(-1, 1)$ into the real line. By contrast, the identity function $f' = f^{-1}$ discussed in the previous paragraph is not continuous, hence f is not a homeomorphism. In fact, two topological spaces cannot be homeomorphic if there is any topological property that holds for one space but not for the other (e.g., for $\mathbb{R}$ but not for $\mathbb{R}_\ell$). Hence the natural question: What exactly are these topological properties? Some of them can be easily grasped at the intuitive level, others require a rather complex mathematical apparatus. A few simple examples are discussed below.

---

[4] The set (a, b) can be written as a countably infinite union of intervals $[a + \frac{1}{n}, b)$.

Connectedness. Intuitively, a space is connected if it consists of a single "piece" and cannot be separated into disjoint open sets. More formally, we define a *separation* of a topological space X as a pair U, V of nonempty, disjoint open sets of X with the property $U \cup V = X$. The space X is said to be *connected* if no separation of X exists. Since it is formulated entirely in terms of the collection of open sets of X, connectedness is a topological property of X. For example, the open interval $(-1, 1)$ (with the standard topology) is connected, while the open interval $[-1, 1)$ (with the lower limit topology) is not connected, since $[-1, 1) = [-1, 0) \cup [0, 1)$, i.e., it has a separation. Similar properties hold for $\mathbb{R}$ and $\mathbb{R}_\ell$, respectively. Since $\mathbb{R}$ is connected and $\mathbb{R}_\ell$ is not, the two spaces cannot be homeomorphic.

Compactness. A collection $\mathcal{A}$ of open subsets is said to be a *covering* of X if the union of the elements of $\mathcal{A}$ is equal to X. The space X is said to be *compact* if every open covering of X contains a finite sub-collection that is a covering of X. Similarly, a subset $K \subseteq X$ is said to be *compact* if every open covering of K has a finite subcovering. For example, the interval $[0, 1] \subset \mathbb{R}$ (with the standard topology[5]) is compact, while the interval $(0, 1]$ is not; the open covering $\mathcal{A} = \{(1/n, 1] | n \in \mathbb{Z}_+\}$ contains no finite sub-collection that covers $(0, 1]$. The real line $\mathbb{R}$ itself is not compact. However, one can construct a compact space by adding one point, $\hat{\mathbb{R}} = \mathbb{R} \cup \{\infty\}$, and expanding the topology with open sets containing ∞, $U_{ab} = \{x \in \mathbb{R} \cup \{\infty\} | x < a \text{ or } x > b\}$, where $a < b$. The space $\hat{\mathbb{R}}$, called the *one-point compactification* of the real line, is homeomorphic with a circle. Similarly, the one-point compactification of $\mathbb{R}^2$ is homeomorphic with a sphere.

There are other topological properties associated with connectedness (e.g., path and local connectedness) and compactness (e.g., limit point compactness), as well as topological properties associated with *separability* (e.g., Hausdorff spaces), *countability* (e.g., first- and second-countable spaces), and *metrizability*. However, none of these topological properties can distinguish between, let's say, a sphere and a torus. In this context, as a note of caution, we emphasize that topological properties should not be understood as always characterizing some sort of "global" feature of the topological space; for example, the distinct ways in which the open sets "fit" inside each other result in different *local* (rather than *global*) structures for $\mathbb{R}$ and $\mathbb{R}_\ell$.

The fundamental group. To address the problem of the sphere–torus distinction, we define a *path* in X from x_0 to x_1 as the continuous map (i.e., function) $f : [0, 1] \to X$ with the property $f(0) = x_0$ and $f(1) = x_1$. Two paths f and f' are said to be *path homotopic* (and we write $f \simeq_P f'$) if they have the same initial and final points (x_0 and x_1, respectively) and there is a continuous map $F : [0, 1] \times [0, 1] \to X$ such that $F(s, 0) = f(s)$, $F(s, 1) = f'(s)$, $F(0, t) = x_0$, and $F(1, t) = x_1$. In other words, F represents a continuous "deformation" of the path f into the path f'. One can show that $\simeq_P$ is an

---

[5]For $X = [0, 1]$ the basis of the standard topology includes open sets (a, b), with $0 < a < b < 1$, as well as $[0, a)$ and $(a, 1]$, with $0 < a < 1$.

equivalence relation, i.e., all paths that are path homotopic with f belong to the same equivalence class, which we will denote by $[f]$. Furthermore, one can create an algebraic structure by defining the *product* $f * g$ of two paths. Specifically, if f is a path in X from x_0 to x_1 and g is a path from x_1 to x_2, then $h = f * g$ is the path from x_0 to x_2 defined by

$$h(s) = \left\{ \begin{array}{ll} f(2s) & \text{for } s \in [0, \frac{1}{2}], \\ g(2s - 1) & \text{for } s \in [\frac{1}{2}, 1]. \end{array} \right. \tag{1.1}$$

One can show that this introduces an operation on the path-homotopy classes defined by $[f] * [g] = [f * g]$. If we consider only paths that start and end at x_0 (called *loops* based at x_0), the corresponding path-homotopy classes with product operation $*$ form a group called the *fundamental group* of X relative to the *base point* x_0 and denoted by $\pi_1(X, x_0)$. Note that the identity element of the group is $[f_0]$ with $f_0(s) = x_0$ (i.e., the class of loops that can be continuously shrunk to a point). One can show that, in certain conditions (specifically, when X is *path-connected*), $\pi_1(X, x_0)$ is independent of x_0 and can be denoted by $\pi_1(X)$. The fundamental group of the Euclidian n-space, $\pi_1(\mathbb{R}^n)$, is the trivial group consisting of the identity alone (i.e., all loops can be continuously shrunk to a point). The same is true for a sphere and, in fact, for all topological spaces that are *simply connected*. By contrast, for a circle $(X = S^1)$, each path-homotopy class consists of loops that wind around the circle n times, where $n \in \mathbb{Z}$ is positive or negative depending on the winding direction. The product of two loops that wind n and m times, respectively, will wind $n + m$ times. Consequently, $\pi_1(S^1) = \mathbb{Z}$, meaning that $\pi_1(S^1)$ is isomorphic with the group of integers endowed with the addition operation. For a torus $(X=T)$, which can be viewed as the direct product of two circles, the fundamental group is $\pi_1(T) = \mathbb{Z} \times \mathbb{Z}$. Hence, the fundamental group of the torus is different from the fundamental group of the sphere and represents a topological property that clearly distinguishes the two objects.

With the fundamental group, we have stepped inside the realm of algebraic topology. Using its tools, one can assign more subtle topological invariants to a given topological space, such as the so-called *homology groups*, a sequence of Abelian groups $H_n(X)$ that, basically, allows one to categorize the different kinds of "holes" associated with X. The classification of topological quantum matter requires the use of some of these rather complex mathematical tools. However, our goal here was just to provide a "taste" of this field and, most importantly, to convey the following core ideas: 1) topology characterizes a class of mathematical objects (topological spaces) regarding how the elements of a set are "packed" together based on information about the neighborhoods (open subsets) of these elements, but regardless of relative distances or directions, which may be meaningless, and 2) this "packing structure" is characterized by a set of properties (topological invariants) that are preserved under transformations (homeomorphisms) that "stretch," "twist," and "deform" the system or completely change the nature of its elements.

Anther important set of mathematical concepts is associated with differential geometry – the natural language of gauge theory, which, in essence, is the study of connections on various types of bundles. Below we provide basic definitions and a few examples as an introduction to our discussion of the mathematical structure of geometric phases (see below, Section 1.5).

Topological manifold. A *manifold* is, in essence, a topological space that locally resembles the Euclidian space $\mathbb{R}^n$. The manifold is obtained by patching together open pieces of $\mathbb{R}^n$, more precisely open subsets that are homeomorphic to open subsets of $\mathbb{R}^n$. Formally, a topological space $(M, \mathcal{T}_M)$ is said to be a *topological manifold* of dimension n if there exists an open covering $\{O_\alpha\}$ of M such that each element of the covering is homeomorphic to an open set U_α of $\mathbb{R}^n$. The homeomorphism

$$\phi_\alpha : O_\alpha \to U_\alpha, \quad \phi(u) = (x_1(u), \ldots, x_n(u)) \tag{1.2}$$

is called a *coordinate system*, while the pair (O_α, ϕ_α) is called a *chart* on M. The complete collection of charts is called an *atlas*. For a pair of overlapping charts, $O_\alpha \cap O_\beta \neq \emptyset$, the homeomorphism $g_{\alpha\beta} : \phi_\beta(O_\alpha \cap O_\beta) \to \phi_\alpha(O_\alpha \cap O_\beta)$ between open subsets of $\mathbb{R}^n$ defined as

$$g_{\alpha\beta} = \phi_\alpha \phi_\beta^{-1}|_{O_\alpha \cap O_\beta} \tag{1.3}$$

is called a *transition map* (*overlap function* or *coordinate transformation*). Note that the topology $\mathcal{T}_M$ of a manifold can be reconstructed from the inverse images $\phi_\alpha^{-1}(U)$ of open subsets of $\mathbb{R}^n$. As a consequence, all n-dimensional manifolds have the same local properties. However, the global topological properties of a manifold depend on how the open subsets $\{O_\alpha\}$ are patched together, which is determined by the transition maps. We note that the mathematical objects relevant to the characterization of topological quantum matter turn out to be topological manifolds. Hence, in this context, the significant topological properties are associated with global features.

Differentiable manifold. Consider the subsets $U \subseteq \mathbb{R}^n$ and $V \subseteq \mathbb{R}^m$. A function $g : U \to V$ is said to be a C^N function $(1 \leq N \leq \infty)$ if g is N-times differentiable. A differentiable *homeomorphism* g is called a C^N *diffeomorphism* if g and its inverse g^{-1} are C^N functions. If the transition functions $g_{\alpha\beta}$ of a manifold M are C^N diffeomorphisms, M is said to be a C^N manifold. In this section, we will consider C^∞ manifolds, also called *smooth manifolds*. Note that the notion of C^N differentiability can be generalized to homeomorphisms between C^N manifolds that satisfy certain properties. This introduces an equivalence relation on the collection of C^N differentiable manifolds.

Example: the sphere S^2. A simple example of a smooth manifold is the space $\mathbb{R}^n$. However, this is a trivial manifold, as it can be covered by a single chart. A nontrivial smooth manifold is, for example, the sphere S^2, which can be covered using at least two charts. To define these charts, one possibility is to use the so-called *stereographic* projection of S^2 on $\mathbb{R}^2$. Let $\mathbf{N}$ and $\mathbf{S}$ denote the north and south poles of the sphere, respectively. We choose a covering of S^2 consisting of the open subsets $O_1 = S^2 - \{\mathbf{S}\}$ and $O_2 = S^2 - \{\mathbf{N}\}$. The homeomorphisms $\phi_i : O_i \to \mathbb{R}^2$, $\phi_i(\mathbf{p}) = \mathbf{r}_i$ are defined by the projection

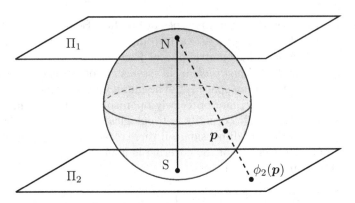

FIGURE 1.3 Stereographic projection of the sphere.

procedure illustrated in Figure 1.3. The planes Π_1 and Π_2 are tangent to the sphere at the north and south poles, respectively. Using Cartesian coordinates on Π_1 and Π_2, we can easily show that the transition function $g_{21} : \mathbb{R}^2 - \{0\} \rightarrow \mathbb{R}^2 - \{0\}$ is

$$g_{21}(x_1, y_1) = \left(\frac{x_1}{x_1^2 + y_1^2}, \frac{y_1}{x_1^2 + y_1^2} \right). \tag{1.4}$$

Note that g_{21} is a smooth diffeomorphism, hence S^2 is a smooth manifold. In fact, the sphere is a representative of an infinite class of smooth manifolds that can be obtained by smooth deformations of S^2, i.e., are related to one another by a diffeomorphism. For example, the surface of an ellipsoid and the sphere are diffeomorphic to one another.

Tangent and cotangent spaces. The stereographic projection discussed above involves the notion of a plane tangent to a sphere embedded in $\mathbb{R}^3$. Differentiability allows us to generalize this concept to arbitrary smooth manifolds. Let $C : [0, T] \longrightarrow M$ be a smooth curve on a smooth manifold M of dimension m embedded in $\mathbb{R}^n$, with $C(0) = p \in M$. The vector $v = \frac{d}{dt}C(t)|_{t=0}$ is called the *tangent vector* to $C \subset M$ at $p \in M$. The set of all tangent vectors to all curves originating at $p \in M$ represents an m-dimensional vector space (isomorphic to $\mathbb{R}^m$) denoted by T_pM and called the *tangent space* of M at p. The space of all linear real-valued functions on T_pM (i.e., the dual vector space of T_pM) is called the *cotangent space* of M at p and is denoted by T_pM^*. The elements of T_pM and T_pM^* are called tangent (or *contravariant*) and cotangent (*covariant*) vectors, respectively. Intuitively, T_pM can be thought of as the space of velocities at $p \in M$ corresponding to all "trajectories" $C(t)$ on the manifold M that pass through p. Also, the action of a cotangent vector (or covector) $\omega_p \in T_pM^*$ on $v_p \in T_pM$ generates a scalar that can be expressed as an inner product, $\omega_p(v_p) = (\omega_p, v_p) = (\omega_p)_i(v_p)^i$, where the *components* of v_p associated with a local coordinate chart (O_α, ϕ_α) that maps the portion of the curve $C(t) \subset O_\alpha$ to the curve $x(t) = (x^1(t), \ldots, x^m(t)) \subset \mathbb{R}^m$ are

$$(v_p)^i = \left.\frac{dx^i(t)}{dt}\right|_{t=0}. \tag{1.5}$$

Consider now two intersecting charts (O_α, ϕ_α) and (O_β, ϕ_β) with $x = \phi_\alpha(p)$ and $x' = \phi_\beta(p) = g_{\beta\alpha}(x)$. The transformation rules for the components of contravariant and covariant vectors are

$$(v'_p)^i = \frac{\partial x'^i}{\partial x^j}(v_p)^j, \qquad\qquad (\omega'_p)_i = \frac{\partial x^j}{\partial x'^i}(\omega_p)_j. \tag{1.6}$$

A convenient way of accounting for these transformation rules is to use the symbols $\frac{\partial}{\partial x^i} = \partial_i$ and dx^i for the local coordinate basis vectors and covectors, respectively. Using this notation, we have $v_p = (v_p)^i \frac{\partial}{\partial x^i}$ and $\omega_p = (\omega_p)_i dx^i$. The duality of the contravariant and covariant basis vectors can be expressed as $dx^i(\partial_j) = (dx^i, \partial_j) = \delta^i_j$.

   One-form. The set of all tangent spaces of a manifold is called the *tangent bundle* of M and is denoted by TM. An element of TM can be represented as a pair of the form (p, v_p), with $p \in M$ and $v_p \in T_pM$. Similarly, the set of all cotangent spaces is called the *cotangent bundle* and is denoted by $TM^* \ni (p, \omega_p)$. A smooth function $V : M \longrightarrow TM$ such that $V(p) = (p, v(p))$ for every $p \in M$ is called a *vector field*. Note that, usually, $V(p)$ is identified with its value $v(p)$. Similarly, a smooth function $\Omega : M \longrightarrow TM^*$ such that $\Omega(p) = (p, \omega(p))$ for every $p \in M$ is called a *differential one-form* (or simply a *one-form*). Typically, $\Omega(p)$ is identified with $\omega(p)$. Since any point $p \in M$ is represented by its coordinates $x = (x^1, \ldots, x^m) = \phi_\alpha(p)$, we can locally (i.e., for $p \in O_\alpha$) express a vector field and a one-form using the notations for the basis vectors and covectors mentioned above,

$$v(x) = v^i(x)\frac{\partial}{\partial x^i}, \qquad\qquad \omega(x) = \omega_i(x)dx^i. \tag{1.7}$$

   Wedge product. The vector fields mentioned above are particular examples of *tensor fields*. An important class of such mathematical objects contains the *totally antisymmetric covariant tensor fields*, also called *differential forms*. In general, the tensor product of two forms is not antisymmetric. The corresponding antisymmetric operation for differential forms is the so-called *wedge product*, or the antisymmetric tensor product. For example, the basis for a differential two-form is given by

$$dx^\mu \wedge dx^\nu = dx^\mu \otimes dx^\nu - dx^\nu \otimes dx^\mu = -dx^\nu \wedge dx^\mu. \tag{1.8}$$

Using this operation, we can write the electromagnetic field tensor as a differential two-form, $F = \frac{1}{2}F_{\mu\nu}dx^\mu \wedge dx^\nu$. Note that the wedge product of two vectors, say $\boldsymbol{A}_1, \boldsymbol{A}_2 \in \mathbb{R}^3$, is related to the signed area of the parallelogram defined by the two vectors, similar to the cross product $\boldsymbol{A}_1 \times \boldsymbol{A}_2$. However, while $\boldsymbol{A}_1 \times \boldsymbol{A}_2 \in \mathbb{R}^3$, the product $\boldsymbol{A}_1 \wedge \boldsymbol{A}_2$ does not belong to the same vector space as the two vectors. Moreover, unlike the cross product, we can generalize the relation between $A_1 \wedge A_2$ and the signed area of the parallelogram to higher dimensions. For example, $A_1 \wedge A_2 \wedge A_3$ is related to the signed volume defined by the three vectors.

Exterior derivative. Consider the space Ω^p of all differential p-forms on a smooth manifold. The *exterior derivative* is the map $d : \Omega^p \longrightarrow \Omega^{p+1}$ defined (locally) by

$$d\omega = \left(\frac{\partial}{\partial x^j} \omega_{i_1 \dots i_p} \right) dx^j \wedge dx^{i_1} \wedge \cdots \wedge dx^{i_p}, \qquad (1.9)$$

where $\omega_{i_1 \dots i_p} dx^{i_1} \wedge \cdots \wedge dx^{i_p} \in \Omega^p$. Note that $d^2 = 0$. In electromagnetism, for example, we can write the differential two-form F as $F = dA$, where $A = A_\mu dx^\mu$ is a one-form with components given by the 4-potential A_μ. The equation $dF = d^2 A = 0$ corresponds to the homogeneous Maxwell equations.

Push-forward and pullback maps. Consider two smooth manifolds M_1 and M_2 and a smooth function $f : M_1 \to M_2$ between them. This function induces certain maps between the vector fields defined on the two manifolds. Specifically, for $p \in M_1$ let $v_p \in T_p M_1$ be an arbitrary (contravariant) vector defined by the curve $C_1(t)$, $v_p = \frac{dC_1}{dt}|_{t=0}$. The image of C_1 under f is the curve $C_2 = f \circ C_1$ in M_2, where $(f \circ C_1)(t) = f(C_1(t))$ represents the composition of the functions C_1 and f. Then, f induces the linear map

$$f_* : T_p M_1 \longrightarrow T_{f(p)} M_2, \qquad f_*(v_p) = \frac{dC_2}{dt}\bigg|_{t=0}, \qquad (1.10)$$

called the *push-forward* or the *differential* map. Similarly, starting from the cotangent vectors and the differential forms of M_2 we can "pull" them "back" on M_1. Let $\omega_{f(p)} \in T_{f(p)} M_2^*$ be an arbitrary cotangent vector at $f(p) \in M_2$. Then, the map

$$f^* : T_{f(p)} M_2^* \longrightarrow T_p M_1^*, \qquad f^*(\omega_{f(p)})[v_p] = \omega_{f(p)}[f_*(v_p)], \qquad (1.11)$$

called the *pullback* map, defines the cotangent vector $f^*(\omega_{f(p)}) \in T_p M_1^*$. In the above equation $f_*(v_p)$ is the push-forward of an arbitrary vector $v_p \in T_p M_1$.

1.3 GEOMETRIC PHASES: EXAMPLES AND OVERVIEW

The Berry phase, a central concept in quantum physics, is rooted in the state space structure of quantum theory and has an observable impact on a wide range of phenomena. The remarkably elegant interpretation of this geometric phase as the *holonomy of a fiber bundle* provides it with a mathematical framework capable of accounting for a broad spectrum of classical and quantum phenomena, such as, for example, the precession of a Foucault pendulum, the parallel transport of a vector on a sphere, the Aharonov–Bohm effect, the gauge theory of molecular physics, the electric polarization and orbital magnetization of solids, and the integer and fractional quantum Hall effects. The Berry phase is a central concept in topological band theory and provides the mechanism responsible for anyonic statistics in correlated many-body systems, a key element that could be used to achieve fault-tolerant quantum computation. In other words, the Berry phase is not just some fashionable idea useful for explaining a few exotic phenomena, but a generic feature of quantum mechanics and an essential ingredient necessary for a coherent understanding of basic quantum phenomena, such as, for example, the electronic properties of materials. It is therefore rather surprising that the full significance of geometric phases was practically ignored for more than fifty years prior to the publication of Berry's work in 1984 [57]. Below we discuss the basic physics that gives rise to these geometric phases, point out the elements that allow us to distinguish different types of holonomies, summarize the basic concepts relevant to understanding the mathematical structure of geometric phases, and mention a few relevant applications. The concepts and the mathematical framework introduced here will be used in subsequent chapters as the basis for describing topological properties of condensed matter systems.

1.3.1 Classical and quantum holonomies

We start our discussion with some simple examples of geometric and topological phases emerging during cyclic evolutions in classical and quantum systems. First, consider the parallel transport of a tangent vector along a loop Γ on the surface of a sphere. During the transport, the vector is always in the plane tangent to the surface of the sphere and is not allowed to rotate with respect to the normal. Yet, after completing the loop, the vector points in a direction that differs from its initial orientation. For example, let us assume that the vector is parallel transported starting from the north pole along a meridian down to the equator, then along the equator and back to the north pole along a different meridian, as shown in Figure 1.4(a). After completing the loop, the vector points in a direction that is rotated by an angle θ_Γ with respect to its initial orientation. This phenomenon is called a *holonomy* and the angle θ_Γ is equal to the solid angle subtended by the surface enclosed by the loop Γ.

Next, consider a Möbius strip and a vector perpendicular to the surface [Figure 1.4(b)]. The vector is transported along closed paths on the surface of the strip. After one circuit, the direction of the vector gets reversed, as illustrated in Figure 1.4(b). On the other hand, if the loop returns to the initial point without completing a full circuit, or if it circles the strip twice, the initial and final directions of the vector coincide. The transport of the perpendicular vector along a closed path Γ on the Möbius strip generates a discrete holonomy with two possible outcomes: $\theta_\Gamma = \pi$ (reversed orientation) and $\theta_\Gamma = 0$ (same orientation). In general, $\theta_\Gamma = \pi n_\Gamma$ (mod 2π), where the winding number n_Γ represents the number of complete circuits around the strip.

There are important differences between these two examples. In the case of the parallel transport over the sphere the holonomy angle, θ_Γ, depends on the details of the path; variations of the loop will generally change the angle, since

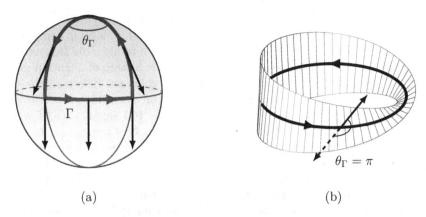

FIGURE 1.4 (a) Parallel transport of a tangent vector on the surface of a sphere. (b) Transport of a perpendicular vector on a Möbius strip.

it is proportional to the area of the enclosed surface. By contrast, in the case of the Möbius strip, arbitrary variations of the path that preserve the number of complete circuits do not affect the holonomy. In the first case, the holonomy is linked to the curvature of the underlying space, i.e., the sphere. Parallel transport of a vector over a plane results in the initial and final orientations being always the same, i.e., no holonomy. In general, the holonomy associated with parallel transport over curved spaces depends on the path geometry and represents a measure of the curvature, which is a geometric property of the underlying space; hence it can be properly called a *geometric holonomy*. In the case of the Möbius strip, on the other hand, the holonomy is due to the nontrivial topological properties of the underlying space and, therefore, it is a *topological holonomy*. Its topological nature is the reason behind the robustness of the holonomy against deformations of the path that do not modify the winding number n_Γ. Note that the (Gaussian) curvature of the Möbius strip is zero. Also note that transport on a topologically trivial surface with zero curvature (e.g., a cylinder) generates no holonomy.

To further emphasize the distinction between geometric and topological holonomies, let us consider again the parallel transport of a vector along a closed path Γ, but this time on the surface of a cone. The curvature is zero everywhere except the apex, where it is undefined. Unlike the parallel transport on a sphere, the holonomy angle does not depend on the details of the path, but only on how many times it circles the apex. Specifically, $\theta_\Gamma = 2\pi \sin(\alpha) n_\Gamma$, where n_Γ is the number of complete circuits around the apex and 2α is the aperture of the cone. In particular, the vector will return to the initial orientation for any loop that does not circle the apex. Hence, the parallel transport of a vector on the surface of a cone generates a topological holonomy.

Turning now to quantum systems, let us consider a spin-$\frac{1}{2}$ particle in a rotating magnetic field $\boldsymbol{B}(t)$. We assume that the particle has a magnetic moment $\boldsymbol{\mu} = g\frac{q\hbar}{2m}\boldsymbol{S}$, where g is the Landé factor, q and m are the charge and mass of the particle, respectively, and $\boldsymbol{S} = \frac{1}{2}\vec{\sigma}$, with $\vec{\sigma} = (\sigma_x, \sigma_y, \sigma_z)$ being the Pauli matrices, is the spin operator in units of $\hbar$. The magnetic field precesses around the z axis at a polar angle θ with constant angular frequency ω, so that the direction of the field at time t is given by the unit vector $\hat{\boldsymbol{R}} = \boldsymbol{B}/B$ defined in spherical coordinates by the angles θ and $\varphi = \omega t$. In the laboratory frame, the Hamiltonian of the system is

$$H(t) = \hbar\omega_B \hat{\boldsymbol{R}}(\theta, \omega t) \cdot \boldsymbol{S}, \tag{1.12}$$

with $\hbar\omega_B = g\frac{q\hbar}{2m}B$. The time evolution of the quantum state vector $|\psi(t)\rangle$ is determined by the Schrödinger equation

$$i\hbar\frac{\partial}{\partial t}|\psi(t)\rangle = H(t)|\psi(t)\rangle. \tag{1.13}$$

We are interested in solutions of the Schrödinger equation (1.13) characterized by *cyclic evolutions*, $|\psi(\tau)\rangle\langle\psi(\tau)| = |\psi(0)\rangle\langle\psi(0)|$, where τ is the period of the cycle. Note that $|\psi(\tau)\rangle$ and $|\psi(0)\rangle$ represent the same quantum state, i.e., they are identical up to a phase factor. Also note that, in general, the period τ does not have to be the same as the period of the Hamiltonian, $T = \frac{2\pi}{\omega}$. However, in this example we will focus on the class of cyclic solutions with $\tau = T$. To find these solutions, it is convenient to change the reference frame by rotating the state vector about the z axis by an angle $\varphi(t) = -\omega t$. In the new frame, the state vector $|\psi(t)\rangle'$, which can be obtained from the original state vector by applying the unitary transformation $U_\omega(t) = e^{i\omega t S_z}$, i.e., $|\psi(t)\rangle' = U_\omega(t)|\psi(t)\rangle$, evolves according to the time-independent Hamiltonian

$$H' = U_\omega H(t)U_\omega^\dagger - i\hbar U_\omega\frac{\partial U_\omega^\dagger}{\partial t} = \hbar\Omega\hat{\boldsymbol{R}}(\theta', 0) \cdot \boldsymbol{S}, \tag{1.14}$$

where

$$\Omega = \omega_B\sqrt{1 - 2\frac{\omega}{\omega_B}\cos\theta + \left(\frac{\omega}{\omega_B}\right)^2}, \tag{1.15a}$$

$$\sin\theta' = \frac{\omega_B}{\Omega}\sin\theta, \qquad \cos\theta' = \frac{\omega_B}{\Omega}\left(\cos\theta - \frac{\omega}{\omega_B}\right). \tag{1.15b}$$

Since H' is time-independent, the Schrödinger equation $i\hbar\frac{\partial}{\partial t}|\psi'(t)\rangle = H'|\psi'(t)\rangle$ can be easily integrated and we have $|\psi'(t)\rangle = e^{-\frac{i}{\hbar}H't}|\psi'(0)\rangle$. Transforming back to the laboratory frame, we find the solution of the Schrödinger equation (1.13) for an arbitrary initial state vector $|\psi(0)\rangle = |\psi'(0)\rangle$ as

$$|\psi(t)\rangle = e^{-i\omega t S_z}e^{-i\Omega t\hat{\boldsymbol{R}}'\cdot\boldsymbol{S}}|\psi(0)\rangle = U^\dagger(t)|\psi(0)\rangle, \tag{1.16}$$

where $\hat{\boldsymbol{R}}' = \hat{\boldsymbol{R}}(\theta', 0)$. A cyclic solution with period τ can be obtained by choosing $|\psi(0)\rangle$ to be an eigenstate of the unitary operator

$U^\dagger(\tau) = e^{-i\omega\tau S_z} e^{-i\Omega\tau \hat{\boldsymbol{R}}'\cdot\boldsymbol{S}}$. Focusing on the case $\tau = T$, one can easily show that the eigenstates of the spin projection along $\hat{\boldsymbol{R}}'$, i.e., the solutions of the eigenvalue problem $\hat{\boldsymbol{R}}' \cdot \boldsymbol{S}|\lambda, \hat{\boldsymbol{R}}(\theta',0)\rangle = \frac{\lambda}{2}|\lambda, \hat{\boldsymbol{R}}(\theta',0)\rangle$, with $\lambda = \pm 1$, are also eigenvectors of the evolution operator $U^\dagger(T)$. Consequently, during a cyclic evolution with period $T = \frac{2\pi}{\omega}$ and initial condition $|\psi(0)\rangle = |\lambda, \hat{\boldsymbol{R}}(\theta',0)\rangle$, the state vector acquires a phase factor $e^{-i\alpha_\lambda}$ with $\alpha_\lambda = \lambda\pi(1 + \Omega/\omega)$. Explicitly,

$$|\psi(T)\rangle = U^\dagger(T)|\psi(0)\rangle = e^{-i\pi\lambda\left(1+\frac{\Omega}{\omega}\right)}|\psi(0)\rangle \qquad (1.17)$$

The appearance of a phase factor as a result of time evolution is not at all surprising. For example, during the time interval $0 \leq t \leq T$ a stationary state with energy E_n acquires a dynamical phase $e^{-i\alpha_n^{\rm dyn}}$ characterized by an angle $\alpha_n^{\rm dyn} = E_n T/\hbar$. For a general cyclic evolution, one would expect a similar dynamical phase. A reasonable generalization of the dynamical phase angle would be

$$\alpha^{\rm dyn} = \frac{1}{\hbar}\int_0^T \langle\psi(t)|H(t)|\psi(t)\rangle\, dt. \qquad (1.18)$$

For the Hamiltonian (1.12), the dynamical phase angle associated with the cyclic state (1.16) with $|\psi(0)\rangle = |\lambda, \hat{\boldsymbol{R}}(\theta',0)\rangle$ is

$$\alpha_\lambda^{\rm dyn} = \lambda\pi\left(\frac{\Omega}{\omega} + \cos\theta'\right), \qquad (1.19)$$

where $\cos\theta'$ is given by Eq. (1.15b). Note, however, that the total phase in Eq. (1.17) does not coincide with the dynamical contribution and we have $e^{-i\alpha_\lambda} = e^{-i\alpha_\lambda^{\rm dyn}} e^{i\gamma_\lambda}$, with the additional contribution given by

$$\gamma_\lambda = -\lambda\pi(1 - \cos\theta') = -\lambda\pi\left(1 - \frac{\omega_B}{\Omega}\cos\theta + \frac{\omega}{\Omega}\right). \qquad (1.20)$$

Let us remark that the quantity $\pi(1 - \cos\theta')$ represents half of the solid angle $\Omega(\theta')$ subtended by a vector $\hat{\boldsymbol{R}}$ precessing around the z axis at a polar angle θ'. Furthermore, for a slowly rotating magnetic field, i.e., in the adiabatic limit characterized by $\omega \ll \omega_B$, we have $\Omega \to \omega_B$, $\theta' \to \theta$ and the additional phase angle $\gamma_\lambda = -\frac{\lambda}{2}\Omega(\theta)$ is solely determined by the geometry of the rotating field. To better understand the significance of this phase, let us consider the adiabatic evolution of a system governed by the Hamiltonian (1.12) that is initially in the low-energy eigenstate $|\psi(0)\rangle = |-, \hat{\boldsymbol{R}}(\theta,0)\rangle$ with energy $E_- = -\frac{1}{2}\hbar\omega_B$ (we assume here that $\omega_B \geq 0$). Since the magnetic field varies slowly in time, the state vector at arbitrary t will be given, up to a phase factor, by the eigenstate corresponding to spin projection $\frac{-\hbar}{2}$ along the direction of field, i.e., $|-, \hat{\boldsymbol{R}}(\theta,\omega t)\rangle$. The states of the system can be represented as points on a Bloch sphere[6] and a cyclic evolution corresponds to a closed loop on this

---

[6]The Bloch sphere is a geometrical representation of the pure state space of a two-level quantum mechanical system. Antipodal points correspond to mutually orthogonal state vectors, e.g., the north and south poles correspond to $|+1\rangle$ and $|-1\rangle$, respectively.

sphere. The dynamical phase angle acquired during a cyclic evolution is $\alpha_-^{\text{dyn}} = -\pi\omega_B/\omega$, i.e., it depends on the period T. Note that the dynamical phase can be eliminated by properly shifting the energy, $H(t) \to H(t) - \hbar\omega\alpha_-^{\text{dyn}}/(2\pi)$. On the other hand, the geometric phase $\gamma_- = \pi(1 - \cos\theta)$ is not affected by the energy shift and is determined solely by the geometry of the problem, more specifically by the solid angle traced by the rotating magnetic field. We conclude that the adiabatic transport of a spin-half particle on the Bloch sphere results in a geometric phase $e^{i\gamma_\lambda^B}$, the *Berry phase*, with γ_λ^B equal to minus half the solid angle enclosed by the rotating field direction,

$$\gamma_\lambda^B = -\lambda\pi(1 - \cos\theta) = -\frac{\lambda}{2}\Omega(\theta). \tag{1.21}$$

This is the quantum analog of the classical holonomy generated by the parallel transport of a vector on the surface of a sphere. The generalization (1.20) of the Berry phase for nonadiabatic evolutions is called the *Aharonov–Anandan phase* [9]. Note the similarity between Eqs. (1.20) and (1.21) and the fact that, in general, $\theta' \neq \theta$. The two angles coincide in the adiabatic limit.

The final example involves the phase associated with the Aharonov–Bohm effect [10]. Let us consider a charged particle moving nonrelativistically in the presence of a magnetic flux Φ generated by an infinitely long and infinitesimally thin impenetrable solenoid that coincides with the z coordinate axis. The corresponding vector potential is $\boldsymbol{A}(\boldsymbol{r}) = \frac{\Phi}{2\pi r^2}(x\hat{\boldsymbol{y}} - y\hat{\boldsymbol{x}})$, where $r = |\boldsymbol{r}|$, while $\hat{\boldsymbol{x}}$ and $\hat{\boldsymbol{y}}$ are unit vectors along the corresponding directions. Note that the magnetic field is zero everywhere outside the infinitesimally thin solenoid, i.e., for any point $\boldsymbol{r}$ that is not on the z axis. The dynamics of the particle is described by the minimal coupling Hamiltonian $H = -\frac{\hbar^2}{2m}(\boldsymbol{\nabla} - i\frac{q}{\hbar}\boldsymbol{A})^2 + V_R(\boldsymbol{r})$, where m and q are the mass and charge of the particle, respectively, and $V_R(\boldsymbol{r})$ is a potential that confines the particle inside a small box centered at $\boldsymbol{R}$. Assume that in the absence of the magnetic field, the confined particle is in a state $\psi_n^0(\boldsymbol{r}')$ of energy E_n, where $\boldsymbol{r}' = \boldsymbol{r} - \boldsymbol{R}$ represents the position of the particle relative to the confining box. In the presence of a magnetic flux, the eigenvector corresponding to E_n will be

$$\psi_n(\boldsymbol{r}, \boldsymbol{R}) = e^{\frac{iq}{\hbar}\int_{\boldsymbol{R}}^{\boldsymbol{r}} \boldsymbol{A}(\boldsymbol{x}) \cdot d\boldsymbol{x}} \, \psi_n^0(\boldsymbol{r} - \boldsymbol{R}). \tag{1.22}$$

If the box containing the particle is adiabatically transported along a closed loop C that does not intersect the (impenetrable) flux tube, the system acquires a geometric phase given by

$$\gamma_{AB}(C) = \frac{q}{\hbar}\oint_C \boldsymbol{A} \cdot d\boldsymbol{R} = \frac{q}{\hbar}\int_{S(C)} \boldsymbol{B} \cdot d\boldsymbol{S} = n_C\frac{q}{\hbar}\Phi, \tag{1.23}$$

where $S(C)$ is the surface bounded by the loop C, $\boldsymbol{B}$ is the magnetic field, and n_C is the number of times C cycles around the flux line. Note that the system acquires a nontrivial phase although the box is moving in a region with no magnetic field. This rather unexpected behavior is known as the Aharonov–Bohm effect [10]. The corresponding phase angle $\gamma_{AB}(C)$ does not depend on the geometric details of the closed path C, being determined solely by the topology of this path (i.e., by how many times it circles around the flux tube) and by the flux Φ. The Aharonov–Bohm phase is an example of *topological phase* and can be viewed as the quantum analog of the (classical) topological holonomy generated by the parallel transport of a vector on the surface of a cone.

1.3.2 Historical overview and conceptual distinctions

Historically, the concept of geometric phase has been recognized as a generic feature of quantum mechanics only relatively recently, particularly after the publication in 1984 of Michael Berry's work [57] on phase factors accompanying the adiabatic evolution of quantum systems with slowly varying external parameters. However, an analog of the quantum geometric phase was first discovered in 1956 by Pancharatnam [401] in the context of polarization optics. This type of geometric holonomy, sometimes called the Pancharatnam phase, represents a phase difference associated with polarized beams passing through crystals. A few years later, in 1959, Aharonov and Bohm showed [10] that the presence of a vector potential can generate phase factors that modify the interference pattern of charged particles even though the electromagnetic field in the region accessible to these particles is identically zero. In the same period, manifestations of the geometric phase were found in molecular systems in the context of the $E \otimes e$ Jahn–Teller problem [245, 342]. It was noticed that the electronic wave function changes sign when the nuclear coordinates traverse a loop in the nuclear coordinate space that encircles a degeneracy point of the potential energy surfaces. About two decades later, in 1979, Mead and Truhlar [359] showed that this double-valuedness of the electronic wave function can be eliminated if one introduces a gauge potential in the effective electronic Hamiltonian. This first concrete derivation of a geometric phase revealed the close relation between the phase and an underlying gauge potential that emerges naturally in a Born–Oppenheimer-type approximation when the nuclear coordinates are treated as slowly-varying quantum observables.

In 1984 Berry [57] carried out an investigation of geometric phases that accumulate during cyclic evolutions of quantum systems with classical environments that change adiabatically. He derived the Mead–Truhlar gauge potential and the corresponding adiabatic geometric phase – the Berry phase – and showed that the Aharonov–Bohm phase represents a special type of geometric phase. Berry's derivation was restricted to adiabatic cyclic evolutions of nondegenerate pure quantum states. This generates Abelian geometric phases that depend on the geometry of the path traced out by the Hamiltonian in the parameter space. All these restrictions were subsequently removed. In the same year, Wilczek and Zee [547] pointed out that adiabatic transport of a degenerate set of quantum states generates a non-Abelian geometric phase. The generalization to the case of nonadiabatic unitary cyclic evolutions was done by Aharonov and Anandan in 1987 [9]. The Aharonov–Anandan phase depends on the geometry of the path in the state space, rather than parameter space. Subsequently, Samuel and Bhandari [444] introduced the notion of noncyclic geometric phases based upon Pancharatnam's work. A few years earlier, in 1986, Uhlmann [515] investigated the geometric phase, as a mathematical concept, in the context of mixed states. This rapid progress was driven, in part, by the remarkably beautiful mathematical structure of the geometric phase. As Simon [479] realized immediately after Berry's introduction of the concept, the geometric phase can be interpreted as the holonomy of a fiber bundle and the corresponding gauge potential can be viewed as a connection on this fiber bundle (see below, Section 1.5).

Today, geometric phases play a central role in many areas of physics, including the study of condensed matter systems. In 1989, Zak [565] introduced the Berry phase across the Brillouin zone – sometimes called the *Zak phase* – associated with the dynamics of electrons in periodic solids. This led to a breakthrough in understanding the electric polarization of crystalline insulators [290, 430] in the framework of adiabatic charge transport. Adiabatic transport in Bloch bands leads naturally to geometric phases, which can successfully explain the anomalous Hall effect [270, 379] and can be directly connected [503] with the topological Chern number associated with the quantized Hall conductance. Beyond Bloch band theory, geometric phases have important manifestations in interacting many-body systems, including fractional quantum Hall systems [511], spin-wave dynamics in itinerant magnets [392], and the fractional statistics of anyons [28].

We conclude this section with a summary intended to help the reader to discriminate among different types of geometric holonomies. First, we have classical versus quantum holonomies. The parallel transport of a vector on the surface of a sphere, or the rotation of the Foucault pendulum are examples of classical holonomies, which are described by the so-called *Hannay angles* [233]. The quantum analogs of these angles are the geometric phases, sometimes generically called Berry phases. Technically, the Berry phase, e.g., that corresponding to the phase angle γ_λ^B given by Eq. (1.21), is associated with adiabatic cyclic evolutions of nondegenerate pure quantum states. Its nonadiabatic generalization is the Aharonov–Anandan phase, e.g., the phase corresponding to γ_λ given by (1.20). Other generalizations of the Berry phase involving noncyclic evolution, non-Abelian structure, or mixed states, were mentioned above. Second, one can distinguish between holonomies that are purely geometrical and holonomies that have a topological nature, such as the transport of a vector on a Möbius band, or the Aharonov–Bohm phase angle $\gamma_{AB}(C)$ from Eq. (1.23). Note that, unlike the more familiar dynamic phase, the geometric phase is independent of the rate of change along the loop in the parameter space or the state space. In addition, the topological phases are invariant with respect to smooth deformations of the loop that do not change its topological properties, e.g., the number of times it circles a certain singularity. Finally, note that there are two distinct concepts that bear the same name: *topological phase*. The first concept refers to geometric phases that are topological in nature, such as the Aharonov–Bohm phase. The second concept refers to *phases of matter* that have nontrivial topological properties (e.g., topological insulators or quantum Hall fluids). The classification and properties of these topological phases of matter are discussed in Chapters 3–9.

1.4 PHASE CHANGES DURING CYCLIC QUANTUM EVOLUTIONS

In this section, we discuss the formal derivation of the quantal phase factors associated with cyclic evolutions focusing on the adiabatic Abelian case originally considered by Berry. We also discuss the main ideas behind the non-Abelian Wilczek–Zee extension and the nonadiabatic Aharonov–Anandan generalization. For a more detailed account of geometric phases see, for example, Shapere and Wilczek [468] and Bohm et al. [64], and references therein.

1.4.1 The Berry phase

We consider a quantum system described by a Hamiltonian $H = H(\boldsymbol{R})$ that depends on a set of parameters $\boldsymbol{R} = (R_1, R_2, \dots)$ characterizing the environment. Each value of $\boldsymbol{R}$ represents a particular environment configuration and corresponds to a point in the parameter space $\mathcal{M}$, which is a smooth manifold. A changing environment is described by a time-dependent set of parameters $\boldsymbol{R}(t)$ and the adiabatic evolution of the system corresponds to $\boldsymbol{R}(t)$ moving slowly along a certain path $\mathcal{C}$ in the parameter space, $\mathcal{C} : [0, T] \to \mathcal{M}$. The initial and final environment configurations are $\boldsymbol{R}(0) = \boldsymbol{R}_0$ and $\boldsymbol{R}(T) = \boldsymbol{R}_T$, respectively.

We postulate that the Hilbert space $\mathcal{H}$ of physical states is the same for all parameters $\boldsymbol{R} \in \mathcal{M}$. Consequently, the basis for the (unique) space of physical state vectors (i.e., the basis for $\mathcal{H}$) can be chosen, for each $\boldsymbol{R} \in \mathcal{M}$, as the orthonormal basis $|n; \boldsymbol{R}\rangle$ given by the eigenstates of $H(\boldsymbol{R})$,

$$H(\boldsymbol{R})|n; \boldsymbol{R}\rangle = E_n(\boldsymbol{R})|n; \boldsymbol{R}\rangle. \tag{1.24}$$

We assume that the observables, including the eigenvalues $E_n(\boldsymbol{R})$ and the projectors $|n; \boldsymbol{R}\rangle\langle n; \boldsymbol{R}|$, are single-valued functions of $\boldsymbol{R}$ over the whole parameter space $\mathcal{M}$. In particular if $\mathcal{C}$ is a closed path, i.e., $\boldsymbol{R}_T = \boldsymbol{R}_0$, we have $E_n(\boldsymbol{R}_T) = E_n(\boldsymbol{R}_0)$ and $|n; \boldsymbol{R}_T\rangle\langle n; \boldsymbol{R}_T| = |n; \boldsymbol{R}_0\rangle\langle n; \boldsymbol{R}_0|$. Note, however, that this does not imply that the state vectors $|\psi(t)\rangle \in \mathcal{H}$ are single-valued functions over the whole $\mathcal{M}$. In general, it is necessary to use different parameterizations over different patches $\mathcal{O}_i \subset \mathcal{M}$ that cover the parameter space, which leads to different functional dependencies. On the other hand, since Eq. (1.24) determines the state vectors up to a phase factor, one can define equivalent sets of solutions related by the phase transformation,

$$|n; \boldsymbol{R}\rangle' = e^{i\zeta_n(\boldsymbol{R})}|n; \boldsymbol{R}\rangle, \tag{1.25}$$

where $\zeta_n(\boldsymbol{R})$ are arbitrary phase angles. We will only consider transformations corresponding to phase factors $e^{i\zeta_n(\boldsymbol{R})}$ that are single-valued functions
over $\mathcal{M}$, which are called *gauge transformations*. In the overlap region $\mathcal{O}_i \cap \mathcal{O}_j$
between two neighboring patches the eigenvectors corresponding to the two
different parameterizations are related by a gauge transformation. To simplify the discussion, we will restrict ourselves to the case when the path $\mathcal{C}$ is
contained inside a single patch $\mathcal{O} \subset \mathcal{M}$, so that the basis vectors $|n; \boldsymbol{R}\rangle$ are
smooth and single-valued functions of $\boldsymbol{R}$.

Next, we focus on the adiabatic evolution of the system as $\boldsymbol{R}(t)$ moves
along the path $\mathcal{C}$. According to the adiabatic theorem [68, 281], if the sys-
tem starts in a state described by $|n; \boldsymbol{R}_0\rangle$, it will remain in the instantaneous
eigenstate corresponding to $E_n(\boldsymbol{R}(t))$. The only degree of freedom in the problem, the phase, can be determined by solving the time-dependent Schrödinger
equation

$$i\hbar \frac{d}{dt}|\psi_n(t)\rangle = H(\boldsymbol{R}(t))|\psi_n(t)\rangle. \qquad (1.26)$$

One can easily verify that the solution of Eq. (1.26) corresponding to the
initial condition $|\psi_n(0)\rangle = |n; \boldsymbol{R}_0\rangle$ is

$$|\psi_n(t)\rangle = e^{i\gamma_n(t)} \exp\left[-\frac{i}{\hbar}\int_0^t dt' \, E_n(\boldsymbol{R}(t'))\right] |n; \boldsymbol{R}(t)\rangle, \qquad (1.27)$$

where the second exponential represents the familiar dynamical phase. The
additional phase is characterized by a phase angle $\gamma_n(t) = \int_0^t d\boldsymbol{R} \cdot \mathcal{A}^n(\boldsymbol{R})$,
where the vector-valued function $\mathcal{A}^n(\boldsymbol{R})$, called the *Berry connection* or the
Mead–Berry vector potential, is given by

$$\mathcal{A}^n(\boldsymbol{R}) = i\langle n; \boldsymbol{R}|\frac{\partial}{\partial \boldsymbol{R}}|n; \boldsymbol{R}\rangle. \qquad (1.28)$$

Note that the Berry connection is a gauge-dependent quantity. Indeed, as a
result of a gauge transformation (1.25) the Berry connection changes according
to $\mathcal{A}^n(\boldsymbol{R}) \to \mathcal{A}^n(\boldsymbol{R}) - \frac{\partial}{\partial \boldsymbol{R}}\zeta_n(\boldsymbol{R})$. Consequently, the additional phase angle
acquired during the evolution along the path $\mathcal{C}$ will also change, $\gamma_n(T) \to$
$\gamma_n(T) + \zeta_n(\boldsymbol{R}_0) - \zeta_n(\boldsymbol{R}_T)$. This property was noticed by V. A. Fock in 1928
[173]. Addressing the case of noncyclic evolutions, he concluded that a suitable
choice of $\zeta_n(\boldsymbol{R})$ can completely eliminate the additional phase accumulated
along the path $\mathcal{C}$. This conclusion induced the idea that γ_n can always be
neglected; the additional phase was overlooked for more than half a century.

It was Berry who discovered that the above argument does not hold for
cyclic evolutions. For a closed path $\mathcal{C}$, i.e., for $\boldsymbol{R}_T = \boldsymbol{R}_0$, the single-valuedness
of the phase factor $e^{i\zeta_n(\boldsymbol{R})}$ associated with a gauge transformation implies
$\zeta_n(\boldsymbol{R}_T) - \zeta_n(\boldsymbol{R}_0) = 2\pi \times$(integer). Consequently, since the phase angle acquired along a closed path can only be changed by an integer multiple of 2π,
it cannot be removed. The additional phase angle $\gamma_n = \gamma_n(T)$, known as the
Berry phase angle or the *geometric phase* angle, is gauge invariant (modulo

2π) and represents a physical observable. The Berry phase angle is given by

$$\gamma_n = \oint_{\mathcal{C}} d\boldsymbol{R} \cdot \mathcal{A}^n(\boldsymbol{R}) \qquad (\text{mod } 2\pi), \tag{1.29}$$

with the Berry connection $\mathcal{A}^n(\boldsymbol{R})$ given by Eq. (1.28).

The Berry curvature. The Berry connection (1.28) is a gauge–dependent quantity analogous to the vector potential from electrodynamics. The circulation of the vector potential along a closed loop $\mathcal{C}$, i.e., an expression similar to Eq. (1.29), represents the magnetic flux through a surface bounded by $\mathcal{C}$. Therefore, exploiting further the analogy with electrodynamics, we introduce a gauge field tensor $\Omega_{\mu\nu}^n$ and express the Berry phase in terms of this gauge invariant quantity. Explicitly, we have

$$\begin{aligned}
\Omega_{\mu\nu}^n(\boldsymbol{R}) &= \nabla_\mu \mathcal{A}_\nu^n(\boldsymbol{R}) - \nabla_\nu \mathcal{A}_\mu^n(\boldsymbol{R}) \tag{1.30}\\
&= i\left[\langle \nabla_\mu(n;\boldsymbol{R})|\nabla_\nu(n;\boldsymbol{R})\rangle - \langle \nabla_\nu(n;\boldsymbol{R})|\nabla_\mu(n;\boldsymbol{R})\rangle\right],
\end{aligned}$$

where we used the notations $\nabla_\mu = \frac{\partial}{\partial R_\mu}$ and $|\nabla_\mu(n;\boldsymbol{R})\rangle = \frac{\partial}{\partial R_\mu}|n;\boldsymbol{R}\rangle$. The gauge invariant field $\Omega_{\mu\nu}^n$ is called the *Berry curvature* tensor. If the parameter space $\mathcal{M}$ is three dimensional, we can define the Berry curvature vector $\boldsymbol{\Omega}^n = \boldsymbol{\nabla} \times \boldsymbol{A}^n(\boldsymbol{R})$, which is related to the Berry curvature tensor by $\Omega_{\mu\nu}^n(\boldsymbol{R}) = \epsilon_{\mu\nu\lambda}\Omega_\lambda^n(\boldsymbol{R})$, where $\epsilon_{\mu\nu\lambda}$ is the antisymmetric Levi–Civita tensor. This provides a powerful intuitive picture of the Berry curvature as a "magnetic field" in parameter space. Furthermore, using the Stokes theorem one can recast Eq. (1.29) as

$$\gamma_n = \int_{\mathcal{S}(\mathcal{C})} d\boldsymbol{S} \cdot \boldsymbol{\Omega}^n(\boldsymbol{R}) \qquad (\text{mod } 2\pi), \tag{1.31}$$

with $\mathcal{S}(\mathcal{C}) \subset \mathcal{M}$ being a surface bounded by the closed path $\mathcal{C}$. In other words, the Berry phase is the flux of the Berry curvature through the surface $\mathcal{S}(\mathcal{C})$.

Returning to the general case of arbitrary $\dim(\mathcal{M})$, one can express $\Omega_{\mu\nu}^n$ in terms of all energy eigenstates as

$$\Omega_{\mu\nu}^n(\boldsymbol{R}) = i\sum_{m\neq n} \frac{\langle n;\boldsymbol{R}|\nabla_\mu H(\boldsymbol{R})|m;\boldsymbol{R}\rangle\langle m;\boldsymbol{R}|\nabla_\nu H(\boldsymbol{R})|n;\boldsymbol{R}\rangle - (\mu \leftrightarrow \nu)}{[E_n(\boldsymbol{R}) - E_m(\boldsymbol{R})]^2}. \tag{1.32}$$

Equation (1.32) can be obtained from Eq. (1.30) using the relation $\langle n;\boldsymbol{R}|\nabla_\mu H|m;\boldsymbol{R}\rangle = \langle \nabla_\mu(n;\boldsymbol{R})|m;\boldsymbol{R}\rangle(E_n - E_m)$. Again, using the Stokes theorem one can rewrite Eq. (1.29) in the form

$$\gamma_n = \int_{\mathcal{S}(\mathcal{C})} dR_\mu \wedge dR_\nu \, \frac{1}{2}\Omega_{\mu\nu}^n(\boldsymbol{R}) \qquad (\text{mod } 2\pi), \tag{1.33}$$

where $\wedge$ is the *wedge product* and $dR_\mu \wedge dR_\nu$ represents the area of an infinitesimal oriented surface.

Several remarks would be appropriate at this point. First, unlike γ_n, which is associated with a closed path, the Berry curvature is a *local* quantity that characterizes certain geometric properties associated with the dependence of the state on the parameter set $\boldsymbol{R}$. Second, the Berry curvature is gauge independent and can be globally defined on $\mathcal{M}$. By contrast, the Berry connection (1.28) has to be calculated using smooth single–valued eigenvectors $|n; \boldsymbol{R}\rangle$ defined on a certain patch in the parameter space. This condition is not required when using Eqs. (1.32) and (1.33) to calculate the geometric phase. Third, Eq. (1.32) suggests that the Berry curvature can be thought of as the result of a "residual interaction" of the nth energy level with the other energy levels that were projected out as a result of the adiabatic approximation. Note that the total Berry curvature $\sum_n \Omega^n_{\mu\nu}(\boldsymbol{R})$ is identically zero. Also note that $\Omega^n_{\mu\nu}(\boldsymbol{R})$ becomes singular at a degeneracy point $\boldsymbol{R}^*$ of two energy bands, $E_n(\boldsymbol{R}^*) = E_m(\boldsymbol{R}^*)$. As shown below, such a degeneracy point behaves like a monopole in the parameter space. Finally, the formalism developed above is valid for a nondegenerate energy level $E_n(\boldsymbol{R})$. In the case of degenerate levels, the dynamics has to be projected onto the subspace spanned by the corresponding states and the Berry connection becomes a non-Abelian matrix of dimension equal to the degeneracy (see below, Section 1.4.2).

Degeneracy points. To illustrate the general formalism, we consider a basic example of significant practical importance: the two-level system. The generic Hamiltonian of a two-level system has the form

$$H(\boldsymbol{R}) = \epsilon(\boldsymbol{R})\sigma_0 + \boldsymbol{d}(\boldsymbol{R}) \cdot \boldsymbol{\sigma}, \tag{1.34}$$

where σ_0 is the 2×2 identity matrix and $\boldsymbol{\sigma} = (\sigma_x, \sigma_y, \sigma_z)$ are the Pauli matrices. A specific example was already discussed in Section 1.3.1, namely the Hamiltonian in Eq. (1.12) describing a spin-$\frac{1}{2}$ particle in a rotating magnetic field. In general, the two energy levels are $E_\pm = \epsilon \pm \sqrt{\boldsymbol{d} \cdot \boldsymbol{d}}$, where the dependence on the parameter set $\boldsymbol{R}$ was omitted for simplicity. Note that the additive term $\epsilon(\boldsymbol{R})$ can be neglected without loss of generality. Also, it is convenient to re-parameterize the Hamiltonian in terms of three independent parameters: $\boldsymbol{d} = (d_x, d_y, d_z)$ or, using spherical coordinates, $\boldsymbol{d} = d(\sin\theta\cos\phi, \sin\theta\sin\phi, \cos\theta)$. Note that the point $\boldsymbol{d}^* = 0$ is a degeneracy point with $E_-(\boldsymbol{d}^*) = E_+(\boldsymbol{d}^*) = 0$. The eigenstates corresponding to $E_\pm = \pm d$ are

$$|-; \boldsymbol{d}\rangle = \begin{pmatrix} \sin\frac{\theta}{2}e^{-i\phi} \\ -\cos\frac{\theta}{2} \end{pmatrix}, \qquad |+; \boldsymbol{d}\rangle = \begin{pmatrix} \cos\frac{\theta}{2} \\ \sin\frac{\theta}{2}e^{i\phi} \end{pmatrix}. \tag{1.35}$$

Using these eigenstates and Eq. (1.28) we can calculate the Berry connection. For the low-energy level, we have

$$\mathcal{A}^{(-)}_d = 0, \qquad \mathcal{A}^{(-)}_\theta = 0, \qquad \mathcal{A}^{(-)}_\phi = \sin^2\frac{\theta}{2}. \tag{1.36}$$

Note that the eigenstates (1.35) are not single-valued for $\theta = 0$, as they have undefined phases. Changing the gauge, e.g., $|\mp; \boldsymbol{d}\rangle \rightarrow e^{\pm i\phi}|\mp; \boldsymbol{d}\rangle$, generates eigenstates that are smooth and single-valued everywhere except at $\theta = \pi$. Using this gauge, we get $\mathcal{A}_d^{(-)} = \mathcal{A}_\theta^{(-)} = 0$ and $\mathcal{A}_\phi^{(-)} = -\cos^2\frac{\theta}{2}$. However, the Berry curvature is gauge independent and we have for any $\boldsymbol{d} \neq 0$

$$\Omega_{\theta\phi}^{(-)} = \nabla_\theta \mathcal{A}_\phi^{(-)} - \nabla_\phi \mathcal{A}_\theta^{(-)} = \frac{1}{2}\sin\theta, \tag{1.37}$$

all other tensor components being zero. For an arbitrary set of parameters $\boldsymbol{R}$, the Berry curvature can be calculated using the relation

$$\Omega_{\mu\nu}^{(-)} = \Omega_{\theta\phi}^{(-)} \frac{\partial(\theta,\phi)}{\partial(R_\mu, R_\nu)} = \frac{1}{2}\frac{\partial(\phi,\cos\theta)}{\partial(R_\mu, R_\nu)}, \tag{1.38}$$

where $\frac{\partial(\theta,\phi)}{\partial(R_\mu,R_\nu)} = \frac{\partial\theta}{\partial R_\mu}\frac{\partial\phi}{\partial R_\nu} - \frac{\partial\theta}{\partial R_\nu}\frac{\partial\phi}{\partial R_\mu}$ is the Jacobian of the corresponding parameter transformation. In the particular case $\boldsymbol{R} = (d_x, d_y, d_z)$, we obtain $\Omega_{ij}^{(-)} = \frac{1}{2d^3}\epsilon_{ijk}d_k$ or, in terms of the Berry curvature vector,

$$\boldsymbol{\Omega}^{(-)} = \frac{1}{2}\frac{\boldsymbol{d}}{d^3}. \tag{1.39}$$

One can recognize $\boldsymbol{\Omega}^{(-)}$ as the field generated by a monopole of strength $1/2$ placed at the origin (in the parameter space), which coincides with the degeneracy point $\boldsymbol{d}^* = 0$. Note that the curvature of the positive-energy band has opposite sign, $\boldsymbol{\Omega}^{(+)} = -\boldsymbol{\Omega}^{(-)}$. If the system is adiabatically transported along a closed loop $\mathcal{C} \subset \mathcal{M}$, the state vector will acquire a geometric phase

$$\gamma_\pm(\mathcal{C}) = \int_{\mathcal{S}(\mathcal{C})} d\boldsymbol{S} \cdot \boldsymbol{\Omega}^\pm = \mp\frac{1}{2}\Omega(\mathcal{C}), \tag{1.40}$$

where $\Omega(\mathcal{C})$ is the solid angle subtended by $\mathcal{C}$ at the degeneracy point. This result confirms Eq. (1.21), where the Berry phase for a $\frac{1}{2}$-spin in a rotating magnetic field was determined directly by calculating the evolution of the wave function. Finally, we note that the integral of the Berry curvature over any closed manifold is quantized in units of 2π. For example, we have $\int d\theta d\phi\, \Omega_{\theta\phi} = 2\pi$. In general, the integer $\nu_\pm = \frac{1}{2\pi}\oint dR_\mu \wedge dR_\nu \frac{1}{2}\Omega_{\mu\nu}^{(\pm)}$, called the *Chern number*, is equal to the net number of monopoles enclosed by the manifold.

The Aharonov–Bohm phase. It would be instructive to briefly revisit the Aharonov–Bohm effect from the perspective of the general formalism developed in this section. We have shown (see Section 1.3.1) that, in the presence of an impenetrable flux line (which we choose to coincide with the z coordinate axis), the nth eigenstate of a particle confined inside a box is given by $\langle \boldsymbol{r}|n; \boldsymbol{R}\rangle = \psi_n(\boldsymbol{r}, \boldsymbol{R})$, where $\boldsymbol{r}$ is the particle position vector, $\boldsymbol{R}$ represents the position of the box, and $\psi_n(\boldsymbol{r}, \boldsymbol{R})$ is given by Eq. (1.22). The Hamiltonian of the system depends parametrically on $\boldsymbol{R}$, so that the parameter space associated with this problem is $\mathcal{M} = \mathbb{R}^3 - \{(0, 0, z) | z \in \mathbb{R}\}$. As the box containing the particle is adiabatically transported along a closed loop $\mathcal{C}$, i.e., $\boldsymbol{R}(t) \in \mathcal{C}$, the state vector acquires a geometric phase γ_n. To determine it, we first calculate the corresponding Berry connection using Eq. (1.28) and the wave function $\psi_n(\boldsymbol{r}, \boldsymbol{R})$. We have

$$
\begin{aligned}
\mathcal{A}^n(\boldsymbol{R}) &= i \int d^3 r \; [\psi_n^0(\boldsymbol{r} - \boldsymbol{R})]^* \left\{ -\frac{iq}{\hbar} A(\boldsymbol{R}) \psi_n^0(\boldsymbol{r} - \boldsymbol{R}) + \nabla \psi_n^0(\boldsymbol{r} - \boldsymbol{R}) \right\} \\
&= \frac{q}{\hbar} A(\boldsymbol{R}),
\end{aligned}
\tag{1.41}
$$

where $\psi_n^0(\boldsymbol{r} - \boldsymbol{R})$ is the nth energy eigenstate of the confined particle in the absence of a magnetic flux and we have used the property $\langle \psi_n^0 | \nabla | \psi_n^0 \rangle = 0$. Note that the Berry connection is proportional to the vector potential associated with the external magnetic field. Again, we emphasize that the magnetic field $\boldsymbol{B}(\boldsymbol{R}) = \nabla \times \boldsymbol{A}(\boldsymbol{R})$ is zero for any $\boldsymbol{R} \in \mathcal{M}$, i.e., everywhere except inside the infinitesimally thin flux tube. Finally, to calculate the geometric phase associated with the adiabatic transport of the box along a closed path $\mathcal{C}$, we use Eq. (1.29) with $\mathcal{A}^n(\boldsymbol{R})$ determined above and the Stokes theorem. We get $\gamma_n = \gamma_{AB}(\mathcal{C})$, where $\gamma_{AB}(\mathcal{C})$ is the Aharonov–Bohm phase given by Eq. (1.23).

The Berry phase in Bloch bands. The geometric phase is a critical concept in the modern theory of crystalline systems. Here we introduce the basic ideas that will enable us to apply the generic Berry phase formalism to band theory. For more details see, for example, Chang and Niu [97] and Xiao et al. [558].

The electronic properties of the crystal are described within the independent particle approximation by the single-particle Hamiltonian

$$
H = \frac{-\hbar^2}{2m_0} \nabla^2 + V(\boldsymbol{r}),
\tag{1.42}
$$

where m_0 is the electron mass and $V(\boldsymbol{r}) = V(\boldsymbol{r} + \boldsymbol{a}_i)$ is a periodic potential, $\boldsymbol{a}_i$ being the primitive lattice vectors. An eigenstate of the Hamiltonian with eigenvalue $E_n(\boldsymbol{q})$ will satisfy the Bloch condition

$$
\psi_{n\boldsymbol{q}}(\boldsymbol{r} + \boldsymbol{a}_i) = e^{i\boldsymbol{q} \cdot \boldsymbol{a}_i} \psi_{n\boldsymbol{q}}(\boldsymbol{r}),
\tag{1.43}
$$

where n is the band index, while $\boldsymbol{q}$ is the Bloch wave number (i.e., $\hbar\boldsymbol{q}$ is the crystal momentum) and takes values inside the Brillouin zone (BZ). Equation (1.43) can be regarded as a $\boldsymbol{q}$-dependent boundary condition. To fully exploit the periodicity of the problem, we reduce the Hilbert space to that associated with a single unit cell. Since the $\boldsymbol{q}$-dependent condition (1.43) will generate a multitude of Hilbert spaces, we perform a unitary transformation and introduce the cell-periodic function

$$
u_{n\boldsymbol{q}}(\boldsymbol{r}) = e^{-i\boldsymbol{q} \cdot \boldsymbol{r}} \psi_{n\boldsymbol{q}}(\boldsymbol{r}), \qquad u_{n\boldsymbol{q}}(\boldsymbol{r} + \boldsymbol{a}_i) = u_{n\boldsymbol{q}}(\boldsymbol{r}).
\tag{1.44}
$$

The price for this is a q-dependent Hamiltonian, $H \to H(q) = e^{-iq \cdot r} H e^{iq \cdot r}$,

$$H(q) = \frac{-\hbar^2}{2m_0}(\nabla + iq)^2 + V(r). \tag{1.45}$$

At this point, we can make the connection with the general Berry phase formalism. We have a parameter-dependent Hamiltonian, $H(R) \to H(q)$, and a single Hilbert space $\mathcal{H}$, so that for each value of the parameter $R \to q$ the set $|n; R\rangle \to |u_n(q)\rangle$ represents an orthonormal basis for $\mathcal{H}$. The parameter space is the Brillouin zone, $\mathcal{M} \to BZ$. A slow cyclic variation of q in the Brillouin zone, which may be caused by an external field that enters the Schrödinger equation as a time-dependent variation of the wave vector, $q \to q + \Delta q(t)$, will result in the wave function acquiring a geometric phase. For example, applying a magnetic field generates cyclotron motion along closed orbits in the BZ. In this case, the Berry phase will be manifested in various magneto-oscillatory effects [364, 365]. If, on the other hand, one applies an electric field, Δq is linear in t and q sweeps the entire Brillouin zone (which has the topology of a torus). The corresponding geometric phase, called the *Zak phase* [565], plays an important role in, for example, adiabatic transport [418, 502] and electric polarization of crystalline solids [290, 398, 430].

To account for these effects, we introduce the Berry connection and the corresponding Berry curvature vector

$$\mathcal{A}^n(q) = i\langle u_n(q)|\nabla_q|u_n(q)\rangle, \tag{1.46}$$

$$\Omega^n(q) = \nabla_q \times \mathcal{A}^n(q) = i\langle \nabla_q u_n(q)| \times |\nabla_q u_n(q)\rangle, \tag{1.47}$$

where $|\nabla_q u_n(q)\rangle = \nabla_q|u_n(q)\rangle$. The corresponding Berry phase is given by Eq. (1.29) or Eq. (1.31). Similarly, the Zak phase can be written as $\gamma_n = \int_{\mathcal{C}_{BZ}} dq \cdot \langle u_n(q)|i\nabla_q|u_n(q)\rangle$, where $\mathcal{C}_{BZ}$ is a path across the Brillouin zone, e.g., $(0, q_y, q_z) \to (2\pi/a, q_y, q_z)$ for a simple cubic lattice. Finally, we note that introducing a slow time-dependence in $V(r)$, so that it preserves the spatial periodicity, results in the Hamiltonian $H(q, t)$ and the eigenstates $|u_n(q, t)\rangle$ becoming explicitly time dependent. In this case, we treat q and t as independent parameters, i.e., we have $R \to (q, t)$. The Berry connection (1.46), which can be viewed as a *geometric vector potential*, becomes time dependent, $\mathcal{A}^n = \mathcal{A}^n(q, t)$. In addition, there is a component associated with the parameter t, which can be interpreted as a *geometric scalar potential*

$$\chi^n(q, t) = i\langle u_n(q)|\frac{\partial}{\partial t}|u_n(q)\rangle. \tag{1.48}$$

All these quantities play key roles in the dynamics of the Bloch electrons under slowly varying external perturbations. For example, the semiclassical equations of motion for a wave packet with crystal momentum $\hbar q_c$ involve an extra contribution, known as the *anomalous velocity*, $\Delta v = -\dot{q}_c \times \Omega^n(q_c) + \frac{\partial \mathcal{A}^n}{\partial t} + \nabla_q \chi^n$. In this equation, the Berry curvature vector $\Omega^n(q_c)$ acts as a "magnetic" field in k-space, while the last two terms are due to the time dependence of the band structure.

1.4.2 The non-Abelian adiabatic phase

In the previous section, we have considered the cyclic adiabatic evolution of a quantum system with nondegenerate energy levels $E_n(\boldsymbol{R})$. In Eq. (1.32), for example, the nondegeneracy condition is a manifest requirement that ensures a finite Berry curvature. Here, we lift this requirement and allow the energy level E_n to be g_n-fold degenerate. For simplicity, we assume that the degeneracy is $\boldsymbol{R}$-independent. This generalization of Berry's construction to the case of degenerate energy levels, first done by Wilczek and Zee [547] in 1984, leads to non-Abelian gauge fields.

For any given value of the parameter $\boldsymbol{R} \in \mathcal{M}$ the set of eigenstates $|n, a; \boldsymbol{R}\rangle$ with $a = 1, 2, \ldots, g_n$ provides an orthonormal basis for the degeneracy subspace $\mathcal{H}_n(\boldsymbol{R})$ associated with the nth energy level $E_n(\boldsymbol{R})$. The basis states are assumed to be single-valued, i.e., for a cyclic evolution with $\boldsymbol{R}(T) = \boldsymbol{R}(0)$ we have $|n, a; \boldsymbol{R}_T\rangle = |n, a; \boldsymbol{R}_0\rangle$. The adiabatic condition ensures that any eigenstate vector $|n, a; \boldsymbol{R}_0\rangle \in \mathcal{H}_n(\boldsymbol{R}_0)$ will evolve in such a way that it remains an element of the degenerate subspace $\mathcal{H}_n(\boldsymbol{R}(t))$ at any $t \in [0, T]$. Note that the subspace $\mathcal{H}_n(\boldsymbol{R})$ is parameter dependent, while the total Hilbert space $\mathcal{H} = \bigotimes_n \mathcal{H}_n(\boldsymbol{R})$ is not. Also note that the choice of basis for $\mathcal{H}_n(\boldsymbol{R})$ is arbitrary, but any two choices are related by a unitary transformation

$$|n, a; \boldsymbol{R}\rangle' = \sum_{b=1}^{g_n} |n, b; \boldsymbol{R}\rangle \, \mathcal{U}_{ba}^n(\boldsymbol{R}), \qquad (1.49)$$

where $\mathcal{U}_{ba}^n(\boldsymbol{R})$ are the matrix elements of a unitary matrix $\mathcal{U}^n(\boldsymbol{R}) \in U(g_n)$. Note that Eq. (1.49) is a generalization of the gauge transformation (1.25), which involves the phase factor $e^{i\zeta_n(\boldsymbol{R})} \in U(1)$, with $U(N)$ designating the unitary group of degree N (for the definition of a *group* see page 45).

Next, we are considering a cyclic evolution of the quantum system corresponding to a closed curve $\mathcal{C}$ in the parameter space $\mathcal{M}$. We assume that the initial state belongs to the nth energy eigenspace, $|\psi_n(0)\rangle \in \mathcal{H}_n(\boldsymbol{R}_0)$. The adiabatic theorem ensures that the state vector will belong to $\mathcal{H}_n(\boldsymbol{R}(t))$ at any later time t and that after one cycle it will return to the initial subspace, since $\mathcal{H}_n(\boldsymbol{R}_T) = \mathcal{H}_n(\boldsymbol{R}_0)$. However, the final state vector $|\psi_n(T)\rangle \in \mathcal{H}_n(\boldsymbol{R}_0)$ is not required to be parallel to $|\psi_n(0)\rangle$. Hence, as the result of its evolution, the state vector undergoes a rotation in the degeneracy subspace $\mathcal{H}_n(\boldsymbol{R}_0)$, in addition to acquiring the standard dynamical phase $e^{-\frac{i}{\hbar} \int_0^T dt E_n(\boldsymbol{R}(t))}$. Formally, this holonomy can be expressed as

$$|\psi_n(T)\rangle = \exp\left[-\frac{i}{\hbar} \int_0^T dt \, E_n(\boldsymbol{R}(t))\right] \hat{\mathcal{U}}_{\mathcal{C}}^n |\psi_n(0)\rangle, \qquad (1.50)$$

where $\hat{\mathcal{U}}_{\mathcal{C}}^n$ is a unitary operator that rotates the basis vectors in the degeneracy subspace $\mathcal{H}_n(\boldsymbol{R}_0)$, more specifically $\hat{\mathcal{U}}_{\mathcal{C}}^n |n, a; \boldsymbol{R}_0\rangle = |n, a; \boldsymbol{R}_0\rangle'$, with the rotated vectors related to the original basis through a unitary matrix

$\mathcal{U}_{\mathcal{C}}^n \in U(g_n)$, as given by Eq. (1.49). Note that the summation in Eq. (1.49) is over the first index of the matrix element. Comparing Eqs. (1.50) and (1.27), we note that the unitary transformation associated with the operator $\hat{\mathcal{U}}_{\mathcal{C}}^n$ (or, equivalently, the matrix $\mathcal{U}_{\mathcal{C}}^n$) represents the non-Abelian generalization of the Berry phase $e^{i\gamma_n} \in U(1)$. The unitary matrix $\mathcal{U}_{\mathcal{C}}^n$ can be expressed in terms of the non-Abelian analog of the Berry connection, which is a vector-valued matrix with matrix elements

$$[\mathcal{A}^n]_{ab}(\boldsymbol{R}) = i\langle n, a; \boldsymbol{R}| \frac{\partial}{\partial \boldsymbol{R}} |n, b; \boldsymbol{R}\rangle. \tag{1.51}$$

One can show that the non-Abelian Berry connection $\mathcal{A}_\mu^n$ is a $g_n \times g_n$ Hermitian matrix that transforms under a gauge transformation (1.49) according to

$$\mathcal{A}_\mu^n \rightarrow [\mathcal{U}^n]^{-1} \mathcal{A}_\mu^n \, \mathcal{U}^n + i[\mathcal{U}^n]^{-1} \frac{\partial}{\partial R_\mu} \mathcal{U}^n. \tag{1.52}$$

The non-Abelian generalization of the Berry phase can be formally expressed in terms of $\mathcal{A}^n$ as a path-ordered exponential,

$$\mathcal{U}_{\mathcal{C}}^n = \mathcal{P} \exp\left(\oint_{\mathcal{C}} d\boldsymbol{R} \cdot \mathcal{A}^n(\boldsymbol{R}) \right), \tag{1.53}$$

where $\mathcal{P}$ is the path-ordering operator. Note that, in general, the matrices $\mathcal{A}_\mu^n(\boldsymbol{R})$ do not commute and their order in the series expansion representing the exponential is determined by $\mathcal{P}$. Also, unlike its Abelian counterpart, the non-Abelian geometric phase is gauge covariant, $\mathcal{U}_{\mathcal{C}}^n \rightarrow [\mathcal{U}^n]^{-1}(\boldsymbol{R}_0)\mathcal{U}_{\mathcal{C}}^n \mathcal{U}^n(\boldsymbol{R}_0)$, hence it is not directly observable. However, quantities such as the eigenvalues of $\mathcal{U}_{\mathcal{C}}^n$ or its trace $\text{Tr}[\mathcal{U}_{\mathcal{C}}^n]$, also called a *Wilson loop*, are gauge invariant and thus experimentally measurable.

As a final remark, we note that this formalism can be easily adapted to the case of degenerate Bloch bands. For example, the Berry connection (1.46) becomes the vector-valued matrix $[\mathcal{A}^n]_{ab}(\boldsymbol{q}) = i\langle u_{na}(\boldsymbol{q})|\boldsymbol{\nabla}_{\boldsymbol{q}}|u_{nb}(\boldsymbol{q})\rangle$. The corresponding Berry curvature, defined as $\boldsymbol{\Omega}^n = \boldsymbol{\nabla}_{\boldsymbol{q}} \times \mathcal{A}^n - i\mathcal{A}^n \times \mathcal{A}^n$, contains a second term that vanishes in the Abelian case. Remarkably, these quantities have the same structure as the gauge potential ($\mathcal{A}^n$) and the gauge field ($\boldsymbol{\Omega}^n$) in a non-Abelian $SU(2)$ gauge theory [547].

1.4.3   The Aharonov–Anandan phase

The Berry phase formalism introduced so far was developed based on the assumption that the quantum system evolves adiabatically. Here, we examine the phenomenon of geometric phase for nonadiabatic cyclic evolutions. This generalization of the Berry phase was proposed in 1987 by Aharonov and Anandan [9]. A specific example – the geometric phase acquired by a spin-$\frac{1}{2}$ particle in a rotating magnetic field – was discussed in Section 1.3.1.

We consider a quantum system described by the Hamiltonian $H(t) = H(\boldsymbol{R}(t))$ with a time-dependence given by the path $\mathcal{C}_R \ni \boldsymbol{R}(t)$, $0 \leq t \leq \tau$, in the parameter space $\mathcal{M}$. The physical states of the system represent equivalence classes (rays) in a complex Hilbert space $\mathcal{H}$. The set of these equivalence classes corresponds to the *projective Hilbert space* $\mathcal{P}(\mathcal{H})$. The density operator $\rho(t) = |\psi(t)\rangle\langle\psi(t)|$, where $\psi \in \mathcal{H}$ is a solution of the Schrödinger equation, carries no redundant phase information and represents an element of $\mathcal{P}(\mathcal{H})$. We define the projection map $\pi : \mathcal{H} \to \mathcal{P}(\mathcal{H})$ by the relation $\pi(|\psi(t)\rangle) = \rho(t)$, which means that all state vectors $|\psi(t)\rangle$ representing a given physical state are projected onto the corresponding density operator $\rho(t) \in \mathcal{P}(\mathcal{H})$.

An arbitrary cyclic evolution of the quantum system with period τ corresponds to a closed path $\mathcal{C} \subset \mathcal{P}(\mathcal{H})$ in the projective Hilbert space. This means that the evolution of the system $\mathcal{C} : \rho(0) \to \rho(t) \to \rho(\tau)$ satisfies the condition $\rho(0) = \rho(\tau)$. Note that $\mathcal{C}$, which is a closed curve in the space of physical states $\mathcal{P}(\mathcal{H})$, is distinct from $\mathcal{C}_R$, which is a path (not necessarily closed) in the parameter space $\mathcal{M}$. During the cyclic evolution along $\mathcal{C}$, the state vector $|\psi(t)\rangle$, which is the solution of the time-dependent Schrödinger equation, evolves along the associated curve $C_\psi \subset \mathcal{H}$. Of course, we have $\pi(C_\psi) = \mathcal{C}$, i.e., at any time t the state vector is mapped onto the corresponding physical state. However, generally the curve C_ψ is not closed, because the condition for cyclic evolutions, $\rho(\tau) = \rho(0)$, does not imply $|\psi(\tau)\rangle = |\psi(0)\rangle$, but rather

$$|\psi(\tau)\rangle = e^{i\alpha}|\psi(0)\rangle. \tag{1.54}$$

To identify the phase $e^{i\alpha}$ acquired by the state vector during the cyclic evolution of the quantum system, we introduce the smooth and singe-valued function $|\phi(t)\rangle = |\phi(\boldsymbol{R}(t))\rangle \in \mathcal{H}$ that satisfies the property $\rho(t) = |\phi(t)\rangle\langle\phi(t)|$. Note that $|\phi(\boldsymbol{R}(t))\rangle$ is defined up to a gauge transformation, $|\phi(\boldsymbol{R}(t))\rangle \to e^{i\zeta(\boldsymbol{R}(t))}|\phi(\boldsymbol{R}(t))\rangle$. Also, since $|\phi(\tau)\rangle) = |\phi(0)\rangle$, as dictated by the single-valuedness condition, the path $C_\phi \subset \mathcal{H}$ defined by $|\phi(t)\rangle$ during the evolution of the system is closed. The curves C_ψ and C_ϕ are called the *dynamical lift* and the *closed lift* of $\mathcal{C}$, respectively, and we have $\pi(C_\phi) = \pi(C_\psi) = \mathcal{C}$. The state vectors $|\psi(t)\rangle$ and $|\phi(t)\rangle$ represent the same physical state, hence they can differ only by a phase factor, $|\psi(t)\rangle = e^{i\alpha(t)}|\phi(t)\rangle$. We introduce this relation into the Schrödinger Eq. and solve for $\alpha(t)$. Using the property $\langle\phi(t)|H(t)|\phi(t)\rangle = \langle\psi(t)|H(t)|\psi(t)\rangle$, we obtain

$$|\psi(t)\rangle = \exp\left[-\frac{i}{\hbar}\int_0^t dt'\langle\psi(t')|H(t')|\psi(t')\rangle\right]e^{i\gamma(t)}|\phi(t)\rangle, \tag{1.55}$$

where the first exponential represents the dynamical phase and the additional phase angle is $\gamma(t) = \int_0^t dt' i\langle\phi(t')|\frac{d}{dt'}|\phi(t')\rangle$. Note that Eq. (1.55) is the nonadiabatic generalization of (1.27). The additional phase acquired during the cyclic evolution corresponding to the closed path $\mathcal{C}$ in $\mathcal{P}(\mathcal{H})$, called the Aharonov–Anandan (AA) phase, is given by the phase angle

$$\gamma(\mathcal{C}) = \oint i\langle\phi(t)|\frac{d}{dt}|\phi(t)\rangle dt, \tag{1.56}$$

where $|\phi(t)\rangle$ corresponds to any of the gauge-dependent closed lifts C_ϕ.

Equation (1.56) represents the nonadiabatic generalization of the Berry phase angle. In the particular case of a spin-$\frac{1}{2}$ particle in a rotating magnetic field (see Section 1.3.1), the closed lift corresponds to the states $|\phi(t)\rangle = |\lambda, \hat{\boldsymbol{R}}(\theta', \omega t)\rangle$ and the AA phase angle is given by Eq. (1.20). Reproducing this result using Eq. (1.56) is left as an exercise. Note that the phase angle $\gamma(\mathcal{C})$ is gauge independent (although $|\phi(t)\rangle$ is not) and independent of the speed with which $|\phi(t)\rangle$ traverses the path C_ϕ. Also, while the Berry phase depends on the geometry of the parameter space $\mathcal{M}$ and on the path $\mathcal{C}_{\boldsymbol{R}}$, the AA phase depends on the curve $\mathcal{C}$ and the geometry of the projective space $\mathcal{P}(\mathcal{H})$. The name *geometric phase* expresses this relation between $e^{i\gamma(\mathcal{C})}$ and the geometric properties of the space of physical states. Finally, we note that the geometric phase introduced here requires the cyclic evolution of the physical state of the system, but not a cyclic Hamiltonian. In other words, we can define the geometric phase even for evolutions with $H(\tau) \neq H(0)$, provided that $\rho(\tau) = \rho(0)$.

1.5 THE MATHEMATICAL STRUCTURE OF GEOMETRIC PHASES

The geometric phase can be interpreted as the holonomy of a fiber bundle over a certain manifold. In this section, we provide an elementary introduction to basic concepts in differential geometry that are relevant in the context of geometric phases. For a more in-depth treatment of this material, the reader is referred to introductory textbooks, for example Refs. [259, 380, 383, 459]. Our goal here is to provide the reader with the basic language and simplest ideas necessary to appreciate the significance, universality, and aesthetic appeal of the mathematical structure underlying the geometric phase. For simplicity, we will limit our discussion to the Abelian case.

1.5.1 Elementary introduction to fiber bundles

In this section, we introduce some basic elements of fiber bundle theory. A thorough treatment of this subject can be found in the literature, for example Refs. [259, 380, 383]. Here, we focus on main concepts and ideas, particularly those that are relevant to the holonomy interpretation of the geometric phase.

A *fiber bundle E* consists of a *total space E* (note that we use the same symbol as for the fiber bundle), a *base manifold M*, and a set of *fibers F_x*, with $x \in M$. All fibers belong to the same class of smooth manifolds, i.e., they are diffeomorphic copies of an N-dimensional manifold F called the *typical fiber*. The total space E can be thought of as the collection of fibers. More specifically, let the projection map $\pi : E \to M$ be a surjective (onto) function so that for any point $p \in E$ we have $\pi(p) = x \in M$ if and only if $p \in F_x$. Then, the fiber F_x over $x \in M$ is the inverse image of x under π, $F_x = \pi^{-1}(x)$ (Figure 1.5). In other words, we associate a fiber F_x to every point x of the base space and the collection of all these fibers represents the total space E.

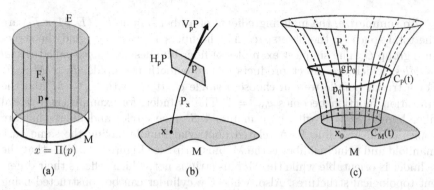

FIGURE 1.5 (a) The cylinder as a (trivial) fiber bundle with base manifold $M = S^1$ (the unit circle) and typical fiber $F = \mathbb{R}$. Note that the Möbius strip (Figure 1.4) is a nontrivial fiber bundle having the same base manifold and typical fiber. (b) Decomposition of the tangent space into vertical (V_pP) and horizontal (H_pP) subspaces. (c) Horizontal lift $C_P(t)$ of a closed curve $C_M(t)$ in the base manifold M. The lift is characterized by a holonomy g.

A fiber bundle is locally diffeomorphic to the Cartesian product of an open subset of M with the typical fiber F. Explicitly, $\pi^{-1}(U_x) \simeq U_x \times F$, where $U_x \subset M$ is an open neighborhood of x, $\pi^{-1}(U_x) \subset E$ is the fiber bundle over U_x, and "$\simeq$" means "diffeomorphic." Moreover, we can cover the base manifold M with a collection $\{U_\alpha\}$ of open subsets such that there exist diffeomorphisms

$$\phi_\alpha : \pi^{-1}(U_\alpha) \longrightarrow U_\alpha \times F. \tag{1.57}$$

The pair (U_α, ϕ_α) is called a *chart*, or a set of *local coordinates*, or a *local trivialization* of the bundle E. Restricting ϕ_α to a point $x \in U_\alpha$, we obtain a diffeomorphism from the fiber F_x to the typical fiber, $\phi_\alpha(x) : \{x\} \times F_x \longrightarrow \{x\} \times F$, or, for simplicity, $\phi_\alpha(x) : F_x \longrightarrow F$.

Similarly to the case of manifolds, the global topological structure of the bundle is determined by how the charts are patched together, i.e., by the *transition functions*

$$g_{\alpha\beta}(x) : F \longrightarrow F, \qquad g_{\alpha\beta}(x) = \phi_\alpha(x)\phi_\beta^{-1}(x)|_{U_\alpha \cap U_\beta}, \tag{1.58}$$

where the product $\phi_\alpha(x)\phi_\beta^{-1}(x)$ denotes the composition of two functions, i.e., $\phi_\alpha(x) \circ \phi_\beta^{-1}(x)$. The transition functions satisfy two important conditions,

$$g_{\alpha\beta}^{-1}(x) = g_{\beta\alpha}(x), \qquad \forall x \in U_\alpha \cap U_\beta \tag{1.59a}$$

$$g_{\alpha\beta}(x)g_{\beta\gamma}(x)g_{\gamma\alpha}(x) = 1, \quad \forall x \in U_\alpha \cap U_\beta \cap U_\gamma \tag{1.59b}$$

where 1 is the identity function on F. As a consequence, the transition functions $g_{\alpha\beta}$ form a group under composition, a subgroup of the so-called *structure group* G of the bundle E, which is a Lie group of diffeomorphisms (smooth transformations) of F.

To summarize, the main ingredients of a fiber bundle $E : (E, M, \pi, G)$ are the total space E, the base space M, the fibers $F_x = \pi^{-1}(x)$ and the structure group G. The simplest examples of fiber bundles are the so-called *trivial bundles*, which are direct products of two smooth manifolds, $E \simeq M \times F$. For a trivial bundle we can choose a single chart ϕ_α with $U_\alpha = M$ and the transition function becomes $g_{\alpha\alpha} = 1$. The cylinder, for example, is a trivial fiber bundle $S^1 \times \mathbb{R}$ with base manifold S^1 (the circle) and fibers that are copies of the real line $\mathbb{R}$. A related nontrivial bundle having the same base manifold and principal fiber is the Möbius strip (see Figure 1.4). Note that the cylinder is orientable while the Möbius strip is not, which reflects their different topological structures. Also, while the cylinder can be constructed using one chart, the Möbius strip requires (at least) two charts, which are patched together via transition functions belonging to the group $\mathbb{Z}_2$. Intuitively, this can be easily understood if we think about the construction of the Möbius strip using two oriented bands that are glued together with the same orientation at one end and the opposite orientation at the other. Finally, note that $\mathbb{Z}_2$ is a subgroup of the general linear group[7] $GL(1, \mathbb{R})$, the group of nonzero real numbers under multiplication, which is the structure group of the Möbius strip.

The cylinder and the Möbius strip are examples of *vector bundles*. These are bundles with the typical fiber being a real (or complex) vector space and the structure group the general linear group, $GL(N, \mathbb{R})$ or $GL(N, \mathbb{C})$. Another important class, the *principal fiber bundles*, are bundles whose typical fiber is identical with the structure group, $F \simeq G$. Here "identical" means that, as smooth manifolds, the two objects are diffeomorphic. As we will discuss in the next section, the geometric phase can be naturally interpreted as a geometric property of certain principal fiber bundles.

The geometric structure of a fiber bundle can be investigated using the notion of parallelism. Consider first a smooth manifold M embedded in a Euclidean space $\mathbb{R}^n$ and two tangent vectors $v_1 \in TM_{x_1}$ and $v_2 \in TM_{x_2}$ at x_1 and x_2, respectively. In what sense can we say that the two vectors are parallel? The idea is to associate with v_1 a vector that belongs to TM_{x_2}, which then can be directly compared with v_2. To determine this vector, we use the following procedure. Let $C(t)$ with $C(0) = x_1$ and $C(T) = x_2$ be a smooth curve that joins x_1 and x_2 and $x_{dt} = C(dt)$ a point nearby x_1. As an element of $\mathbb{R}^n$, the vector at x_{dt} parallel to v_1 is uniquely defined but, in general, does not belong to $TM_{x_{dt}}$. We define the vector $\bar{v}_1^C(dx) \in TM_{x_{dt}}$ associated with v_1 as the tangent component of this parallel (in the sense of $\mathbb{R}^n$) vector. The procedure is carried out successively for all points $x \in C$ between x_1 and x_2 to determine the vectors $\bar{v}_1^C(x) \in TM_x$ associated with v_1. Finally, we compare v_2 with $\bar{v}_1^C(x_2)$. The vector v_1 is said to be *parallel transported* to x_2 along the

---

[7]The *general linear group* $GL(N, X)$ is the set of $N \times N$ invertible matrices with entries from X (e.g., real numbers, $X = \mathbb{R}$, or complex numbers, $X = \mathbb{C}$) together with the the the operation of matrix multiplication. The *special linear group*, $SL(N, X)$, is the subgroup of $GL(N, X)$ consisting of matrices with determinant equal to 1.

curve C. Note that this notion of parallelism, unlike the familiar Euclidean case, is not absolute, but depends on the particular choice of the curve C. Nonetheless, similar to parallelism in Euclidean geometry, parallel transport along a curve defines an equivalence relation on the set of tangent vectors $v_x \in TM_x$ with $x \in C$. All vectors $\bar{v}_1^C(x)$ constructed as described above belong to the same equivalence class, i.e., they are *parallel* to each other.

Next, we apply this notion of parallelism to principal fiber bundles. To better understand the idea of parallel transport on a principal fiber bundle $P : (P, M, \pi, G)$, we first introduce the concepts of *vertical* and *horizontal vectors*. Consider the smooth curves $C : [0, T] \longrightarrow P$ with $C(0) = p$ that lie entirely on the fiber P_x. The tangent vectors to these curves, $v_p = \frac{d}{dt}C(t)|_{t=0}$, are called *vertical vectors* and form a subspace V_pP of the tangent space T_pP called the *vertical subspace*. The tangent vectors that have a zero vertical component are called *horizontal vectors* and form a subspace H_pP called the *horizontal subspace* (see Figure 1.5). The vector space T_pP is the direct sum of the vertical and horizontal subspaces,

$$T_pP = V_pP \oplus H_pP. \tag{1.60}$$

This decomposition of the tangent space is unique.

Consider now a smooth curve $C_M \subset M$ with $C(0) = x_0$ and a point $p_0 \in P_{x_0}$. The concept of parallel transport on P implies finding a smooth curve in the total space, $C_P \subset P$, that satisfies the following properties:
i) C_P is projected onto C_M under the projection map π, i.e., $\pi(C_P(t)) = C_M(t)$, and $C_P(0) = p_0$.
ii) The tangent vectors $v_p \in TP$ to C_p are projected onto tangent vectors $u_x \in TM$ to C_M under the push-forward map $\pi_* : TP \longrightarrow M$, i.e., $\pi_*(v_p) = u_x$.
iii) The tangent vectors $v_p \in TP$ to C_p are *horizontal vectors*, i.e., $v_p \in H_pP$.

The structure that enables one to uniquely determine C_P is called a *geometry* or a *connection* on the principal fiber bundle P, while the curve C_P is called the *horizontal lift* of C_M associated with this connection (see Figure 1.5). We note that there are several equivalent definitions of a connection on a principal fiber bundle [64, 380, 383] and many developments based on this notion, some directly related to physical problems of significant current interest. Here, we will not address any of these technical aspects.

In essence, a *connection* on a principal fiber bundle P is the collection of horizontal subspaces H_pP. In other words, the connection describes the "bending" of the fibers, i.e., the geometry of the bundle, by defining the horizontal subspace at each point $p \in P$. In practice, we need a mathematical object that yields these horizontal subspaces algebraically. This object, the so-called *connection one-form* ω, is a differential one-form with values in the *Lie algebra*[8] $\mathfrak{g}$ of the structure group G that has the property $\omega(v_p) = 0$ if

---

[8]A Lie algebra-valued one-form is a linear function that maps tangent vectors to elements of a Lie algebra [304]. Consider, for example, a principal bundle P with $G = U(N)$ or $G = SU(N)$. Every element $h \in G$ can be obtained as the exponential of some element

(and only if) $v_p \in H_p P$. Hence, the horizontal subspace at p can be defined as the set of all tangent vectors $v_p \in T_p P$ that satisfy $\omega(v_p) = 0$.

Furthermore, the one-form ω can be represented *locally* by a one-form on M. More specifically, let $s_\alpha : U_\alpha \longrightarrow P$ be a smooth function such that $\pi(s(x)) = x$ for all $x \in U_\alpha \subseteq M$, where (U_α, ϕ_α) is an open chart (the function s is called a *local section* of the fiber bundle). Then, the connection one-form ω is represented by the so-called *local connection one-form* $\mathcal{A}_\alpha = s_\alpha^* \omega$, where s_α^* is the pullback map induced by π. Acting on P with a group element $g(x)$ transforms the local section, $s(x) \to s'(x) = s(x) \cdot g(x)$ and, consequently, the corresponding local connection one form, $\mathcal{A}_\alpha \to \mathcal{A}_\alpha'$. Explicitly, we have

$$\mathcal{A}_\alpha'(x) = g^{-1}(x) \cdot \mathcal{A}_\alpha(x) \cdot g(x) + g^{-1}(x) \cdot dg(x). \tag{1.61}$$

Also, for two overlapping charts (U_α, ϕ_α) and (U_β, ϕ_β) we have $\mathcal{A}_\beta(x) = g_{\alpha\beta}^{-1}(x) \cdot \mathcal{A}_\alpha(x) \cdot g_{\alpha\beta}(x) + g_{\alpha\beta}^{-1}(x) \cdot dg_{\alpha\beta}(x)$, where $x \in U_\alpha \cap U_\beta$. In terms of its components, we can write the local connection one-form as $\mathcal{A}_\alpha = A_i dx^i$. Note that physicists define the "connection" as $i\mathcal{A}_\alpha = A = A_i dx^i$. Finally, let us mention that the related two-form $\mathcal{F} = d\mathcal{A} - i\mathcal{A} \wedge \mathcal{A}$, called the *local curvature two-form* $\mathcal{F}$, represents the mathematical counterpart of the field strength in gauge theories. In terms of its components, the curvature two-form can be written as $\mathcal{F} = \frac{1}{2}\mathcal{F}_{ij} dx^i \wedge dx^j$ with

$$\mathcal{F}_{ij} = \frac{\partial \mathcal{A}_j}{\partial x_i} - \frac{\partial \mathcal{A}_i}{\partial x_j} - i[\mathcal{A}_i, \mathcal{A}_j]. \tag{1.62}$$

Consider now a closed curve $C_M(t)$ in the base manifold M with $C_M(T) = C_M(0) = x_0$. Its horizontal lift $C_P(t)$ staring at $C_P(0) = p_0 \in P_{x_0}$ is uniquely defined and, in general, open (see Figure 1.5). More specifically, we have

$$C_P(T) = C_P(0) \cdot g, \qquad \text{with } g \in G. \tag{1.63}$$

In other words, the end point $p_T = C_P(T) \in P_{x_0}$ and the starting point p_0 differ by an element of the structure group G. Note that both points belong to the same fiber P_{x_0}. The element $g \in G$ is called the *holonomy element* associated with the point p_0, the connection ω, and the curve C_M. The set of all holonomy elements associated with different curves $C_M \ni p_0$ is called the *holonomy group* of the connection ω associated with the point p_0. We can express the holonomy g in terms of the local connection one-form as

$$g = \mathcal{P} \exp \left[i \oint_{C_M} \mathcal{A} \right], \tag{1.64}$$

where $\mathcal{P}$ is the path-ordering operator and we have used the physicist's convention for $\mathcal{A}$. In the Abelian case, we have $g = e^{i\gamma}$ with $\gamma = \oint_{C_M} \mathcal{A}$.

---

of the associated Lie algebra, $h = \exp(X)$, where $X \in \mathfrak{g}$ can be represented as an $N \times N$ *anti-Hermitian matrix* (e.g., for $G = SU(2)$, $X = i\lambda_j \sigma_j$, with σ_j being Pauli matrices). Note that the vector spaces $\mathfrak{g}$ and $T_e P$, where $e \in G$ is the identity element, are isomorphic.

1.5.2 Holonomy interpretations of geometric phases

The elementary introduction presented in the previous subsection, although rather technical and largely incomplete, gives us a flavor of fiber bundle theory, a very powerful formalism that captures important aspects of the physical reality, such those underlying gauge theories. Here, we go back to the concept of geometric phase from the perspective of this general formalism and interpret it as the holonomy of a certain connection.

The first holonomy interpretation of the Berry phase, proposed by Simon [479], is based on a fiber bundle constructed over a parameter space $\mathcal{M}$. Consider a quantum system described by the parameter-dependent Hamiltonian $H(\boldsymbol{R})$ undergoing a cyclic evolution with $\boldsymbol{R}(t) \in \mathcal{C}$, where $\mathcal{C} \subset \mathcal{M}$ is a closed path, as discussed in Section 1.4.1. We assume that the eigenvalues $E_n(\boldsymbol{R})$ and the corresponding eigenstates $|n; \boldsymbol{R}\rangle$ are smooth functions of $\boldsymbol{R} \in \mathcal{M}$. In addition, we assume that $E_n(\boldsymbol{R})$ are nondegenerate for all $\boldsymbol{R} \in \mathcal{M}$ and that the time dependence of the Hamiltonian, $H(t) = H(\boldsymbol{R}(t))$, is consistent with the adiabatic approximation. At $t = 0$, the system is assumed to be in a state $|n; \boldsymbol{R}(0)\rangle\langle n; \boldsymbol{R}(0)|$ corresponding to the nth energy level.

We construct a fiber bundle λ^n over $\mathcal{M}$ by attaching to each point $\boldsymbol{R}$ of the base manifold a fiber defined by

$$\lambda_{\boldsymbol{R}}^n = \{|\psi\rangle \in \mathcal{H} : |\psi\rangle = e^{i\varphi}|n; \boldsymbol{R}\rangle, \varphi \in [0, 2\pi]\}, \tag{1.65}$$

where $|n; \boldsymbol{R}\rangle$ are single-valued normalized eigenvectors. Of course, the correspondence between the parameter set $\boldsymbol{R}$ and the fiber $\lambda_{\boldsymbol{R}}^n$ can be expressed in terms of a projection map, $\pi(|\psi\rangle) = \boldsymbol{R}$ if and only if $|\psi\rangle \in \lambda_{\boldsymbol{R}}^n$. Note that the fibers are normalized rays in the Hilbert space $\mathcal{H}$ associated with the quantum system. Since any two vectors $|\psi\rangle, |\psi'\rangle \in \lambda_{\boldsymbol{R}}^n$ differ by a phase factor, the fiber $\lambda_{\boldsymbol{R}}^n$ represents a copy of the unit circle or the group $U(1)$. In other words, as manifolds, these mathematical objects are diffeomorphic. Also note that the single-valued vectors $|n; \boldsymbol{R}\rangle$ are only defined locally, i.e., on open neighborhoods $U_\alpha \subset \mathcal{M}$, while the fiber $\lambda_{\boldsymbol{R}}^n$ is uniquely determined by the eigenvalue $E_n(\boldsymbol{R})$. Let $|n; \boldsymbol{R}, \alpha\rangle$ and $|n; \boldsymbol{R}, \beta\rangle$ designate the single-valued eigenvectors associated with the overlapping neighborhoods U_α and U_β, respectively. For any $\boldsymbol{R} \in U_\alpha \cap U_\beta$ the two eigenvectors may only differ by a phase factor,

$$|n; \boldsymbol{R}, \beta\rangle = e^{i\zeta_{\beta\alpha}(\boldsymbol{R})}|n; \boldsymbol{R}, \alpha\rangle. \tag{1.66}$$

We define these phase factors to be the transition functions of λ^n, $g_{\beta\alpha}(\boldsymbol{R}) = e^{i\zeta_{\beta\alpha}(\boldsymbol{R})} \in U(1)$. Since the typical fiber and the structure group $U(1)$ are diffeomorphic (hence "identical"), we conclude that the fiber bundle that we have constructed is a $U(1)$ principal bundle over the parameter space $\mathcal{M}$, $\lambda^n : (\lambda^n, \mathcal{M}, \pi, U(1))$.

Consider now the cyclic evolution along $\mathcal{C} \in \mathcal{M}$ with the initial condition $|\psi(0)\rangle = |n; \boldsymbol{R}(0)\rangle$. For simplicity, let us assume that the dynamical phase angle associated with this cyclic evolution is zero, $\alpha_n^{dyn} = 0$ (which can always be obtained by properly shifting the energy). The final state of

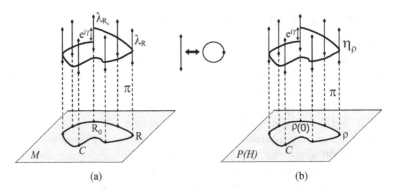

FIGURE 1.6 Schematic representation of the (a) Berry–Simon and (b) Aharonov–Anandan fiber bundles. The base manifolds are the parameter space $\mathcal{M}$ and the projective Hilbert space $\mathcal{P}(\mathcal{H})$ (i.e., the space of physical states), respectively. The fibers $\lambda_{\boldsymbol{R}}$ and η_ρ defined by Eqs. (1.65) and (1.67), respectively, are copies of the unit circle, but are represented as line segments.

the quantum system belongs to the same fiber, $\lambda^n_{\boldsymbol{R}(0)}$, as $|\psi(0)\rangle$, but differs from the initial state by an element of the structure group $g = e^{i\gamma} \in U(1)$, as illustrated in Figure 1.6(a). This holonomy, which basically characterizes the horizontal lift of the closed curve $\mathcal{C} \in \mathcal{M}$, can be expressed in terms of a local connection one-form as given by Eq. (1.64). One can show that this connection is, in fact, related to the Mead–Berry vector potential (1.28). More specifically, we can express this potential as the local connection one-form $\mathcal{A}^n = i\langle n; \boldsymbol{R}|\nabla_j|n; \boldsymbol{R}\rangle dx^j = i\langle n; \boldsymbol{R}|d|n; \boldsymbol{R}\rangle$. Hence, the Berry phase $e^{i\gamma_n}$ with γ_n given by Eq. (1.29) can be identified as the holonomy of the (local) connection $\mathcal{A}^n$ over the principal fiber bundle λ_n.

Another holonomy interpretation of the geometric phase, suggested by Aharonov and Anandan [9], is based on a fiber bundle η over the projective Hilbert space $\mathcal{P}(\mathcal{H})$. We consider the nonadiabatic cyclic evolution of a quantum system, as discussed in Section 1.4.3. The physical states of the system are represented by $\rho(t) = |\psi(t)\rangle\langle\psi(t)| \in \mathcal{P}(\mathcal{H})$, where $|\psi(t)\rangle$ is a (normalized) solution of the Schrödinger equation. To each point ρ of the base manifold $\mathcal{P}(\mathcal{H})$ we attach a fiber

$$\eta_\rho = \{|\psi\rangle : |\psi\rangle\langle\psi| = \rho, \langle\psi|\psi\rangle = 1\}. \tag{1.67}$$

The projection map $\pi(|\psi\rangle) = \rho$ if and only if $|\psi\rangle \in \eta_\rho$ expresses the fact that all the vectors that belong to a given fiber represent the same physical state, as shown schematically in Figure 1.6(b). As in the Berry–Simon construction, the fiber η_ρ represents a copy of the unit circle or of the group $U(1)$ (i.e., they are diffeomorphic). The total space $\mathcal{S}(\mathcal{H})$ is the set of all normalized state vectors, which, as a manifold, is identical with the unit sphere S^{2N-1}, where $N \leq \infty$ is the dimension of the Hilbert space $\mathcal{H}$. Finally, the structure group of the fiber bundle is again $U(1)$. Consequently, the Aharonov–Anandan (AA)

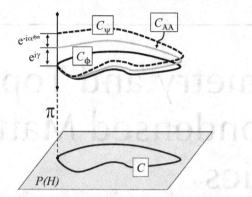

FIGURE 1.7 Closed path $\mathcal{C}$ in the space of physical states and its lifts: the closed lift $\mathcal{C}_\phi$, the Aharonov–Anandan lift $\mathcal{C}_{AA}$, and the dynamical lift $\mathcal{C}_\psi$.

construction results in a $U(1)$ principal bundle over the projective Hilbert space, $\eta_\rho : (\mathcal{S}(\mathcal{H}), \mathcal{P}(\mathcal{H}), \pi, U(1))$.

The evolution of a cyclic state corresponding to the closed loop $\mathcal{C} \in \mathcal{P}(\mathcal{H})$ with $\rho(T) = \rho(0)$ [see Figure 1.6(b)] results in a holonomy $g = e^{i\gamma}$ of the horizontal lift of $\mathcal{C}$ that is determined by a local connection one–form. We identify this connection as the integrand in Eq. (1.56), i.e., the AA local connection one-form $\mathcal{A} = i\langle\phi|\frac{d}{dt}|\phi\rangle dt = i\langle\phi|d|\phi\rangle$. Note that $\mathcal{A}$ depends on the single-valued function $|\phi(t)\rangle$, which represents a local section s of the fiber bundle η_ρ. Moreover, under a gauge transformation $|\phi(t)\rangle \rightarrow e^{i\zeta(t)}|\phi(t)\rangle$, the transformation rule for $\mathcal{A}$, i.e., $\mathcal{A} \rightarrow \mathcal{A} - d\zeta$, represents a special case of the transformation (1.61) for a local connection one-form corresponding to the Abelian case $G = U(1)$. We conclude that the Aharonov–Anandan phase $e^{i\gamma(\mathcal{C})}$ with $\gamma(\mathcal{C})$ given by Eq. (1.56) can be identified as the holonomy of the (local) connection $\mathcal{A}$ over the principal fiber bundle η_ρ. For the relationship between the Berry–Simon and the AA constructions in the limit of the adiabatic approximation we refer the reader to Bohm et al. [64].

As a final note, let us stress that the holonomy $g = e^{i\gamma}$ (i.e., the Aharonov–Anandan phase) is an element of the structure group $U(1)$ of the principal fiber bundle η_ρ that connects the initial and final points of the horizontal lift of the closed path $\mathcal{C}$ — the so-called *Aharonov–Anandan lift*

$$\mathcal{C}_{AA} : |\tilde\psi(0)\rangle \rightarrow |\tilde\psi(t)\rangle \rightarrow |\tilde\psi(T)\rangle = e^{i\gamma}|\tilde\psi(0)\rangle. \tag{1.68}$$

The open path $\mathcal{C}_{AA}$ is a curve in the total space $\mathcal{S}(\mathcal{H})$ of the fiber bundle that projects onto $\mathcal{C}$ in the base manifold $\mathcal{P}(\mathcal{H})$. There are two other important curves that project onto $\mathcal{C}$: the *dynamical lift* $\mathcal{C}_\psi$ defined by the state vector $|\psi(t)\rangle$ and the *closed lift* $\mathcal{C}_\phi$ defined by the single-valued function $|\phi(t)\rangle$ (see Figure 1.7). Note that $|\tilde\psi(t)\rangle = e^{i\gamma(t)}|\phi(t)\rangle$, where $\gamma(t) = \int_0^t dt' i\langle\phi(t')|\frac{d}{dt'}|\phi(t')\rangle$, while the relation between the state vector $|\psi(t)\rangle$ and the single-valued function $|\phi(t)\rangle$ is given by Eq. (1.55). Of course, we have $|\phi(t)\rangle\langle\phi(t)| = |\tilde\psi(t)\rangle\langle\tilde\psi(t)| = |\psi(t)\rangle\langle\psi(t)| = \rho(t)$.

CHAPTER 2

Symmetry and Topology in Condensed Matter Physics

U NIFYING THEMES, CONCEPTS, STRUCTURES, AND IDEAS are the cornerstones of understanding physical reality. The remarkable success of condensed matter physics is intimately related to the existence of a few powerful paradigms that operate throughout a vast field with rather blurry boundaries stretching from the realm of atomic physics to the gates of astrophysics and from fundamental problems that, typically, constitute the object of particle physics or cosmology to applied problems addressed, for example, by materials science and biophysics. Condensed matter, this Middle Earth of physics, is a rich and diverse field that provides wide avenues for transferring themes among different areas and serves as the battleground of reality for many great ideas. This is also an extremely successful field due to the experimental accessibility of condensed matter systems, which promotes a fruitful synergy between theory and experiment. The knowledge acquired in this field enables exquisite control of materials properties and constitutes the bedrock of present–day information technology. Historically, the ability to control and exploit condensed matter systems was always intimately linked to the technological progress of humankind, from the Stone Age to the "Silicon Age." Recent developments in condensed matter physics clearly illustrate the antinomy between our desire for complete understanding and the joy of discovering something new. As a result of these developments, the researcher's status of "being on the road" will be preserved, at least in the foreseeable future. This should be great news, for who would really want to reach the end of the road, the theory of everything, the One Ring?

DOI: 10.1201/9781003226048-2

2.1 THEMES IN MANY-BODY PHYSICS

The main task in condensed matter physics is to describe, predict, and understand the physical properties of many-body systems. The problem involves a macroscopically large number of degrees of freedom, a feature that is also a defining attribute of quantum field theories in particle physics. Thus, not surprisingly, the two areas share many key components of their theoretical frameworks. But the link between the two fields is, perhaps, much deeper. The key question is what does it mean to understand the physics of a system containing 10^{23} particles? Finding the solution of the corresponding Schrödinger equation would be not only a practical impossibility, but an utterly meaningless enterprise, since the resulting predictions cannot be experimentally tested in any detail. Understanding the system means, in fact, being able to describe and predict the correlations among its observable properties. These correlations can be captured by effective theories containing degrees of freedom typically associated with the low-energy excitations of the system. In a broad sense, every theory in physics is an effective theory that captures the correlations among various observable properties of a system corresponding to a set of measurements that operate at certain length and time scales, within a certain energy range. To predict the trajectory of a crystal moving in a gravitational field there is no need to incorporate into the theory information about the constituent atoms or about the lattice structure of the crystal. Similarly, to understand the electrical transport properties of that crystal one does not have to consider nuclear forces, which are completely irrelevant in this context.

The paradigmatic effective theory in traditional condensed matter physics is the Landau Fermi liquid theory [4, 312, 313, 338]. Landau's main idea is that the low-energy excitations of an interacting Fermi system – the so-called *quasiparticles* – are in a one-to-one correspondence with the excited states of a free fermion system. A quasiparticle can be viewed as a fermion "dressed" by interactions, which preserve the particle spin and momentum but renormalize the effective mass and the effective particle–particle interaction. The Landau Fermi liquid theory is, probably, the most successful renormalized field theory ever constructed. The theory accounts for the low-temperature properties of most normal metals and, in combination with band theory, the concept of *quasiparticle* sits at the core of our understanding of semiconductors, superconductors, and superfluids. This theoretical framework has great appeal because it reduces a complicated interacting many-body problem to the familiar free-fermion picture of "filling the bands."

The *quasiparticle* concept can be generalized to include the low-energy excitations of arbitrary systems. In general, quasiparticles have properties that are very different from the properties of the basic constituents. For example, the lowest-energy excitations in a crystal are collective modes, called phonons, that represent noninteracting bosonic quasiparticles with an energy-independent velocity. These quasiparticles are strikingly different from the atoms or molecules that form the system by interacting with each other through specific and, typically, rather complicated interactions. By contrast, the quasiparticles are governed by much simpler and aesthetically appealing laws. One can view the phonons as an example of *emergent physics*. Quite generally, many-body systems acquire qualitatively new properties – *emergent properties* – that have no correspondent at the microscopic level. A familiar example is temperature, a macroscopic physical observable that becomes meaningless at the microscopic level. By contrast, the momentum of a macroscopic object can be viewed as the sum of the momenta of all its microscopic constituents. The whole is, in some sense, more than the sum of its parts – this is, in essence, the principle of emergence, one of the fundamental principles of many-body physics. From this perspective, constructing effective theories to describe the low-energy physics of a macroscopic system is quite natural, since a system characterized by a large number of degrees of freedom has low-energy excitations (quasiparticles) that are governed by basic laws different from and, typically, simpler than those governing the original constituents of the system. As a result, it is often the case that a hopeless-looking problem involving complicated interactions among the original microscopic constituents of the system can be mapped onto a manageable effective theory containing degrees of freedom associated with the low-energy quasiparticles.

Another main theme in condensed matter physics concerns the classification of different states of matter and the characterization of systems in the vicinity of a phase transition. A state (or *phase*) of matter is defined as a distinct form in which the constituents of a many-body system are assembled and organized, which corresponds to a qualitatively distinct set of physical properties. For example, a (homogeneous) gas is a compressible fluid that takes the shape of its container, occupies the entire available volume, and is characterized by a position-independent probability of finding a constituent particle at a given position. By contrast, a crystalline solid has a definite shape and a definite volume and its constituents are spatially ordered following a well-defined repeating pattern. Within a given phase, the properties of a system as functions of parameters such as temperature, external fields, and microscopic coupling constants vary smoothly, while transitions among different phases are signaled by discontinuities in the functional dependence of certain properties on some of these parameters. The key question is how many different phases of matter exist and, most importantly, what are the principles that would enable one to classify them? The classical answer is based on the observation that the properties of a macroscopic system are governed by conservation laws and broken symmetries. This observation sits at the core of Landau's symmetry breaking theory [203]. In essence, two states of a many-body system that have different symmetries are manifestly distinct and, consequently, represent different phases. For example, upon lowering the temperature, a rotationally invariant paramagnetic system can undergo a transition to a ferromagnetic state characterized by a net magnetization oriented along a specific direction. In the ferromagnetic state, the constituent spins are preferentially oriented along that specific direction and, as a result, the system breaks rotational symmetry. The magnetization, which is nonzero in the low symmetry, ordered phase (i.e., the ferromagnetic phase) and vanishes in the high symmetry, disordered phase (paramagnetic), can be viewed as a local *order parameter* that represents a measure of the "degree of order" corresponding to the two phases.

The "Standard Model" of condensed matter theory is based on the Landau Fermi liquid theory and the Landau theory of symmetry breaking. The main challenges to this "Standard Model," which has dominated many-body physics for several decades, were expected to come from disorder and strong correlations. For example, glassy phases have nonzero shear rigidity, like crystalline solids, but no long-range order; consequently, these phases have the same symmetry as their liquid counterparts. Glass is not in thermodynamic equilibrium and can be viewed as practically "frozen" in a metastable state. Strong correlations, on the other hand, call into question Landau's concept of fermionic quasiparticle. For example, a one-dimensional system of interacting fermions, which is described by the Tomonaga–Luttinger model [357], is characterized by two types of low-energy excitations, one carrying spin and the other one charge (the so-called *spin-charge separation*), instead of Landau-type quasiparticles (which are charged spin-$\frac{1}{2}$ fermions). Mott insulators [374] – systems that according to band theory are characterized by partially filled bands and, consequently, should be good conductors – provide another striking example.

It was a class of strongly correlated systems discovered in the early 1980s – the fractional quantum Hall (FQH) liquids [511] – that decisively opened the way toward a new paradigm in condensed matter physics. The study of FQH systems has shown that the classification based on symmetry breaking is not complete and has revealed that phases of matter should also be distinguished based on their *topological* properties. In conformity with Landau's theory, if two phases cannot be continuously connected without crossing a phase boundary, they *must* have different symmetries. Nonetheless, different FQH states have the same symmetry, despite being separated by phase transitions. This discovery demands a new principle for classifying distinct phases of matter. It turns out that the distinctiveness of FQH states (and of many other states of matter) stems from *topology*, rather than *symmetry*.

In the remainder of this chapter, we first summarize the main ideas behind Landau's theory of symmetry breaking – the "standard" framework for classifying different phases of matter. Then, we consider the new paradigm (based on topology) and address a few general questions that every newcomer into the realm of topological quantum matter probably has in mind. What do we mean by "topological" when talking about quantum states? How important is quantum mechanics in the topological world? How can one distinguish different topological states and how do we classify them? What kind of emergent physics can one find in this topological "New World"? Are there implications beyond condensed matter physics? Here, we merely sketch the contour lines of the answers to these questions, focusing on generic, qualitative aspects. In addition, since the study of topological matter is a developing field and the terminology is not yet standardized, we provide some clarifications (to avoid possible confusions). Technical details and more in-depth discussions will be provided in the subsequent chapters.

2.2 LANDAU THEORY OF SYMMETRY BREAKING

A phase transition [207, 314] between two states of matter corresponds to singularities in the functional dependence of the system properties on parameters such as temperature, external fields, and coupling strengths. Considering the nature of these singularities, one can distinguish two types of transitions: *first-order* phase transitions, which involve energy exchange as latent heat, and *second-order* or *continuous* phase transitions, which are characterized by continuous first-order derivatives of the thermodynamic potential and divergent susceptibilities. Further insight can be gained by considering the symmetries of the states (i.e., phases) separated by the phase transition.

One possibility is that the two phases have the same symmetry. As we will see below, this case also includes transitions between states of matter with different topological properties. For now, we only consider topologically trivial phases that can be addressed within the traditional framework for understanding critical phenomena. A familiar example of a transition between same-symmetry states is the liquid–gas phase transition, a first-order transition corresponding to a line in the temperature-pressure plane that ends at a critical point. In this case, there is a *weak* distinction between the two phases, in the sense that the liquid can always be smoothly connected with the gas (i.e., without crossing the phase transition line) by going around the critical point (see Figure 2.2 on page 51). Since there is no *qualitative* distinction between the two phases, in some sense one can view them as a single state.

A second possibility is that the two phases have different symmetries. Consider the situation when one of the phases has the same symmetry as the Hamiltonian describing the many-body system, for example the high-temperature paramagnetic phase of the Heisenberg ferromagnet. Upon varying a certain parameter (in this case lowering the temperature), the system undergoes a phase transition from the high-symmetry disordered state (paramagnet) to a lower-symmetry ordered state (ferromagnet). This is a so-called *spontaneous symmetry breaking* phase transition. Unlike weakly distinct isosymmetric phases, states with different symmetries are distinct in the *strong* sense and cannot be smoothly connected, i.e., without going through a phase transition. Spontaneous symmetry breaking, which represents a major pillar of both condensed matter and high-energy physics, sits at the core of Landau's paradigm.

Landau's approach includes three key steps. First, one has to identify an appropriate local *order parameter* $\psi(r)$ as a measure of the "degree of order" in the broken symmetry phase. This quantity is zero in the disordered phase and nonzero in the ordered phase, hence clearly enables one to distinguish the two phases. The order parameter can be defined as the average of a certain local operator and, within Landau's theory, is assumed to be small and uniform near the phase transition. Familiar examples include the magnetization in a ferromagnet and the amplitudes of the Fourier components of the density in a crystal. Second, one assumes that the equilibrium thermodynamics is completely determined by a free energy obtained by minimizing the functional

$$F[\psi(\boldsymbol{r}), T] = F_0(T) + F_L[\psi(\boldsymbol{r}), T] \tag{2.1}$$

with respect to the order parameter. In Eq. (2.1) $F_0(T)$ represents a smooth function of temperature and the Landau functional F_L is assumed to be an analytic function of ψ and T with $F_L[0, T] = 0$. Close to the phase transition, where the order parameter is small, F_L is typically written in terms of a polynomial expansion in ψ.

The third step involves the construction of the free energy functional. Landau used a phenomenological approach based on the observation that the functional has to be invariant under all possible symmetry transformations associated with the symmetry group[1] $\mathcal{G}$ of the Hamiltonian. At this stage, the powerful machinery of group theory provides valuable insight into the physics of the phase transition and the very principle for classifying different phases. For example, if the disordered phase has symmetry group $\mathcal{G}$, the same as the Hamiltonian, a spontaneous symmetry breaking transition will reduce the symmetry of the system to one of the possible subgroups $\mathcal{U}_i$ of $\mathcal{G}$. We note that, upon varying the external parameters (e.g., lowering the temperature), the system may undergo successive symmetry breaking transitions corresponding to the symmetry sequence $\mathcal{G} \supset \mathcal{U}_i \supset \mathcal{V}_j$. For example, upon lowering the temperature a fluid crystallizes and the symmetry of the system is reduced from a continuous translation invariance to invariance with respect to a discrete set of translations. At a lower temperature, the full spin-rotation symmetry of the paramagnetic crystal may be spontaneously broken and reduced to a symmetry with respect to rotation about an axis parallel to the magnetization, while the system becomes a ferromagnet.

---

[1] A *group* $\mathcal{G}$ is a collection of elements equipped with an operation • (called the *group law* of $\mathcal{G}$) that satisfies the following properties (known as the *group axioms – closure, associativity, identity element,* and *inverse element*):
(a) If g_1 and g_2 are elements of $\mathcal{G}$ (i.e., $g_1, g_2 \in \mathcal{G}$), then $g_1 \bullet g_2 \in \mathcal{G}$
(b) For all $g_1, g_2, g_3 \in \mathcal{G}$, $(g_1 \bullet g_2) \bullet g_3 = g_1 \bullet (g_2 \bullet g_3)$
(c) There exists an identity element $e \in \mathcal{G}$ such that $g \bullet e = e \bullet g = g$ for every $g \in \mathcal{G}$
(d) For each $g \in \mathcal{G}$, there exists an inverse element g^{-1}, such that $g \bullet g^{-1} = g^{-1} \bullet g = e$
Two groups $(\mathcal{G}, \bullet)$ and $(\mathcal{H}, \circ)$ are said to be *isomorphic* if there exists a bijective (one-to-one) mapping $f : \mathcal{G} \to \mathcal{H}$ with the property $f(g_1 \bullet g_2) = f(g_1) \circ f(g_2)$ for all $g_1, g_2 \in \mathcal{G}$ (i.e., the two groups have the same "multiplication table"). Isomorphic groups are usually considered to be "the same."
A *symmetry operation* is a transformation that leaves a certain object (e.g., a geometric figure, a physical system, or the Hamiltonian that describes it) invariant. Two symmetry operations can be *composed* (by applying them successively), the resulting transformation being another symmetry operation. The *symmetry group* of an object is the group of all its symmetry operations with composition as the group law.

In addition to providing tools for constructing the Landau free energy functional and "selection rules" for identifying the allowed broken-symmetry phase transitions, group theory gives valuable information concerning the properties of the low-energy excitations in the ordered phase. Without going into details, we emphasize the importance of distinguishing i) between continuous and discrete broken symmetries and ii) between global and local symmetries. For example, translation and rotation symmetries are continuous symmetries that are spontaneously broken in a liquid-solid transition, while time reversal and inversion symmetries are examples of discrete symmetries that are broken in, e.g., a ferromagnetic phase. All these are examples of global symmetries, in the sense that the corresponding Hamiltonians and disordered phases are invariant under any such symmetry operation acting identically at every point in space. By contrast, local symmetries, such as gauge symmetries, impose the stricter constraint of invariance under local group transformations. For example, the Hamiltonian describing a system of spin-$\frac{1}{2}$ charged fermions coupled to an electromagnetic field is characterized by a local gauge symmetry that becomes spontaneously broken when the system undergoes a transition to a superconducting state.

The importance of correctly identifying the nature of the broken symmetry is nicely illustrated by Goldstone's theorem [212, 381]. The theorem states that for each spontaneously broken *continuous global* symmetry there is a low-energy gapless mode, i.e., a massless quasiparticle, called a *Goldstone boson* (see below, Section 2.2.2). Physically, one can view Goldstone modes, such as phonons and spin waves (magnons), as long-wavelength fluctuations of the corresponding order parameter that attempt to restore the full symmetry of the system. We emphasize that Goldstone's theorem does not apply when *discrete* symmetries or *local* symmetries are broken.

2.2.1 Construction of the Landau functional

To better understand the main ideas underlying Landau's theory, as well as the intimate connection with the field theories of particle physics, it is instructive to briefly discuss a few examples. Consider first an Ising ferromagnet [260] described by the Hamiltonian

$$H_{\mathrm{I}} = -\frac{1}{2} \sum_{i,j} J_{ij} S_i S_j, \qquad (2.2)$$

where J_{ij} are short-range exchange couplings, S_i are classical "spin" variables taking values $S_i = \pm 1$, and the summations are over all lattice sites. Under time reversal $S_i \to -S_i$ and the Hamiltonian (2.2) remains invariant, hence the system is characterized by a $\mathbb{Z}_2$ symmetry.[2] At low temperatures, the spins will spontaneously break this symmetry and will preferentially point either

---

[2]A group $\mathcal{G}$ is called *cyclic* if there exists an element $g_1 \in \mathcal{G}$ such that *every* element g_m of the group can be generated as $g_m = g_1^m \equiv g_1 \bullet g_1 \bullet \cdots \bullet g_1$ (m times). The identity is $e \equiv g_0 = g_1^0$. For example, the set $\{0, 1, \ldots, n-1\}$ with *addition* modulo n is a cyclic

up or down. The natural order parameter associated with this transition is the average spin value $\psi = \langle S_i \rangle$. Note that under time reversal $\psi \to -\psi$. The phenomenological Landau functional can be constructed in terms of a polynomial expansion in ψ that is invariant under a $\mathbb{Z}_2$ symmetry operation, i.e., an expansion containing only even powers of ψ. To gain further insight, instead of pursuing this phenomenological approach, we sketch a different route that starts from the microscopic lattice Hamiltonian (2.2). The main steps are: i) Construct a lattice field theory. Such a theory will accurately describe the physics at all length scales and arbitrarily far away from the phase transition, but may be technically challenging. ii) Obtain a continuum field theory by *coarse graining* the lattice theory. The result is an effective theory that captures the long wavelength physics arbitrarily far away from the phase transition. iii) Perform a mean-field approximation and a power expansion in the order parameter. This provides the Landau functional, including the dependence of the expansion coefficients on model and external parameters (e.g., J_{ij} and temperature).

One can construct a lattice field theory for the Ising model (or any other Hamiltonian that can be expressed in a quadratic form) by writing a functional integral representation of the partition function $Z = \mathrm{Tr} e^{-\beta H_I}$ using the Hubbard–Stratonovich transformation

$$e^{-\beta H_I} = A \prod_i \int d\psi_i \exp\left[-\frac{1}{2} \sum_{i,j} \psi_i K_{ij} \psi_j + \sum_{i,j} \psi_i K_{ij} S_j \right], \qquad (2.3)$$

with $K_{i,j} = \beta J_{ij}$ and $A = (2\pi)^{N/2} (\det K)^{-1/2}$, where N is the number of lattice sites. The linear (rather than quadratic) dependence on S_j in the exponent allows for an exact evaluation of the trace over spins. Adding an interaction term $-\sum_i h_i S_i$ that couples the spins to an external field and evaluating the trace gives

$$Z = A \prod_i \int d\psi_i \exp\left[-\frac{1}{2} \sum_{i,j} \psi_i K_{ij} \psi_j + \sum_i \ln[2\cosh(\overline{K}\psi_i + \beta h_i)] \right], \quad (2.4)$$

where $\overline{K} = \sum_j K_{ij}$. Note that the average of the constrained variable $S_i = \pm 1$ is equal to the average of the unconstrained field ψ_i, $\langle S_i \rangle = \partial \ln Z / \partial \beta h_i = \langle \psi_i \rangle$, as one can see by changing the derivative with respect to βh_i into a derivative with respect to $\sum_j K_{ij} \psi_j$, followed by an integration by parts. To obtain the continuum limit, we first express Eq. (2.4) in terms of the Fourier components

---

group with $g_1 = 1$ and $e = 0$; this group is isomorphic with the group of complex numbers $z_m = \exp(2\pi i m/n)$, where $0 \leq m \leq n - 1$, (i.e., the solutions of the equation $z^n = 1$) with multiplication, or with the group of rotations of a regular n-sided polygon; these isomorphic groups are usually denoted by $\mathbb{Z}_n$. In particular, the $\mathbb{Z}_2$ symmetry group is the set $\{e, g\}$ (with composition $\circ$) that contains the identity e (i.e., no transformation) and one symmetry operation g (e.g., inversion or time reversal) with the property $g^2 \equiv g \circ g = e$.

of the ψ field. Then, after retaining only the lowest order contributions in k, we rewrite the functional integral in terms of a continuum field $\psi(x)$. This approximation amounts to neglecting short-range fluctuations of the field, i.e., imposing a short-range (large wavenumber) cutoff $k_{\max} \sim 2\pi\Lambda^{-1/d}$, where Λ is a certain volume larger than the volume of the unit cell (v_0) but much smaller than the volume of the d-dimensional macroscopic system. Consequently, one can view the continuum field $\psi(x)$ as the average of ψ_i over all the lattice sites inside a domain Λ centered at x. We have

$$\sum_i f[\psi_i] \to \frac{1}{v_0} \int d^d x f[\psi(x)], \tag{2.5}$$

where $f[\psi]$ is an arbitrary function. Also, approximating the Fourier transform of the coupling as $K(k) \approx \overline{K} - wk^2$, where $w = (1/2d) \sum_j R_{0j}^2 K_{0j}$, with R_{0j} being the position vector of site j relative to the origin ($j = 0$), we have

$$\sum_{i,j} \psi_i K_{ij} \psi_j \approx \int \frac{d^d k}{(2\pi)^d} (\overline{K} - wk^2)|\psi(k)|^2 \to \frac{1}{v_0} \int d^d x \left[\overline{K}\psi^2 + w(\boldsymbol{\nabla}\psi)^2 \right]. \tag{2.6}$$

Using these relations, we can write an approximate partition function for the Ising system in terms of a functional integral over the continuum field $\psi(x)$,

$$Z = \int \mathcal{D}\psi \, e^{-\beta F[\psi(x)]}, \tag{2.7}$$

where the functional $F[\psi(x)]$ (often called *effective Hamiltonian*) has the expression

$$F[\psi(x)] = \frac{1}{\beta v_0} \int d^d x \left[\frac{1}{2} w(\boldsymbol{\nabla}\psi)^2 + \frac{1}{2}\overline{K}\psi^2 - \ln[2\cosh(\overline{K}\psi + \beta h)] \right]. \tag{2.8}$$

The approximate free energy $F(T)$ of the Ising system can be obtained by evaluating the functional integral in Eq. (2.7) using a *saddle point* approximation, i.e., replacing the integral by a constant times the maximum value of the exponential, which is given by the minimum of the free energy functional $F[\psi]$ with respect to ψ. Explicitly, $F(T) = F_0(T) + \min_\psi\{F[\psi(x)]\}$. Comparison with Eq. (2.1) suggests that $F[\psi]$ in Eq. (2.8) represents, in fact, the Landau functional, while the field $\psi(x)$ is the order parameter. In the vicinity of the phase transition the order parameter is small and, assuming weak external fields, one can expand the logarithm from Eq. (2.8) in a Taylor series. After relabeling the coefficients, the Landau functional takes the familiar form

$$F[\psi(x)] = \int d^d x \left[\frac{1}{2} w(\boldsymbol{\nabla}\psi)^2 + f_L(\psi) \right], \quad \text{with} \quad f_L(\psi) = \frac{1}{2}a\psi^2 + \frac{1}{4}b\psi^4 - h\psi, \tag{2.9}$$

where $a = \overline{K}(1 - \overline{K})/\beta v_0$, $b = \overline{K}^4/3\beta v_0$, $w/\beta v_0 \to w$, and $\overline{K}h/v_0 \to h$.

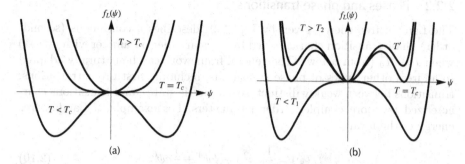

FIGURE 2.1 Evolution of the Landau free energy $f_L(\psi)$ across a continuous (second-order) phase transition (a) and a discontinuous (first-order) phase transition (b). For $T_1 < T < T_2$ f_L has local minima corresponding to metastable states (e.g., superheated liquid at $T = T'$).

This construction reveals several important points. The order parameter $\psi(x)$ represents the average spin value over all the sites within a region Λ around point x, $\psi(x) = \langle S_i \rangle_{i \in \Lambda(x)}$. At low temperatures, in the absence of an external field ($h = 0$), the order parameter will fluctuate around one of the two possible mean values corresponding to the nonzero minima of f_L, $\psi_0 = \pm\sqrt{-a/b}$. These values characterize the broken symmetry states with spins oriented preponderantly up or down, respectively, and require the condition $a < 0$. Note that $a \sim T - T_0$, where $T_0 = \frac{1}{k_B}\sum_j J_{0j}$ is the mean field transition temperature. Typically, T_0 has a value larger than the ferromagnetic transition temperature T_c, as the mean field treatment becomes inaccurate below a certain (upper) critical dimension[3] $d_c^{(u)}$, i.e., for $d < d_c^{(u)}$. This problem also affects the critical exponents and is properly addressed within the framework of the renormalization group. So, below the critical temperature, $a < 0$ and $f_L(\psi)$ has two nonzero minima corresponding to the two possible values of the average spin in the broken symmetry phase, as illustrated in Figure 2.1(a). By contrast, for $a > 0$ (i.e., when $T > T_c$), the unique minimum of f_L obtains for $\psi_0 = 0$; the system is in a disordered (paramagnetic) phase. Also note that for $h = 0$ (i.e., in the presence of time reversal symmetry) the Landau free energy density $f_L(\psi)$ contains only even powers of the order parameter, as required by the invariance under time-reversal operations ($\psi \to -\psi$). Finally, we observe that the contribution to $F[\psi]$ proportional to w has the role of suppressing the spatial variations of the order parameter. In fact, Landau's construction assumes a uniform order parameter near the phase transition.

---

[3]There is also a *lower* critical dimension, $d_c^{(\ell)}$, below which (i.e., for $d \leq d_c^{(\ell)}$) there is no phase transition. The critical dimensions have specific values that characterize all systems belonging to a given *universality class*.

2.2.2 Phases and phase transitions

The Landau free energy given by Eq. (2.9) describes a continuous (second-order) phase transition characterized by a scalar order parameter with $\psi_0 \to 0$ when $T \to T_0$ (from below). The general framework can be naturally adapted to address other types of phase transitions, including first order transitions, transitions between weakly distinct isosymmetric states, and transitions characterized by more complex order parameters. For example, a Landau free energy of the form

$$f_L(\psi) = \frac{1}{2}a\psi^2 - \frac{1}{4}b\psi^4 + \frac{1}{6}c\psi^6, \qquad (2.10)$$

with $b > 0$ (and $c > 0$), will describe a first-order phase transition. The ψ^6 term was included to ensure that f_L is bounded from below (i.e., has a finite absolute minimum). The dependence of f_L on ψ for different temperatures is illustrated in Figure 2.1(b). Deep inside the broken symmetry phase f_L has two nonzero minima, while for high-enough temperature (i.e., deep inside the disordered phase) there is a unique minimum at $\psi = 0$. However, in contrast with the continuous transition [Figure 2.1(a)], there is a certain temperature range $T_1 < T < T_2$ in which the Landau free energy has three local minima. For $T_1 < T < T_c$ the lowest minima, which characterize the thermodynamically stable state, correspond to $\psi_0 \neq 0$ (i.e., an ordered state), but there is a third minimum at $\psi = 0$ that corresponds to a metastable (disordered) state of the system. For $T_c < T < T_2$ the stable (disordered) state corresponds to the minimum at $\psi = 0$, while the higher local minima at $\psi_0 \neq 0$ describe metastable ordered states. Note that for $T \lesssim T_c$ the order parameter [i.e., ψ_0 corresponding to the lowest degenerate minima of $f_L(\psi)$ in Figure 2.1(b)] does not vanish upon approaching the critical temperature (from below), as is the case in a continuous phase transition [see Figure 2.1(a)].

The case of weakly distinct isosymmetric states (e.g., the liquid-gas transition) can be incorporated within this framework by identifying an appropriate order parameter that does not break the symmetry of the system. For the liquid-gas transition, a possible choice is $\psi = n - n_c$, where n_c is the particle density at the critical point. We emphasize that the condition $\psi = 0$ at high-enough temperature is not a symmetry requirement in this case. The situation is analogous to that of an Ising ferromagnet in the presence of an external field, as illustrated in Figure 2.2. When $h \neq 0$, the "up" and "down" states of the system have the same symmetry as the Hamiltonian and can be smoothly connected by going around the critical point.

The Ising ferromagnet discussed above is a system with discrete $\mathbb{Z}_2$ symmetry. Consider now, as an example of a system characterized by a continuous symmetry, the so-called classical xy model

$$H_{\mathrm{XY}} = -\frac{1}{2}\sum_{i,j} J_{ij}\vec{S}_i \cdot \vec{S}_j, \qquad (2.11)$$

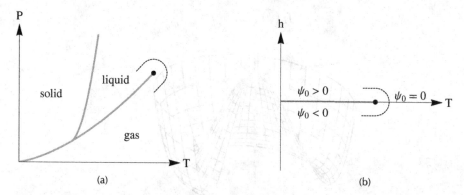

FIGURE 2.2 Schematic phase diagrams for (a) a classical fluid (in the temperature-pressure plane) and (b) an Ising ferromagnet (in the temperature-magnetic field plane). The full lines represent first-order (discontinuous) phase transitions and the black dots are critical points. Phases with the same symmetry, i.e., the liquid and the gas in (a) and states with opposite magnetization in (b), can be smoothly connected along the dashed lines without encountering a phase transition.

where $\vec{S}_i$ are unit vectors in the $x - y$ plane. The order parameter, given again by the average "spin," $\vec{\psi}(x) = \langle \vec{S}_i \rangle_{i \in \Lambda(x)}$, is now a two-component vector, while the Hamiltonian has global $O(2)$ symmetry, i.e., it is invariant with respect to in-plane rotations and reflections of the "spins."[4] The Landau free energy has to be invariant with respect to the symmetry operations and, consequently, will only contain even terms,

$$f_L(\vec{\psi}) = \frac{1}{2} a\, \vec{\psi} \cdot \vec{\psi} + \frac{1}{4} b (\vec{\psi} \cdot \vec{\psi})^2 + \dots . \qquad (2.12)$$

In the disordered phase $f_L(\vec{\psi})$ has a unique minimum at $\vec{\psi} = 0$, while in the broken symmetry phase $f_L(\vec{\psi})$ has a "Mexican hat" profile with minima located along a circle in the $\psi_x - \psi_y$ plane, as shown in Figure 2.3. The implications are major. In particular, there are two types of excitations: those associated with fluctuations of $|\vec{\psi}|$ (the magnitude of the order parameter) and those corresponding to fluctuations of the direction of $\vec{\psi}$. Since the energy cost to change the direction of the order parameter (i.e., move $\vec{\psi}$ around the circle of minima of f_L) is zero, the corresponding excitations ("spin" waves) dominate the low-energy physics and represent gapless Goldstone bosons associated

---

[4]The *orthogonal group*, $O(n)$, is the group of $n \times n$ orthogonal matrices with matrix multiplication as the group operation. Equivalently, $O(n)$ is the group of distance-preserving transformations (e.g., rotations and reflections) of an n-dimensional Euclidean space. An orthogonal matrix M is a real matrix with the property $M^T = M^{-1}$; hence, $\det(M) = \pm 1$. The matrices with determinant $+1$ form a subgroup called the *special orthogonal group* (or the *rotation group*), $SO(n)$.

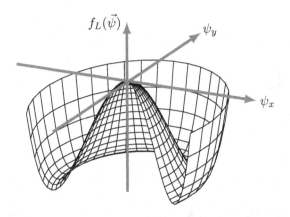

FIGURE 2.3 Landau free energy $f_L(\vec{\psi})$ for a system with spontaneously broken $O(2)$ symmetry. Long wavelength fluctuations that move $\vec{\psi}$ around the circle of minima involve no energy cost (Goldstone bosons).

with the broken $O(2)$ symmetry. Note that the zero mass of the Goldstone bosons can be associated with the vanishing curvature of $f_L(\vec{\psi})$ along the azimuthal direction at the minima. By contrast, the excitations associated with magnitude fluctuations are gapped and the corresponding quasiparticles have a mass proportional to the curvature of $f_L(\vec{\psi})$ in the radial direction.

At this point, a comment about the fate of Goldstone bosons in systems with gauge symmetry would be appropriate. Consider first the case of a system with global $U(1)$ symmetry.[5] The order parameter is now a complex function, $\psi(\boldsymbol{x}) = |\psi(\boldsymbol{x})|e^{i\theta(\boldsymbol{x})}$, and the free energy functional has the form $F(\psi) = \int d^d x \left[\frac{1}{2}|\boldsymbol{\nabla}\psi|^2 + f_L(\psi)\right]$, with $f_L = \frac{1}{2}a|\psi|^2 + \frac{1}{4}b|\psi|^4$. Note that f_L has the same form as Eq. (2.12), as one can easily establish using the parameterization $\psi = \psi_1 + i\psi_2$ (with ψ_1 and ψ_2 being real functions of position). The Goldstone boson corresponds to (long wavelength) fluctuations of the phase θ. The Landau free energy density f_L has local $U(1)$ symmetry [i.e., it is invariant to local changes of $\theta(\boldsymbol{x})$], but the term containing the derivatives is only invariant to global changes of the phase, $\theta(\boldsymbol{x}) \to \theta(\boldsymbol{x}) + \theta_0$.

Next, let us assume that the system consists of charged particles interacting electromagnetically. We can introduce the electromagnetic field through the substitution $\boldsymbol{\nabla} \to \boldsymbol{\nabla} - iq\boldsymbol{A}$ (with $\hbar = 1$), where $\boldsymbol{A}$ is the vector potential. For $q = 2e$ the resulting effective theory describes a superconductor. Once the

---

[5]The *unitary group*, $U(n)$, is the group of $n \times n$ unitary matrices with matrix multiplication as the group operation. Equivalently, it is the group of transformations that preserve the standard inner product on $\mathbb{C}^n$. Since a unitary matrix satisfies the relation $M^\dagger = M^{-1}$, its determinant is, in general, a complex number with absolute value 1, i.e., $\det(M) = e^{i\phi}$. The unitary matrices with $\det(M) = 1$ form a subgroup called the *special unitary group*, $SU(n)$.

charged particles are coupled to the (electromagnetic) gauge field, the phase $\theta(\boldsymbol{x})$ of the complex order parameter no longer represents an actual degree of freedom, as it can be "absorbed" into the vector field through the gauge transformation $\boldsymbol{A} - 1/q\boldsymbol{\nabla}\theta \to \boldsymbol{A}$. As a consequence, the term in the free energy containing the derivatives, $\frac{1}{2}|(\boldsymbol{\nabla} - iq\boldsymbol{A})\psi|^2$, becomes $\frac{1}{2}(\boldsymbol{\nabla}|\psi|)^2 + \frac{1}{2}q^2|\psi|^2\boldsymbol{A}^2$, i.e., independent of the phase $\theta(\boldsymbol{x})$. Note that in the broken symmetry phase the contribution $\frac{1}{2}q^2|\psi_0|^2\boldsymbol{A}^2$, where $|\psi_0|$ is the magnitude of the order parameter (i.e., the radius of the circle of minima of the Mexican hat), is nonzero. This quadratic contribution adds a Proca mass term [354] to the free field Lagrangian density of the electromagnetic field,[6] which becomes

$$\mathcal{L} = -\frac{1}{4}F_{\mu\nu}F^{\mu\nu} + \frac{1}{2}q^2|\psi_0|^2 A_\mu A^\mu. \tag{2.13}$$

Consequently, in the broken symmetry phase the interaction among the charged particles becomes short-range, since it is mediated by a massive gauge boson (i.e., a massive "photon") of mass $m = q|\psi_0|$. In the case of superconductivity, the physical consequence of this short-range electromagnetic interaction is that the magnetic field is expelled out of the superconductor (the Meissner effect). We conclude that, in a gauge theory, the phase angle $\theta(\boldsymbol{x})$ of a complex (scalar) order parameter combines with the redundant gauge degrees of freedom and, instead of generating a gapless Goldstone boson, results in the gauge boson (e.g., the photon) acquiring mass. This is the famous Higgs mechanism,[7] which is key to explaining the mass of gauge bosons in particle physics.

Landau's theory, which is equivalent to a mean field approximation, as argued above, does not provide accurate values for the transition temperature and the critical exponents. While transition temperatures are system-specific and depend on the details of the interaction, critical exponents describing the behavior of the system near a continuous phase transition (e.g., near critical points) may be the same for completely different physical systems, which are said to be in the same *universality class*. In fact, critical exponents only depend on the spatial dimension (d), the symmetry of the order parameter (e.g., ψ), and the symmetry (and range) of the interaction (e.g., J_{ij}). For example, the liquid-gas, the binary fluid, and the uniaxial (anti)ferromagnetic transitions belong to the Ising universality class and are characterized by the same set of critical exponents.

---

[6]The Lagrangian density of the free electromagnetic field can be written in the gauge-invariant form $\mathcal{L} = -\frac{1}{4}F_{\mu\nu}F^{\mu\nu}$, where $F^{\mu\nu} = \partial^\mu A^\nu - \partial^\nu A^\mu$ is the so-called Maxwell tensor (or electromagnetic field strength), with $A^\mu : (\phi/c, \boldsymbol{A})$ being the four-vector gauge field (or four-vector potential). This gauge field generates long-range interactions among charged particles mediated by massless bosons (photons). Adding a Proca mass makes the gauge bosons massive and the corresponding interaction short-range.

[7]In the case of superconductivity, the massive gauge boson is the "photon" of mass $m = q|\psi_0|$, while the so-called Higgs boson corresponds to fluctuations of the order parameter magnitude, $|\psi(\boldsymbol{x})|$, with respect to the value corresponding to the minimum of the free energy, $|\psi_0|$.

In essence, this remarkable behavior stems from the fact that the correlation length, i.e., the characteristic length scale ξ associated with correlated fluctuations of the order parameter, diverges as the system approaches a second-order phase transition. Consequently, the system becomes *scale invariant* at the critical point, i.e., its properties appear to be the same regardless of the scale at which one examines them. This suggests that valuable information about the system could be obtained if one steps back from the microscopic level and examines the physics at larger and larger length scales. The idea led to the development of the *renormalization group* by L. P. Kadanoff and K. G. Wilson. The approach, which has been implemented in various forms, involves a sequence of transformations consisting in a coarse graining (thinning) of the degrees of freedom followed by a rescaling of the relevant variables. The renormalization group illustrates the power of the "effective theory" concept and, in fact, can be viewed as a method for calculating the evolution of the effective theory with the spatial length scale on which we examine the system. As a final note, we point out that within this framework there is no fundamental difference between classical (or thermal) phase transitions and quantum (i.e., zero temperature) phase transitions. For example, the scaling of the order parameter near the critical temperature, $\psi_0 \sim (T_c - T)^\beta$, is similar to the behavior at $T = 0$ near a critical value of some coupling constant $\psi_0 \sim (g_c - g)^{\bar{\beta}}$, where β and $\bar{\beta}$ are critical exponents that depend on the universality classes of the classical and quantum phase transitions, respectively. In general, a quantum phase transition in d dimensions has a field-theoretical description that looks the same as the description of a classical phase transition in $d + 1$ dimensions, imaginary time representing the extra dimension.

These highlights on Landau's theory of symmetry breaking, which barely scratch the surface of a vast, well-established field, are intended to emphasize the novelty of the topological quantum world, but also to underline its formal links with the field theories initially developed within the traditional paradigm.

2.3 TOPOLOGICAL ORDER, SYMMETRY, AND QUANTUM ENTANGLEMENT

In Chapter 1, we have briefly introduced one manifestation of topological quantum matter associated with space–time processes involving localized quasiparticles emerging in so-called topological quantum systems. But, more generally, what do we mean by "topological" in the context of condensed matter phases and phase transitions? The term has several different meanings, some of them not yet completely standardized. In this section, we sketch a broad view of the "topological quantum world" focusing on several key aspects of its "geography": neighbors (what was left out), regions (main classes of topological quantum states discussed in this book), and toponyms (some basic terminology).

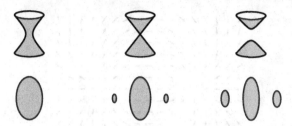

FIGURE 2.4 Evolution of the Fermi surface in a Lifshitz phase transition: neck collapsing (top) and pocket formation (bottom).

There are many-body phenomena that involve topological aspects, yet they are not included in our "topological world." For example, *Lifshitz transitions* [337] are continuous quantum phase transitions not associated with symmetry breaking that are characterized by a change of the Fermi surface topology, more specifically, a change of the Fermi surface connectedness. There are two types of Lifshitz transitions: *neck-collapsing* and *pocket-vanishing* (or *pocket-formation*) transitions, as illustrated in Figure 2.4. Originally proposed for noninteracting fermions, this type of "topological" quantum phase transition has more recently been considered in the context of interacting electron systems, such as electron-doped iron–arsenic superconductors, heavy fermion systems, $Na_x CoO_2$, and underdoped cuprates. Another example is the Kostelitz–Thouless transition – a vortex unbinding transition that occurs without spontaneous symmetry breaking in two-dimensional systems with O(2) or U(1) symmetry [52, 308]. As we have seen in Section 2.2, states with a broken continuous symmetry are characterized by low-energy fluctuations (Goldstone bosons) that tend to restore the full symmetry of the system. In fact, it was shown that the symmetry of a many-body system cannot be spontaneously broken at any finite temperature in dimensions $d \leq 2$. This result, which holds in systems with sufficiently short-range interactions, is called the *Mermin–Wagner theorem*. However, the emergence of Goldstone bosons is not the only mechanism for restoring the symmetry of the system; an alternative mechanism involves the creation of *topological defects*, i.e., finite energy distortions of the order parameter field (such as *vortices* in superfluid helium, *dislocations* in periodic crystals, and *disclinations* in nematic liquid crystals) that cannot be eliminated by any continuous change of the order parameter. For example, the low-temperature phase of the xy model (2.11) is characterized by finite spin stiffness (i.e., nonzero magnitude of the order parameter, $|\vec{\psi}_0| \neq 0$) and by the presence of bound vortex-antivortex pairs (see Figure 2.5), which results in a power-law decay of the spin-spin correlation (i.e., quasi long-range order). At the Kostelitz–Thouless transition, the pairs unbind and the vortices proliferate resulting in a state with no spin rigidity ($|\vec{\psi}_0| = 0$) and exponentially decaying correlations.

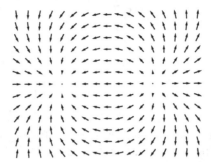

FIGURE 2.5 Vortices in a two-dimensional xy model.

At this point, it is instructive to take a closer look at the characterization of topological defects. In the case of the two-dimensional xy model, a vortex is characterized by a position-dependent order parameter that vanishes at the vortex core and rises to the bulk magnitude $|\vec{\psi}_0|$ over a certain finite length scale. To further characterize the vortex configuration, it is convenient to introduce the *order parameter space* $\mathcal{M}$ — the space of parameters that define all the values $\vec{\psi}_0$ of the order parameter corresponding to the ground state (i.e., given by the minima of the Landau free energy). For the xy model, this manifold is the unit circle S^1, which is in a one-to-one correspondence with the minima of the "Mexican hat." The direction of the order parameter along any contour Γ surrounding the vortex core defines a closed path (i.e., a *loop*) in $\mathcal{M}$ (i.e., a continuous map from Γ to $\mathcal{M}$ having the same initial and final points). Moreover, two different contours Γ and Γ^1 that surround the vortex correspond to *path homotopic* loops. As discussed in the previous chapter (Section 1.2), the corresponding *path homotopy* classes form the *fundamental group* of $\mathcal{M}$. For the circle, we have $\pi_1(S^1) = \mathbb{Z}$. We conclude that the vortices of the two-dimensional xy model are classified by an integer topological number $\nu \in \mathbb{Z}$. Physically, this *winding number* corresponds to the number of times a loop (defined by the direction of the spin as one goes around the vortex core) wraps around the unit circle (i.e., the order parameter space). For example, going (counterclockwise) around either of the two vortices shown in Figure 2.5 results in the direction of the spin cycling the unit circle once (e.g., starting at the rightmost side of the figure and going around the right vortex, the spin points right–up–left–down–right, i.e., one cycle in the order parameter space, which means $\nu = 1$).

Physical phenomena such as the Lifshitz and the Kostelitz–Thouless phase transitions clearly involve elements that can be characterized as topological. However, we are not concerned with this type of phenomena. Instead, our "topological quantum world" will primarily include *fully gapped (bulk) systems* (with the notable exception of *Weyl semimetals* discussed in Chapter 4) in equilibrium at *zero temperature* (except the periodically driven *Floquet insulators*, also discussed in Chapter 4) and possessing nontrivial topological properties that essentially stem from the properties of the ground state. In other words, we focus on systems with many-body ground states that are separated from the excited states by finite energy bulk gaps; these systems may or may not support gapless boundary modes, depending on the specific topological phase to which they belong. Each topological phase includes all ground states than can be smoothly deformed into one another (without closing the bulk gap and while preserving certain symmetries) and thus belong to an *equivalence class* characterized by a specific set of topological invariants. The transitions between different phases, called *topological quantum phase transitions*, are associated with the vanishing of the (bulk) energy gap. These quantum states, which evade the standard classification based on Landau's symmetry breaking paradigm and cannot be described using *local* order parameters, have recently burst into the arena of condensed matter physics with the discovery of *topological insulators*. However, topological insulators represent a rather small region inside a vast and diverse territory; they are a sort of "garden variety" topological state – not the first to be discovered, nor the most interesting, not even possessing full "topological protection," but, probably, the simplest to understand. So, what are the main types of (gapped) topological quantum states?

Historically, the incompleteness of the theoretical framework based on the symmetry breaking paradigm has been revealed by the discovery of the quantum Hall effect (QHE) [301, 511]. Different quantum Hall states, although clearly distinct, have the same symmetry as the Hamiltonian and cannot be described using local order parameters. Discriminating between two quantum Hall states cannot be done based on symmetry, but requires considering the topological properties of the ground states, i.e., those properties that remain invariant under sufficiently small but otherwise arbitrary adiabatic deformations of the Hamiltonian. The remarkably well-defined value of the Hall conductance is a consequence of the relationship between this quantity and the topological invariant that characterizes the quantum Hall state [503] (see Chapter 3, page 89).

The topological properties that define different phases of matter are intrinsically quantum mechanical and have a rather subtle relationship with the concept of *quantum entanglement*. There are two main categories of topologically nontrivial quantum phases: 1) states characterized by patterns of long-range quantum entanglement and possessing a so-called intrinsic *topological order* [532], and 2) states with short-range quantum entanglement (i.e., no topological order). Topologically ordered phases emerge in *strongly interacting* many-body systems and are protected against *any* type of adiabatic deformation of the Hamiltonian. Topological order is revealed by the degeneracy of the ground state in systems defined on compact (multiply connected) manifolds (e.g., on a torus) and is characterized by the emergence of (bulk) excitations with fractional statistics and fractional quantum numbers. The standard example of topologically ordered quantum states is provided by the fractional quantum Hall effect (FQHE) [511]. By contrast, topologically nontrivial states with short-range quantum entanglement are supported by both interacting and noninteracting many-body systems and require the presence of certain symmetries. These symmetry-protected short-range entangled states are characterized by topological properties that are robust only against adiabatic deformations of the Hamiltonian that preserve the symmetry and have unique ground states on all closed manifolds. Typical examples of topologically nontrivial short-range entangled states include the quantum spin Hall effect [347], and the three dimensional topological insulators [54, 239]. Note that many of the the symmetry-protected topological states have the same symmetry as the Hamiltonian, in contrast with the Landau symmetry breaking states, which are also short-range entangled states, but are topologically trivial and break the symmetry of the Hamiltonian. Also note that, in general, symmetry breaking can combine with either long-range or short-range entanglement leading to the emergence of a variety of symmetry breaking topological quantum phases.

In Chapters 3, 4, and 5, we will return to the problem of classifying topological quantum phases with details concerning various subclasses and criteria used in the classification, including the characterization of several topological invariants. The diagrams in Figure 5.1 (page 156) provide a quick general idea about the basic structure of this topological quantum world. Also, in Chapter 4, we will generalize the discussion of noninteracting topological phases by introducing Weyl semimetals (which are characterized by a gapless bulk spectrum) and Floquet insulators (which are periodically driven nonequilibrium systems).

In a spin system, for example, a short-range entangled quantum state can be connected to a real space direct product state through a set of smooth local deformations.[8] By contrast, the long-range entangled states do not have this property. Note that this behavior has no relation with the correlation length of local operators. For example, a gapped spin liquid is a topologically ordered state characterized by long-range entanglement and by exponentially decaying correlation functions for all local operators, which makes different topological

---

[8]We should warn the reader that the terminology in this field is rather fluid. For example, there are two different definitions of short-range entangled (SRE) states: i) gapped states that can be smoothly deformed into a trivial product state without a phase transition (and without breaking any symmetry during the deformation), and ii) gapped states corresponding to a unique ground state on any closed space manifold. In this context, one should also bear in mind that the concept of entanglement is defined in relation to a *partition* of the system (e.g., a real space or a particle-based partition). In terms of particle-based partitions, fermionic states are always entangled and cannot be reduced to direct product states due to their intrinsic antisymmetry. In other words, a topologically trivial band insulator has an entagled ground state represented by a Slater determinant. However, with respect to a real space partition, there is no entabglement (also see discussion on page 152). When we talk about short- or long-range entangled fermionic states we refer to this type of entaglement. Another example concerns the concept of *topological order*, which is used both in a special sense (to designate topological states with long-range entanglement), or generally for any quantum state that has nontrivial topological properties (including symmetry-protected SRE states).

phases locally indistinguishable. By contrast, a Landau symmetry breaking state (e.g., a ferromagnet) has short-range entanglement, but the correlation length of certain local operators (e.g., spin) diverges. A useful theoretical tool for characterizing the entanglement of a many-body quantum state is the entanglement entropy [295, 328]. This quantity can be defined by dividing the system into complementary subsystems A and B and tracing out the degrees of freedom in B. Specifically, if $|\Psi\rangle$ is the normalized ground state wave function and $\rho_A = Tr_B|\Psi\rangle\langle\Psi|$ is the reduced density matrix for subsystem A, the entanglement entropy is defined as the von Neumann entropy

$$S = -\text{Tr}\left[\rho_A \log \rho_A\right]. \tag{2.14}$$

The leading term in Eq. (2.14) is proportional to the area of subsystem A, a property called the area law. However, the entanglement entropy of a topologically ordered state acquires a universal sub-leading contribution. For a disc with circumference L the entanglement entropy has the form

$$S = \alpha L - \gamma + \mathcal{O}(L^{-1}), \tag{2.15}$$

where α is a nonuniversal coefficient and $\gamma \neq 0$ is a universal constant that (partially) characterizes the topologically ordered phase. Note that γ, also called topological entanglement entropy, does not provide a unique criterion for classifying topologically ordered states, but vanishes in all short-range entangled phases. Hence, the topological entanglement entropy clearly discriminates between the two main classes of topological quantum states, i.e., the long-range and short-range entangled states.

The general classification of topologically ordered phases represents an outstanding open question, although significant progress has been made in understanding short-range entangled phases with symmetry-protected topological properties, particularly the topological phases of free fermions (e.g., the topological insulators). We note that, in the presence of interactions, topological phases with short-range entanglement, which can be viewed as generalizations of the topological insulators, can be realized in fermionic systems as well as bosonic and spin systems. An example of interacting short-range entangled topological state is the Haldane phase of a spin $S = 1$ Heisenberg chain [228]. We emphasize that topological phases with short-range entanglement require symmetry protection. If the relevant symmetries are broken, the phase can be adiabatically connected to a topologically trivial state (e.g., an atomic insulator). We also note that topological phases are often characterized as states with a gapped bulk and a gapless boundary. However, gapless boundary modes are typically present in (almost) noninteracting topologically nontrivial systems with internal (nonspatial) symmetries, but they are not generically required in interacting systems, or in systems with spatial (crystalline) symmetries.

In conclusion, we distinguish two main categories of topological quantum states: (1) topologically ordered phases and (2) topological states with short-range entanglement. Category (1) contains (strongly) interacting systems and its members are characterized by long-range entanglement, which results in a nonvanishing topological entanglement entropy, degenerate ground states for systems defined on multiply connected manifolds, and nontrivial bulk excitations. Typical examples are fractional quantum Hall fluids and gapped spin liquids. Category (1) states are robust against *any* type of perturbation and do not require symmetry protection. However, in the presence of symmetry, each class of category (1) states will be divided into distinct subclasses of so-called *symmetry enriched* topological phases. Category (2) contains both interacting (e.g., the Haldane phase of a spin-1 chain) and noninteracting systems (e.g., the topological insulators). Category (2) states have vanishing topological entanglement entropy, nondegenerate ground states (on simply and multiply connected manifolds), and trivial bulk excitations; oftentimes, however, they support gapless boundary modes. These states – also called *symmetry protected* topological states – require symmetry protection, i.e., they are robust only against perturbations that preserve the symmetry of the Hamiltonian.

A few final notes. We will use the term *topological phases* (or *states*) for both category (1) and category (2) quantum systems. In the literature, the term *topological phase* is sometimes used to specifically denote category (1) phases. Topological order (*intrinsic* or without any qualification) applies to category (1) phases. Again, sometimes category (2) states are said to possess *symmetry protected* topological order, but we will try to avoid this term. The condition requiring the existence of a gap can be relaxed; in Weyl semimetals, for example, topological phases can exist in the absence of a bulk energy gap. We will briefly discuss this type of system in Chapter 4. Also, long-range entanglement is not a feature that characterizes only the topologically ordered gapped phases; gapless spin liquids and string-net condensates are also long-range entangled non-symmetry-breaking phases. The larger family of non-symmetry-breaking phases are sometimes said to possess *quantum order*. Intrinsic and symmetry-protected topological orders are specific types of quantum orders. Finally, we reiterate that a general principle for classifying topological quantum matter has not yet been identified. There are different proposed theoretical frameworks that successfully capture the topological properties of specific categories of topological phases (e.g., the topological phases of noninteracting fermions), but are not generally applicable. Throughout this book (and, more generally, within this field) the reader should always have in mind that various statements and results may have a limited domain of validity. Defining topological quantum matter as consisting of systems with a gapped bulk and gapless surface (boundary) states is an obvious example (the boundary states may be absent in interacting topological systems and even in some noninteracting systems – see Chapter 4). Another example would be the classification of a "regular" three-dimensional s-wave superconductor: when viewed as a noninteracting systems (i.e., at the mean-field level, as described by, e.g., a Bogoliubov–de Gennes Hamiltonian), it is topologically trivial; however, as a system of charged fermions with a dynamical $U(1)$ gauge field the superconductor has intrinsic topological order.

2.4 TOPOLOGY AND QUANTUM COMPUTATION

At the time of this writing (summer 2023), quantum computing (QC) is a highly technical area actively present in the public consciousness, with attitudes ranging from enthusiastic expectations of a revolutionary technology that will solve some of the most challenging current problems to serious concerns regarding a hype-fueled bubble. While betting one's future on any short- or medium-term economic success of QC technology would probably be unwise, the basic science underlying it is beautiful, fascinating, and rewarding. We will discuss basic concepts and ideas regarding quantum computation in Chapter 11 and specific aspects concerning topological quantum computation in Chapter 12. Here, we will simply argue that leveraging the properties of topological quantum matter is a good strategy for realizing a quantum computer. Before making the point, it is useful to put some flesh on the relation between topological quasiperticles and knot invariants sketched in Section 1.1.

We have seen that the quantum amplitudes associated with space–time processes involving the creation and annihilation of quasiparticles in certain "topological" quantum systems are related to knot invariants. Specific examples of systems that support such quasiparticles include quantum Hall liquids. While for many quantum Hall states the corresponding knot invariants are rather trivial, the fractional quantum Hall states with a low-energy physics described by, e.g., a $SU(2)_k$ non-Abelian Chern–Simons field theory[9] are characterized by quantum amplitudes corresponding to Kauffman bracket invariants with $A = ie^{-\frac{i\pi}{2(2+k)}}$. For example, the fractional quantum Hall fluid with filling factor $\nu = 12/5$ belongs to the $SU(2)_4$ class (with Kauffman bracket invariant corresponding to $A = i^{5/6}$), while the quantum Hall fluid with $\nu = 5/2$ belongs to the $SU(2)_2$ class ($A = i^{3/4}$). We note that the $SU(2)_2$ class, which also includes two-dimensional p-wave superconductors (see Chapter 6, page 194) and ^{3}He superfluid films, supports quasiparticles (e.g., Majorana zero modes) that are similar to the so-called "Ising anyons," which we discuss in Chapter 12.

Before continuing our discussion, let us clarify a technical point regarding the knots corresponding to quasiparticle world lines. We should imagine the knotted strings as consisting of (finite width) ribbons lying in the plane (the so-called *blackboard framing*). Within this convention, a string with a loop corresponds to a twisted string (i.e., ribbon), as illustrated in Figure 2.6. Furthermore, using rule I for evaluating the Kauffman bracket invariant (see Figure 1.2), we can easily show that the twisted ribbon is related to a straight ribbon by a factor of $Ad + A^{-1} = -A^3$ (Figure 2.6). Mathematically, we say that the equivalence relation between a looped string and a straight string is an *ambient isotopy*[10] (which, basically, ignores self twists); however, the two objects are not *regular isotopic*. Physically, a self-twisted string represents a

---

[9]See Chapter 6 (page 204) and Chapter 12 (page ??) for details on Chern–Simons field theories. An $SU(2)_k$ theory involves gauge fields that are elements of the $SU(2)$ group, with the integer k (called the *level* of the theory) being a coupling constant.

[10]Two knots are *ambient isotopic* if they can be continuously deformed into one another without cutting the strings. Two framed knots are *regular isotopic* if they can be continuously deformed into one another without cutting or twisting the strings.

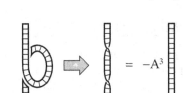 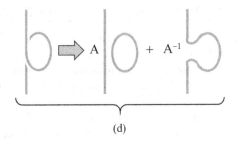

(a)   (b)   (c)

(d)

FIGURE 2.6 (a) A piece of a knot containing a loop represented as a ribbon lying flat in the plane (blackboard framing). (b) Pulling straight the loop generates a twist. (c) The twisted ribbon is equivalent to a straight string multiplied by a factor $-A^3$. (d) Eliminating the loop using rule I. Since a simple loop can be replaced by d (rule II), the resulting diagrams correspond to a straight string multiplied by a factor $Ad + A^{-1} = -A^3$.

particle having a fixed position in space that rotates around an axis by an angle 2π. The factor $-A^3$ corresponds to the phase accumulated during this rotation by a particle that carries spin. We note that one can construct a knot invariant that takes the same value for all ambient isotopic knots by simply constructing the Kauffman bracket invariant and removing a factor of $-A^3$ for each self twist. The resulting quantity is the so-called *Jones polynomial* knot invariant [267].

Next, let us express the quantum amplitudes discussed in Chapter 1 as inner products of quantum states at a given time t using Dirac bra–ket notation. Consider a state corresponding to a particle–antiparticle pair that was created sometime in the past, each particle evolving along a certain world line up to the time slice t when it reaches a specified position. For example, one can associate a state, say $|\alpha\rangle$, to the half-loop diagram in Figure 2.7(a), up to a normalization factor $\mathcal{N}$ (to be determined). Furthermore, we can associate the bra $\langle\alpha|$ to the time-reversed diagram shown in Figure 2.7(b). The inner product $\langle\alpha|\alpha\rangle$ corresponds to "gluing together" the two diagrams, which results in a simple loop. If we consider a topological system with quantum amplitudes given by the Kauffman bracket invariant, the simple loop corresponds to an invariant equal to d (rule II in Figure 1.2) and the normalization factor is $\mathcal{N} = 1/\sqrt{d}$. This construction can be generalized for an arbitrary number of particle–antiparticle pairs. Consider, for example, two particle–antiparticle pairs and two states, $|0\rangle$ and $|1\rangle$, characterized by the same measurable *local* properties at time t (e.g., location of the particles, spin values, etc.), but having different space–time histories, as illustrated in Figure 2.7(c) and (d). The normalization factors are determined by the condition $\langle 0|0\rangle = \langle 1|1\rangle = 1$, after evaluating the the Kauffman bracket invariants characterizing the knots obtained by "gluing together" the diagrams representing the kets and those representing the corresponding (time-reversed) bras. Thus, the diagram representing $\langle 0|0\rangle$ consists of two simple loops, which, according to rule II give a

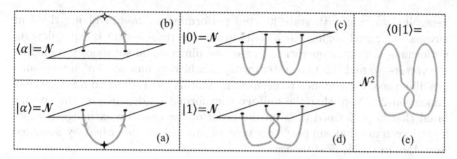

FIGURE 2.7 (a) Diagram representing the state $|\alpha\rangle$ of a particle–antiparticle pair at time t. The particles have specified positions (small black disks) and specific space–time histories. The normalization factor is $\mathcal{N} = 1/\sqrt{d}$. (b) The bra $\langle\alpha|$ associated with state $|\alpha\rangle$ can be obtained by time-reversing the diagram in (a). The inner product $\langle\alpha|\alpha\rangle$ corresponds to "gluing together" the two diagrams. (c) Diagram representing the state of two particle–antiparticle pairs (with no particle exchange). (d) State of two particle–antiparticle pairs with the same local properties at time t as in (c), but different space–time histories. (e) Knot representing the inner product $\langle 0|1\rangle$.

contribution equal to d^2 (hence, $\mathcal{N} = 1/d$). Similarly, the diagram representing $\langle 1|1\rangle$ is a knot consisting of two linked loops with four crossings. After eliminating the crossings (using rule I), we obtain 16 weighted diagrams and a Kauffman bracket invariant equal to d^2 (hence, again, $\mathcal{N} = 1/d$). Note that this is a concrete example of inequivalent knots characterized by the same value of the knot invariant.[11]

Let us now evaluate the inner product $\langle 0|1\rangle$. The corresponding diagram, shown in Figure 2.7(e), is equivalent (i.e., regular isotopic) to the knot in Figure 1.1(b), up to the overall $1/d^2$ factor. Consequently, we have $\langle 0|1\rangle = 2/d-d$. If we consider the specific case of Ising anyons, we have $A = i^{3/4}$ and $d = -A^2 - A^{-2} = \sqrt{2}$, which implies $\langle 0|1\rangle = 0$. This is a remarkable result showing that the two states are orthogonal; they are not uniquely determined by the locally measurable properties of the quasiparticles, but depend on their space–time histories. The physical reason for this unique property is that the topological vacuum is not an "inert environment," but a "medium" that encodes nonlocally the information associated with the space–time history of the quasiparticles. Thus, performing two exchanges of a pair of Ising anyons (i.e., moving one anyon around the other one) generates a state ($|1\rangle$) that is orthogonal on the state with a space–time history involving no exchange ($|0\rangle$). In other words, the probability for the anyons to fuse (pairwise) back to vacuum vanishes after two exchanges that return the anyons to their original positions, whereas this probability is one if no exchange is performed. More

---

[11]All knots within a given equivalence class have the same value of the knot invariant. However, there can be different equivalence classes having the same value of the knot invariant.

generally, starting with state $|0\rangle$ and performing a *braid* involving the four anyons (e.g., exchanging them pairwise in a certain order) is equivalent to performing a unitary operation on the two dimensional subspace spanned by the vectors $|0\rangle$ and $|1\rangle$. In a system with an arbitrary number n of anyon pairs, braiding them corresponds to performing unitary operations on a subspace of dimension 2^n. Note that the unitary transformation depends on the specific braid that is performed (i.e., on the order of the pairwise exchanges); topological systems that support this type of anyons are described by so-called *non-Abelian* topological theories.

We are now ready to sketch the main ideas underlying topological quantum computation. Consider the state $|0\rangle$ corresponding to n well-separated particle–antiparticle pairs created out of the vacuum of a (non-Abelian) topological system. Braiding these quasiparticles corresponds to performing unitary operations on a 2^n-dimensional vector subspace, $|0\rangle \rightarrow U_{\text{braid}}|0\rangle$, while evaluating the quantum amplitude $\langle 0|U_{\text{braid}}|0\rangle$ corresponds to calculating the Kauffman bracket invariant of the knot associated with the braided world lines. Assume that the knot contains N crossings. Calculating the knot invariant using a classical computer can be done by evaluating 2^N diagrams (obtained after applying rule I)[12] – a computational task that becomes practically impossible for values of N larger than a certain limit of order 10^2. On the other hand, determining the Kauffman bracket invariant would become a relatively easy task if one could controllably braid anyons in a (Kauffman type) topological quantum system and physically measure the corresponding quantum amplitudes. Hence, the topological quantum system could provide computational abilities beyond those of a classical computer. Furthermore, as we will discuss in detail in Chapter 11, a key requirement for a general-purpose quantum computer is to be able to perform generic unitary operations (or, more precisely, to approximate them with a given precision) on a 2^n-dimensional Hilbert space, where n (the number of *qubits*) is "sufficiently large." Could such unitary operations be realized by braiding non-Abelian anyons? It turns out that for certain types of anyons (e.g., the so-called *Fibonacci anyons*) the answer is affirmative. Thus, one can do (general-purpose) quantum computation using non-Abelian anyons hosted by a topological quantum system. Our final point is that doing quantum computation with topological systems is not only possible; it actually represents an excellent strategy for overcoming the fundamental challenge of quantum computation – quantum decoherence. Since the coupling between a quantum computer and its environment is never exactly zero, the qubits will be affected by local perturbations, e.g., phonons or photons entering the system. In a "standard" (i.e., nontopological) quantum computer information is encoded locally and the unitary operations are controlled by local variables, which makes them extremely susceptible to local perturbations that generate quantum decoherence and, ultimately, errors.

---

[12]More efficient classical algorithms can be designed, but the problem remains computationally *hard*. See Chapter 11 for an elementary introduction to computational complexity.

While quantum error correction is possible (if the error rate is below a certain level), it involves a huge overhead cost (i.e., using many *physical* qubits to realize a single *logical* qubit). By contrast, in a topological quantum computer quantum information is encoded nonlocally and the unitary operations are determined by the topology of the anyon world lines, rather than their geometric (local) details. Locally perturbing an anyon may significantly affect its world line but this will not generate quantum errors, as long as the topology of the braid remains the same. Hence, topological quantum computation enables an inherent (topological) protection against quantum errors generated by local perturbations.

2.5 TOPOLOGY AND EMERGENT PHYSICS

We end this chapter with a brief discussion about the potential impact of topology on the physics of emergent behavior. Is there anything new that topological quantum matter brings to the table? But, first, what do we mean by emergent physics? Consider, for example, a system of atoms that, at low-enough temperature, condense and form an insulating crystal. The low-energy physics of the crystal is governed by the longitudinal and transverse acoustic phonons. We distinguish two levels: the microscopic level (L_1), which can be described in terms of "elementary" particles, e.g., electrons and, say, nuclei, that obey the laws of quantum mechanics and interact via the Coulomb interaction, and the "macroscopic" level (L_2), which can be characterized in terms of noninteracting phonons. The "elementary" particles of L_2 and the laws that govern them can be viewed as *emergent* phenomena. Using the laws of L_1 to derive the "elementary" particles and the "fundamental" laws of L_2 (i.e., solving the Schrödinger equation for the interacting many-body system) is a practical impossibility. On the other hand, we know that the emergent behavior (i.e., the existence of low-energy phonons) is a consequence of the fact that the L_1 particles are organized in a structure that breaks a continuous symmetry, which results in the emergence of gapless Goldstone bosons.

This simple example can be interpreted in several different registers, ranging from "strong emergentism" (the physics at level L_2 is fundamentally irreducible to the properties of the L_1 constituents; there are extra laws/principles that do not operate at and do not originate from level L_1) to "reductionism" (the physics at level L_2 is in principle determined by the fundamental laws that govern L_1 and can be derived from them; it may be difficult to reconstruct theoretically the phenomenology of L_2 starting from the fundamental principles, but there is nothing really "new" operating at L_2). We will not dwell on these interpretations or on the possible variations and nuances, nor will we mention any of the fascinating philosophical implications. After all, stepping into this territory may be a "lone theorist's" temptation. Instead, let us notice that one of the main differences between the two levels is that the entities and the laws of L_1 are simple, robust, and beautiful, while those of L_2 may be simple, but not as robust and, consequently, not as beautiful as the "fundamental" ones. This is where topological quantum matter comes into the picture. Topological protection endows the low-energy quasiparticles of a topologically ordered state with a robustness qualitatively similar to that enjoyed by their "fundamental" counterparts. Moreover, while in the "Landau world" the collective excitations (e.g., the phonons) are always scalar bosons, the discovery of the fractional quantum Hall effect and the subsequent developments have revealed other possibilities, such as elementary excitations with fractional quantum numbers and fractional or non-Abelian statistics.

An intriguing proposal by Levin and Wen suggests the possibility of having fermions and boson-mediated gauge interactions as emergent phenomena in a purely bosonic long-range entangled system – the so-called string-net condensed liquid [329]. In essence, one can start with a spin model characterized by interactions that i) allow strong fluctuations of large nets formed by intersecting and/or overlapping closed strings of spins and ii) inhibit local spin flips or the formation of open strings. The fluctuations of closed string-nets condensed in the ground state are gauge bosons, while the topological defects associated with the ends of open strings are fermions carrying the charge of the corresponding gauge field. On a more speculative note, this suggests the possibility that photons and electrons have a unified origin and represent the low-energy excitations of a highly entangled vacuum. Then, maybe all "elementary" particles and the laws the govern their interactions are nothing but emergent phenomena in a "vacuum" whose constituents are inaccessible below some extremely large energy scale. Ultimately, the standard model may be just a phenomenological description of emerging low-energy behavior, but this will remain an intriguing hypothesis as long as we do not have clear falsifiable predictions and experimental access to the underlying microscopic reality. Meanwhile, the wonders and mysteries of the topological quantum world await closer to us in condensed matter systems, our "Universe in a box."

CHAPTER 3

Topological Insulators and Superconductors

THE CLASSIFICATION OF DISTINCT PHASES OF MATTER is a central theme in condensed matter physics. A phase of matter can be viewed as an equivalence class of physical states that share a certain set of properties. In the case of topological quantum matter, these properties are associated with topological invariants that take specific values for each equivalence class. Perturbing a gapped state that corresponds to a given topological phase (i.e., smoothly deforming the Hamiltonian of the system) does not affect the topological invariant, as long as the bulk energy gap remains finite. The value of this invariant changes only if the system undergoes a quantum phase transition, which is signaled by the collapse of the gap. Full topological protection can only emerge in interacting systems and requires the presence of long-range entanglement, which is at the origin of intrinsic topological order. However, topological properties protected by symmetries emerge even in noninteracting systems and, perhaps surprisingly, can be described within the framework of good old band theory of solids. Systems belonging to this category are generically called topological insulators, after the class of materials that is largely responsible for the recent surge in research on topological matter. In addition, topological band theory considerations can also be applied to superconductors (and superfluids), as described at the mean-field level using Bogoliubov–de Gennes theory. This chapter summarizes the key ideas behind the topological classification of noninteracting systems and discusses the basic properties of various classes of topological insulators and superconductors. For in-depth discussions and related developments the reader is referred to a number of books [54, 175, 469] and review papers [23, 239, 240, 421] specifically dedicated to this subject.

DOI: 10.1201/9781003226048-3

3.1 INTRODUCTION

Topological insulators (TIs) represent a subset of the set of noninteracting insulating phases, which are equivalence classes of fermionic many-body quantum states characterized by a bulk energy gap. According to the band theory of solids, the insulating gap separates the occupied valence-band states from the empty conduction-band states. The simplest example of an insulator, the atomic insulator, is characterized by narrow, nearly flat bands with energies corresponding to the electronic spectrum of an isolated atom. Note that, in general, the ground state properties of an insulator do not depend on the core bands. Adding or subtracting a number of trivial core bands generates an equivalent insulating state that belongs to the same phase as the original system. With this in mind, we introduce the following equivalence relation on the set of insulating states. Consider the insulating states A and B corresponding to the ground states of Hamiltonians H_A and H_B, respectively. The two insulators are said to be topologically equivalent if (and only if) by tuning the Hamiltonian one can continuously interpolate between the ground states A and B (modulo trivial core bands) without closing the energy gap.

According to this definition, solid argon and solid xenon belong to the same insulating phase. Moreover, these atomic insulators are topologically equivalent with all the other conventional insulators and semiconductors, such as, for example, silicon, as one can continuously interpolate (modulo trivial core bands) from one state to another without closing the energy gap. Interpolations are done by smoothly deforming the Hamiltonian of the system, for example by changing the inter-atomic distance. We also note that conventional insulators are topologically equivalent to the vacuum, which can be viewed as an insulator characterized by a conduction band (electrons), a valence band (positrons), and a large energy gap (associated with pair production).

Are there any other classes of insulating states? In other words, are there noninteracting gapped electronic states that are not topologically equivalent to the vacuum? The answer is yes, provided we introduce a constraint in the definition of topological equivalence. Specifically, instead of interpolating between two states by considering *arbitrary* continuous deformations of the Hamiltonian, we require these deformations to preserve certain *symmetries*. Then, preserving time-reversal symmetry, for example, one cannot continuously interpolate between a quantum spin Hall (QSH) insulator and a trivial two-dimensional insulator without closing the energy gap, i.e., without going through a quantum phase transition. However, the two states can be continuously connected without closing the gap if deformations of the Hamiltonian that break time-reversal symmetry are allowed. Hence, with a definition of topological equivalence that includes the requirement of preserving time-reversal symmetry, the QSH insulator and the trivial two-dimensional insulator belong to different topological classes. In fact, the QSH state is an example of a topological insulator, i.e., a noninteracting symmetry-protected topological state. In general, TIs are defined as the noninteracting insulating phases representing the symmetry-protected equivalence classes topologically distinct from the trivial insulator (i.e., the atomic insulator or the vacuum).

Below we discuss the main ideas behind the classification of topological insulators and superconductors. The basic question is the following. Given a certain spatial dimension and a set of generic symmetries (e.g., time-reversal and particle-hole symmetries), how many distinct topological phases (i.e., symmetry-protected equivalence classes) are there? Before presenting the basic ideas and the main results, let us make a few observations.

First, one can view a superconductor (SC), described at the *mean-field level*, as a gapped system of noninteracting fermionic (Bogoliubov) quasiparticles endowed with particle-hole (or charge conjugation) symmetry. Consequently, the scheme used for classifying band insulators can also be used for the classification of superconductors. In particular, two superconducting states are said to be topologically equivalent if and only if one can smoothly interpolate between them (modulo certain trivial core bands) without closing the superconducting quasiparticle gap and while preserving the appropriate symmetries (e.g., the particle-hole symmetry).

Second, all TIs and topological SCs protected by nonspatial symmetries (e.g., time-reversal and particle-hole) are characterized by topologically protected gapless boundary modes. Since this feature is generic for these systems, it is often used in the definition of topological insulators (superconductors). More specifically, they are defined as gapped phases of noninteracting fermions characterized by gapless boundary modes that are robust against any symmetry-preserving perturbation of the Hamiltonian that does not close the bulk gap. However, we warn the reader that, unlike "standard" TIs and topological SCs, noninteracting topological phases protected by spatial symmetries (see Chapter 4) and interacting topological systems (Chapter 5) may not have gapless modes on some boundaries (e.g., the boundaries that break the spatial symmetries), or may not support gapless boundary states at all.

Third, since preserving certain symmetries of the Hamiltonian is a key requirement in the definition of topological equivalence, classifying the TIs and the topological SCs is a twofold task. First, one has to classify the single-particle Hamiltonians in terms of the presence or absence of general symmetries, such as time-reversal and particle-hole symmetries. Then, within each of the symmetry classes, one has to determine the topologically distinct sectors corresponding to the (topological) equivalence classes of insulating quantum ground states. Different equivalence classes (i.e., topological phases) are characterized by different values of a suitable topological invariant (e.g., the Chern number in the case of the integer quantum Hall effect). Note that, in view of the bulk-boundary correspondence mentioned above, one can alternatively characterize the topological properties of a system by looking at its boundary states, which holographically reflect the topological features of the bulk.

Fourth (and last), there are several important caveats regarding the vanishing of the bulk spectral gap characterizing a gapped system. We have already mentioned several times that topologically distinct (gapped) phases are separated by topological quantum phase transitions (TQPTs), which are associated with the closing (and reopening) of the *bulk energy gap*. However, not all closings of the bulk spectral gap are associated with TQPTs. Considering first the case of a clean, translation invariant system (i.e., a system with no disorder), we point out that the emergence of a topological phase may be associated with a "band inversion" at certain (high symmetry) points in the Brillouin zone. The TQPT is associated with the vanishing of the spectral gap at those specific (high symmetry) momenta; the vanishing (and possible reopening) of the bulk gap at an *arbitrary* point in the Brillouin zone is *not* associated with a TQPT. Also, the transition between two topologically distinct gapped phases as a function of a certain control parameter does not always happen at a critical *point* (where the bulk gap vanishes), but may involve a finite parameter range, i.e., an intermediate *gapless phase*. In certain conditions, this gapless phase can be topologically nontrivial, e.g., a Weyl semimetal (see Chapter 4).

Consider now a system with disorder. The presence of weak-enough disorder does not destroy a topological phase, but induces low-energy (sub-gap) states; in particular, these disorder-induced states may have zero energy. However, the vanishing of the spectral gap due to the presence of disorder-induced states is not associated with a TQPT. The key point is that the disorder-induced low-energy states are *localized*, typically within the bulk. The TQPT is always associated with a *delocalized* zero energy state that, in a finite system, couples to all boundary modes. Intuitively, if one views the presence of gapless boundary modes as a direct consequence of connecting a trivial region to a region with "inverted bands," eliminating these modes (i.e., "undoing" the band inversion) requires a zero-energy state that extends throughout the whole system. Localized, disorder-induced states do not satisfy this requirement. Consequently, in disordered systems a TQPT is not necessarily associated with the closing an reopening of the *spectral* gap, but rather with the closing and reopening of the *mobility* (i.e., transport) gap. A prime example is the integer quantum Hall system; a brief discussion of disorder effects in one-dimensional topological SCs (i.e., "Majorana wires") can be found in Chapter 8. Throughout this book, when referring to the closing (and reopening) of the *bulk gap* we will implicitly assume the closing (and reopening) of the mobility gap, i.e., the presence of a *delocalized* zero energy state (rather than a *localized*, disorder induced zero energy state), unless stated otherwise.

3.2 SYMMETRY CLASSIFICATION OF GENERIC NONINTERACTING HAMILTONIANS

The topological properties of the ground state of a gapped Hamiltonian characterized by certain symmetries are not affected by perturbations that preserve these symmetries, as long as the bulk gap remains finite. Since we are particularly interested in topological features that are robust against disorder, we need to identify the symmetry classes of Hamiltonians that are *not* translation invariant. Furthermore, in the presence of "ordinary" symmetries, which are represented by unitary operators that commute with the Hamiltonian, the (first-quantized) Hamiltonian can be written in a block-diagonal form. The symmetry classification that concerns us refers to these "irreducible" blocks. The symmetry classes are determined by the spatial dimensionality and by the presence or absence of *time-reversal* and *charge conjugation* (or *particle-hole*) symmetries. Specifically, following Zimbauer and Altland [18, 573], there are ten distinct symmetry classes of random matrices, which can be interpreted as first-quantized Hamiltonians of certain noninteracting fermionic systems. Below, we review the main implications of having time-reversal (TR) or particle-hole (PH) symmetries and summarize the generic properties of the ten symmetry classes of noninteracting Hamiltonians.

3.2.1 Time-reversal symmetry

Time-reversal (TR) symmetry is a fundamental property of a system associated with the invariance of the Hamiltonian with respect to a time-reversal transformation, $\mathcal{T} : t \longrightarrow -t$, i.e., reversing the arrow of time. Formally, TR invariance is expressed by the condition

$$[H, \mathcal{T}] = 0. \tag{3.1}$$

This implies that, if $|\psi\rangle$ is an eigenstate of H with energy E, then $\mathcal{T}|\psi\rangle$ is also an eigenstate of H corresponding to the same eigenvalue E. The answer to the question regarding whether or not $|\psi\rangle$ and $\mathcal{T}|\psi\rangle$ represent the same physical state depends on the spin carried by the constituents of the many-body system described by H. Before discussing this point, let us notice that the TR operator commutes with the position operator, $\mathcal{T}\hat{x}\mathcal{T}^{-1} = \hat{x}$, but flips the sign of the momentum operator, $\mathcal{T}\hat{p}\mathcal{T}^{-1} = -\hat{p}$. Consequently, we have $\mathcal{T}[\hat{x}, \hat{p}]\mathcal{T}^{-1} = -[\hat{x}, \hat{p}]$, or $\mathcal{T}i\mathcal{T}^{-1} = -i$, which shows that $\mathcal{T}$ is proportional to the complex conjugation K. Moreover, one can show that the scalar product of any two states $|\phi\rangle$ and $|\psi\rangle$ transforms under TR into its complex conjugate, $\langle\phi'|\psi'\rangle = \langle\phi|\psi\rangle^*$, where $|\phi'\rangle = \mathcal{T}|\phi\rangle$ and $|\psi'\rangle = \mathcal{T}|\psi\rangle$. We conclude that $\mathcal{T}$ is an *anti-unitary* operator. In general, we can represent an anti-unitary operator as the product of a unitary operator and the complex conjugation. Explicitly, we have

$$\mathcal{T} = U_t K, \tag{3.2}$$

with $U_t^\dagger = U_t^{-1}$. Note that applying the TR operator twice should leave the system invariant, i.e., we have $\mathcal{T}^2 = U_t K U_t K = U_t U_t^* = e^{i\phi}$. This implies that $U_t = e^{i\phi} U_t^T = e^{i\phi}(e^{i\phi} U_t^T)^T = e^{2i\phi} U_t$, which means that $e^{i\phi} = \pm 1$ and, consequently, the TR operator squares to plus or minus the identity, $\mathcal{T}^2 = \pm 1$.

Consider now the simple case of a spinless particle in a quantum state determined by the Schrödinger equation $i\hbar\frac{\partial}{\partial t}|\psi(t)\rangle = H|\psi(t)\rangle$. Under TR we have $t \longrightarrow -t$, $H \longrightarrow \mathcal{T}H\mathcal{T}^{-1}$ and $|\psi(t)\rangle \longrightarrow \mathcal{T}|\psi(-t)\rangle$. To identify a suitable *representation* of the TR operator, let us first consider the position representation. In this "conventional" representation the Hamiltonian H and the state vector $|\psi(t)\rangle$ are represented by the operator $\mathcal{H}(x,p)$ (with $p = -i\frac{\partial}{\partial x}$) and the wave function $\psi(x,t)$, respectively. The TR operator acts on a position eigenvector as $\mathcal{T}|x\rangle = |x\rangle$ and on a generic state vector as

$$\mathcal{T}|\psi\rangle = \int dx \ \psi^*(x)|x\rangle. \tag{3.3}$$

This definition, which implies that in the position representation we have $\mathcal{T} = K$, ensures that $\mathcal{T}$ is anti-unitary – $\langle \mathcal{T}\phi|\mathcal{T}\psi\rangle = \int dx \ \phi(x)\psi^*(x) = \langle\phi|\psi\rangle^*$ – and satisfies the basic commutation relations with the position and momentum operators – $KxK^{-1} = x$ and $KpK^{-1} = K(-i\hbar \, \partial/\partial x)K^{-1} = -p$. Consequently, for a spinless particle we can represent the TR operator as the identity operator times the complex conjugation, $\mathcal{T} = K$. Note that we have $KH(x,p)K^{-1} = H(x,-p)$ and $K\psi(x,-t) = \psi^*(x,-t)$, which represent the canonical relations for the TR transformed Hamiltonian and wave function, respectively. Also note that for spinless particles we have $\mathcal{T}^2 = K^2 = 1$, i.e., the TR operator squares to the identity operator.

The "conventional" representation discussed above is not the only possible representation. It is illustrative to consider the alternative representation based on a discrete basis $|\mu\rangle$. For example, let us assume that the state vectors $|\mu\rangle$ correspond to localized Wannier-type states associated with a particle moving in the presence of a periodic potential. Then, the single-particle Hamiltonian is represented by the $N \times N$ matrix $\mathcal{H}_{\mu\nu} = \langle\mu|H|\nu\rangle$, where the $\mu = (i,\alpha)$ and $\nu = (j,\beta)$ represent combined labels for the lattice sites i and j and (possibly) some additional orbital degrees of freedom (α and β). It is straightforward to show that $(U_t)_{\mu\nu} = \langle\mu|\mathcal{T}|\nu\rangle = \int dx \ \langle\mu|x\rangle\langle\nu|x\rangle$ is a symmetric unitary matrix, $U = U^T = (U^*)^{-1}$. Hence, in this new representation we have $\mathcal{T} = U_t K$. Note that the property $\mathcal{T}^2 = UU^* = 1$ holds, i.e., for spinless particles the TR operator squares to the identity operator independent of representation. More generally, one can show that $\mathcal{T}^2 = 1$ for any quantum system with *integer* total spin.

Next, let us consider the case of a spin-$\frac{1}{2}$ particle. In addition to the transformations discussed above, the time-reversal operation flips the spin, $\mathbf{S} \longrightarrow \mathcal{T}\mathbf{S}\mathcal{T}^{-1} = -\mathbf{S}$, where $\mathbf{S} = \frac{\hbar}{2}(\sigma_x, \sigma_y, \sigma_z)$ is the spin operator expressed in terms of Pauli matrices σ_α. The change in the spin direction is equivalent to a rotation around an arbitrary axis – say, for concreteness, the y-axis – by an angle π, which is represented by the operator $\exp[-\frac{i}{\hbar}\pi S_y] = -i\sigma_y$. With this choice, the TR operator for a spin-$\frac{1}{2}$ particle is represented by

$$\mathcal{T} = -i\sigma_y K. \tag{3.4}$$

Indeed, one can easily check that $\mathcal{T}\mathbf{S}\mathcal{T}^{-1} = \sigma_y \mathbf{S}^* \sigma_y = -\mathbf{S}$. Note that in this case $\mathcal{T}^2 = -i\sigma_y i\sigma_y^* K^2 = -1$, where "1" designates the identity matrix. More generally, one can show that the TR operator of a quantum system with *half-integer* total spin squares to minus the identity operator.

For a particle moving in a periodic potential, translation symmetry allows us to block-diagonalize the Hamiltonian. For a block labeled by the crystal momentum $\boldsymbol{k}$, i.e., the Bloch Hamiltonian $\mathcal{H}(\boldsymbol{k})$, time-reversal invariance is expressed by the condition

$$\mathcal{T}\mathcal{H}(\boldsymbol{k})\mathcal{T}^{-1} = U_t \mathcal{H}^*(\boldsymbol{k}) U_t^{-1} = \mathcal{H}(-\boldsymbol{k}), \tag{3.5}$$

where $\mathcal{T}^2 = U_t(U_t^T)^{-1} = 1$ for spinless particles (or systems with integer total spin) and $\mathcal{T}^2 = -1$ for spin-$\frac{1}{2}$ particles (or half-integer total spin). If $\psi(\boldsymbol{k})$ is an eigenstate of $\mathcal{H}(\boldsymbol{k})$ with energy $E(\boldsymbol{k})$, $\mathcal{T}\psi(\boldsymbol{k})$ is an eigenstate of $\mathcal{H}(-\boldsymbol{k})$ with energy $E(-\boldsymbol{k}) = E(\boldsymbol{k})$. If $\boldsymbol{k}$ and $-\boldsymbol{k}$ are distinct momenta, TR symmetry implies that the energy $E = E(\boldsymbol{k}) = E(-\boldsymbol{k})$ is (at least) double degenerate. However, for TR invariant momenta $\boldsymbol{k}_{TR}$, such as $(0,0,0)$ or $(\frac{\pi}{a}, 0, 0)$ (for a cubic lattice), i.e., when $\boldsymbol{k}_{TR}$ and $-\boldsymbol{k}_{TR}$ differ by a reciprocal lattice vector, the states $\psi(\boldsymbol{k}_{TR})$ and $\mathcal{T}\psi(\boldsymbol{k}_{TR})$ are not necessarily distinct. Specifically, for a spinless system with $\mathcal{T}^2 = 1$ the state with energy $E(\boldsymbol{k}_{TR})$ is, generally, nondegenerate. In particular, the system can have a single band with $E(\boldsymbol{k}) = E(-\boldsymbol{k})$ and a minimum (or maximum) at $\boldsymbol{k} = \boldsymbol{k}_{TR}$. By contrast, the condition $\mathcal{T}^2 = -1$ ensures that $E(\boldsymbol{k})$ is at least double degenerate for any value of the momentum. This property is a consequence of the so-called *Kramers theorem*, which states that the ground state of a TR invariant quantum system with $\mathcal{T}^2 = -1$ (i.e., a system with half-integer total spin) is at least twofold degenerate. Consequently, in the presence of TR symmetry, the spectrum of a spin-$\frac{1}{2}$ particle is necessarily (at least) double degenerate, which, in the presence of translation invariance, results in the twofold degeneracy of $E(\boldsymbol{k})$. Furthermore, the ground state of any (noninteracting or interacting) TR invariant system with an odd number of electrons, i.e., with half-integer total spin, is at least double degenerate. The Kramers theorem is a direct consequence of the property $\mathcal{T}^2 = -1$. Indeed, let $|\psi\rangle$ represent an energy E eigenstate of a TR invariant Hamiltonian that describes a system with half-integer total spin. The state $\mathcal{T}|\psi\rangle$ will also be an eigenstate with the same energy E. Since $\mathcal{T}$ is anti-unitary, we have $\langle\psi|\mathcal{T}\psi\rangle = \langle\mathcal{T}\psi|\mathcal{T}^2\psi\rangle^* = -\langle\psi|\mathcal{T}\psi\rangle = 0$, i.e., $|\psi\rangle$ and $\mathcal{T}|\psi\rangle$ are orthogonal. Hence, the energy level E is (at least) twofold degenerate.

3.2.2 Particle-hole and chiral symmetries

The other symmetry that plays a critical role in the classification of single-particle Hamiltonians is the particle-hole (PH) (or charge conjugation) symmetry. In particle physics, charge conjugation is a transformation that changes a particle into its antiparticle. For example, in the case of a system described by the Dirac equation [145], charge conjugation transforms a spinor $\psi(x)$ that describes the motion of a particle with charge e and mass m_0 in a potential $A_\mu(x)$ into the spinor $\psi_c(x)$ describing a particle with charge $-e$ and mass m_0 moving in the same potential $A_\mu(x)$.

In condensed matter physics, PH can emerge as an approximative symmetry between electrons (i.e., occupied states above the Fermi level) and holes (empty states below the Fermi level) that holds within a certain energy window. Within this energy range, the system can be modeled using a Dirac-type Hamiltonian, as, for example, in the case of graphene [94]. Furthermore, PH symmetry is an intrinsic property of the mean-field theory of superconductivity. The single-particle s-wave pairing mean-field Hamiltonian – the so-called Bogoliubov–de Gennes (BdG) Hamiltonian – has the generic form $H = \frac{1}{2}\hat{\Psi}^\dagger \mathcal{H}_{BdG}\hat{\Psi}$, where

$$\mathcal{H}_{BdG} = \begin{pmatrix} \mathcal{H}_0 & -i\sigma_y\Delta \\ i\sigma_y\Delta^* & -\mathcal{H}_0^T \end{pmatrix} \tag{3.6}$$

is the (first quantized) BdG Hamiltonian, while the four-component Nambu spinor $\hat{\Psi}^\dagger = (\hat{\psi}^\dagger, \hat{\psi}) = (\hat{\psi}_\uparrow^\dagger, \hat{\psi}_\downarrow^\dagger, \hat{\psi}_\uparrow, \hat{\psi}_\downarrow)$ represents the (second quantized) field operator and $\Delta = \langle\hat{\psi}_\uparrow\hat{\psi}_\downarrow\rangle$ is the pair potential. One can interpret $\mathcal{H}_{BdG}$ as the Hamiltonian that governs the dynamics of the Bogoliubov quasiparticles, $i\hbar\frac{\partial}{\partial t}\Psi = \mathcal{H}\Psi$, with $\Psi = (u, v)^T = (u_\uparrow, u_\downarrow, v_\uparrow, v_\downarrow)^T$, where $u(x,t)$ and $v(x,t)$ are the particle and hole components, respectively. Note that the BdG theory contains an intrinsic particle-hole redundancy, in the sense that each eigenfunction Ψ_E with $E > 0$ has a negative energy correspondent, $\Psi_{-E} = \tau_x\Psi_E^*$, where the Pauli matrix τ_x acts on the particle-hole space and switches the electron and hole components. At the level of the BdG Hamiltonian, this particle-hole symmetry can be expressed in terms of the charge conjugation operator $\mathcal{C} = \tau_x K$ as the condition

$$\mathcal{C}\mathcal{H}_{BdG}\mathcal{C}^{-1} = \tau_x\mathcal{H}_{BdG}^*\tau_x = -\mathcal{H}_{BdG}. \tag{3.7}$$

The charge conjugation operator $\mathcal{C}$ is anti-unitary and squares to $+1$. In the presence of $SU(2)$ spin symmetry, we have $\mathcal{H}_0 = \text{diag}\{h_0, h_0\}$ and the Hamiltonian (3.6) can be rearranged into block-diagonal form, $\mathcal{H}_{BdG} = \text{diag}\{\mathcal{H}_+, \mathcal{H}_-\}$, with the blocks

$$\mathcal{H}_\pm = \begin{pmatrix} h_0 & \pm\Delta \\ \pm\Delta^* & -h_0^T \end{pmatrix} \tag{3.8}$$

acting on $(u_\uparrow, v_\downarrow)$ and $(u_\downarrow, v_\uparrow)$, respectively. In this case, the charge conjugation operator can be represented as $\mathcal{C} = i\tau_y K$ and we have $\mathcal{C}^2 = -1$.

To summarize, the time-reversal and particle-hole transformations can be represented by anti-unitary operators that square to either $+1$ or -1. We describe a noninteracting fermionic system by the single-particle Hamiltonian $\mathcal{H}$, which represents a $pN \times pN$ Hermitian matrix corresponding to N site and orbital degrees of freedom and p spin and particle-hole degrees of freedom. If the system is translation-invariant, we describe it using the Bloch Hamiltonian $\mathcal{H}(\boldsymbol{k})$, which is a $pM \times pM$ Hermitian matrix, with M representing the number of orbital degrees of freedom. The time-reversal operation is represented by $\mathcal{T} = U_t K$, with K being the complex conjugation and U_t a $pN \times pN$ (or $pM \times pM$, with translation invariance) unitary matrix. Similarly, the charge-conjugation operation is represented by $\mathcal{C} = U_c K$, with U_c being a $pN \times pN$

$(pM \times pM)$ unitary matrix. The presence of time-reversal and particle-hole symmetries is expressed by the conditions

$$U_t \mathcal{H}^* U_t^{-1} = +\mathcal{H}, \qquad U_t U_t^* = \mathcal{T}^2 = \pm 1, \qquad (3.9\text{a})$$
$$U_c \mathcal{H}^* U_c^{-1} = -\mathcal{H}, \qquad U_c U_c^* = \mathcal{C}^2 = \pm 1, \qquad (3.9\text{b})$$

where 1 represents the $pN \times pN$ unit matrix. For a translation invariant system these conditions become

$$U_t \mathcal{H}^*(\boldsymbol{k}) U_t^{-1} = +\mathcal{H}(-\boldsymbol{k}), \qquad U_t U_t^* = \mathcal{T}^2 = \pm 1, \qquad (3.10\text{a})$$
$$U_c \mathcal{H}^*(\boldsymbol{k}) U_c^{-1} = -\mathcal{H}(-\boldsymbol{k}), \qquad U_c U_c^* = \mathcal{C}^2 = \pm 1, \qquad (3.10\text{b})$$

where all quantities are now $pM \times pM$ matrices. Finally, we note that the presence of TR symmetry implies Kramers degeneracy (for half-integer spin) and, for translation invariant systems, a symmetric spectrum with $E(\boldsymbol{k}) = E(-\boldsymbol{k})$. In addition, PH symmetry implies an energy spectrum that is symmetric about zero. In particular, every state with wave vector $\boldsymbol{k}$ and positive energy $E_+(\boldsymbol{k}) > 0$ has a correspondent at $-\boldsymbol{k}$ with energy $E_-(-\boldsymbol{k}) = -E_+(\boldsymbol{k})$.

When the Hamiltonian has both TR symmetry and PH symmetry, the product $\mathcal{S} = \mathcal{T} \cdot \mathcal{C}$ corresponds to the unitary operation $U_t U_c^*$ acting on $\mathcal{H}$. Note, however, that $\mathcal{S}$ does not represent an "ordinary" symmetry since it does not commute but rather *anti-commutes* with the Hamiltonian,

$$\mathcal{S} \mathcal{H} \mathcal{S}^{-1} = U_t U_c^* \, \mathcal{H} \, U_c^T U_t^{-1} = -\mathcal{H}, \qquad \mathcal{S}^2 = 1. \qquad (3.11)$$

This unitary operation is called *chiral* or *sublattice* (SL) symmetry and represents an additional key ingredient for the classification of the Hamiltonian blocks. Similarly to PH symmetry, the presence of SL symmetry implies a symmetric spectrum. In particular, every state with wave vector $\boldsymbol{k}$ and positive energy $E_+(\boldsymbol{k}) > 0$ has a negative energy correspondent with $E_-(\boldsymbol{k}) = -E_+(\boldsymbol{k})$.

It is worth noting that the presence of two chiral symmetries $\mathcal{S}$ and $\mathcal{S}'$, i.e., two operations that satisfy Eq. (3.11), allows us to construct the conserved quantity $\mathcal{S} \mathcal{S}'$, $[\mathcal{H}, \mathcal{S} \mathcal{S}'] = 0$. This implies that $\mathcal{H}$ can be block-diagonalized and that we can apply our classification scheme to each block. Consequently, it is enough to consider only one SL-type symmetry. Similarly, we can assume without loss of generality that there is a single TR-type operator $\mathcal{T}$ and a single PH-type operator $\mathcal{C}$. As in the case of SL-type symmetry, the presence of two charge-conjugation (or time-reversal) operations allows us to construct the unitary symmetry operator $U_{c_1} U_{c_2}^*$ (or $U_{t_1} U_{t_2}^*$) that commutes with the Hamiltonian, which, consequently, can be block-diagonalized. Also, we note that the presence of both TR and PH symmetries automatically implies the presence of SL symmetry, while the presence of TR (PH) symmetry and the absence of PH (TR) symmetry automatically implies the absence of chirality. However, the absence of both TR and PH symmetries allows for chiral symmetry to be either present or absent.

TABLE 3.1 The ten symmetry classes of single-particle Hamiltonians classified according to their behavior under time-reversal symmetry (TRS), particle-hole symmetry (PHS), and chiral (sublattice) symmetry (SLS).

	Cartan label	TRS	PHS	SLS
Standard classes	A (unitary)	0	0	0
(Wigner–Dyson)	AI (orthogonal)	+1	0	0
	AII (symplectic)	−1	0	0
Chiral classes	AIII (chiral unitary)	0	0	1
(sublattice)	BDI (chiral orthogonal)	+1	+1	1
	CII (chiral symplectic)	−1	−1	1
BdG classes	D	0	+1	0
	C	0	−1	0
	DIII	−1	+1	1
	CI	+1	−1	1

3.2.3 Classification of random Hamiltonians

After these considerations, it is easy to understand the basic idea behind the symmetry classification of generic single-particle Hamiltonians according to their behavior under time-reversal symmetry (TRS), charge-conjugation (or particle-hole) symmetry (PHS), and chiral (or sublattice) symmetry (SLS). There are three distinct possibilities for a system to respond to the time-reversal operation: i) no time-reversal invariance (TRS = 0), ii) time-reversal invariance with a TR operator $\mathcal{T}$ that squares to the identity operator (TRS = +1), and iii) time-reversal invariance with a TR operator $\mathcal{T}$ that squares to minus the identity operator (TRS = −1). Similarly, there are three distinct possibilities for a system to respond to the particle-hole operation, which we denote by PHS = 0 (no symmetry), PHS = +1 (PH symmetry with $\mathcal{C}^2 = +1$), and PHS = −1 (PH symmetry with $\mathcal{C}^2 = −1$). Hence, there are $3 \times 3 = 9$ distinct ways for the Hamiltonian $\mathcal{H}$ to respond to the combination of time-reversal and charge conjugation (particle-hole) symmetries. In addition, a system with TRS = 0 and PHS = 0 may possess chiral (sublattice) symmetry (SLS = 1) or not (SLS = 0). Consequently, there are ten symmetry classes corresponding to the distinct ways in which single-particle Hamiltonians behave under TRS, PHS, and SLS. The existence of precisely ten symmetry classes, as listed in Table 3.1, is a fundamental result due to Altland and Zirnbauer [18, 573]. This completes an earlier classification done by Wigner and Dyson [149] in the context of random matrix theory, which contains the *unitary, orthogonal*, and *symplectic* symmetry classes.

The ten generic symmetry classes of single-particle Hamiltonians provide the framework for the classification scheme of topological insulators and superconductors, as we will discuss in the next section. These classes are typically grouped in three categories, i.e., standard (Wigner–Dyson), chiral, and BdG classes, as shown in Table 3.1. Note, however, that classes CI and DIII are closely related to the chiral classes, since they are characterized by SLS = 1. Furthermore, in certain conditions a Hamiltonian that belongs to class A or class AIII can be thought of as a BdG superconductor. Indeed, consider a superconductor described by a BdG Hamiltonian that is invariant under spin rotations about the z axis, $[\mathcal{H}_{BdG}, S_z] = 0$. In the presence of this $U(1)$ symmetry, the Hamiltonian becomes block-diagonal and one can show [104] that, up to a term proportional to the identity matrix, a (second quantized) Hamiltonian block can be written in the form

$$H = (\hat{\psi}_\uparrow^\dagger, \hat{\psi}_\downarrow) \, \mathcal{H} \begin{pmatrix} \hat{\psi}_\uparrow \\ \hat{\psi}_\downarrow^\dagger \end{pmatrix}, \qquad \mathcal{H} = \begin{pmatrix} h_\uparrow & \Delta \\ \Delta^\dagger & -h_\downarrow^T \end{pmatrix}, \qquad (3.12)$$

where $h_\sigma^\dagger = h_\sigma$. Without further constraints, $\mathcal{H}$ is a member of the unitary symmetry class (class A). A spinful 2D chiral $p \pm ip$ wave superconductor with an order parameter $\Delta(\boldsymbol{k}) = \Delta_0(k_x + ik_y)$ represents a specific physical realization of this class. If we impose the additional constraints $\Delta^\dagger = \Delta$ and $h_\downarrow^T = h_\uparrow$, $\mathcal{H}$ becomes a member of the chiral unitary class (AIII). The chiral symmetry $\mathcal{S}$ is represented by the Pauli matrix r_y associated with the spin-up particle – spin-down hole space and we have $r_y \mathcal{H} r_y = -\mathcal{H}$ and $r_y^2 = 1$. The spinful p wave superconductor is a specific physical realization of this class.

As a final remark, let us emphasize the relation between the ten symmetry classes of single-particle Hamiltonians and the *compact symmetric spaces* (of arbitrarily large dimension) identified by Élie Cartan in 1926. More specifically, if $\mathcal{H}$ is a member of a certain symmetry class, the quantum mechanical time-evolution operator $\exp[it\mathcal{H}]$ is an element of the symmetric space designated by the corresponding Cartan label. The compact symmetric spaces consist of the orthogonal, unitary, and symplectic groups, $O(N)$, $U(N)$, and $Sp(2N)$, respectively, as well as cosets,[1] such as $U(N)/O(N)$ or $U(N+M)/U(N) \times U(M)$ [244]. If, for example, the system has no symmetry, (TRS, PHS, SLS) = $(0, 0, 0)$, $\mathcal{H}$ is a generic $N \times N$ Hermitian matrix and, consequently, $\exp[it\mathcal{H}]$ will be an element of the unitary group $U(N)$, which is a symmetric space corresponding to the Cartan label A. Imposing time-reversal symmetry, e.g., for spinless particles (TRS = $+1$), results in the existence of a basis in which $\mathcal{H}$ is represented by a real symmetric $N \times N$ matrix. The corresponding time-evolution operator is an element of the coset $U(N)/O(N)$, the symmetric space with the Cartan label AI. Similar considerations apply for the remaining symmetry classes [344].

---

[1] If H is a subgroup of G and g is an element of G, we define a (left) coset H in G as $gH = \{gh | h \in H\}$. Note that any two cosets $g_1 H$ and $g_2 H$ are either equal (as sets) or disjoint. We define G/H as the set of all (left) cosets. Also, we define the index of H in G as the number of distinct cosets of H in G.

3.3 TOPOLOGICAL CLASSIFICATION OF BAND INSULATORS AND SUPERCONDUCTORS

In the previous section, we have classified the Hamiltonians of noninteracting fermion systems according to their behavior under certain generic symmetry operations. Consider now a d-dimensional fermionic system with specified symmetry properties that is characterized by the (first quantized) Hamiltonian $\mathcal{H}$ belonging to the appropriate symmetry class. In general, $\mathcal{H}$ possesses either a gapless or a gapped ground state. The question that we address is the following: given the symmetry class of $\mathcal{H}$ and the spatial dimension d, how many *distinct gapped phases* (i.e., insulating or superconducting) are there?

First, let us clarify the expression "distinct gapped phases." As we already mentioned in Section 3.1, we define two gapped states as topologically equivalent if (and only if) they can be adiabatically connected without closing the bulk gap (up to the addition of a number of trivial bands). The "adiabatic connection" that defines this equivalence relation involves *only* deformations of the Hamiltonian that preserve its *symmetry*. The resulting equivalence classes are precisely the "distinct gapped phases" we want to identify and classify. Assume that the Hamiltonian $\mathcal{H}$ is parameterized by two sets of parameters, $\{p_1, p_2, \ldots, p_n\}$ and $\{q_1, q_2, \ldots, q_m\}$, so that $\mathcal{H}(p_1, \ldots, p_n; 0, \ldots, 0)$ belongs to a given symmetry class for arbitrary values of $\{p_i\}$, but corresponds to some other class if any of the parameters q_j is nonzero. In other words, $\{q_j\}$ parameterize perturbations that break the symmetry of the Hamiltonian. We define the equivalence classes of gapped ground states (i.e., the distinct insulating and superconducting phases) under the condition $q_j = 0$, with $1 \leq j \leq m$.

Anticipating the result, let us note that there are three generic possibilities characterizing the structure of the equivalence classes, i.e., the phase diagram of topological band insulators and SCs. More specifically, for any given spatial dimension d, the topological phase diagram of a system described by $\mathcal{H}(p_1, \ldots, p_n; 0, \ldots, 0)$ will correspond to one of the situations illustrated schematically in Figure 3.1. Panel (a) corresponds to the situation when all possible gapped ground states can be continuously connected without closing the gap. In this case, there is a single equivalence class (the trivial insulating phase Φ_0) and all gapped states are equivalent to the atomic insulator (or the vacuum). Panel (b) illustrates the case when two distinct gapped phases are possible: Φ_0, which corresponds to the trivial insulator/superconductor, and Φ_1, which includes all gapped ground states that cannot be continuously connected with the Φ_0 states without closing the gap. For example, by varying the parameter p_1 the system can be driven from $A \in \Phi_1$ to $B \in \Phi_0$ through the gapless state X, which corresponds to a topological phase transition. As we will discuss below, the two phases are labeled by a $\mathbb{Z}_2$ topological invariant, i.e., a quantity that takes one value (e.g., $\nu = 0$) for all the states in Φ_0 and another value (e.g., $\nu = 1$) for all the states belonging to the other equivalence class. The quantum spin Hall states (in 2D) and the 3D topological insulators are specific examples of Φ_1-type phases. The third possibility is illustrated in

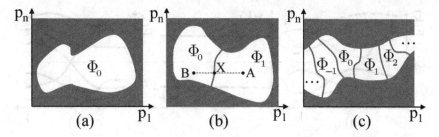

FIGURE 3.1 Schematic topological phase diagram in the parameter space $\{p_1, \ldots, p_n\}$ of the Hamiltonian. White domains and gray regions/lines correspond to gapped and gapless ground states, respectively. Two Hamiltonians that belong to the same connected region Φ_ν can be continuously deformed into each other without closing the gap, while Hamiltonians from distinct regions, e.g., $\mathcal{H}_A$ and $\mathcal{H}_B$ in panel (b), can only be connected by going through a quantum phase transition associated with a vanishing gap (e.g., point X). (a) All insulators are topologically trivial. (b) Two distinct phases labeled by a $\mathbb{Z}_2$ topological invariant. (c) Gapped phases with $\mathbb{Z}$ topological classification.

panel (c). In this case, there are an infinite number of equivalence classes (i.e., distinct gapped phases) that can be labeled by a $\mathbb{Z}$ topological invariant, i.e., a quantity that takes a specific integer value for all the states belonging to a given equivalence class (e.g., $\nu = 0, \pm 1, \pm 2, \ldots$). Again, the system can be driven from one phase to another only through topological phase transitions characterized by a vanishing energy gap. The best-known example of $\mathbb{Z}$-type gapped phases are the integer quantum Hall states.

In the remainder of this section, we first discuss the origin of topology in band insulators (superconductors), then we sketch the main ideas behind the topological classification of band insulators and superconductors by focusing on one of the possible equivalent approaches, the bulk-boundary correspondence. We close with a discussion of some of the caveats and implications of this topological classification.

3.3.1 The origin of topology in gapped noninteracting systems

To understand the key elements that lead to the appearance of topological properties in insulators (and SCs), it is instructive to begin with the translationally invariant case. In the presence of translation invariance, a band insulator can be described in terms of Bloch states satisfying the eigenvalue equation

$$\mathcal{H}(\mathbf{k})|u_{a\mathbf{k}}\rangle = E_a(\mathbf{k})|u_{a\mathbf{k}}\rangle, \tag{3.13}$$

where the momentum $\mathbf{k}$ takes values inside the Brillouin zone (BZ) and a labels different bands. The Fermi level E_F lies inside an energy gap that separates the filled (valence) bands from the empty (conduction) bands, as shown schematically in Figure 3.2. The many-body ground state of the system

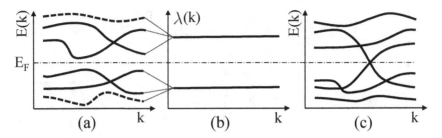

FIGURE 3.2 (a) Schematic band structure of a band insulator represented by a 4×4 Hamiltonian $\mathcal{H}(\boldsymbol{k})$ (full lines). Adding two trivial bands (dashed lines) results in a 6×6 Hamiltonian $\mathcal{H}'(\boldsymbol{k})$ that describes the same insulating phase. (b) Spectrum of the reduced Hamiltonian $Q(\boldsymbol{k})$ obtained by continuously deforming $\mathcal{H}$ (or $\mathcal{H}'$). (c) Gapless spectrum associated with a quantum phase transition, e.g., point X in Figure 3.1(b).

is determined by the filled Bloch states; hence it corresponds to the map $\boldsymbol{k} \longrightarrow P(\boldsymbol{k})$ from the BZ to the space of projection operators

$$P(\boldsymbol{k}) = \sum_{a}^{\text{(filled)}} |u_{a\boldsymbol{k}}\rangle \langle u_{a\boldsymbol{k}}|. \tag{3.14}$$

To better capture the bare bones of the underlying topological structure, it is convenient to introduce the operator

$$Q(\boldsymbol{k}) = 1 - 2P(\boldsymbol{k}), \tag{3.15}$$

where 1 denotes the identity operator. The Q operator, also called the "Q matrix" or the "projector," has the following properties

$$Q^{\dagger} = Q, \qquad Q^2 = 1, \qquad \text{Tr}[Q] = m - n, \tag{3.16}$$

where m and n represent the number of empty and occupied bands, respectively. The Q matrix has n negative eigenvalues, $\lambda_{\nu} = -1$, and m positive eigenvalues, $\lambda_{\nu} = +1$. Note that $Q(\boldsymbol{k})$, which can be viewed as a "simplified Hamiltonian" having the same ground state as the original Hamiltonian, can be obtained by continuously deforming $\mathcal{H}$ without closing the gap (Figure 3.2).

Consider now a band insulator in the symmetry class A (i.e., no additional constraints imposed on the Hamiltonian) so that the Hamiltonian $\mathcal{H}$ is represented by a generic $n + m$ Hermitian matrix. Consequently, the simplified Hamiltonian Q is represented by the unitary matrix constructed using the set of $n + m$ eigenvectors of $\mathcal{H}$, $Q = \{-\phi_1, \dots, -\phi_n, \phi_{n+1}, \dots, \phi_{n+m}\}^T$, each eigenvector ϕ_{ν} being an $(n + m)$-dimensional vector. Since rotations within the degenerate subspaces spanned by the filled and the empty states do not change the physics, there is a natural gauge symmetry represented by block diagonal unitary matrices that are elements of $U(n) \times U(m)$. Consequently, the simplified Hamiltonian is an element of the coset $U(n+m)/U(n) \times U(m)$, which is also called a *Grassmannian* and denoted $G_{n,n+m}(\mathbb{C})$,

$$Q \in U(n+m)/U(n) \times U(m). \tag{3.17}$$

Coming back to our original question concerning the existence of inequivalent phases, we note that answering it amounts to finding the equivalence classes of maps $\boldsymbol{k} \longrightarrow Q(\boldsymbol{k})$ from the Brillouin zone to the Grassmannian, $Q : BZ \longrightarrow G_{n,n+m}(\mathbb{C})$. Indeed, if two maps $Q(\boldsymbol{k})$ and $Q'(\boldsymbol{k})$ can be continuously deformed into each other, then the original Hamiltonians $\mathcal{H}(\boldsymbol{k})$ and $\mathcal{H}'(\boldsymbol{k})$ can be continuously connected and, consequently, the corresponding systems are in the same phase. Mathematically, the problem reduces to finding the *homotopy group* of the map $\boldsymbol{k} \longrightarrow Q(\boldsymbol{k})$, i.e., $\pi_d[G_{n,n+m}(\mathbb{C})]$, where d is the dimension of the Brillouin zone.

As a specific example, in two dimensions ($d = 2$) we have $\pi_2[G_{n,n+m}(\mathbb{C})] = \mathbb{Z}$, which means that there is an infinite set of inequivalent gapped phases (i.e., insulators that cannot be continuously deformed into each other without crossing a quantum phase transition) labeled by an integer topological number, as in Figure 3.1(c). The integer quantum Hall insulators, which are ground states of a 2D Hamiltonian belonging to the symmetry class A (no TRS, PHS, or SLS), are physical realizations of these distinct topological phases. Note that the quantized Hall conductance σ_{xy} is proportional to the $\mathbb{Z}$ topological invariant (see Section 3.4). If instead we consider the three-dimensional case ($d = 3$), the homotopy group $\pi_3[G_{n,n+m}(\mathbb{C})] = \{1\}$ is the trivial group consisting of only one element. This means that all the maps can be continuously deformed into each other, i.e., in $d = 3$ spatial dimensions there is only one phase (the trivial insulator) in symmetry class A, as in Figure 3.1(a).

Next, let us impose some additional discrete symmetry on the Hamiltonian. For example, let us assume that the system has chiral symmetry, which means that $\mathcal{H}$ belongs to the symmetry class AIII (see Table 3.1). In a system defined on a bipartite lattice, this chiral symmetry can be realized as a sublattice symmetry that is represented by the Pauli matrix λ_z associated with the sublattice degrees of freedom and we have $\lambda_z \mathcal{H} \lambda_z = -\mathcal{H}$. It is straightforward to show that, as a consequence, $\mathcal{H}$ takes a block off-diagonal form when represented using a local basis with the first n entries corresponding to sublattice A and the other n to sublattice B. Furthermore, in this basis we have

$$Q(\boldsymbol{k}) = \begin{pmatrix} 0 & q(\boldsymbol{k}) \\ q^\dagger(\boldsymbol{k}) & 0 \end{pmatrix}, \qquad \text{with } qq^\dagger = q^\dagger q = 1. \tag{3.18}$$

For Hamiltonians in the symmetry class AIII, $q(\boldsymbol{k})$ are arbitrary unitary matrices, i.e., elements of the unitary group $U(n)$. Hence, the number of inequivalent phases is given by the homotopy group of the map $q : BZ \longrightarrow U(n)$. In two dimensions, for example, there are no topologically nontrivial phases, $\pi_2[U(n)] = \{1\}$. By contrast, in $d = 3$ there is an infinite sequence of distinct phases indexed by an integer topological invariant,[2] $\pi_3[U(n)] = \mathbb{Z}$. Finally, we note that for Hamiltonians in symmetry classes other than A and AIII, the matrix $Q(\boldsymbol{k})$ (or the off-diagonal block $q(\boldsymbol{k})$ if SLS= 1) has to satisfy additional conditions. For example, in the symplectic class (AII), the projector $Q(\boldsymbol{k}) \in G_{2n,2n+2m}(\mathbb{C})$ has to satisfy the TRS condition $\sigma_y Q^*(\boldsymbol{k})\sigma_y = Q(-\boldsymbol{k})$. Again, identifying topologically distinct phases amounts to determining Q maps that cannot be continuously deformed into each other. However, because of the additional condition, the number of inequivalent Q maps cannot be obtained directly from the homotopy group of the Grassmannian and one has to use a different strategy, as discussed below.

Let us summarize. Two gapped states are defined as equivalent (i.e., belonging to the same phase) if the corresponding Hamiltonians (which are in the same symmetry class) can be continuously deformed into each other without closing the gap and while preserving the symmetry. Consequently, each connected component Φ_ν of the manifold Φ of gapped Hamiltonians corresponds to a distinct phase (see Figure 3.1). Adding a number of trivial bands generates an equivalent state that belongs to the same phase (see Figure 3.2). To determine the topology of Φ in the presence of translation invariance, it is convenient to continuously deform the Bloch Hamiltonian $\mathcal{H}(\boldsymbol{k})$ into the projector $Q(\boldsymbol{k})$: the problem of finding inequivalent insulating phases amounts to identifying how many different maps $\boldsymbol{k} \longrightarrow Q(\boldsymbol{k})$ there are that cannot be continuously deformed into each other. In other words, one has to determine the homotopy group of the map from the Brillouin zone to the space of projectors. For Hamiltonians in symmetry class A, $Q(\boldsymbol{k})$ is an element of the Grassmannian, $G_{n,n+m}(\mathbb{C})$ and the relevant homotopy group is $\pi_d[G_{n,n+m}(\mathbb{C})]$, which is the trivial group $\{1\}$ for odd spatial dimensions and $\mathbb{Z}$ for d even. Consequently, integer quantum Hall-type states exist only in spaces with even dimensionality (e.g., $d = 2$). By contrast, for Hamiltonians in the chiral symmetry class AIII, the relevant homotopy group is $\pi_d[U(n)]$, which is $\{1\}$ for d even and $\mathbb{Z}$ for d odd. Adding time-reversal symmetry and/or charge conjugation symmetry imposes additional constraints on $Q(\boldsymbol{k})$. A general strategy for determining the homotopy classes is discussed below.

3.3.2 Classification of topological insulators and superconductors

There are several methods [293, 442, 454] for establishing the classification of noninteracting topological insulators and superconductors, in particular the approach proposed by Kitaev [293] based on K-theory and the scheme proposed by Schnyder et al. [454] that exploits the bulk-boundary correspondence. Remarkably, all these methods, which are ultimately equivalent, establish a link between the existence of distinct topological phases and the nontrivial homotopy groups of symmetric spaces.

The K-theory classification [293] is based on a systematic study of the homotopical structure of gapped bulk Hamiltonians. Roughly speaking, the

---

[2]The explicit form of this invariant, called the winding number, is given in Section 3.4.

key idea in topological K-theory is to classify vector bundles not just up to a homotopy equivalence, but up to a *stable equivalence*, which essentially means that, when comparing two objects, it is allowed to augment them by (direct sum) trivial bundles. Consequently, bundles of different rank, which cannot be directly deformed into each other, can be in the same equivalence class after augmentation. Physically, this corresponds to the addition of trivial bands discussed above (see, for example, Figure 3.2). These trivial bands, e.g., those associated with inner atomic shells, always exist, but they are often ignored when considering finite dimensional Hamiltonians. It is common to have situations when two Hamiltonians cannot be continuously deformed into each other, but such a deformation becomes possible after the augmentation (i.e., addition of trivial bands). In these situations, the Hamiltonians must physically describe the same phase.

Without going into technical details, we note that for each symmetry class the zero-dimensional simplified Hamiltonian (3.15), basically $Q(0)$, defines a so-called *classifying space*. The "periodic table" of topological insulators follows from the homotopy group of the classifying spaces and the insight that it obeys a certain periodicity that originates from the *Bott periodicity* in K-theory [69]. Remarkably, the classifying spaces are exactly the ten symmetric spaces that appear in the symmetry classification of generic Hamiltonians. For example, in symmetry class A, i.e., for quantum mechanical systems with a time evolution operator $\exp(it\mathcal{H})$ that is an element of the symmetric space $U(N) \times U(N)/U(N)$, the reduced Hamiltonian $Q(0)$ is, according to Eq. (3.17), an element of the symmetric space $V_A = U(n+m)/U(n) \times U(m)$. Similarly, the classifying space associated with a class AIII Hamiltonian is $V_{AIII} = U(N) \times U(N)/U(N)$. The complete list is given in Table 3.2.

Having identified the classifying space for each Altland–Zirnbauer symmetry class, one can determine the number of equivalence classes (i.e., topologically distinct phases) corresponding to a zero-dimensional Hamiltonian $\mathcal{H}$ that belongs to a given symmetry class using the homotopy group $\pi_0(V)$ of the corresponding classifying space V. Note that $\pi_0(V)$ indexes the path-connected subspaces of V, each subspace corresponding to a distinct phase. For example, in symmetry class A we have $\pi_0(V_A) = \mathbb{Z}$, i.e., there is an infinite number of $d=0$ distinct phases classified by an integer topological invariant. On the other hand, in symmetry class AIII $\pi_0(V_{AIII}) = \{1\}$, i.e., there is only the trivial phase. The complete list of homotopy groups of classifying spaces corresponds to the column $d=0$ in Table 3.3. Extending the classification to arbitrary dimensions $d \geq 1$ reveals a pattern that originates in the Bott periodicity of the homotopy groups [293]. K-theory predicts that there are two families of *complex* classifying spaces, $C_0 = V_A$ and $C_1 = V_{AIII}$, corresponding to Hamiltonians without time-reversal or particle-hole symmetries, and eight families of *real* classifying spaces, i.e., $R_0 = V_{AI}, \ldots, R_7 = V_{CI}$, which correspond to Hamiltonians having at least one reality condition ($\mathcal{T}$ or $\mathcal{C}$). According to Bott periodicity [69], the dependence on q for the complex (C_q) and real (R_q) classifying spaces enters modulo two and modulo eight,

TABLE 3.2 The ten Cartan *symmetric spaces* in the context of i) basic quantum mechanics, where they describe the time evolution operator $\exp(itH)$, ii) K-theory, where they characterize the *classifying spaces*, and iii) NLσM field theories, where they describe the *target spaces*.

	Time evolution operator $\exp(it\mathcal{H})$	Classifying space	Fermionic replica NLσM target space
A	$U(n)\times U(n)/U(n)$	$U(n+m)/U(n)\times U(m)$	$U(2n)/U(n)\times U(n)$
AIII	$U(n+m)/U(n)\times U(m)$	$U(n)\times U(n)/U(n)$	$U(n)\times U(n)/U(n)$
AI	$U(n)/O(n)$	$O(n+m)/O(n)\times O(m)$	$Sp(2n)/Sp(n)\times Sp(n)$
BDI	$O(n+m)/O(n)\times O(m)$	$O(n)\times O(n)/O(n)$	$U(2n)/Sp(2n)$
D	$O(n)\times O(n)/O(n)$	$O(2n)/U(n)$	$O(2n)/U(n)$
DIII	$SO(2n)/U(n)$	$U(2n)/Sp(2n)$	$O(n)\times O(n)/O(n)$
AII	$U(2n)/Sp(2n)$	$Sp(n+m)/Sp(n)\times Sp(n)$	$O(2n)/O(n)\times O(n)$
CII	$Sp(n+m)/Sp(n)\times Sp(m)$	$Sp(n)\times Sp(n)/Sp(n)$	$U(n)/O(n)$
C	$Sp(2n)\times Sp(2n)/Sp(2n)$	$Sp(2n)/U(n)$	$Sp(2n)/U(n)$
CI	$Sp(2n)/U(n)$	$U(n)/O(n)$	$Sp(2n)\times Sp(2n)/Sp(2n)$

respectively,

$$C_{q+2} = C_q, \qquad\qquad R_{q+8} = R_q. \qquad (3.19)$$

Furthermore, for each symmetry class Λ and given space dimension d the classifying space $V_\Lambda(d)$ corresponds to the solution of an *extension problem* of a Clifford algebra[3] [293]. Specifically, one can show that for the complex case $V_A(d) = C_{0+d}$ and $V_{AIII}(d) = C_{1+d}$, while for the real case $V_{AI}(d) = R_{0-d}$, $V_{BDI}(d) = R_{1-d}$, etc. Finally, the topological classification is done in terms of the zeroth homotopy group[4] of the corresponding classifying spaces $\pi_0(V_\Lambda(d))$. The resulting "periodic table" of topological insulators and superconductors is listed in Table 3.3. Below, we will comment on the structure of this table, after briefly discussing a different strategy for obtaining it. Here, we note that the physical meaning of the higher homotopy groups $\pi_n(V)$ with $n \geq 1$ is related to the classification of defects. Specifically, the classification of defects that can

---

[3]Let K be a field and A a vector space over K equipped with an additional binary operation (which we denote by $\cdot$). A is said to be an algebra over K (or a K-algebra) if for all vectors $x, y, z \in A$ and scalars $a, b \in K$ we have:

(i) Right distributivity: $(x + y) \cdot z = x \cdot z + y \cdot z$

(ii) Left distributivity: $x \cdot (y + z) = x \cdot y + x \cdot z$

(iii) Compatibility with scalars: $(ax) \cdot (by) = (ab)(x \cdot y)$

A *Clifford algebra* [115] is a type of associative K-algebra that generalizes the concepts of real numbers, complex numbers, and quaternions. A familiar example of Clifford algebra is the algebra generated by the Dirac matrices.

[4]Recall that the homotopy group $\pi_n(X)$ is the set of homotopy classes of maps $f : S^n \longrightarrow X$ between the unit sphere S^n and the topological space X. Note that $S^0 = \{-1, 1\}$ and $\pi_0(X)$ indexes the distinct subspaces of X that are path-connected.

be surrounded by a D-dimensional sphere S^D (e.g., point, line, and surface defects in d-dimensional space corresponding to $D = d - 1$, $D = d - 2$, and $D = d - 3$, respectively), is done by the homotopy group $\pi_D(V_\Lambda(d))$. Also note that these homotopy groups satisfy the relation $\pi_D(V_\Lambda(d)) = \pi_0(V_\Lambda(d - D))$.

An alternative derivation of Table 3.3 proposed by Schnyder et al. [454] is based on the so-called bulk-boundary correspondence. A *defining* property of topological insulators and superconductors is that gapless states *necessarily* exist at the boundary (interface) between the system and a "topologically trivial" state, such as, e.g., vacuum. There is a one-to-one correspondence between the topological properties of the bulk and the properties of the gapless boundary degrees of freedom. Hence, topological insulators are inherently "holographic" systems: the topological nature of the bulk state is "revealed" by the gapless boundary modes.

The existence of robust gapless states at the boundary of a TI is a nontrivial property, as in general quantum states tend to become localized in the presence of disorder because of the phenomenon of Anderson localization [21]. Hence, establishing a classification of d-dimensional topological insulators and superconductors is equivalent to classifying the systems of noninteracting fermions that completely evade the phenomenon of Anderson localization at the $d - 1$ dimensional boundary. The absence of Anderson localization can be understood using a variety of theoretical approaches [151, 321]. In general, the problem is described by a random Hamiltonian that belongs to one of the symmetry classes listed in Table 3.1. In the low-energy, long wavelength limit, the physics of the disordered system is captured by an effective theory that can be formulated in terms of a *nonlinear sigma model*[5] (NLσM) [174, 199]. Within this effective theory, the property of the boundary degrees of freedom to completely evade Anderson localization can be incorporated by adding certain extra terms of topological origin to the action of the NLσM. There are two such terms which contain no adjustable parameter: the Wess-Zumino-Witten (WZW) term and the $\mathbb{Z}_2$ topological term [164]. Hence, a topological insulator (superconductor) exists in d dimensions whenever including a WZW or a topological term in the action of the $(d - 1)$ dimensional NLσM is allowed. Finally, whether adding such an extra term is allowed or not depends on the spatial dimensionality $\bar{d} = d - 1$ of the boundary and on the target space G/H of the NLσM, more specifically on their homotopy groups. A WZW term can be included when $\pi_{\bar{d}+1}(G/H) = \mathbb{Z}$, while adding a topological term is allowed when $\pi_{\bar{d}}(G/H) = \mathbb{Z}_2$.

To summarize, the approach proposed by Schnyder et al. [454] exploits the holographic nature of topological insulators and maps the problem of classifying them into a problem concerning the existence of robust gapless

---

[5]A simple example of NLσM is the $O(3)$ nonlinear sigma model with Lagrangian density $\mathcal{L} = \frac{1}{2}\partial^\mu \hat{n} \cdot \partial_\mu \hat{n}$, which describes the classical statistical mechanics of Heisenberg magnets. In general, the unit vector $\hat{n} \in S^2 = U(2)/U(1) \times U(1)$ (note that S^2 is an example of a symmetric space) is replaced by an element of one of the symmetric spaces listed in the last column of Table 3.2, which is called the *target space* of the NLσM and is denoted by G/H.

TABLE 3.3 The "Periodic Table" of topological insulators and superconductors. All possible topologically nontrivial (gapped) quantum ground states of noninteracting Hamiltonians are listed as a function of symmetry class (left column) and spatial dimension (d). The symbols $\mathbb{Z}$ and $\mathbb{Z}_2$ indicate that the topologically distinct phases within a given symmetry class and spatial dimension are characterized by an integer or a $\mathbb{Z}_2$ topological invariant, respectively. The label $2\mathbb{Z}$ corresponds to topological phases characterized by an even-integer topological invariant, indicating the presence of an even number of protected gapless boundary modes. The symbol "0" indicates that all ground states belong to the same (topologically trivial) phase.

Cartan	d	0	1	2	3	4	5	6	7	8
A		$\mathbb{Z}$	0	$\mathbb{Z}$	0	$\mathbb{Z}$	0	$\mathbb{Z}$	0	$\mathbb{Z}$
AIII		0	$\mathbb{Z}$	0	$\mathbb{Z}$	0	$\mathbb{Z}$	0	$\mathbb{Z}$	0
AI		$\mathbb{Z}$	0	0	0	$2\mathbb{Z}$	0	$\mathbb{Z}_2$	$\mathbb{Z}_2$	$\mathbb{Z}$
BDI		$\mathbb{Z}_2$	$\mathbb{Z}$	0	0	0	$2\mathbb{Z}$	0	$\mathbb{Z}_2$	$\mathbb{Z}_2$
D		$\mathbb{Z}_2$	$\mathbb{Z}_2$	$\mathbb{Z}$	0	0	0	$2\mathbb{Z}$	0	$\mathbb{Z}_2$
DIII		0	$\mathbb{Z}_2$	$\mathbb{Z}_2$	$\mathbb{Z}$	0	0	0	$2\mathbb{Z}$	0
AII		$2\mathbb{Z}$	0	$\mathbb{Z}_2$	$\mathbb{Z}_2$	$\mathbb{Z}$	0	0	0	$2\mathbb{Z}$
CII		0	$2\mathbb{Z}$	0	$\mathbb{Z}_2$	$\mathbb{Z}_2$	$\mathbb{Z}$	0	0	0
C		0	0	$2\mathbb{Z}$	0	$\mathbb{Z}_2$	$\mathbb{Z}_2$	$\mathbb{Z}$	0	0
CI		0	0	0	$2\mathbb{Z}$	0	$\mathbb{Z}_2$	$\mathbb{Z}_2$	$\mathbb{Z}$	0

boundary states. Remarkably, as in the K-theory approach, the solution of this problem involves the homotopy groups of symmetric spaces. A topological insulator (superconductor) of either i) $\mathbb{Z}$-type or ii) $\mathbb{Z}_2$-type exists in a given symmetry class in $d = \bar{d} + 1$ spatial dimensions if and only if the target space G/H of the NLσM on the $\bar{d}$-dimensional boundary allows the addition of either i) a WZW term, which is possible when $\pi_{\bar{d}+1}(G/H) = \mathbb{Z}$, or ii) a $\mathbb{Z}_2$ topological term, which is the case when $\pi_{\bar{d}}(G/H) = \mathbb{Z}_2$. For example, the NL$\sigma$M target space for a system in symmetry class A is $G/H = U(2n)/U(n) \times U(n)$ (see Table 3.2), which has a nontrivial homotopy group $\pi_{\bar{d}+1}(G/H) = \pi_d(G/H) = \mathbb{Z}$ in each even spatial dimension $d = 0, 2, \ldots$ (i.e., odd boundary dimension $\bar{d} = d - 1$). For d odd, we have $\pi_d(G/H) = \{1\}$. Consequently, in even dimensions a Hamiltonian belonging to symmetry class A has robust gapless boundary states protected by a WZW term, hence it has nontrivial (bulk) topological phases classified by an integer invariant. By contrast, in odd dimensions all class A insulators are topologically trivial.

Consider now the "real case" category, for example class AII. The target space is $G/H = O(2n)/O(n) \times O(n)$ and we have $\pi_0(G/H) = \mathbb{Z}$ and $\pi_1(G/H) = \pi_2(G/H) = \mathbb{Z}_2$. Consequently, a WZW term is allowed for $\bar{d} + 1 = 0$, while a $\mathbb{Z}_2$ topological term can be included in the action of the NLσM on boundaries with $\bar{d} = 1$ and $\bar{d} = 2$. Hence, a $\mathbb{Z}$-type TI exists in zero dimensions ($\bar{d} + 1 = 0$), while $\mathbb{Z}_2$ TIs exist in two and three spatial dimensions (with $\bar{d} = 1$ and $\bar{d} = 2$, respectively). The complete list of topological insulators (superconductors) corresponding to each symmetry class and spatial dimension is given in Table 3.3.

A few comments on the structure of Table 3.3 are appropriate. We note that the symmetry classes have been reordered as compared to the list given in Table 3.1. This reorganization reveals the underlying periodicity [293] originating from the Bott periodicity of K theory [69], i.e., the alternating pattern (period 2) of the "complex" classes (A and AIII) and the 8-fold periodicity in d of the remaining "real classes." We also note that there are exactly five classes of topological insulators (superconductors) in each spatial dimension: three characterized by integer ($\mathbb{Z}$) topological invariants and two having $\mathbb{Z}_2$ topological numbers. The columns in Table 3.3 corresponding to $d > 3$ may become relevant in situations when the Hamiltonian depends on external parameters that can be changed adiabatically (e.g., along closed loops in parameters space – a process referred to as "adiabatic pumping"); these parameters can be interpreted as additional momentum components, so that the system has an "effective dimension" $d > 3$.

In any "real case" symmetry class, a $\mathbb{Z}_2$ TI appears as part of a triplet consisting of a d-dimensional $\mathbb{Z}$ TI and a pair of $\mathbb{Z}_2$ TIs of dimensions $d-1$ and $d-2$. This structure reflects a connection among the members of the triplet, which can be exploited to derive the $\mathbb{Z}_2$ classification in $d-1$ and $d-2$ spatial dimensions from a d-dimensional "parent" $\mathbb{Z}$ topological insulator through a procedure called *dimensional reduction* [419].

Table 3.3 can also be used to determine whether or not localized zero-energy modes are supported by r-dimensional topological defects inside d-dimensional topological insulators (superconductors) [442]. A topological defect hosted by a system that belongs to a certain symmetry class can support zero modes if the entry of Table 3.3 at the intersection of the corresponding row with the column $d = r + 1$ is different from zero. For example, point-like defects ($r = 0$) hosted by a system in symmetry class D (e.g., vortex cores in a two-dimensional $p \pm ip$ superconductor) can bind zero-energy modes (Majorana bound states). Similarly, a dislocation line ($r = 1$) in a three-dimensional lattice hosting a so-called *weak topological insulator* – basically a stack of two-dimensional topological insulator (superconductor) layers – in symmetry classes A, D, DIII, AII, or C is predicted [423, 442] to bind an extended gapless mode.

We emphasize that Table 3.3 contains the classification of topological insulators and superconductors with generic symmetries such as time-reversal (TR) and particle-hole (PH). However, solid state system often have additional spatial symmetries, such as parity, reflection and discrete rotations. In this case, the topological quantum states can be protected by a combination of generic symmetries (e.g., TR) and spatial symmetries (e.g., inversion). The classification of topological insulators and superconductors protected by spatial symmetries or combinations involving spatial and generic symmetries can be determined based on the general ideas sketched in this section. For example, consider inversion symmetry and another discrete symmetry (e.g., TR or PH) [442]. When the Hamiltonian is invariant under inversion, $\mathcal{H}(\boldsymbol{k})$ and $\mathcal{H}(-\boldsymbol{k})$ can be related by a k-independent unitary transformation,

$$U\mathcal{H}(\boldsymbol{k})U^{-1} = \mathcal{H}(-\boldsymbol{k}). \tag{3.20}$$

We combine the inversion operation with a discrete symmetry satisfying the conditions given by Eq. (3.10), i.e., $V\mathcal{H}^*(\boldsymbol{k})V^{-1} = \epsilon_V\mathcal{H}(-\boldsymbol{k})$ with $V = U_t$, $\epsilon_V = 1$ or $V = U_c$, $\epsilon_V = -1$. More specifically, we consider systems that are invariant under the combined transformation $W = UV$,

$$W\mathcal{H}(\boldsymbol{k})W^{-1} = \epsilon_W\mathcal{H}(-\boldsymbol{k}), \tag{3.21}$$

where $\epsilon_W = \epsilon_V$. Since $WW^* = \eta_W = \pm 1$, we have four distinct symmetry classes of noninteracting Hamiltonians that are symmetric under a W-type symmetry labeled by $(\epsilon_W, \eta_W) = (\pm 1, \pm 1)$. Note that the system may not be symmetric under the *separate* action of U (inversion) and V (TR or PH). One can show that the projector (or the Q-matrix) associated with a Hamiltonian satisfying Eq. (3.21) is an element of a symmetric space, e.g., of the real Grassmannian, $G_{m,n+m}(\mathbb{R}) = O(n + m)/O(n) \times O(m)$ for $(\epsilon_W, \eta_W) = (+1, +1)$. Consequently, the topological classification of d-dimensional insulators and superconductors protected by a W-type symmetry with $(\epsilon_W, \eta_W) = (+1, +1)$ will be given [442] by the homotopy group $\pi_d(G_{m,n+m}(\mathbb{R}))$. For $d = 0, 1, 2$, and 3 the homotopy group that yields the classification is $\mathbb{Z}$, $\mathbb{Z}_2$, $\mathbb{Z}_2$, and $\{1\}$, respectively. Similar considerations apply for the other symmetry classes. A more in-depth discussion of the impact of spatial symmetries on the topological classification of noninteracting systems and on the stability of gapless boundary modes can be found in Chapter 4.

As a final note, we reiterate that the topological insulators (superconductors) classified in Table 3.3 are robust against disorder, i.e., against perturbations that break the translational symmetry of the underlying crystal lattice. Quantum states that possess this property are often referred to as *strong* topological insulators (superconductors). In the presence of translation invariance, additional topological states – the so-called *weak* topological insulators (superconductors) – become protected by translation symmetry. Of course, a *weak* TI becomes trivial in the presence of disorder. An example of weak TI is the 3D weak integer quantum Hall insulator [305], which, basically, consists of layered 2D integer quantum Hall states. Similarly, a weak 3D topological insulator in class AII can be viewed as a $k = 1$ dimensional array of $d = 2$ dimensional strong TIs (i.e., spin quantum Hall states). In general, a weak TI of "codimension" k exists in a given symmetry class and in d spatial dimensions whenever there is a $d - k$ dimensional strong TI in the same class. Specific cases can be identified using the "Periodic Table" 3.3. A more detailed discussion of weak topological phases is deferred to Chapter 4.

3.4 TOPOLOGICAL INVARIANTS: CHERN NUMBERS, WINDING NUMBERS, AND Z_2 INVARIANTS

The topological invariants of a topological space X are properties that depend only on the topology of the space and, consequently, are shared by all topological spaces that are homeomorphic to X. Examples include compactness, connectedness, orientability, as well as algebraic invariants such as the homology and homotopy groups. There is no unique "formula" for calculating the topological invariants associated with various topological insulators and superconductors. The procedure for determining these algebraic invariants may depend, for example, on the symmetries of the Hamiltonian, the spatial dimensionality, and the presence or absence of disorder. Moreover, for a given class of topological insulators (superconductors) there may be several equivalent approaches for determining the topological invariant. We will not analyze all these issues in detail; instead, we will discuss the basic ideas and provide a few examples. For simplicity, we focus on systems without disorder, i.e., systems with translation invariance. In this case, the eigenstates $|u_{nk}\rangle$ corresponding to the occupied (valence) bands (or the corresponding projectors $|u_{nk}\rangle\langle u_{nk}|$) can be viewed as defining a fiber bundle over the Brillouin zone. Topological insulators (superconductors) correspond to nontrivial fiber bundles and the powerful mathematical machinery sketched in Chapter 1 can be used for calculating the topological invariants associated with these bundles.

3.4.1 Hall conductance and the Chern number

We start with a classic example: the topological invariant that characterizes the integer quantum Hall (IQH) states, i.e., the two-dimensional TI in class A. Historically, the first insight into the topological nature of the IQH states was provided in 1982 by Thouless, Khomoto, Nightingale, and den Nijs (TKNN) who realized that the quantum number n in the Hall conductivity $\sigma_{xy} = n\frac{e^2}{h}$ is, in fact, the first Chern number associated with a principal $U(1)$ bundle over the (magnetic) Brillouin zone [503]. To better understand the main idea, let us consider a 2D electron system of size $L \times L$ in the presence of a perpendicular magnetic field $\boldsymbol{B}$ (oriented along the z axis) corresponding to a rational number p/q of magnetic flux quanta per unit cell and an in-plane electric field $\boldsymbol{E}$ (oriented along the y axis). Working in the Landau gauge, one can show [503] that k_x and k_y are good quantum numbers with values in the magnetic Brillouin zone, which corresponds to an enlarged unit cell with an integer number (p) of magnetic flux quanta going through. The Schrödinger equation for the 2D electrons in a uniform magnetic field can be written in the form

$$H(\boldsymbol{k})|u_{nk}\rangle = E_n|u_{nk}\rangle, \tag{3.22}$$

with E_n being the energy of the nth Landau level and

$$H(\boldsymbol{k}) = \frac{1}{2m}\left(-i\hbar\boldsymbol{\nabla} + \hbar\boldsymbol{k} + e\boldsymbol{A}\right)^2 + U(\boldsymbol{r}), \tag{3.23}$$

where $U(r)$ describes the position dependence of a periodic potential over one unit cell and $B = \nabla \times A$. The eigenfunctions $u_{nk}(r) = \langle r|u_{nk}\rangle$, with r in the magnetic unit cell, satisfy generalized periodic boundary conditions [503]. For a small applied electric field $E = E_y\hat{y}$ corresponding to the potential $V(r) = -eEy$, a perturbed eigenstate can be approximated as

$$|u_\alpha\rangle_E = |u_\alpha\rangle + \sum_\beta^{(\beta \neq \alpha)} \frac{\langle u_\beta|(-eEy)|u_\alpha\rangle}{E_\alpha - E_\beta}|u_\beta\rangle + \ldots, \tag{3.24}$$

where $\alpha = (n, k)$ and $\beta = (m, k')$. Equation (3.24) can be used to express the expectation value of the current density along the x axis in the presence of the perturbation, $\langle j_x\rangle_E = \sum_\alpha f(E_\alpha)\,_E\langle u_\alpha|\frac{ev_x}{L^2}|u_\alpha\rangle_E$, in terms of unperturbed matrix elements $(v)_{\alpha\beta} = \langle u_\alpha|v|u_\beta\rangle$ of the velocity operator. We note that the Heisenberg equation of motion $v_y = \frac{1}{i\hbar}[y, H]$ leads to $(v_y)_{\beta\alpha} = \frac{1}{i\hbar}(E_\beta - E_\alpha)\langle u_\beta|y|u_\alpha\rangle$. By putting together all these ingredients we arrive at the linear response (Kubo) formula for the Hall conductance $\sigma_{xy} = \langle j_x\rangle_E/E_y$. Explicitly,

$$\sigma_{xy} = \frac{e^2\hbar}{iL^2} \sum_\alpha^{(E_\alpha < E_F)} \sum_\beta^{(\beta \neq \alpha)} \frac{(v_x)_{\alpha\beta}(v_y)_{\beta\alpha} - (v_y)_{\alpha\beta}(v_x)_{\beta\alpha}}{(E_\alpha - E_\beta)^2}, \tag{3.25}$$

where E_F is the Fermi energy. Furthermore, in Eq. (3.25) we can replace the components of velocity operator $v = (-i\hbar\nabla + eA)/m$ by partial derivatives of the Hamiltonian (3.23), since we are considering only off-diagonal matrix elements,

$$(v_j)_{\beta\alpha} = \frac{1}{\hbar}\langle u_\beta|\frac{\partial H}{\partial k_j}|u_\alpha\rangle = (E_{nk} - E_{mk'})\langle u_{mk'}|\frac{\partial}{\partial k_j}|u_{nk}\rangle. \tag{3.26}$$

Using these identities and replacing the summation over k by an integral over the Brillouin zone, $\sum_k \rightarrow \frac{L^2}{(2\pi)^2}\int d^2k$, one can rewrite Eq. (3.25) as

$$\sigma_{xy} = \frac{e^2}{h} \sum_n^{(E_n < E_F)} \int \frac{d^2k}{2\pi i}\left[\frac{\partial}{\partial k_x}\langle u_{nk}|\frac{\partial}{\partial k_y}|u_{nk}\rangle - \frac{\partial}{\partial k_y}\langle u_{nk}|\frac{\partial}{\partial k_x}|u_{nk}\rangle\right]. \tag{3.27}$$

The integral runs over the entire Brillouin zone and the summation is over the occupied bands. We notice that the quantity inside the parentheses is the Berry curvature[6] vector $\Omega^n(k) = \nabla \times \mathcal{A}^n(k)$ corresponding to the Berry connection $\mathcal{A}^n(k) = i\langle u_{nk}|\nabla_k|u_{nk}\rangle$. Hence, the Hall conductivity reduces to

$$\sigma_{xy} = \frac{e^2}{h}\nu, \tag{3.28}$$

---

[6]See Chapter 1 for details, particularly Section 1.4.1 starting on page 22 and the discussion leading to Eqs. (1.46) and (1.47).

where $\nu = \sum_n \nu_n$ and the contribution ν_n from the nth occupied band is related to the Berry phase associated with encircling the Brillouin zone,

$$\nu_n = -\frac{1}{2\pi} \int_{BZ} d^2k \, \hat{e}_3 \cdot \mathbf{\Omega}^n(\mathbf{k}) = -\frac{1}{2\pi} \oint_{\partial_{BZ}} d\mathbf{k} \cdot \mathcal{A}^n(\mathbf{k}) = -\frac{\gamma_n[\partial_{BZ}]}{2\pi}, \quad (3.29)$$

where $\hat{e}_3$ is the unit vector perpendicular to the (k_x, k_y) plane and ∂_{BZ} is the boundary of the Brillouin zone. Because of the single-valued nature of the wave function, the Berry phase factor associated with encircling the Brillouin zone has to be an integer multiple of 2π, i.e., $\gamma_n[\partial_{BZ}] = 2\pi m$ with $m \in \mathbb{Z}$. The integer m – the so-called *first Chern number* – given by the integral of the Berry curvature over the Brillouin zone (divided by 2π) is a topological invariant. We conclude that the Hall conductance is quantized to integer multiples of e^2/h, with the integer ν (also called the TKNN invariant) being the topological invariant that characterizes the IQH system.

3.4.2 Chern numbers and winding numbers

It is instructive to move this discussion about the Chern invariant into a broader perspective. First, let us consider another example, the two-level system described by the Hamiltonian $H(\mathbf{k}) = \mathbf{h}(\mathbf{k}) \cdot \boldsymbol{\sigma}$, which was already discussed in Section 1.4.1. For concreteness, let $\mathbf{h}(\mathbf{k}) = (\sin k_x, \sin k_y, m + \cos k_x + \cos k_y)$, which corresponds to a tight-binding model with nearest-neighbor hopping on a square lattice. Note that for $m = 0, \pm 2$ the vector $\mathbf{h}(\mathbf{k})$ vanishes at certain values of $\mathbf{k}$, e.g., $\mathbf{h}^* = \mathbf{h}(0,0) = 0$ for $m = -2$, which results in degeneracy points with $E_+(\mathbf{h}^*) = E_-(\mathbf{h}^*) = 0$. When $m \neq 0, \pm 2$, $\mathbf{h}(\mathbf{k}) \neq 0$ over the entire Brillouin zone and we can define a flat band model with $E_\pm(\mathbf{k}) = \pm 1$ through the substitution $\mathbf{h}(\mathbf{k}) \rightarrow \mathbf{d}(\mathbf{k}) = \mathbf{h}(\mathbf{k})/|\mathbf{h}(\mathbf{k})|$. Note that $\mathbf{d}(\mathbf{k})$ can be viewed as a map from the Brillouin torus to the unit sphere, $\mathbf{d} : \mathbb{T}^2 \longrightarrow \mathbb{S}^2$. As shown in 1.4.1, the Berry curvature vector for the low-energy band is $\mathbf{\Omega} = \frac{1}{2}\frac{\mathbf{h}}{|\mathbf{h}|^3}$ (or $\frac{1}{2}\mathbf{d}$ for the flat-band model) and the corresponding curvature two-form[7] can be written as

$$\mathcal{F} = \frac{1}{2}\epsilon^{ijk}\Omega_i dh_j \wedge dh_k = \frac{1}{4}\epsilon^{ijk}\frac{h_i}{h^3}\frac{\partial h_j}{\partial k_\alpha}\frac{\partial h_k}{\partial k_\beta}dk_\alpha \wedge dk_\beta. \quad (3.30)$$

A similar expression holds for the flat-band model. The first Chern number, defined as the integral of the curvature two-form divided by 2π, reads

$$C_1 = \frac{1}{2\pi} \int_{BZ} \mathcal{F} = \frac{1}{4\pi} \int_{\mathbb{T}^2} \mathbf{d} \cdot \left(\frac{\partial \mathbf{d}}{\partial k_x} \times \frac{\partial \mathbf{d}}{\partial k_y}\right) dk_x \wedge dk_y. \quad (3.31)$$

Geometrically, when $\mathbf{k}$ covers the Brillouin torus, $\mathbf{d}(\mathbf{k})$ describes a closed surface Σ on the unit sphere. The Chern number C_1 given by Eq. (3.31) represents the number of times the surface Σ wraps around the unit sphere (which

---

[7]See Section 1.4.1.

is sometimes called a *winding number*). Mathematically, this represents the homotopy class of Σ in the punctured space $\mathbb{R}^3 - \{0\}$. Alternatively, one can view the Berry curvature vector $\boldsymbol{\Omega}$ as the field generated by a monopole of strength $1/2$ placed at the degeneracy point $\boldsymbol{h}^* = 0$ (or $\boldsymbol{d}^* = 0$, i.e., the center of the sphere). Then, C_1 represents the flux of this field through the surface Σ. Note that for the specific model considered above we have $C_1 = -1$ for $-2 < m < 0$, $C_1 = +1$ for $0 < m < 2$, and $C_1 = 0$ for $|m| > 2$.

The bulk-boundary correspondence dictates the existence of gapless states at interfaces between two insulators characterized by different values of the topological invariant. Consider, for example, a semi-infinite strip of a Chern insulator (described, say, by the so-called Haldane model[8] [231]), as shown in Figure 3.3. A *chiral* (i.e., propagating in one direction only) edge state is bound to the interface between the Chern insulator (with $\nu \equiv C_1 = +1$) and vacuum ($\nu \equiv C_1 = 0$). Note that the edge state corresponding to an insulator with $C_1 = -1$ will propagate in the opposite direction (i.e., left). These chiral edge states are robust against disorder because there are no states available for backscattering, a property that is intrinsically linked to the quantized electronic transport of IQH systems. If the system is perturbed near the boundary, the dispersion of the edge mode can change leading to the appearance of counter-propagating modes at the Fermi energy, as shown in Figure 3.3 (right panel). However, the difference $N_R - N_L$ between the number of right and left moving modes remains the same, as it is uniquely determined by the topological properties of the bulk. In fact, the bulk-boundary correspondence is formally expressed by the relation

$$N_R - N_L = \Delta\nu, \tag{3.32}$$

where $\Delta\nu$ is the difference in Chern number across the interface. In general, $\nu = \sum_n C_1^{(n)}$, the summation being over the occupied bands.

From a more abstract perspective, the Chern number can be viewed as an "obstruction" to Stokes theorem over the whole Brillouin zone (BZ). In Eq. (3.29) we implicitly treated the BZ as an open surface with a boundary (e.g., the square $-\pi/a \leq k_x, k_y \leq \pi/a$). However, considering the periodic dependence of physical quantities on the reciprocal lattice vectors, the BZ is, in fact, a torus. Since the torus has no boundary, applying the Stokes theorem to the integral over the BZ gives $\nu_n = 0$. Nonetheless, this can be done only if the Berry connection $\mathcal{A}^n(\boldsymbol{k})$ is smooth over the entire torus, which is not the case when $\nu_n \neq 0$. For example, gluing the open BZ of Eq. (3.29) into a torus necessarily results in discontinuities of $\mathcal{A}^n(\boldsymbol{k})$ across the juncture lines whenever $\nu_n \neq 0$. In general, the Berry connection is a gauge-dependent quantity that can be smoothly defined on subsets $O_\alpha \in \mathbb{T}^2$ representing an open covering of the torus.[9] Let us assume that $\{O_1, O_2\}$ represents such a covering with the eigenstates $|u_{n\boldsymbol{k}}\rangle_1$ and $|u_{n\boldsymbol{k}}\rangle_2$ smoothly defined over O_1 and

---

[8]For a detailed discussion of the Haldane model see Section 6.2.1 starting on page 189.

[9]See Chapter 1 for a discussion of the Berry connection in the context of fiber bundles.

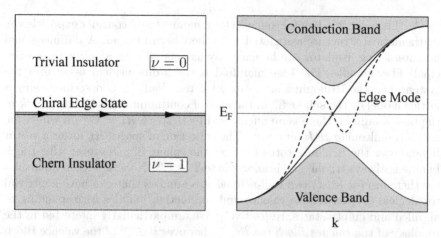

FIGURE 3.3 *Left*: Semi-infinite strip of Chern insulator (with Chern number $\nu = +1$) having a chiral edge mode at the boundary with a trivial insulator (vacuum). *Right*: Schematic energy spectrum of the Chern insulator strip showing the bulk bands and the chiral (right-propagating) edge mode. Perturbing the system may create counter-propagating modes (dashed line), but the difference $N_R - N_L$ between left and right movers is uniquely determined by the topological properties of the bulk (see main text).

O_2, respectively. In the overlap region $O_1 \cap O_2$ the two states are related by a gauge transformation

$$|u_{nk}\rangle_2 = e^{i\chi(k)}|u_{nk}\rangle_1. \tag{3.33}$$

As we argue below, the state vectors $|u_{nk}\rangle$ can be viewed as representing a fiber bundle over the BZ, with $g_{21}(\boldsymbol{k}) = e^{i\chi(k)} \in U(1)$ being the transition functions that patch together the two charts. The relation between the smoothly defined Berry connections in the overlap region is

$$\mathcal{A}_2^n(\boldsymbol{k}) = \mathcal{A}_1^n(\boldsymbol{k}) - \boldsymbol{\nabla}\chi(\boldsymbol{k}). \tag{3.34}$$

Consider now the closed path $\mathcal{C} \subset O_1 \cap O_2$ that separates the torus into two regions $M_1 \subset O_1$ and $M_2 \subset O_2$. The integral of the Berry curvature over the BZ can be written using the Stokes theorem as

$$\int_{\mathbb{T}^2} d^2k \, \hat{e}_3 \cdot \boldsymbol{\Omega}^n(\boldsymbol{k}) = \int_{M_1} \boldsymbol{\nabla} \times \mathcal{A}_1^n(\boldsymbol{k}) + \int_{M_2} \boldsymbol{\nabla} \times \mathcal{A}_2^n(\boldsymbol{k}) = \oint_{\mathcal{C}} d\boldsymbol{k} \cdot [\mathcal{A}_1^n(\boldsymbol{k}) - \mathcal{A}_2^n(\boldsymbol{k})]. \tag{3.35}$$

Hence, the Chern number of the nth band can be written as the winding number of the transition function around the path separating regions of the BZ where the Berry connection can be smoothly defined

$$\nu_n = \frac{1}{2\pi} \oint_{\mathcal{C}} d\boldsymbol{k} \cdot \boldsymbol{\nabla}\chi(\boldsymbol{k}). \tag{3.36}$$

At this point it is helpful to pause for a moment and consider explicitly the mathematical structure associated with the Chern number. A d-dimensional noninteracting insulator can be naturally associated with a vector (or a principal) fiber bundle. The base manifold is the d-dimensional k-space of the system, i.e., the Brillouin zone torus $\mathbb{T}^d$ if the insulator is described using a lattice model or the sphere $\mathbb{S}^d$ in the case of continuum models. For each $\boldsymbol{k}$ in the base manifold, there is an effective Hilbert space $\mathcal{H}_{\boldsymbol{k}} \cong \mathbb{C}^m$ on which the Bloch Hamiltonian $H(\boldsymbol{k})$ operates. The collection of spaces $\mathcal{H}_{\boldsymbol{k}}$ forms a vector bundle over the Brillouin torus (or over the sphere $\mathbb{S}^d$). However, this Bloch bundle is always trivial, i.e., isomorphic to $\mathbb{T}^d \times \mathbb{C}^m$. Nonetheless, in an insulator there are (at least) two well-defined sub-bundles that can have nontrivial topological properties: the valence and conduction bundles corresponding to all filled and empty states, respectively. We are particularly interested in the topology of the *valence Bloch bundle*. A fiber over $\boldsymbol{k} \in \mathbb{T}^d$ of the valence Bloch bundle corresponds to the space of *occupied states* or, alternatively, the *projector* $P(\boldsymbol{k})$ defined by Eq. (3.14). With no further symmetry conditions, the structure group of the bundle is $U(n)$, where n is the dimension of the occupied subspace. Hence, the occupied states of a class A insulator form a Bloch bundle $B_n : (B_n, \mathbb{T}^d, \pi, U(n))$ with fibers $\pi^{-1}(\boldsymbol{k}) = P(\boldsymbol{k})$, where $\boldsymbol{k} \in \mathbb{T}^d$ is a point in the Brillouin zone. Determining whether or not the insulator is topologically nontrivial amounts to determining the topological properties of B_n. For example, if the base manifold can be covered with one open set, the fiber is trivial. For nontrivial fiber bundles the base manifold can only be covered by two or more open sets with transition functions in the overlap regions that are different from the identity function.

For $d = 2$ (i.e., for IQH systems), the topology of the (valence) Bloch bundle is described by the first Chern number. More generally, the $\mathbb{Z}$ topological invariant of a class A insulator in *even* spatial dimensions $d = 2\nu$ is the Chern number

$$C_\nu = \frac{1}{\nu!(2\pi)^\nu} \int_{BZ} \mathrm{Tr}\left(\mathcal{F}^\nu\right), \qquad (3.37)$$

where $\mathcal{F}$ is the curvature two-form of the corresponding Bloch bundle and the integrand $\mathrm{ch}_\nu(\mathcal{F}) = \frac{1}{\nu!}\left(\frac{\mathcal{F}}{2\pi}\right)^\nu$ represents the νth *Chern character*.[10] When $d = 2$ this expression reduces to the TKNN invariant C_1 given by Eq. (3.31), while the second Chern number C_2 describes class A topological insulators in $d = 4$, or certain "adiabatic pumping" processes in $d < 4$ spatial dimensions. Finally, we note that four of the "real" symmetry classes in Table 3.3, specifically those classes that break chiral symmetry (AI, D, AII, and C), have $\mathbb{Z}$ topological insulators in *even* spatial dimensions. The topological invariants that characterize these insulators are the Chern numbers given by Eq. (3.37).

The winding number. What about odd-dimensional base manifolds? For example, how do we characterize the $\mathbb{Z}$-type TIs in class AIII or in a "real"

---

[10]Here we use the physicist's definitions of connection and curvature (see Chapter 1). The corresponding mathematical quantities can be obtained by taking $\mathcal{A} \to i\mathcal{A}$ and $\mathcal{F} \to i\mathcal{F}$.

symmetry class with chiral symmetry, i.e., class BDI, DIII, CII, or CI (see Table 3.1 and Table 3.3)? In the presence of chiral (sublattice) symmetry, one can find a unitary matrix Γ that anticommutes with the Hamiltonian, $\{\mathcal{H}(\boldsymbol{k}), \Gamma\} = 0$. As a consequence, the spectrum is symmetric with respect to zero energy and, using a basis in which Γ is diagonal, the Hamiltonian, as well as the Q matrix (i.e., the projector), can be brought into block off-diagonal form, e.g., Eq. (3.18). For a Hamiltonian in symmetry class AIII, the off-diagonal component $q(\boldsymbol{k})$ defines a map from the BZ onto the group of unitary matrices $U(N)$, where N is the number of occupied bands (equal to the number of empty bands). Consequently, the topological classification of class AIII insulators is given by the homotopy group $\pi_d(U(N))$, which is nontrivial (and equal to $\mathbb{Z}$) in odd spatial dimensions and trivial in even dimensions. The corresponding $\mathbb{Z}$ topological invariant, which can be defined in odd spatial dimensions $d = 2n + 1$, is the *winding number*

$$\nu_{2n+1}[q] = \int_{BZ} \omega_{2n+1}[q], \tag{3.38}$$

where the winding number density is given by

$$\omega_{2n+1}[q] = \frac{(-1)^n n!}{(2n+1)!} \left(\frac{i}{2\pi}\right)^{n+1} \epsilon^{\alpha_1 \alpha_2 \dots \alpha_{2n+1}} \operatorname{Tr}[q^{-1}\partial_{\alpha_1} q \cdot q^{-1}\partial_{\alpha_2} q \dots] d^{2n+1}k. \tag{3.39}$$

The number $\nu_{2n+1}[q]$ counts the nontrivial winding (wrapping) of the map $q: BZ \longrightarrow U(N)$, where BZ is either the Brillouin torus $\mathbb{T}^{2n+1}$ (for lattice models), or the sphere $\mathbb{S}^{2n+1}$ (for continuum models). We note that in class AIII the winding number $\nu_{2n+1}[q]$ can be any integer, while in the case of "real" classes with chiral symmetry additional constraints may impose the realization of specific integer values, e.g., even integers for the entries labeled $2\mathbb{Z}$ in Table 3.3. Similar constraints apply for the Chern number characterizing "real" classes with no chiral symmetry in even spatial dimensions. However, we point out that the groups of integers ($\mathbb{Z}$) and even integers ($2\mathbb{Z}$) are isomorphic.

3.4.3 The $\mathbb{Z}_2$ topological invariant

We turn our attention to a different kind of TI, the $\mathbb{Z}_2$ topological insulator, focusing on class AII systems in two and three dimensions. This type of time-reversal invariant TI was discovered in 2005 by Kane and Mele [274, 275], who considered a model of spin-$\frac{1}{2}$ particles with spin-orbit interaction on a honeycomb lattice.[11] The corresponding 2D phase, dubbed quantum spin Hall effect (QSHE), is characterized by pairs of counter-propagating edge states (see Figure 3.4) that preserve time-reversal symmetry (TRS). From the perspective of the bulk-boundary correspondence, the $\mathbb{Z}_2$ index represents the parity of the number of edge state *pairs*. The first experimental realization of the QSHE

---

[11]This model corresponds to doubling the Haldane model [231] of a Chern insulator by introducing spin into the problem. See Chapter 6 for details.

was achieved by the Molenkamp group in 2007 using HgTe quantum wells [55, 282]. The same year, it was discovered by three independent theoretical groups [189, 370, 434] that, unlike Chern topological insulators, TIs characterized by a $\mathbb{Z}_2$ invariant also exist in 3D. The discovery was subsequently confirmed experimentally [254, 557] and triggered intense research activity.

Similarly to the $\mathbb{Z}$ invariant, the $\mathbb{Z}_2$ index can be constructed in several different ways and can be understood using different perspectives. Before discussing specific constructions, let us emphasize the key role played by the additional constraint imposed by TRS. First, we note that for time-reversal symmetric 2D systems the Bloch bundle B_n associated with the occupied states is trivial. Indeed, one can easily show that in the presence of TRS the Berry connections $\mathcal{A}(\boldsymbol{k})$ and $\mathcal{A}(-\boldsymbol{k})$ differ by a gauge transformation $\nabla\chi(\boldsymbol{k})$, while the Berry curvature satisfies the condition

$$\mathcal{F}(-\boldsymbol{k}) = -\mathcal{F}(\boldsymbol{k}). \tag{3.40}$$

Consequently, the Chern character is subjected to the constraint $\mathrm{ch}_n[\mathcal{F}(-\boldsymbol{k})] = (-1)^n \mathrm{ch}_n[\mathcal{F}(\boldsymbol{k})]$ and the Chern number (3.37) vanishes in $4n - 2$ dimensions (i.e., for $\nu = 2n - 1$), but not in $4n$ dimensions ($\nu = 2n$). Hence, for a $d = 2$ time-reversal invariant system the first Chern number vanishes, $C_1 = 0$, and there is no obstruction to globally define the eigenstates $|u_{n\boldsymbol{k}}\rangle$ over the Brillouin torus $\mathbb{T}^2$, i.e., the Bloch bundle B_n is trivial. However, TRS imposes additional constraints on the eigenstates and it is when these constraints are considered that nontrivial topology emerges. Specifically, the Bloch Hamiltonians at $\boldsymbol{k}$ and $-\boldsymbol{k}$ satisfy the relation $\mathcal{T}H(\boldsymbol{k})\mathcal{T}^{-1} = H(-\boldsymbol{k})$, which implies that the time-reversal image of any eigenstate of the Bloch Hamiltonian at $\boldsymbol{k}$, $\mathcal{T}|u_{n\boldsymbol{k}}\rangle$, is an eigenstate of the Bloch Hamiltonian at $-\boldsymbol{k}$ having the same energy (Kramers theorem). Note that, as a consequence of $\mathcal{T}^2 = -1$ (TR operator for spin-$\frac{1}{2}$ particles), the two Kramers partners are orthogonal, i.e., the matrix

$$M_{ij} = \langle u_{i\boldsymbol{k}}|\mathcal{T}|u_{j\boldsymbol{k}}\rangle = -\langle u_{j\boldsymbol{k}}|\mathcal{T}|u_{i\boldsymbol{k}}\rangle \tag{3.41}$$

contains only off-diagonal elements. In the presence of TRS the band structure is characterized by the existence of Kramers pairs, as illustrated schematically in Figure 3.4. While there always exists a global basis $|u_{i\boldsymbol{k}}\rangle$ for the valence bands that is smoothly defined over the whole Brillouin torus, it is impossible to continuously define *Kramers pairs* on the whole Brillouin torus when the insulator is nontrivial. In other words, there is no global basis that satisfies $\mathcal{T}|u_{i,\boldsymbol{k}}\rangle = e^{-i\chi(\boldsymbol{k})}|u_{j,-\boldsymbol{k}}\rangle$ with $e^{-i\chi(\boldsymbol{k})}$ being a smooth function of $\boldsymbol{k}$ over the Brillouin torus (e.g., $\chi(\boldsymbol{k}) = 0$ everywhere). This obstruction is, in essence, the source of nontrivial topology in TR invariant insulators. To capture it, one has to explicitly consider the Kramers constraint. In the language of fiber bundles, one has to replace the (trivial) bundle B_n (having fibers corresponding to the subspace spanned by the n occupied states $|u_{i\boldsymbol{k}}\rangle$) with the rank 2 TRS Bloch bundle $B_n^{\mathcal{T}}$ with fibers defined as the subspaces spanned by the (valence band) Kramers pairs $|\Psi_i(\boldsymbol{k})\rangle = (|u_{i\boldsymbol{k}}\rangle, \mathcal{T}|u_{i\boldsymbol{k}}\rangle)$. Note that the

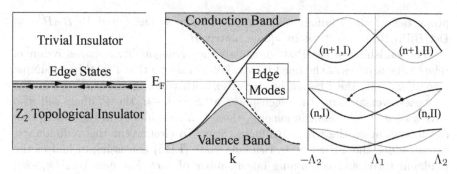

FIGURE 3.4 *Left*: Semi-infinite strip of $\mathbb{Z}_2$ topological insulator supporting counter-propagating edge modes at the boundary with a trivial insulator (e.g., vacuum). *Center*: Schematic representation of the energy spectrum of the semi-infinite strip showing the bulk bands (shaded regions) and the counter-propagating edge modes. *Right*: Typical band structure of a spin-$\frac{1}{2}$ TR-invariant system. The small black circles represent Kramers pairs, while Λ_1 and Λ_2 are TR-invariant momenta (TRIM).

subspace spanned by $|\Psi_i(\boldsymbol{k})\rangle$ can be viewed as a *quaternionic* Hilbert space with quaternionic elements i, $j = \mathcal{T}$, and $k = i\mathcal{T}$. [12]

As pointed out by Freed and Moore [177], the $\mathbb{Z}_2$ index can be defined as a topological invariant of the $B_n^{\mathcal{T}}$ bundle. Without going into technical details, we note that this construction of the $\mathbb{Z}_2$ index involves the orientations of a so-called *determinant line bundle* at the TR invariant points of the Brillouin zone. The product of these orientations is constrained by the topology of the TR symmetric bundle $B_n^{\mathcal{T}}$ [177]. The TR invariant points, usually called high-symmetry points or time-reversal invariant momenta (TRIM), $\boldsymbol{\Lambda} = -\boldsymbol{\Lambda} + \boldsymbol{G}$, with $\boldsymbol{G}$ a reciprocal lattice vector, play an important role in TR invariant systems. Note, for example, that at a TRIM the Kramers partners have the same momentum and the spectrum is (at least) double degenerate.

The zeros of the Pfaffian. The original Kane and Mele construction [275] of the $\mathbb{Z}_2$ index is based on the observation that the parity of the number of zeros of the Pfaffian Pf$[M]$, where the matrix M is given by Eq. (3.41), in half of the BZ is a topological invariant. In general, the Pfaffian Pf$[A]$ of a $2n \times 2n$ *skew-symmetric* matrix A is defined as

$$\mathrm{Pf}[A] = \frac{1}{2^n n!} \sum_{\sigma \in \Pi_{2n}} \mathrm{sign} \prod_{i=1}^n A_{\sigma(2i-1),\sigma(2i)}, \tag{3.42}$$

where Π_{2n} is the permutation group of $2n$ elements. For example, for a 2×2 skew-symmetric matrix $A = \begin{pmatrix} 0 & a \\ -a & 0 \end{pmatrix}$, we have Pf$[A] = a$. Among the

---

[12] A generic quaternion reads $q = a + bi + cj + dk$ with a, b, c, and d real numbers and $i^2 = j^2 = k^2 = ijk = -1$.

properties of the Pfaffian we mention $(\text{Pf}A)^2 = \text{Det}A$ and $\text{Pf}[BAB^T] = \text{Det}[B]\text{Pf}[A]$ for any $2n \times 2n$ square matrix B.

The vanishing of the Pfaffian signals the orthogonality between Kramers related eigenspaces of the filled bands. Let us assume that $\text{Pf}[M]$ has a simple zero at $\boldsymbol{k}_\alpha$, i.e., the Pfaffian has a vortex with winding number[13] ± 1 at $\boldsymbol{k}_\alpha$. As time-reversal maps the eigenspaces at $\boldsymbol{k}$ and $-\boldsymbol{k}$, the Pfaffian will also have a zero at $-\boldsymbol{k}_\alpha$ and, in general, there is a redundancy in the description of the system on the whole Brillouin torus. To circumvent this redundancy, we introduce the TR effective Brillouin zone (EBZ) consisting of half of the Brillouin torus and containing one member of each Kramers pair $(\boldsymbol{k}, -\boldsymbol{k})$, except at the boundary. Note that the EBZ has the topology of a torus with a boundary consisting of two homotopic TR invariant closed curves connecting two TRIM.

At TR invariant points, $|u_{i\Lambda}\rangle$ and $\mathcal{T}|u_{i\Lambda}\rangle$ are orthogonal eigenstates and $M(\Lambda)$ is a unitary matrix containing only off-diagonal elements. Consequently, $|\text{Pf}[M(\Lambda)]| = 1$. On the other hand, $\text{Pf}[M(\boldsymbol{k}_\alpha)] = \text{Pf}[M(-\boldsymbol{k}_\alpha)] = 0$, i.e., the Pfaffian has vortices at $\boldsymbol{k} = \pm\boldsymbol{k}_\alpha$. Note that if there is only one pair of vortices it is not possible to remove it, as the only points where $\boldsymbol{k}_\alpha = -\boldsymbol{k}_\alpha$ (i.e., points where the vortices could annihilate) are TR invariant points characterized by $|\text{Pf}[M(\Lambda)]| = 1$. For two pairs of vortices, on the other hand, it is always possible to continuously deform the Hamiltonian and remove them by merging the vortices with opposite vorticity at $\boldsymbol{k}_\alpha = \pm\boldsymbol{k}_\beta$ (i.e., at points on the boundary of the EBZ that are not TRIM). This merging procedure results in the parity of the number of vortices in the EBZ being an invariant. Hence we can define the $\mathbb{Z}_2$ index in terms of the total vorticity associated with the zeros of the Pfaffian in half of the BZ as

$$\nu = \frac{1}{2\pi i} \oint_{\partial \text{EBZ}} d\boldsymbol{k} \cdot \boldsymbol{\nabla} \log \text{Pf}[M] \qquad \text{(modulo 2)}, \qquad (3.43)$$

where ∂EBZ is the boundary of the effective BZ. The relation between the zeros of the Pfaffian and the $\mathbb{Z}_2$ topological invariant is illustrated in Figure 3.5. Note that breaking TRS removes the constraint $|\text{Pf}[M(\Lambda)]| = 1$ and the zeros are no longer prevented from annihilating at TRIM, i.e., the topological distinction between $\nu = 1$ and $\nu = 0$ is lost. In the presence of TRS the parity of the number of zeros of the Pfaffian can only change through phase transitions in which the bulk gap collapses.

The sewing matrix invariant. An alternative construction of the $\mathbb{Z}_2$ index, introduced by Fu and Kane [185], provides additional physical insight by formulating the problem in the context of the modern theory of polarization [431]. The construction uses the so-called *sewing matrix* defined as

$$w_{ij}(\boldsymbol{k}) = \langle u_{i,-\boldsymbol{k}}|\mathcal{T}|u_{j,\boldsymbol{k}}\rangle. \qquad (3.44)$$

---

[13]Let $f(z) = r(z)e^{i\theta(z)}$ be a complex function with $r(z_0) = 0$. The winding of the phase as one goes around z_0 on a closed path $\mathcal{C}$ is given by the integer $W[\mathcal{C}] = \frac{1}{2\pi i} \oint_{\mathcal{C}} d\log[f(z)]$.

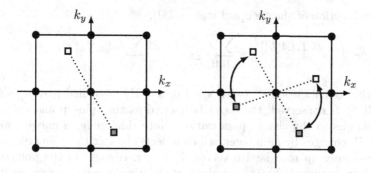

FIGURE 3.5 Vortices of the Pfaffian in a topological insulator (left) and in a trivial insulator (right). The empty and filled squares represent vortices with opposite winding numbers (e.g., +1 and −1, respectively). In the trivial case one can move (arrows) and annihilate the vortices pairwise without crossing the TRIM (black circles).

This matrix is unitary, $w^{\dagger}(\mathbf{k})w(\mathbf{k}) = 1$, and relates the eigenvectors at $-\mathbf{k}$ to the Kramers partners of the eigenstates at $\mathbf{k}$,

$$|u_{i,-\mathbf{k}}\rangle = \sum_{j} w_{ij}^{*}(\mathbf{k})\mathcal{T}|u_{j,\mathbf{k}}\rangle. \tag{3.45}$$

The sewing matrix has the property $w_{ij}(\mathbf{k}) = -w_{ji}(-\mathbf{k})$ and at the TRIM it becomes antisymmetric, $w_{ij}(\mathbf{\Lambda}) = -w_{ji}(\mathbf{\Lambda})$. Also, considering the set of Berry connection matrices $\mathcal{A}_{ij}(\mathbf{k}) = i\langle u_{ik}|\nabla_{\mathbf{k}}|u_{jk}\rangle$ one can show that

$$\mathrm{Tr}[\mathcal{A}(-\mathbf{k})] = \mathrm{Tr}[\mathcal{A}(\mathbf{k})] + i\,\mathrm{Tr}[w^{\dagger}(\mathbf{k})\nabla_{\mathbf{k}}w(\mathbf{k})]. \tag{3.46}$$

Equation (3.46) can be obtained using the relation $\langle \mathcal{T}\psi|\mathcal{T}\phi\rangle = \langle\phi|\psi\rangle$ and the properties of the sewing matrix.

Next, let us introduce the charge and *time-reversal* polarizations. Consider a one-dimensional TR invariant system of spin-$\frac{1}{2}$ particles on a lattice with lattice constant $a = 1$ and suppose that there are no degeneracies other than those required by TRS. The bands corresponding to a Kramers pair are labeled (n, I) and (n, II), as in Figure 3.4, and we have

$$\mathcal{T}|u_{n,k}^{\mathrm{II}}\rangle = e^{-i\chi_n(k)}|u_{n,-k}^{\mathrm{I}}\rangle, \qquad \mathcal{T}|u_{n,k}^{\mathrm{I}}\rangle = -e^{-i\chi_n(-k)}|u_{n,-k}^{\mathrm{II}}\rangle. \tag{3.47}$$

Note that the sewing matrix w is block-diagonal with 2×2 blocks of the form

$$w_n(k) = \begin{pmatrix} 0 & e^{i\chi_n(k)} \\ -e^{i\chi_n(-k)} & 0 \end{pmatrix}. \tag{3.48}$$

The charge polarization can be calculated as an integral over the BZ of the

Berry connection of the occupied states [431],

$$P_\rho = \frac{1}{2\pi} \int_{-\pi}^{\pi} dk\, \text{Tr}[\mathcal{A}(k)] = \sum_{s \in \{\text{I,II}\}} \frac{i}{2\pi} \int_{-\pi}^{\pi} dk \sum_n \langle u_{nk}^s | \nabla_k | u_{nk}^s \rangle = P^\text{I} + P^\text{II},$$
(3.49)

where the contributions P^s (with $s = \text{I}$ or $s = \text{II}$) are called *partial polarizations*. If S_z is conserved, these can be viewed as the spin-up and spin-down contributions to the charge polarization. Note that under a gauge transformation P_ρ changes by an integer, which reflects the fact that the polarization is defined only up to a lattice vector. However, changes in the polarization induced by cyclic adiabatic evolutions of the Hamiltonian are gauge invariant. For example, let us assume that the Hamiltonian depends on an extra parameter $R(t)$ that changes between $R(0) = R_0$ and $R(T) = R_T$. We have

$$P_\rho(T) - P_\rho(0) = \frac{1}{2\pi} \left(\oint_{C_T} dk\, \text{Tr}[\mathcal{A}(k, R_T)] - \oint_{C_0} dk\, \text{Tr}[\mathcal{A}(k, R_0)] \right), \quad (3.50)$$

where C_T and C_0 are loops corresponding to $-\pi \leq k \leq \pi$ for $t = T$ and $t = 0$, respectively. Following the general ideas discussed in Chapter 1 in the context of Berry phases, we can view k and R as the components of a vector belonging to a two-dimensional parameter space $\mathcal{M}$. The corresponding (diagonal) components of the Berry connection are $i\langle u_{nkR}^s | \nabla_k | u_{nkR}^s \rangle$ and $i\langle u_{nkR}^s | \nabla_R | u_{nkR}^s \rangle$, while the associated Berry curvature vector is $\Omega_n^s(k, R) = i[\langle \nabla_R u_{nkR}^s | \nabla_k u_{nkR}^s \rangle - c.c.]$. Using the Stokes theorem, we can write Eq. (3.50) in the form $P_\rho(T) - P_\rho(0) = \frac{1}{2\pi} \int_\mathcal{M} dk dR\, \Omega(k, R)$, where $\Omega = \sum_{n,s} \Omega_n^s$. For a cyclic evolution, $R_T = R_0$ and the parameter space $\mathcal{M}$ becomes a torus, e.g., the Brillouin torus if the parameters represent the components of a two-dimensional momentum, $(k, R) \rightarrow (k_x, k_y)$. Hence, the difference $P_\rho(T) - P_\rho(0)$ representing the charge pumped across the system during one cycle is given by the integral of the Berry curvature over the parameter space torus, which is nothing but the first Chern number.

For a TR invariant system $C_1 = 0$. However, instead of considering the charge pumping process we can consider the spin pumping associated with the change during an adiabatic cyclic evolution of the so-called *time-reversal polarization* $P_\mathcal{T} = P^\text{I} - P^\text{II}$. We note that

$$\mathcal{A}^\text{I}(-k) = \mathcal{A}^\text{II}(k) - \sum_n \nabla_k \chi_n(k), \qquad \mathcal{A}^s(k) = i \sum_n \langle u_{nk}^s | \nabla_k | u_{nk}^s \rangle, \quad (3.51)$$

and the fact that the sewing matrix is antisymmetric for $k = 0$ and $k = \pi$, so that from Eq. (3.48) we have

$$\frac{\text{Pf}[w(\pi)]}{\text{Pf}[w(0)]} = \exp \left\{ i \sum_n [\chi_n(\pi), -\chi_n(0)] \right\}. \quad (3.52)$$

Using these properties, we can express the partial polarization as

$$P^\text{I} = \frac{1}{2\pi} \left[\int_0^\pi dk\, \text{Tr}[\mathcal{A}(k)] + i \ln \frac{\text{Pf}[w(\pi)]}{\text{Pf}[w(0)]} \right]. \quad (3.53)$$

A similar expression holds for P^{II}. The TR polarization $P_T = 2P^I - P_\rho$ has a term of the form $\frac{1}{2\pi} \int_0^\pi \{\text{Tr}[\mathcal{A}(k)] - \text{Tr}[\mathcal{A}(-k)]\}$ that can be written using Eq. (3.46) as $\frac{1}{2\pi i} \int_0^\pi dk \, \text{Tr}[w^\dagger(k) \nabla_k w(k)]$. Since $w(k)$ is block-diagonal, brute force calculation gives $\text{Tr}[w^\dagger \nabla_k w] = \nabla_k \ln \text{Det}[w(k)]$. Finally, using $\text{Det}[w] = (\text{Pf}[w])^2$, we obtain the TR polarization as

$$P_T = \frac{1}{\pi i} \log \left[\frac{\sqrt{\text{Det}[w(\pi)]}}{\text{Pf}[w(\pi)]} \frac{\text{Pf}[w(0)]}{\sqrt{\text{Det}[w(0)]}} \right]. \tag{3.54}$$

Since the argument of the logarithm is $+1$ or -1, P_T is 0 or 1 (mod 2). Note that in Eq. (3.54) the sewing matrix is evaluated at the 1D TRIM points $\Lambda_1 = 0$ and $\Lambda_2 = \pi$.

Consider now a cyclic adiabatic process of period T controlled by some parameter $R(t)$ that is *odd* under time-reversal, $R(-t) = -R(t)$. Eventually, we will interpret this parameter as the second momentum component for a 2D system. Since $H[k, R + R_T] = H[k, R]$, the points $(k, R) \in \mathcal{M}$ corresponding to the values $\Lambda_1 = (0,0)$, $\Lambda_2 = (\pi, 0)$, $\Lambda_3 = (0, R_T/2)$, and $\Lambda_4 = (\pi, R_T/2)$ are TR invariant points. The $\mathbb{Z}_2$ invariant can be defined as the change in TR polarization in half a cycle, $\Delta = P_T(R_T/2) - P_T(0)$ (mod 2). Explicitly, we have

$$(-1)^\Delta = \prod_i \frac{\text{Pf}[w(\Lambda_i)]}{\sqrt{\text{Det}[w(\Lambda_i)]}}. \tag{3.55}$$

We note that in this formula the sewing matrix is to be calculated using wave functions that are smoothly defined over the parameter space torus, which is always possible because the Chern number is equal to zero, hence there is no obstruction. For a $d = 2$ dimensional system in class AII we have $(k, R) \rightarrow (k_x, k_y)$, $R_T/2 \rightarrow \pi$ and Eq. (3.55), with Λ_i being the TRIM, gives the $\mathbb{Z}_2$ invariant for a quantum spin Hall system. The same expression applies to 3D topological insulators (class AII), with Λ_i representing the eight TR-invariant momenta, $(0,0,0)$, $(\pi, 0, 0), \ldots, (\pi, \pi, \pi)$.

CHAPTER 4

Extensions of the Noninteracting Topological Classification

TOPOLOGICAL INSULATORS AND SUPERCONDUCTORS are subsets of (almost) noninteracting symmetry protected topological (SPT) phases. More generally, SPT phases represent equivalence classes of (equilibrium) gapped quantum states with short-range entanglement that are protected by certain symmetries. Two quantum states are "equivalent" if they can be adiabatically deformed into one another without closing the bulk gap. Thus, quantum states belonging to different SPT phases cannot be smoothly deformed into each other without a phase transition, as long as the underlying symmetry is preserved. However, upon removing the symmetry all SPT states can be adiabatically deformed to a trivial product state (e.g., an atomic insulator) without undergoing a phase transition. In this chapter, we generalize our discussion of (almost) noninteracting topological states by removing some of the explicit and implicit constraints underlying the classification and properties of topological insulators and superconductors discussed in Chapter 3. In Section 4.1, we enlarge the type of symmetries that protect the topological phases by supplementing internal symmetries (e.g., time-reversal and particle-hole symmetries) with crystalline symmetries (e.g., mirror and rotation symmetries) and discuss *topological crystalline insulators and superconductors*, a subsets of SPT phases protected by internal and crystalline symmetries. In Section 4.2, we remove the requirement of having a finite (bulk) gap and discuss *Weyl semimetals* – crystalline systems characterized by topologically protected bulk band degeneracy points separated in momentum space, which

    DOI: 10.1201/9781003226048-4

support low energy excitations that are massless chiral fermions, i.e., Weyl fermions. Finally, in Section 4.3, we remove the condition of thermodynamic equilibrium and consider topological quantum states in periodically driven systems, focusing on the so-called *Floquet topological insulators*, which exhibit a richer topological structure than their static counterparts. We note that aspects regarding the effects of particle-particle interactions, including the emergence of topological order in systems with long-range entanglement, will be addressed in Chapter 5. We also point out that the areas summarized in this chapter have grown tremendously in recent years and continue to grow; even a brief mentioning of all recent developments would be impossible within the constraints of a book chapter. Instead, we point out the key elements that determine the nontrivial topology of the quantum systems mentioned above, discuss specific manifestations of their topological properties (e.g., the presence of protected gapless states), and provide representative examples.

4.1 TOPOLOGICAL CRYSTALLINE INSULATORS AND SUPERCONDUCTORS

In the previous chapter, we have discussed the topological classification of non-interacting fermion systems described by single-particle Hamiltonians that are invariant under rather generic internal symmetries, such as time–reversal. A natural question concerns the fate of this classification in the presence of more stringent symmetry constraints that include crystalline symmetries. More specifically, let us assume a crystalline d-dimensional system that is symmetric under the elements of a certain symmetry group G. An element $g \in G$ is called an *internal* symmetry if it acts trivially on all real space vectors, $gr = r$. On the other hand, we call g a *crystalline* symmetry if there are vectors on which it acts nontrivially, i.e., $\exists r \in \mathbf{R}^d$ such that $gr \neq r$. The group G may include both internal symmetries, such as $SU(2)$ spin rotation symmetry, particle–hole symmetry, and time–reversal symmetry, as well as crystalline symmetries, e.g., translation, rotation, and inversion symmetries. In this section, we focus on aspects regarding the classification of topological phases of matter and the characterization of their phenomenology that are determined by the presence of crystalline symmetries.

First, the presence of crystalline symmetries augments (significantly) the "bestiary" of symmetry protected topological phases. Identifying and classifying these new phases require extending the classification schemes (e.g., the classification scheme based on K-theory) and defining new topological invariants. In this context, we note that the sheer number of different crystalline symmetry groups (e.g., the 230 symmetry groups associated with three-dimensional crystals) promotes the problem of classifying the topological phases to a new level of complexity. Here, we do not address the issues regarding the generic classification of topological phases protected by crystalline symmetries; instead, we provide a few examples that illustrate the new aspects associated with the presence of these symmetries (see Section 4.1.2). Second, the bulk–boundary correspondence becomes less straightforward in the presence of crystalline symmetries. Gapless boundary modes require the boundary on which they are localized to preserve the symmetry of the bulk. While topological phases protected solely by internal symmetries automatically satisfy this condition (and, consequently, always support gapless boundary modes), crystalline symmetries can be broken by some (or all) surfaces/edges of the system, which, consequently, will not support gapless states. In other words, topological phases protected by crystalline symmetries are characterized by (robust) gapless boundary modes only on the boundaries that preserve the symmetry of the bulk. An interesting class of topological crystalline phases are the so-called *higher order topological phases*, which support gapless modes on corners and hinges. More specifically, a d-dimensional nth-order topological insulator (or superconductor) is a system characterized by protected gapless modes on boundaries of dimension $d - n$ (with $n > 1$), while higher dimensional boundaries are gapped. A brief introduction to higher-order topological phases is provided in Section 4.1.3. However, before starting our main discussion, it is useful to clarify a few subtleties regarding the classification of topological quantum phases, which become particularly relevant in the presence of crystalline symmetries. These issues are presented in Section 4.1.1

4.1.1 Weak topological phases and fragile topology

The symmetry protected topological phases of noninteracting gapped fermionic systems are defined as equivalence classes of insulating (or superconducting) ground states, with the natural notion of adiabatic (or continuous) deformation – which corresponds to the mathematical concept of homotopy – representing the equivalence relation. Thus, we can define a topological insulator as a gapped (free fermionic) system that cannot be adiabatically connected to an atomic insulator, i.e., a system with electrons occupying orbitals exponentially localized near the lattice points. To distinguish among different topologically nontrivial phases and to classify them (in the presence of some generic internal symmetries) one can use K-theory [293], which is based on the notion of *stable* homotopy equivalence, i.e., homotopy equivalence up to addition of (trivial) valence bands. We note that homotopy equivalence results in a finer topological classification as compared to the classification based on stable homotopy equivalence. Also, in the presence of translation invariance (associated with a given crystalline structure), the topological K-theory classification of Chapter 3 gets enriched and includes a new type of topological phase called *weak topological insulator (superconductor)*, which is unstable against disorder (i.e., against breaking the translation symmetry). Finally, a warning is warranted: The results based on K-theory are strictly applicable if the number of bands is large enough, i.e., in the so-called *stable regime*. Systems with a low number of bands (i.e., in the nonstable regime) require additional considerations.

To clarify these points, let us consider a translation invariant (crystalline) d-dimensional system described by a Bloch single-particle Hamiltonian $\mathcal{H}(\boldsymbol{k})$ that is symmetric under the elements of the symmetry group G. Note that invariance under a symmetry transformation $g \in G$ implies the existence of a (unitary or anti-unitary) *projective representation*[1] with the property

$$U(g)\,\mathcal{H}\,(\boldsymbol{k})\,U^{\dagger}(g) \quad = \quad \epsilon(g)\,\mathcal{H}(g\boldsymbol{k}), \qquad \text{(unitary)} \qquad (4.1)$$

$$U(g)\,\mathcal{H}^{*}(\boldsymbol{k})\,U^{\dagger}(g) \quad = \quad \epsilon(g)\,\mathcal{H}(g\boldsymbol{k}), \qquad \text{(antiunitary)} \qquad (4.2)$$

where $\epsilon(g) = \pm 1$ and $U(g)$ is a unitary matrix. If $\epsilon(g) = -1$ the element g is sometimes called an *anti-symmetry*. Specific examples of internal (anti-)symmetries and of corresponding (projective) representations have been discussed in the previous chapter. Two Bloch Hamiltonians, $\mathcal{H}_0(\boldsymbol{k})$ and $\mathcal{H}_1(\boldsymbol{k})$ are said to be *homotopy equivalent*, $\mathcal{H}_1(\boldsymbol{k}) \simeq \mathcal{H}_2(\boldsymbol{k})$, if there exists a path $\mathcal{H}(\boldsymbol{k};t)$, $t \in [0,1]$, such that $\mathcal{H}(\boldsymbol{k};0) = \mathcal{H}_0(\boldsymbol{k})$, $\mathcal{H}(\boldsymbol{k};1) = \mathcal{H}_1(\boldsymbol{k})$, and $\mathcal{H}(\boldsymbol{k};t)$ is gapped and symmetric under G for all $t \in [0,1]$. In particular, a Bloch Hamiltonian $\mathcal{H}(\boldsymbol{k})$ is homotopy equivalent to the "simplified" flat band Hamiltonian $Q(\boldsymbol{k})$ defined in Chapter 3 (see Section 3.3.1). This implies that the classification of gapped Bloch Hamiltonians is equivalent to the classification of (noninteracting) many-body ground states viewed as vector bundles over the Brillouin zone torus T^d. Homotopy equivalence provides the finest classification of gapped Bloch Hamiltonians, but the resulting classification does not follow a regular pattern and may involve technically complicated aspects.

Assuming that the addition of trivial valence bands (representing, e.g., strongly bound bands far below the Fermi energy) does not affect the topological properties of the system, we can simplify the problem by considering the concept of *stable* homotopy equivalence [104, 293]. Two Bloch Hamiltonians are said to be *stable* homotopy equivalent, $\mathcal{H}_1(\boldsymbol{k}) \sim \mathcal{H}_2(\boldsymbol{k})$, if $\mathcal{H}_1(\boldsymbol{k}) \oplus Y_1 \simeq \mathcal{H}_2(\boldsymbol{k}) \oplus Y_2$ for certain matrices Y_1 and Y_2 representing topologically trivial objects (e.g., flat bands of strongly bound orbitals). Here, "$\oplus$" (which denotes the direct sum) incorporates mathematically the

---

[1]The physical states of a quantum system correspond to *rays* in a Hilbert space H, i.e., sets of vectors that differ by a (nonzero) complex number. For example, the ray associated with $|\psi\rangle \in H{-}\{0\}$ is $\overline{\psi} = \{|\phi\rangle \in H | \exists \lambda \in \mathbb{C}{-}\{0\} \text{ such that } |\phi\rangle = \lambda|\psi\rangle\}$. A *unit ray* contains vectors of norm 1, which differ by a phase factor, $\lambda = e^{i\alpha}$. The space of (unit) rays is known as the *projective Hilbert space*, $\mathbb{P}(H)$. We can define the *ray product* of $\overline{\psi}$ and $\overline{\phi}$ in terms of the Hilbert space inner product of two representative vectors as $\overline{\psi} \cdot \overline{\phi} = |\langle\psi|\phi\rangle|/||\psi||\,||\phi||$. Note that $(\overline{\psi} \cdot \overline{\phi})^2$ gives the transition probability between the two states (according to the Born rule). A bijective ray transformation $\overline{\psi} \mapsto T\overline{\psi}$ is called a *symmetry transformation* if and only if $T\overline{\psi} \cdot T\overline{\phi} = \overline{\psi} \cdot \overline{\phi}$ for all $\overline{\psi}, \overline{\phi} \in \mathbb{P}(H)$. A transformation $U : H \to H$ of the Hilbert space is said to be *compatible* with the transformation T of the ray space if $U|\psi\rangle \in T\overline{\psi}$ for all $|\psi\rangle \in H{-}\{0\}$. *Wigner's theorem* states that, given a symmetry transformation T, there exists a unitary or antiunitary transformation $V : H \to H$ that is compatible with T. If V_1 and V_2 are both compatible with T, they only differ by a phase factor. Thus, if G is a symmetry group and $f, g \in G$, then the corresponding ray transformations satisfy $T(f)T(g) = T(fg)$, while the compatible representatives satisfy the relation $V(f)V(g) = \omega(f,g)V(fg)$, with $\omega(f,g)$ being a phase factor. If $\omega(f,g) = 1$ for all pairs $f, g \in G$, V is called a *representation*; otherwise, V is called a *projective representation*.

augmentation of the system by trivial bands. Note that the "ten-fold way" classification discussed in the previous chapter can be obtained by constructing the stable homotopy equivalence classes for pairs of (same-size) Hamiltonians, $[\mathcal{H}(\boldsymbol{k}); \mathcal{H}'(\boldsymbol{k})]$, based on the stable equivalence relation

$$[\mathcal{H}_1(\boldsymbol{k}); \mathcal{H}_1'(\boldsymbol{k})] \sim [\mathcal{H}_2(\boldsymbol{k}); \mathcal{H}_2'(\boldsymbol{k})] \quad \text{if} \quad \mathcal{H}_1(k) \oplus \mathcal{H}_2'(k) \sim \mathcal{H}_2(k) \oplus \mathcal{H}_1'(k). \quad (4.3)$$

For example, the stable equivalence in Eq. (4.3) is satisfied by two pairs of Chern insulators with $\mathbb{Z}$ invariants $[\nu_1; \nu_1'] = [2; 0]$ and $[\nu_2; \nu_2'] = [3; 1]$. However, one may assume that the second band structure in each pair, e.g., $\mathcal{H}_1'(\boldsymbol{k})$ and $\mathcal{H}_2'(\boldsymbol{k})$, is trivial, without loss of generality [293]. Indeed, a band structure of the form $H' \oplus (-H')$, which corresponds to a matrix with diagonal blocks $H'(\boldsymbol{k})$ and $-H'(\boldsymbol{k})$, is always trivial [293] and we have $[H, H'] \sim [H \oplus (-H'), H' \oplus (-H')]$. The set of equivalence classes defined by the stable equivalence relation (4.3), $\{[[H(\boldsymbol{k}), H'(\boldsymbol{k})]]\}$, can be promoted to an Abelian group K [104] – the K-theory of the associated vector bundles over the Brillouin zone torus – by defining the addition and subtraction of equivalence classes as

$$[[H_1(\boldsymbol{k}), H_1'(\boldsymbol{k})]] + [[H_2(\boldsymbol{k}), H_2'(\boldsymbol{k})]] := [[H_1(\boldsymbol{k}) \oplus H_2(\boldsymbol{k}), H_1'(\boldsymbol{k}) \oplus H_2'(\boldsymbol{k})]], \quad (4.4)$$

$$[[H_1(\boldsymbol{k}), H_1'(\boldsymbol{k})]] - [[H_2(\boldsymbol{k}), H_2'(\boldsymbol{k})]] := [[H_1(\boldsymbol{k}) \oplus H_2'(\boldsymbol{k}), H_1'(\boldsymbol{k}) \oplus H_2(\boldsymbol{k})]]. \quad (4.5)$$

Note that the sum of two equivalence classes corresponds to an equivalence class involving pairs or augmented band structures. Each stable equivalence class can be identified by a set of topological invariants $\nu_i = \nu_i(\mathcal{H}) - \nu_i(\mathcal{H}')$, with $\nu_i(\mathcal{H})$ being defined on the noncontractible loops, surfaces, etc. of T^d. We emphasize that (i) the topological invariants ν_i characterize the "relative topology" of the band structures $\mathcal{H}$ and $\mathcal{H}'$ and (ii) the results based on this approach are applicable to real systems that have a "sufficiently large" number of bands (i.e., systems in the so-called *stable regime*). In the presence of (only) internal symmetries (e.g., in systems with translation symmetry broken by disorder), the relevant topological invariant, ν_0, is defined on a d-dimensional manifold equivalent to the sphere S^d; the trivial phase can be unambiguously identified, e.g., $\nu_0 = 0$ for all pairs $[\mathcal{H}; \mathcal{H}']$ of atomic insulators, and we obtain the (stable regime) classification discussed in Chapter 3.

Consider now a system with internal *and* translation symmetries. Using K-theory one can show [285, 293] that the topological invariants defined on T^d factorize into products of $\binom{d}{\ell}$ independent invariants defined over the spheres S^ℓ, with ℓ running from 1 to d for symmetry classes A, AI and AII and from 0 to d for the other symmetry classes. For example, consider three-dimensional insulators in class AII [189]. Based on the results in Table 3.3, there exist nontrivial ($\mathbb{Z}_2$) topological invariants defined over S^3 and S^2 (corresponding to $d = 3$ and $d = 2$, respectively), hence, the invariant defined over T^3 has a $\mathbb{Z}_2 \times (\mathbb{Z}_2 \times \mathbb{Z}_2 \times \mathbb{Z}_2)$ structure and can be represented as $(\nu_0; \nu_1, \nu_2, \nu_3)$, with ν_i taking values 0 or 1. Consequently, there are 16 topologically distinct phases: one trivial phase, $(0; 0, 0, 0)$, eight so-called *strong* topological insulators, $(1; \nu_1, \nu_2, \nu_3)$, and seven *weak* topological insulators, $(0; \nu_1, \nu_2, \nu_3)$ with

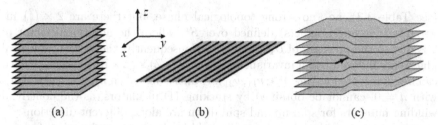

(a)   (b)   (c)

FIGURE 4.1 Weak topological phases adiabatically connected to stacked lower-dimensional (strong) topological systems. (a) Weak 3D integer quantum Hall system obtained by stacking 2D quantum Hall layers in the z direction. The $x - z$ and $y - z$ surfaces support gapless boundary modes, while the $x - y$ (top and bottom) surfaces are gapped. (b) Weak 2D superconductor in the symmetry class BDI obtained by stacking topological wires. The edges parallel to the y axis (i.e., perpendicular to the wires) support gapless modes. (c) Gapless mode along a line dislocation in a 3D quantum Hall system.

at least one ν_i different from zero. If translation invariance is broken (e.g., in the presence of disorder), the strong TIs collapse into a single (topologically nontrivial) phase, while the weak TIs become trivial. In other words, the stability of topological phases associated with $\nu_0 = 0$ and nontrivial invariants over noncontractible manifolds of dimension $\ell < d$ (i.e., the stability of *weak* topological phases) relies on translation invariance. By contrast, ν_0 is robust against perturbations that break the translation symmetry (but do not break internal symmetries and do not close the bulk gap).

One way of constructing weak topological phases is by *stacking* lower dimensional (strong) topological systems. For example, one can obtain a weak 3D integer quantum Hall state by layering two dimensional integer quantum Hall systems, or we can build a 2D weak topological superconductor in class BDI by stacking topological SC wires (see Figure 4.1). Here, we should note that weak topological phases may not support stable gapless modes on all boundaries. For example, the systems sketched in panels (a) and (b) of Figure 4.1 have gapless modes on the $(d-1)$-dimensional boundaries that contain the $(d-2)$-dimensional boundaries of the stacked (strong) topological phases. On the other hand, in weak topological phases dislocations and other topological defects also carry anomalous states [104, 423, 499]. For example, Figure 4.1(c) illustrates the emergence of a chiral fermion mode along a line dislocation in a weak 3D integer quantum Hall system. Finally, we should point out that weak topological phases can be "truly d-dimensional," i.e., not realizable by stacking lower dimensional strong topological systems. As an example [285], consider a two-dimensional $4N$-band insulator in symmetry class AIII having $2N$ occupied bands. We assume that the S_z spin component is conserved, which implies the presence of a U_1 symmetry that commutes with the chiral (anti-)symmetry operator. There is no topological invariant defined over S^2

(see Table 3.3), i.e., no strong topological phase, but there are $2 \times \binom{2}{1}$ invariants (winding numbers) defined over S^1, where the additional factor of 2 is due to the presence of U_1 symmetry. Consequently, the weak topological phases are indexed by an invariant having a $(\mathbb{Z} \times \mathbb{Z}) \times (\mathbb{Z} \times \mathbb{Z})$ structure, which can be represented as $(\nu_{1\uparrow}, \nu_{2\uparrow}; \nu_{1\downarrow}, \nu_{2\downarrow})$. A phase described by, e.g., $(n, 0; 0, n)$, with $n \neq 0$, cannot be obtained by stacking 1D insulators, as the nontrivial winding numbers for spin up and spin down are along different directions.

The above considerations generally hold in the stable regime, i.e., for systems having a sufficiently large number of bands. In this regime strong topological phases can be viewed as "truly d-dimensional," as they are characterized by a nontrivial topological invariant, ν_0, defined over S^d. However, in the nonstable regime one can show [285] that phases having $\nu_0 \neq 0$ can be obtained by stacking lower dimensional systems, unless all other (weak) invariants, $\nu_1, \nu_2, \ldots$, are trivial. Thus, in the nonstable regime stackable phases with $\nu_0 \neq 0$ and at least one $\nu_i \neq 0$ (with $i \geq 1$) should be viewed as *weak*, as they are protected by translation symmetry.

While translation symmetry leads to the emergence of weak topological phases, point group[2] crystalline symmetries can protect additional types of topological phases [190] and lead to new physical complications. One problem concerns the possibility of having mutually distinct product states [191]. Here, a "product state" corresponds to an atomic insulator, i.e., a set of completely occupied, strictly localized orbitals. Such a phase does not possess symmetry–protected quantum entanglement and could be naturally viewed as trivial [75, 84, 410]. However, it is possible to obtain nontrivial K-theoretic topological indexes that describe the "mutual topology" between atomic insulators [311, 410], signaling that the two phases cannot be adiabatically connected without closing the bulk gap, as long as the symmetries of the system are preserved. To address this issue, it is useful to consider the concept of *Wannier functions* – a complete set of orthogonal functions representing the "molecular orbitals" of the crystalline system. Thus, we say that a given band structure is *Wannier-representable* if it can be represented in terms of *exponentially localized* Wannier functions that *preserve all symmetries* [411]. In this case the band structure is defined as *trivial* and the corresponding ground state represents a (trivial) atomic insulator. By contrast, "true" topological phases correspond to bands that are not Wannier-representable, i.e., bands that feature a so-called *Wannier obstruction*.

To gain further intuition, let us consider a simple example involving the Su–Schrieffer–Heeger model [495], initially proposed as a model for polyacetylene. The model describes a 1D lattice with two atoms per unit cell (A and B – see Figure 4.2) and alternating hopping with intra-cell and and inter-cell strengths $-t'$ and $-t$, respectively. For concreteness we assume $t > 0$. The

---

[2] A point group is the group of symmetry operations that do not modify the position of (at least) one point in the crystal. There are 32 crystallographic point groups.

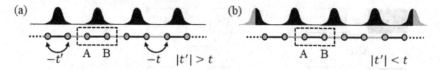

FIGURE 4.2 Su-Schrieffer-Heeger model consisting of a 1D lattice with a two-atom unit cell and nearest-neighbor hopping of alternating strengths. (a) In the trivial state ($|t'| > t$) the Wannier functions are localized near the center of the unit cells. (b) In the "topological phase" ($|t'| < t$) the Wannier functions are localized within the links between unit cells.

corresponding Bloch Hamiltonian has the form

$$H(k) = -\begin{pmatrix} 0 & t' + te^{ik} \\ t' + te^{-ik} & 0 \end{pmatrix} = \sigma_1[m + t(1 - \cos k)] + \sigma_2 t \sin k, \quad (4.6)$$

where σ_1 and σ_2 are Pauli matrices and $t' = -(t + m)$. The model is invariant under time–reversal $\mathcal{T}$ (with $U_t = 1$), particle–hole conjugation $\mathcal{C}$ (with $U_c = \sigma_3$), and chiral (sublattice) symmetry $\mathcal{S}$ (with $U_t U_c^* = \sigma_3$). In addition, the system has inversion symmetry $\mathcal{I}$ (with $U_I = \sigma_1$). Note that Eq. (4.6) can also be viewed as the BdG Hamiltonian of a 1D superconductor with a one-atom unit cell (the "Kitaev chain" [298]). If we enforce $\mathcal{C}$ or $\mathcal{S}$, i.e., if we allow no perturbation that breaks these symmetries, the system is in a topological phase with zero energy end modes (e.g., Majorana zero modes) for $-2t < m < 0$ (i.e., for $|t'| < t$). On the other hand, if $\mathcal{C}$ and $\mathcal{S}$ are broken, but we enforce inversion symmetry, the zero energy modes are no longer protected, but the system has anomalous half-integer "end charges" for $|t'| < t$ [318]. Nonetheless, the corresponding band structure is Wannier-representable both in the trivial phase ($|t'| > t$), when the (symmetry-preserving) Wannier functions are localized near the center of the unit cells, and in the "topological phase" ($|t'| < t$), when the Wannier functions are localized on the inter-cell links (see Figure 4.2). Hence, both phases can be viewed as ("trivial") atomic insulators. However, one cannot continuously change the configuration in Figure 4.2(a) into the configuration in Figure 4.2(b) without breaking the inversion symmetry and, consequently, the two atomic insulators are topologically distinct. In general, we refer to such topologically-distinct atomic-limit insulators as "symmetry obstructed atomic insulators" [75, 410].

Thus, in the presence of crystalline symmetries a nontrivial K-theoretic topological index can describe either (i) the mutual topology between (trivial) atomic insulators with Wannier-representable band structures, or (ii) a true (strong or weak) topological phase. In the second case a Wannier obstruction is necessarily present. However, if a Wannier obstruction is present, it does not necessarily imply a "true" topological phase in the sense of K-theory. More specifically, one can find examples [411] of band structures that are not Wannier-representable (and, consequently, are not *homotopy equivalent*

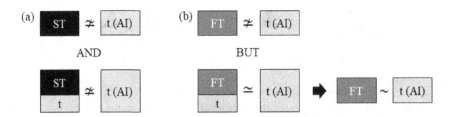

FIGURE 4.3 Stable versus fragile topology. (a) A *stable* topological phase (ST) cannot be adiabatically connected to a trivial (t) atomic insulator (AI) and remains nontrivial upon adding any set of trivial bands (t). (b) A *fragile* topological phase (FT) cannot be adiabatically connected to a trivial atomic insulator, but the obstruction is removed by adding trivial bands. This implies that the system is stable homotopy equivalent ($\sim$) to a trivial insulator.

to an atomic insulator), but can be made Wannier-representable by adding trivial bands (i.e., they are *stable homotopy equivalent* to an atomic insulator). This scenario is illustrated in Figure 4.3. The corresponding topology of the Wannier-obstructed bands, which is weaker than that of stable topological phases, was dubbed "fragile topology." Fragile topological phases are expected to arise in systems with rich enough spatial symmetries.

In summary, gapped band structures with internal symmetries and broken translation symmetry (i.e., in the presence of weak disorder) can be classified into trivial and (strong) topological phases using, e.g., K-theory (i.e., based on the concept of stable homotopy equivalence), as discussed in Chapter 3. In the presence of translation symmetry, each strong topological phase expands into distinct (strong) topological phases (characterized by a nontrivial "strong" invariant and different "weak" invariants) and new weak topological phases (having a trivial "strong" invariant) emerge. These topological phases are not Wannier-representable, i.e., they cannot be adiabatically connected to an atomic-limit insulator, and support topologically protected gapless modes on the whole boundary (in the case of strong topological phases) or on boundaries with certain orientations (weak topological phases). The K-theoretical results are strictly applicable in the stable regime, i.e., when the number of bands is large enough. In the nonstable regime, the "weak" and "strong" invariants may not be independent of each other and more careful considerations (e.g., based on homotopy, rather than stable homotopy) are necessary.

This K-theoretic approach can be generalized in the presence of additional (point group) crystalline symmetries, as discussed in the subsequent sections. A notable difference is that the bulk–boundary correspondence becomes more complex, since the boundary may break the spatial symmetries and, consequently, a nontrivial topological bulk is not necessarily associated with the presence of gapless boundary modes. In addition, in the presence of spatial symmetries there can be atomic-limit (i.e., Wannier-representable) insulators that are topologically-distinct (i.e., stable homotopy nonequivalent) and represent different elements of the K group. These "symmetry obstructed atomic insulators" do not have gapless boundary modes, but they may support anomalous boundary charges. Conversely, not being Wannier-representable (i.e., not being homotopy equivalent to an atomic insulator) does not necessarily imply being topologically nontrivial according to K-theory, as the system may be stable homotopy equivalent to an atomic insulator. The corresponding band structures are said to possess fragile topology.

4.1.2 Crystalline topological phases

Consider a noninteracting gapped system with internal symmetries[3] *and* crystalline symmetries, including point group symmetries (e.g., reflection and rotation). In general, the effect of imposing additional (spatial) symmetries is threefold: (i) The topological classification is modified, being enriched with new (strong) topological phases protected by point group symmetries [24, 104, 121, 473, 485] and additional weak topological phases, whose existence relies on the presence of lattice translation symmetry. (ii) There are new types of topological invariants. (iii) The phenomenology associated with the presence of gapless boundary modes is more complex. To illustrate these points, we will consider two specific examples: topological quantum phases protected by mirror symmetry and topological crystalline insulators with TR and C_4 rotation symmetry. Here, we define the topological phases based on stable homotopy equivalence, in the spirit of K-theory. A classification based on homotopy equivalence, which includes fragile topological phases, can be found in Ref. [85]. Also, we note that determining the topological invariants that identify various topological phases in the presence of crystalline symmetries is a rather complicated task. One approach is based on calculating Wilson loops[4] defined on noncontractible one-dimensional lines (or higher dimensional manifolds) in the Brillouin zone [13, 44, 70, 335, 474]. A practical approach suitable for investigating large sets of materials based on electron band structure calculations uses only the data from a few high symmetry points in the Brillouin zone to determine so-called *symmetry-based indicators* [75, 311, 410, 485].

Before presenting our examples, let us sketch the main ideas underlying the construction of symmetry-based indicators. The technical details of this approach can be found in the references cited above. A fundamental theorem states that at high-symmetry points Λ_j in the Brillouin zone each band (or set of degenerate bands) corresponds to an irreducible representation of the symmetry subgroup under the action of which the momentum is invariant up to a reciprocal lattice vector (i.e., an irreducible representation of the so-called

---

[3]More specifically, with so-called Altland–Zirnbauer (AZ) symmetries, time-reversal (TR), particle-hole (PH), and chiral symmetries.

[4]The Wilson loop is the non-Abelian generalization of the Berry phase factor; see Section 1.4.2 for details.

little group at $\mathbf{\Lambda}_j$). Note that $\mathbf{\Lambda}_j$ is a high-symmetry point if its little group is larger that the little group of any point in the neighborhood. The *symmetry data* of a band structure is an integer vector with elements corresponding to the number of appearances of each irreducible representation associated with valence bands at the high-symmetry points. Furthermore, one can define the addition of two sets of symmetry data corresponding to the superposition of two band structures. For gapped systems, the symmetry data must satisfy certain constraints called *compatibility relations*, which are expressed as a set of linear equations. The symmetry data satisfying these constraints form a linear space called the *band structure space*, $\{BS\}$. We can also define the linear space $\{AI\} \subseteq \{BS\}$ associated with the symmetry data of atomic insulators. Finally, we consider the quotient space $X_{BS} = \{BS\}/\{AI\}$. Note that each element of X_{BS} corresponds to an infinite set of band structures that only differ from each other by the stacking of an atomic insulator. In particular, the linear space $\{AI\}$ collapses into the trivial element of X_{BS}, while the nontrivial elements correspond to band structures that cannot be reduced to atomic insulators by adding (trivial) AI bands. Hence, X_{BS} (or its generators) are symmetry-based indicators of band topology. We emphasize that any two gapped band structures having different sets of symmetry indicators are topologically distinct, while any two band structures having different symmetry data but the same set of symmetry indicators differ by an atomic insulator band structure.

Our first example addresses the problem of classifying crystalline topological phases with AZ and mirror symmetries. The internal (AZ) symmetries are represented by the following (anti)unitary matrices: $\mathcal{T} = U_t K$ – time-reversal (TR) symmetry; $\mathcal{C} = U_c K$ – charge-conjugation or particle-hole (PH) symmetry; $\mathcal{S} = \mathcal{T} \cdot \mathcal{C}$ – chiral or sublattice (SL) symmetry. The properties of these matrices were discussed in the previous chapter (see Sections 3.2.1 and 3.2.2). In addition, we assume that the system possesses mirror (or reflection) symmetry, M, which is a spatial symmetry. We note that, in general, spatial symmetry operations form a group called the *space group* and consist of *translations* and *point group* operations (which leave at least one point in space unchanged). Examples of point group symmetry operations include inversion, reflections, and n-fold rotations. Note that only rotations with $n = 2, 3, 4$ and 6 are consistent with the translation symmetry of a crystal. A mirror (reflection) symmetry in the direction x_ℓ transforms the position vector $\boldsymbol{r} = (x_1, \ldots, x_\ell, \ldots, x_d)$ into $\tilde{\boldsymbol{r}}_\ell = (x_1, \ldots, -x_\ell, \ldots, x_d)$ and the momentum $\boldsymbol{k} = (k_1, \ldots, k_\ell, \ldots, k_d)$ into $\tilde{\boldsymbol{k}}_\ell = (k_1, \ldots, -k_\ell, \ldots, k_d)$. The action of the reflection symmetry on the Bloch Hamiltonian is given by the unitary matrix $U_{M_\ell}(\boldsymbol{k})$ as

$$U_{M_\ell} \mathcal{H}(\boldsymbol{k}) U_{M_\ell}^\dagger = \mathcal{H}(\tilde{\boldsymbol{k}}_\ell). \tag{4.7}$$

Note that this is a specific case of Eq. 4.1 corresponding to $g = \mathcal{M}_\ell$. Also note that the reflection $\mathcal{M}_\ell$ can be viewed as the product of spatial inversion and a twofold rotation around the x_ℓ-axis (i.e., the axis perpendicular to the plane of reflection). Thus, for spinless particles two consecutive reflections imply no change, i.e., $U_{M_\ell}^2 = 1$, while for spin-1/2 systems this corresponds to a spin rotation by 2π, which involves a phase factor of -1, i.e., $U_{M_\ell}^2 = -1$. More specifically, since spin is invariant under inversion, the spin structure of the reflection matrix for spin-1/2 particles corresponds to a rotation by π around the x_ℓ-axis, $U_{M_\ell} = \exp\left(i\frac{\pi}{2}\sigma_\ell\right) = i\sigma_\ell$, where σ_ℓ are Pauli matrices, and the relation $U_{M_\ell}^2 = -1$ is manifestly satisfied. The eigenvalues of U_{M_ℓ} are either $+i$ or $-i$ (for a spinless system the corresponding eigenvalues are $+1$ or -1). We can already anticipate the emergence of new topological phases protected by mirror symmetry. For a two dimensional crystal with mirror symmetry, for example, we can choose $|u_{k\eta}\rangle$, with $\eta = \pm$, to be eigenstates of $\mathcal{H}(k)$ and $U_{M_\ell}(k)$ for all k and we can define Chern numbers ν_η corresponding to each class of states with a given mirror eigenvalue, η. The total Chern number, $\nu = \nu_+ + \nu_-$, will determine the quantized Hall conductance, as discussed in the previous chapter. However, even if $\nu = 0$, we can define a new invariant called mirror Chern number, $\nu_M = (\nu_+ - \nu_-)/2$, which can be a nonzero integer characterizing a topological crystalline phase protected by mirror symmetry.

We have established that reflection symmetry (in the x_ℓ direction) can be represented by a unitary matrix U_{M_ℓ} acting on the Bloch Hamiltonian as given by Eq. (4.7) and satisfying $U_{M_\ell}^2 = 1$ (for spinless systems) or $U_{M_\ell}^2 = -1$ (for spin-1/2 particles). However, it is convenient to choose a representation of the reflection matrix that squares to 1 (i.e., the identity); for example, we can define $R_\ell = U_{M_\ell}$ for spinless particles and $R_\ell = -iU_{M_\ell}$ for spin-1/2 fermions. While Eq. (4.7) is not affected by this phase choice, the commutation relation of the reflection matrix with the internal (AZ) symmetry matrices is. More specifically, while the physical reflection operator always commutes with internal symmetries, with this phase choice we have

$$\mathcal{T}R_\ell = \eta_t R_\ell \mathcal{T}, \quad \mathcal{C}R_\ell = \eta_c R_\ell \mathcal{C}, \quad \mathcal{S}R_\ell = \eta_s R_\ell \mathcal{S}, \tag{4.8}$$

where $\eta_{t,c,s} = \pm 1$. Thus, the ten-fold symmetry classification of single-particle Hamiltonians (see Table 3.1) gets enriched in the presence of mirror symmetry with sub-classes determined by the commutation relations between R_ℓ and the internal symmetries. For example, systems in the complex class A have no internal symmetries, hence in this class there is only one possible type ($t = 0$) of reflection symmetry (which we denote R). By contrast, in the complex symmetry class AIII we have two types of reflection symmetries: $t = 0$, which corresponds to R_ℓ commuting with the sublattice symmetry (i.e., $\eta_s = +1$) and is denoted R_+, and $t = 1$, which corresponds to R_ℓ anticommuting with the sublattice symmetry (i.e., $\eta_s = -1$) and is denoted R_-. Similarly, each real symmetry class with one internal symmetry (TR or PH), i.e., class AI, AII, C, or D, supports two types of reflection symmetry denoted R_+ and R_-, which correspond to $\eta_{t(c)} = +1$ and $\eta_{t(c)} = -1$, respectively. Finally, each of the real symmetry classes with both TR and PH symmetries, i.e., class BDI, CI, CII, or DIII, supports four types of reflection symmetries denoted R_{++}, R_{+-}, R_{-+}, and R_{--}, which correspond to the four possible values of (η_t, η_c). We conclude that in the presence of AZ and reflection symmetries there are 27 different symmetry classes [105, 371, 473] (see Table 4.1). The possible types of reflection symmetries are labeled $t = 0, 1$ for the complex AZ classes and $t = 0, 1, 2, 3$ for the real AZ classes. We note that this type of symmetry classification can be realized for any unitary or anti-unitary order-two spatial symmetry S (e.g., inversion and two-fold rotation) having the property $[S^2, \mathcal{H}(\boldsymbol{k})] = 0$ [473].

The next step is to establish the classification of topological crystalline insulators and superconductors corresponding to each symmetry class. We will use a generalization of the K-theory approach developed in the context of single-particle Hamiltonians in the ten AZ symmetry classes (see Section 3.3). We first note that the classification of topological insulators and superconductors in AZ symmetry classes can be summarized by the following relations:

$$K_{\mathbb{C}}(s; d) = K_{\mathbb{C}}(s - d; 0) = \pi_0(\mathcal{C}_{s-d}), \qquad s = 0, 1 \ (\text{mod } 2) \tag{4.9}$$

$$K_{\mathbb{R}}(s; d) = K_{\mathbb{R}}(s - d; 0) = \pi_0(\mathcal{R}_{s-d}), \qquad s = 0, 1, \ldots, 7 \ (\text{mod } 8) \tag{4.10}$$

where $K_{\mathbb{C}}(s; d)$ and $K_{\mathbb{R}}(s; d)$ denote the K-groups of maps from $\boldsymbol{k} \in S^d$ to the complex $(\mathcal{C}_s)$ and real $(\mathcal{R}_s)$ classifying spaces of the s symmetry class, respectively. The classifying spaces are explicitly given in Table 3.2. The first equality in Eqs. (4.9) and (4.10) establishes the relation between the topological properties of a d-dimensional system and those of a 0-dimensional system, while the second equality explicitly gives the topological classification in zero dimensions for complex symmetry class $s - d$ (mod 2) and real symmetry class $s - d$ (mod 8). Together, these relations generate the classification shown explicitly in the "Periodic Table" 3.3. In the presence of an additional reflection symmetry, R, the K-groups (denoted $K_{\mathbb{C}}^R$ and $K_{\mathbb{R}}^R$) also depend on the type t of mirror symmetry characterizing each (complex or real) symmetry class. It has been shown [473] that the topological properties of a d-dimensional system can be obtained from those in 0-dimension using the relations

$$K_{\mathbb{C}}^R(s, t; d) = K_{\mathbb{C}}^R(s - d, t - 1; 0), \quad s, t = 0, 1 \ (\text{mod } 2) \tag{4.11}$$

$$K_{\mathbb{R}}^R(s, t; d) = K_{\mathbb{R}}^R(s - d, t - 1; 0). \quad s = 0, \ldots, 7 \ (\text{mod } 8);$$

$$t = 0, \ldots, 3 \ (\text{mod } 4) \tag{4.12}$$

Furthermore, after identifying the classifying spaces of the 0-dimensional K-group one obtains [473]

$$K_{\mathbb{C}}^R(s, t = 0; 0) = \pi_0(\mathcal{C}_s \times \mathcal{C}_s) = \pi_0(\mathcal{C}_s) \oplus \pi_0(\mathcal{C}_s), \tag{4.13}$$

$$K_{\mathbb{C}}^R(s, t = 1; 0) = \pi_0(\mathcal{C}_{s+1}), \tag{4.14}$$

$$K_{\mathbb{R}}^R(s, t = 0; 0) = \pi_0(\mathcal{R}_s \times \mathcal{R}_s) = \pi_0(\mathcal{R}_s) \oplus \pi_0(\mathcal{R}_s), \tag{4.15}$$

$$K_{\mathbb{R}}^R(s, t = 1; 0) = \pi_0(\mathcal{R}_{s-1}), \tag{4.16}$$

$$K_{\mathbb{R}}^R(s, t = 2; 0) = \pi_0(\mathcal{C}_s), \tag{4.17}$$

$$K_{\mathbb{R}}^R(s, t = 3; 0) = \pi_0(\mathcal{R}_{s+1}), \tag{4.18}$$

where $\mathcal{C}_s$ [$s = 0, 1 \ (\text{mod } 2)$] and $\mathcal{R}_s$ [$s = 0, 1, \ldots, 7 \ (\text{mod } 8)$] represent the classifying spaces of the complex and real AZ classes, respectively. The system of equations (4.11-4.18) gives the topological classification of the 27 symmetry classes of crystalline insulators and superconductors with reflection symmetry in arbitrary spatial dimension d. Let us examine a few example. Consider first a d-dimensional system in symmetry class A with reflection symmetry R (i.e., the complex symmetry class with $s = 0$, $t = 0$). According to Eqs. (4.11) and (4.14), we have $K_{\mathbb{C}}^R(0,0;d) = K_{\mathbb{C}}^R(d \,(\text{mod } 2),1;0) = \pi_0(\mathcal{C}_{d+1 \,(\text{mod } 2)})$. Based on the results summarized in Table 3.3, we have $\pi_0(\mathcal{C}_0) = \mathbb{Z}$ and $\pi_0(\mathcal{C}_1) = 0$. Thus, we conclude that a system with only reflection symmetry (i.e., in symmetry class $A + R$) supports reflection-symmetry-protected topological phases classified by a $\mathbb{Z}$ invariant in odd spatial dimensions, while being topologically trivial in even spatial dimensions. Consider now a system in the real symmetry class AII + R_+ ($s = 4$, $t = 0$). Using Eqs. (4.12) and (4.18) we have $K_{\mathbb{R}}^R(4,0;d) = K_{\mathbb{R}}^R(4 - d \,(\text{mod } 8),3;0) = \pi_0(\mathcal{R}_{5-d \,(\text{mod } 8)})$. In other words, a d-dimensional crystalline system with reflection symmetry in symmetry class AII–R_+ supports topological phases with the same classification as those corresponding to a 0-dimensional system with AZ symmetries in real symmetry class $s = 5 - d \,(\text{mod } 8)$. Again, using the results in Table 3.3 (from the column $d = 0$), we conclude that the system is topologically trivial in spatial dimensions $d = 0, 2, 6, 7$ and supports topological phases classified by a $\mathbb{Z}$ invariant in $d = 1, 5$ and by a $\mathbb{Z}_2$ invariant in $d = 3, 4$. The analysis of the topological classification of the remaining symmetry classes proceeds along similar lines. The results are summarized in Table 4.1. The first column indicates the Cartan label (CL) of the AZ symmetry class, the second column specifies the type of mirror symmetry (MS) characterizing the system, while the third column summarizes this information in terms of the corresponding indices s and t, with $s, t = 0, 1 \,(\text{mod } 2)$ for the complex symmetry classes (top three lines) and $s = 0, 1, \ldots, 7 \,(\text{mod } 8)$, $t = 0, 1, \ldots, 3 \,(\text{mod } 4)$ for the real symmetry classes (next 24 lines). Columns labeled $0, 1, \ldots, 7$ list the topologically-nontrivial reflection-symmetry-protected crystalline phases in $d \,(\text{mod } 8)$ spatial dimensions and specify the corresponding topological invariants. Finally, we note that the classification summarized in Table 4.1 can be generalized to systems with generic order-two symmetry in the presence of topological defects [473].

Our next question concerns the nature of the topological invariants that label the nontrivial topological phases supported by the crystalline system with mirror symmetry. In general, we distinguish four types of invariants: (i) Chern or winding number $\mathbb{Z}$ invariants and $\mathbb{Z}_2$ invariants of the original AZ classification (marked $\mathbb{Z}$ and $\mathbb{Z}_2$ in Table 4.1); (ii) mirror Chern or winding numbers ($M\mathbb{Z}$) and mirror $\mathbb{Z}_2$ invariants ($M\mathbb{Z}_2$); (iii) translation-protected $\mathbb{Z}_2$ invariants ($T\mathbb{Z}_2$); (iv) combined mirror and original AZ invariants ($M\mathbb{Z} \oplus \mathbb{Z}$ and $M\mathbb{Z}_2 \oplus \mathbb{Z}_2$). Note that in Table 4.1 we use the compact notation $M\mathbb{Z} \oplus \mathbb{Z} \rightarrow M\mathbb{Z} \oplus$, (or $2M\mathbb{Z} \oplus 2\mathbb{Z} \rightarrow 2M\mathbb{Z} \oplus$) and $M\mathbb{Z}_2 \oplus \mathbb{Z}_2 \rightarrow M\mathbb{Z}_2 \oplus$. We remind the reader that the factors of 2 (e.g., $2\mathbb{Z}$ and $2M\mathbb{Z}$) indicate invariants that only take even integer values and describe topological phases that support an even number of boundary modes. Below we briefly describe each type of topological

TABLE 4.1 Classification of mirror-symmetric topological crystalline phases.

CL	MS	(s,t)	0	1	2	3	4	5	6	7
A	R	(0,0)	0	$M\mathbb{Z}$	0	$M\mathbb{Z}$	0	$M\mathbb{Z}$	0	$M\mathbb{Z}$
AIII	R_+	(1,0)	$M\mathbb{Z}$	0	$M\mathbb{Z}$	0	$M\mathbb{Z}$	0	$M\mathbb{Z}$	0
AIII	R_-	(1,1)	0	$M\mathbb{Z}\oplus$	0	$M\mathbb{Z}\oplus$	0	$M\mathbb{Z}\oplus$	0	$M\mathbb{Z}\oplus$
AI	R_+	(0,0)	$M\mathbb{Z}_2$	$M\mathbb{Z}$	0	0	0	$2M\mathbb{Z}$	0	$M\mathbb{Z}_2$
BDI	R_{++}	(1,0)	$M\mathbb{Z}_2$	$M\mathbb{Z}_2$	$M\mathbb{Z}$	0	0	0	$2M\mathbb{Z}$	0
D	R_+	(2,0)	0	$M\mathbb{Z}_2$	$M\mathbb{Z}_2$	$M\mathbb{Z}$	0	0	0	$2M\mathbb{Z}$
DIII	R_{++}	(3,0)	$2M\mathbb{Z}$	0	$M\mathbb{Z}_2$	$M\mathbb{Z}_2$	$M\mathbb{Z}$	0	0	0
AII	R_+	(4,0)	0	$2M\mathbb{Z}$	0	$M\mathbb{Z}_2$	$M\mathbb{Z}_2$	$M\mathbb{Z}$	0	0
CII	R_{++}	(5,0)	0	0	$2M\mathbb{Z}$	0	$M\mathbb{Z}_2$	$M\mathbb{Z}_2$	$M\mathbb{Z}$	0
C	R_+	(6,0)	0	0	0	$2M\mathbb{Z}$	0	$M\mathbb{Z}_2$	$M\mathbb{Z}_2$	$M\mathbb{Z}$
CI	R_{++}	(7,0)	$M\mathbb{Z}$	0	0	0	$2M\mathbb{Z}$	0	$M\mathbb{Z}_2$	$M\mathbb{Z}_2$
BDI	R_{+-}	(1,1)	$M\mathbb{Z}_2\oplus$	$M\mathbb{Z}\oplus$	0	0	0	$2M\mathbb{Z}\oplus$	0	$M\mathbb{Z}_2\oplus$
DIII	R_{-+}	(3,1)	0	$M\mathbb{Z}_2\oplus$	$M\mathbb{Z}_2\oplus$	$M\mathbb{Z}\oplus$	0	0	0	$2M\mathbb{Z}\oplus$
CII	R_{+-}	(5,1)	0	$2M\mathbb{Z}\oplus$	0	$M\mathbb{Z}_2\oplus$	$M\mathbb{Z}_2\oplus$	$M\mathbb{Z}\oplus$	0	0
CI	R_{-+}	(7,1)	0	0	0	$2M\mathbb{Z}\oplus$	0	$M\mathbb{Z}_2\oplus$	$M\mathbb{Z}_2\oplus$	$M\mathbb{Z}\oplus$
AI	R_-	(0,2)	0	0	0	$2M\mathbb{Z}$	0	$T\mathbb{Z}_2$	$\mathbb{Z}_2$	$M\mathbb{Z}$
BDI	R_{--}	(1,2)	$M\mathbb{Z}$	0	0	0	$2M\mathbb{Z}$	0	$T\mathbb{Z}_2$	$\mathbb{Z}_2$
D	R_-	(2,2)	$\mathbb{Z}_2$	$M\mathbb{Z}$	0	0	0	$2M\mathbb{Z}$	0	$T\mathbb{Z}_2$
DIII	R_{--}	(3,2)	$T\mathbb{Z}_2$	$\mathbb{Z}_2$	$M\mathbb{Z}$	0	0	0	$2M\mathbb{Z}$	0
AII	R_-	(4,2)	0	$T\mathbb{Z}_2$	$\mathbb{Z}_2$	$M\mathbb{Z}$	0	0	0	$2M\mathbb{Z}$
CII	R_{--}	(5,2)	$2M\mathbb{Z}$	0	$T\mathbb{Z}_2$	$\mathbb{Z}_2$	$M\mathbb{Z}$	0	0	0
C	R_-	(6,2)	0	$2M\mathbb{Z}$	0	$T\mathbb{Z}_2$	$\mathbb{Z}_2$	$M\mathbb{Z}$	0	0
CI	R_{--}	(7,2)	0	0	$2M\mathbb{Z}$	0	$T\mathbb{Z}_2$	$\mathbb{Z}_2$	$M\mathbb{Z}$	0
BDI	R_{-+}	(1,3)	0	$2\mathbb{Z}$	0	$2M\mathbb{Z}$	0	$2\mathbb{Z}$	0	$2M\mathbb{Z}$
DIII	R_{+-}	(3,3)	0	$2M\mathbb{Z}$	0	$2\mathbb{Z}$	0	$2M\mathbb{Z}$	0	$2\mathbb{Z}$
CII	R_{-+}	(5,3)	0	$2\mathbb{Z}$	0	$2M\mathbb{Z}$	0	$2\mathbb{Z}$	0	$2M\mathbb{Z}$
CI	R_{+-}	(7,3)	0	$2M\mathbb{Z}$	0	$2\mathbb{Z}$	0	$2M\mathbb{Z}$	0	$2\mathbb{Z}$

invariant, as well as the boundary modes supported by the corresponding topological phases.

(i) *Original AZ invariants* ($\mathbb{Z}$ *and* $\mathbb{Z}_2$). These integer (Chern or winding number) and $\mathbb{Z}_2$ invariants, which were described in Chapter 3, characterize certain topological phases in real symmetry classes with at least one AZ symmetry that anticommutes with R, more specifically symmetry classes with type $t = 2$ and $t = 3$ reflection (see Table 4.1). The corresponding topological phases, which are AZ phases that "survive" in the presence of mirror symmetry (see, for comparison, Table 3.3), support (topologically-protected) gapless modes on all boundaries, regardless of their orientation.

(ii) *Mirror invariants* ($M\mathbb{Z}$ *and* $M\mathbb{Z}_2$). These topological invariants are defined on the hyperplanes of the Brillouin zone that are invariant under reflection. Assume, for concreteness, that the system has mirror symmetry in the x_ℓ direction and that the hyperplanes $k_\ell = 0$ and $k_\ell = \pi$ are reflection-invariant, i.e., $\tilde{k}_\ell = k$ for $k_\ell = 0, \pi$. According to Eq. (4.7), the Hamiltonian $\mathcal{H}(k)|_{k_\ell=0,\pi}$ commutes with U_{M_ℓ} (and with R_ℓ) and can be block diagonalized with respect to the eigenvalues $\lambda_R = \pm 1$ of R. Furthermore, each of the two blocks can be characterized by an integer or a $\mathbb{Z}_2$ invariant of the original AZ classification in dimension $d - 1$. For example, consider that the block with reflection eigenvalue λ_R is characterized at $k_\ell = 0, \pi$ by a Chern or winding number $\nu_{k_\ell}^{\lambda_R}$. Also, let us assume that the weak topological index vanishes, which implies that the system cannot be understood as a stack of $d - 1$ dimensional topologically nontrivial layers and requires $\nu_{k_\ell}^+ + \nu_{k_\ell}^- = 0$. Then, we can define the integer mirror invariant as [105]

$$\nu_{MZ} = \mathrm{sgn}(\nu_0^+ - \nu_\pi^+) \left(|\nu_0^+| - |\nu_\pi^+|\right). \tag{4.19}$$

Similarly, if the $\lambda_R = \pm 1$ Hamiltonian block is characterized at $k_\ell = 0, \pi$ by a $\mathbb{Z}_2$ invariant $n_{k_\ell}^{\lambda_R} = 0, 1$ and we assume vanishing weak topological index, $n_{k_\ell}^+ + n_{k_\ell}^- \,(\mathrm{mod}\,2) = 0$, one can define the mirror $\mathbb{Z}_2$ invariant of the system as [105]

$$n_{MZ_2} = n_0^+ + n_\pi^+ \,(\mathrm{mod}\,2). \tag{4.20}$$

The structure of the invariant reflects the fact that a process like $(n_0^+, n_\pi^+) = (0, 0) \longrightarrow (n_0^+, n_\pi^+) = (1, 0)$ involves the closing of the bulk band gap at $k_\ell = 0$ (hence, a topological quantum phase transition), while the process $(0, 0) \longrightarrow (1, 1)$ does not imply the closing of gap at $k_\ell = 0$ and $k_\ell = \pi$ as the corresponding pair of bulk bands can be coupled and gapped out by a symmetry-preserving perturbation, e.g., a density wave (hence, it does not involve a phase transition).

We note that for symmetry classes with R_+ and R_{++} mirror symmetry the internal symmetries (which commute with R) can be block diagonalized simultaneously with the Hamiltonian and, consequently, the $\lambda_R = \pm 1$ blocks can be viewed as $d-1$ dimensional systems with the same internal symmetries as the original (d dimensional) system. In other words, all mirror invariants in symmetry classes with $t = 0$ and dimension d correspond to invariants of the original AZ classification in dimension $d - 1$ (see Table 3.3). For symmetry classes with R_- or R_{--} mirror symmetry, on the other hand, neither time-reversal nor particle-hole symmetry is block-diagonal simultaneously with the Hamiltonian, hence the $\lambda_R = \pm 1$ blocks do not possess these internal symmetries. However, if d is even a system supporting topological phases characterized by an $M\mathbb{Z}$ invariant has chiral symmetry that commutes with R and can be block-diagonalized. Hence, the $\lambda_R = \pm 1$ Hamiltonian blocks possess chiral symmetry and can be viewed as $d - 1$ dimensional systems in symmetry class AIII. The corresponding winding numbers can be used to evaluate ν_{MZ} based on Eq. (4.19). Similarly, if d is odd the $\lambda_R = \pm 1$ Hamiltonian blocks belong to symmetry class A (in $d - 1$ dimensions). In this case, the $M\mathbb{Z}$ invariant is evaluated using the corresponding Chern numbers and Eq. (4.19). Based

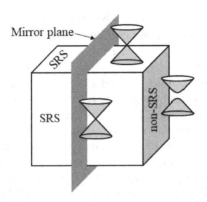

FIGURE 4.4 Protected gapless boundary modes (schematically represented by gapless Dirac cones) for a topological crystalline insulator or superconductor with mirror symmetry. Only the self-reflected surfaces (SRSs) support gapless modes, while the non-self-reflected surfaces are, in general, gapped.

on these considerations, it is clear that in symmetry classes with R_- or R_{--} mirror symmetry one cannot define $\mathbb{Z}_2$ topological numbers for the $\lambda_R = \pm 1$ Hamiltonian blocks, i.e., there are no $M\mathbb{Z}_2$ topological phases. However, some of the $\mathbb{Z}_2$ phases of the original AZ classification "survive" in the presence of mirror symmetry (see Table 4.1).

Finally, we point out that nontrivial values of the integer or $\mathbb{Z}_2$ mirror invariants indicate the presence of protected gapless boundary modes on reflection-symmetric boundaries, i.e., on surfaces or edges that are perpendicular to the reflection hyperplane $x_\ell = 0$. By contrast, boundaries that break the spatial symmetry are, in general, gapped. This behavior, is illustrated schematically in Figure 4.4. We emphasize that, in general, topological crystalline insulators and superconductors support protected boundary modes only on the surfaces that are left invariant by the crystal symmetries. The absence of gapless modes at boundaries that break spatial symmetries does not indicate a trivial bulk topology. Furthermore, while the boundary modes of ordinary topological insulators and superconductors are robust against spatial disorder, the delocalized surface states of topological crystalline systems are protected by spatial symmetries that are typically broken by disorder. However, if disorder preserves the spatial symmetries *on average*, the boundary modes may evade Anderson localization [142, 372]. Examples of 3D topological crystalline insulators in symmetry classes A+R and AII+R_- characterized by nonzero mirror Chern numbers include SnTe, $Pb_{1-x}Sn_x$, and $Pb_{1-x}Sn_x$Te [150, 255, 498].

(iii) *Translation-protected $\mathbb{Z}_2$ invariants ($T\mathbb{Z}_2$).* In the presence of type $t = 2$ reflection symmetry half of the $\mathbb{Z}_2$ topological phases, which are labeled by $T\mathbb{Z}_2$ invariants (see Table 4.1), are stable only in the presence of translation symmetry [105]. The technical details associated with the definition of

$T\mathbb{Z}_2$ invariants can be found in Ref. [473]. In systems with $T\mathbb{Z}_2$ topological phases, the gapless boundary modes are protected by the internal symmetries, reflection, and translation and can be gapped by reflection-preserving perturbations that break translation symmetry (e.g., density wave perturbations) [105]. However, in the presence of random perturbations that are spatially uniform on average the surface gapless states remain critical [371].

(iv) *Composite invariants* ($M\mathbb{Z}\oplus\mathbb{Z}$ and $M\mathbb{Z}_2\oplus\mathbb{Z}_2$). System with chiral symmetry and type $t = 1$ mirror symmetry support topological phases described by combinations of a $\mathbb{Z}$ (or $\mathbb{Z}_2$) invariant of the original AZ classification and a mirror invariant $M\mathbb{Z}$ (or $M\mathbb{Z}_2$), the two indices being independent of one another. Thus, the $\mathbb{Z}$ ($\mathbb{Z}_2$) topological classification of the chiral AZ symmetry classes expands into additional sub-classes labeled by the mirror invariant $M\mathbb{Z}$ ($M\mathbb{Z}_2$). For even (odd) spatial dimension d the $\mathbb{Z}$ index is the Chern number (winding number), while $M\mathbb{Z}$ is the mirror winding number (mirror Chern number). The topological phases characterized by this type of invariant support a number of protected gapless states given by $\max\{|\nu_{MZ}|, |\nu_z|\}$ at boundaries that are perpendicular to the mirror plane [105].

Our second example consists of a three-dimensional topological crystalline insulator with four-fold (C_4) rotational symmetry. To gain some intuition, it is useful to consider a specific, simple model and explicitly calculate the boundary modes. Following Ref. [190], we consider a system of spinless fermions occupying p_x and p_y (or d_{xz} and d_{yz}) orbitals on a tetragonal lattice with a unit cell with two atoms, A and B, along the c-axis [see Figure 4.5(a)]. The Bloch Hamiltonian consists of intra-layer blocks, $H^A(\boldsymbol{k})$ and $H^B(\boldsymbol{k})$, and an inter-layer coupling $H^{AB}(\boldsymbol{k})$:

$$H(\boldsymbol{k}) = \begin{pmatrix} H^A & H^{AB} \\ (H^{AB})^\dagger & H^B \end{pmatrix}. \tag{4.21}$$

By taking into account the symmetry of the system and including only nearest and next-nearest neighbor intra-layer and inter-layer hoppings we have

$$H^\alpha(\boldsymbol{k}) = 2t_1^\alpha \begin{pmatrix} \cos k_x & 0 \\ 0 & \cos k_y \end{pmatrix} + 2t_2^\alpha \begin{pmatrix} \cos k_x \cos k_y & \sin k_x \sin k_y \\ \sin k_x \sin k_y & \cos k_x \cos k_y \end{pmatrix}, \tag{4.22}$$

$$H^{AB}(\boldsymbol{k}) = \left[t_1' + 2t_2'(\cos k_x + \cos k_y) + t_z' e^{ik_z} \right] I, \tag{4.23}$$

where $\alpha = A, B$ and I is the 2×2 identity matrix corresponding to the two-dimensional subspace associated with the p_x, p_y orbitals.

The system has C_4 rotation symmetry around the z axis [i.e., the c axis in Figure 4.5 (a)], which is implemented by the unitary operator $U_{C_4} = \exp(-i\hat{L}_z/2\pi)$, with $\hat{L}_z$ being the angular momentum operator, and time-reversal symmetry (for spinless fermions), which is implemented by the anti-unitary operator $\mathcal{T} = K$, with K being the complex conjugation (see Section 3.2.1). The action of these symmetries on the Bloch Hamiltonian is given by

$$\mathcal{T} H(k_x, k_y, k_z) \mathcal{T}^{-1} = H(-k_x, -k_y, -k_z), \tag{4.24}$$

$$U_{C_4} H(k_x, k_y, k_z) U_{C_4}^{-1} = H(-k_y, k_x, k_z). \tag{4.25}$$

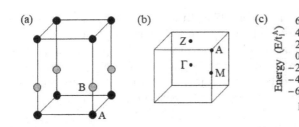

FIGURE 4.5 (a) Unit cell of the tetragonal lattice with two atoms, A and B. (b) Brillouin zone and the four high symmetry points. (c) Bulk band structure along lines between high symmetry points.

Note that $\mathcal{T}^2 = 1$, so time-reversal symmetry does not result in a two-fold degeneracy (i.e., Kramers theorem does not apply). However, for $\tilde{\boldsymbol{k}} = (k_x, k_y, k_z)$ with $k_x = k_y = 0, \pi$ (i.e., along the ΓZ and MA directions in the Brillouin zone) we have $-k_x = k_x = k_y = -k_y$ and, according to Eq. (4.25), the Bloch Hamiltonian commutes with U_{C_4}. Moreover, at the high symmetry points $\Gamma = (0,0,0)$, $Z = (0,0,\pi)$, $M = (\pi,\pi,0)$ and $A = (\pi,\pi,\pi)$ the Hamiltonian also commutes with $\mathcal{T}$ [see Eq. (4.24)]. Thus, the eigenstates of $H(\tilde{\boldsymbol{k}})$ are also eigenstates of U_{C_4} with possible eigenvalues 1, -1, i, and $-i$, since $U_{C_4}^4 = 1$. Next, we assume that the relevant bands belong to the subspace spanned by the eigenstates with $\pm i$ eigenvalues and show that these bands are double degenerate. The model that we consider here, which is based on p_x and p_y (or d_{xz} and d_{yz}) orbitals, is explicitly within that subspace. The physical assumption is that (single) bands associated with, e.g., p_z orbitals do not cross with the "relevant" (doublet) bands.

A direct consequence of U_{C_4} having $\pm i$ eigenvalues is that $U_{C_4}^2 = -1$. Also, since $\mathcal{T}\hat{L}_z\mathcal{T}^{-1} = -\hat{L}_z$, we have $\mathcal{T}U_{C_4}\mathcal{T}^{-1} = U_{C_4}$ and, consequently, $(U_{C_4}\mathcal{T})^2 = -1$. In addition, at the high symmetry points ($\boldsymbol{k} = \Lambda_i = \Gamma, Z, M, A$) the Bloch Hamiltonian $H(\Lambda_i)$ is real and commutes with both U_{C_4} and $\mathcal{T}$. These symmetry properties protect the double degeneracy of the bands for $\boldsymbol{k} = \tilde{\boldsymbol{k}}$, in particular at the high symmetry points. Indeed, for $\boldsymbol{k} = \tilde{\boldsymbol{k}}$ the Hamiltonian $H(\tilde{\boldsymbol{k}})$ consists of two decoupled identical blocks, one involving the p_x orbitals and the other the p_y orbitals of the A and B atoms. Basically, a $\pi/2$ rotation transforms the blocks into one another (but leaves the Hamiltonian unchanged, since the blocks are identical). If we re-order the orbital basis as $(p_x^A, p_x^B, p_y^A, p_y^B)$, the tight-binding Hamiltonian becomes block-diagonal, with blocks of the form

$$h(\tilde{\boldsymbol{k}}) = \begin{pmatrix} 2(t_2^A \pm t_1^A) & t_1' \pm 4t_2' + t_z' e^{ik_z} \\ t_1' \pm 4t_2' + t_z' e^{ik_z} & 2(t_2^B \pm t_1^B) \end{pmatrix}, \qquad (4.26)$$

where the "+" and "−" signs correspond to $k_x = k_y = 0$ and $k_x = k_y = \pi$, respectively. Let $|v_n(\widetilde{\boldsymbol{k}})\rangle$ be an eigenstate of $h(\widetilde{\boldsymbol{k}})$ with eigenvalue $E_n(\widetilde{\boldsymbol{k}})$. The full Hamiltonian is characterized by a double degenerate eigenvalue $E_n(\widetilde{\boldsymbol{k}})$ corresponding to the eigenstates $|u_{n\pm}(\widetilde{\boldsymbol{k}})\rangle = (|v_n(\widetilde{\boldsymbol{k}})\rangle, \pm i |v_n(\widetilde{\boldsymbol{k}})\rangle)^T$. Since under a C_4 rotation we have $p_x \to p_y$ and $p_y \to -p_x$, the states $|u_{n\pm}(\widetilde{\boldsymbol{k}})\rangle$ are also eigenstates of U_{C_4} with eigenvalues $\pm i$, while time-reversal transforms the degenerate bands into one another, $\mathcal{T}|u_{n+}(\widetilde{\boldsymbol{k}})\rangle = |u_{n+}^*(\widetilde{\boldsymbol{k}})\rangle \propto |u_{n-}(-\widetilde{\boldsymbol{k}})\rangle$.

At the high symmetry points, for generic Hamiltonians with the same symmetries as $H(\boldsymbol{k})$ we have

$$H(\Lambda_i) |u_{n+}(\Lambda_i)\rangle = E_n(\Lambda_i) |u_{n+}(\Lambda_i)\rangle, \tag{4.27}$$
$$U_{C_4} |u_{n+}(\Lambda_i)\rangle = +i |u_{n+}(\Lambda_i)\rangle. \tag{4.28}$$

Applying the time-reversal operator $\mathcal{T} = K$ to these equations and taking into account that $\mathcal{T} U_{C_4} \mathcal{T}^{-1} = U_{C_4}$, $\mathcal{T} H(\Lambda_i) \mathcal{T}^{-1} = H(\Lambda_i)$, and $\mathcal{T}|u_{n+}(\Lambda_i)\rangle = |u_{n+}^*(\Lambda_i)\rangle$ gives

$$H(\Lambda_i) |u_{n+}^*(\Lambda_i)\rangle = E_n(\Lambda_i) |u_{n+}^*(\Lambda_i)\rangle, \tag{4.29}$$
$$U_{C_4} |u_{n+}^*(\Lambda_i)\rangle = -i |u_{n+}^*(\Lambda_i)\rangle. \tag{4.30}$$

Equation (4.29) shows that $|u_{n+}^*(\Lambda_i)\rangle$ is an eigenstate of $H(\Lambda_i)$ with eigenvalue $E_n(\Lambda_i)$, while equation (4.30) implies that $|u_{n+}^*(\Lambda_i)\rangle$ is an eigenstate of U_{C_4} with eigenvalue $-i$. In other words, $|u_{n+}^*(\Lambda_i)\rangle \propto |u_{n-}(\Lambda_i)\rangle$ and the energy spectrum is double degenerate. These electronic states can be used to define a $\mathbb{Z}_2$ topological invariant that characterizes the system with time-reversal and four-fold rotation symmetries [190]. For convenience, let us change the basis of the double degenerate subspaces at high-symmetry points, $\{|u_{n+}\rangle, |u_{n-}\rangle\} \to \{|u_{2n-1}\rangle, |u_{2n}\rangle\}$, $n = 1, 2, \ldots, N$, and choose a real gauge, $\mathcal{T}|u_m\rangle = |u_m\rangle$. Of course, the real eigenstates $|u_m(\Lambda_i)\rangle$ of $H(\Lambda_i)$ are *not* eigenstates of U_{C_4}. We define the invariants $\nu_{\Gamma M}$ and ν_{AZ} characterizing the band structures in the 2D momentum spaces $k_z = 0$ and $k_z = \pi$, respectively as

$$(-1)^{\nu_{\Lambda_i \Lambda_j}} = \mathrm{Pf}[w(\Lambda_i)]/\mathrm{Pf}[w(\Lambda_j)], \tag{4.31}$$

where $\mathrm{Pf}[\ldots]$ is the Pfaffian and the antisymmetric matrix $w(\Lambda_i)$ is given by

$$w_{nm}(\Lambda_i) = \langle u_n(\Lambda_i)|U\mathcal{T}|u_m(\Lambda_i)\rangle. \tag{4.32}$$

Note that $w_{nm}(\Lambda_i)$ is a $2N \times 2N$ antisymmetric matrix in any gauge, because $U\mathcal{T}$ is anti-unitary and $(U\mathcal{T})^2 = -1$. One can show [190] that $(-1)^{\nu_{\Lambda_i \Lambda_j}} = \pm 1$, i.e., they can be viewed as $\mathbb{Z}_2$ "weak" indices. Finally, the $\mathbb{Z}_2$ invariant characterizing the 3D system with time-reversal and four-fold rotation symmetries is defined as $\nu_0 = \nu_{\Gamma M} + \nu_{AZ} \pmod 2$ [190]. If $\nu_0 = 1$ the system is topologically nontrivial, while $\nu_0 = 0$ corresponds to a trivial phase adiabatically connected to an atomic insulator.

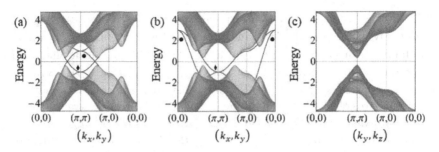

FIGURE 4.6 (a) Energy spectrum for a system in the $(0, 0, 1)$ slab geometry with an integer number of AB bilayers, i.e., A and B terminations on the bottom and top surfaces, respectively. The gapless modes on the bottom and top surfaces (marked by diamond and disk symbols, respectively) have quadratic degeneracies at $(k_x, k_y) = (\pi, \pi)$. (b) Energy spectrum for a system in the $(0, 0, 1)$ slab geometry with A termination on both top and bottom surfaces. The quadratic degeneracy of the top surface mode occurs at $(k_x, k_y) = (0, 0)$. (c) Energy spectrum for a system in the $(1, 0, 0)$ slab geometry. The symmetry-breaking surfaces do not support protected gapless modes.

To corroborate these results and gain further insight into the bulk–boundary correspondence, we solve numerically the tight-binding model given by Eqs. (4.21–4.23) for the following set of parameters: $t_1^A = -t_1^B = 1$ (i.e., we choose t_1^A as the energy unit), $t_2^A = -t_2^B = 0.5$, $t_1' = 2.5$, $t_2' = 0.5$, and $t_z' = 2$. The corresponding bulk spectrum is gapped, as illustrated in Figure 4.5 (c), and is characterized by a topological invariant $\nu_0 = 1$. Based on our general discussion of topological crystalline insulators, we expect the presence of gapless boundary states (only) on surfaces that preserve the spatial C_4 symmetry, i.e., on $(0, 0, 1)$ surfaces. To verify this phenomenology, we consider a slab geometry with the system being finite either in the z direction – with $(0, 0, 1)$ symmetry-preserving surfaces – or the x-direction – with $(1, 0, 0)$ symmetry-breaking surfaces. The numerical results are shown in Figure 4.6. For a system containing an integer number of AB bilayers, the top and bottom $(0, 0, 1)$ surfaces support gapless modes that span the entire bulk gap and have a quadratic (rather than linear) degeneracy at the $(k_x, k_y) = (\pi, \pi)$ point, as shown in Figure 4.6(a). We emphasize that this degeneracy is protected by the combination of time-reversal symmetry (for spinless fermions) and four-fold rotation symmetry with $U_{C_4}^2 = -1$. Breaking the symmetries, or having single bands (with $U_{C_4}^2 = 1$) near the Fermi energy can open a gap and destroy the stability of the gapless surface states. On the other hand, details of the surface that do not affect its symmetry properties can substantially alter the dispersion of the boundary mode, but do not modify its gapless nature. For example, as shown in Figure 4.6(b), adding an A layer on the top surface dramatically changes the dispersion of the top surface mode. However, it still spans the bulk gap and has a symmetry-protected (quadratic) degeneracy point [at $(k_x, k_y) = (0,0)$]. Finally, Figure 4.6(c) shows explicitly that symmetry-breaking surfaces, e.g., the $(1,0,0)$ surfaces, do not support (topologically-protected) gapless boundary modes.

4.1.3 Higher order topological phases

In Chapter 3, we have classified gapped band structures with internal symmetries into different topological equivalence classes (referred to as "topological phases") and identified the "anomalous" boundary modes – gapless states that only exist in the presence of a topologically nontrivial bulk and are robust against symmetry-preserving perturbations that do not close the bulk gap – as a defining property of topological insulators and superconductors (the so-called bulk–boundary correspondence). In the previous two subsections, we have pointed out that the general picture of noninteracting topological phases is more complex. First, the topological classification of gapped band structures depends on the precise definition of topological equivalence, (e.g., defining it in terms of homotopy equivalence or stable homotopy equivalence; requiring the presence of translation symmetry or not). These considerations lead to concepts such as fragile topology and weak topology. Second, while the bulk–boundary correspondence of topological phases with internal symmetries is complete – each d-dimensional topological band structure is associated with a specific type of $(d-1)$-dimensional gapless boundary mode,[5] topological crystalline phases protected by spatial symmetries support gapless modes only on $(d-1)$-dimensional boundaries that preserve these symmetries. In this subsection, we generalize this bulk–boundary correspondence by introducing the so-called *higher order topological phases*, or *higher order topological insulators* (HOTIs) and *superconductors* (HOTSCs) [44, 287, 316, 449, 484, 509]. We define a d-dimensional nth-order topological phase as an equivalence class of systems with certain internal and crystalline symmetries that support anomalous (gapless) modes on boundaries of dimension $d - n$ (i.e., *codimension n*, with $n \geq 1$) when the boundary, as a whole, respects the crystalline symmetries. For example, the "standard" topological insulators and superconductors (without spatial symmetries) are first order topological phases, as they support boundary modes with codimension $n = 1$. To illustrate the main ideas, we will consider two paradigmatic examples of second-order topological band structures: a two-dimensional system that supports gapless corner states and a three-dimensional system with anomalous edge (i.e., hinge) modes. But first, we sketch the generic qualitative picture and point out a few caveats.

One common approach to studying topological phases is to consider the system in the vicinity of a phase transition characterized by a quasi-linear

---

[5]Here, we assume an effectively noninteracting system. The presence of interactions may lead to the emergence of a topological phase with (intrinsic) topological order on the boundary, which becomes gapped.

band gap closing and approximately describe the low-energy physics using a Dirac model.[6] The topological quantum phase transition corresponds to the vanishing of the Dirac mass, which has opposite signs within the trivial and topological phases. Similar low-energy Dirac models (in $d-1$ dimensions) can be used to describe gapless (or weakly gapped) boundary modes. Assume a d-dimensional topologically-nontrivial crystalline system with a termination of codimension $n = 1$ (e.g., edge for $d = 2$ or surface for $d = 3$) that remains invariant (locally) under the relevant spatial symmetries. The boundary hosts gapless modes, which can be described using a massless Dirac model. Upon deforming the boundary so that it remains invariant only under "global" symmetry transformations, which corresponds to generic orientations of the edges or surfaces (see Figure 4.7), the codimension-one boundary states become gapped and eventually disappear. However, if the system represents a second-order topological phase, this boundary change will leave behind gapless boundary modes of codimension $n = 2$ (at the corners or along the hinges). In terms of the effective Dirac model description, changing the orientation of the edge (surface) corresponds to the emergence of a mass term (m) that generates a nonzero gap. For a first-order topological phase the sign of the mass term will be independent of the edge (surface) orientation. However, for a second-order topological phase edges (surfaces) with different orientations can be characterized by opposite sings of the mass terms (see Figure 4.7). Thus, the codimension-two states can be understood as gapless modes emerging at domain walls (i.e., corners or hinges) separating faces with negative and positive mass terms. We note that for certain symmetries (e.g., inversion) there is no (locally) invariant boundary orientation. However, one can extend the domain wall argument using a "trick" proposed in Ref. [196], which involves considering a $(d + 1)$-dimensional system with spatial symmetry operations acting on the first d coordinates, while the $d + 1$ coordinate is left unchanged (hence, hyper-surfaces perpendicular to the $d + 1$ direction are manifestly symmetry-invariant).

It is useful to make a distinction between higher order phases characterized by anomalous boundary states rooted in the topology of the bulk band structure – called *intrinsic* – and phases with higher-order boundary states originating from a nontrivial topological phase located on the crystal boundary – called *extrinsic* [196, 316, 480]. The intrinsic higher-order boundary modes are robust against symmetry-preserving perturbations that do not close the bulk gap. For example, one can modify the boundary termination of a 2D system by "decorating" the edges with topological 1D chains, while preserving the (global) symmetry (see Figure 4.7). More specifically, in Figure 4.7(c) we consider second–order topological phase with a $\mathbb{Z}_2$ classification, which supports zero–energy modes at the symmetry-preserving corners A and A'. These modes are robust against (symmetry–preserving) changes of the boundary termination, since this could only add pairs of corner modes (which become

---

[6]See Chapter 6 for an explicit construction of a low-energy Dirac model.

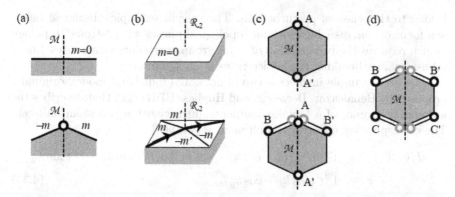

FIGURE 4.7 (a) *Top*: Symmetry–preserving edge of a 2D second–order topological crystal with mirror symmetry. The edge supports gapless boundary modes described by a massless Dirac model. *Bottom*: Generic termination that is invariant under a "global" mirror transformation. The gapped ($m \neq 0$) edges with different orientations are characterized by opposite Dirac masses, while the (mirror–symmetric) corner supports a gapless mode (open circle). (b) *Top*: Symmetry–preserving surface of a 3D second–order topological crystal with twofold rotation symmetry. *Bottom*: Generic termination with "global" twofold rotation symmetry. The symmetry–related surfaces have opposite Dirac masses, while the edge separating regions with opposite mass terms supports a gapless chiral mode (we assume $m \cdot m' > 0$). (c) *Top*: Corner states at the symmetry–preserving corners of a 2D crystal with mirror symmetry. *Bottom*: Changing the boundary termination by "decorating" two edges with topological 1D chains. At the A corner one mode remains gapless. (d) Extrinsic 2D second–order topological system. The gapless corner modes (at B, B', C, and C') can be removed by "decorating" the vertical edges with topological 1D chains.

gapped). By contrast, zero–energy modes at the B, B' (or C, and C') corners are not robust against symmetry-preserving perturbations of the boundary. Of course, none of the corner modes are robust against symmetry-breaking perturbations – e.g., "decorating" the three edges on the right side of the system with topological 1D chains in Figure 4.7(c). The extrinsic anomalous boundary modes, on the other hand, are determined solely by the properties of the crystal termination. For example, the system in Figure 4.7(d) has a topologically–trivial bulk, but supports zero–energy modes at the B, B', C, and C' corners. These modes can be removed by a symmetry-preserving perturbation of the boundary (e.g., "decorating" the vertical edges with topological 1D chains). Nonetheless, they are robust against symmetry-preserving perturbations that do not close the gaps in the bulk or on the boundary. Note that "adding" a $(d - 1)$-dimensional topological system on a trivial boundary involves the closing and reopening of the boundary gap, so it does not belong to this class of perturbations. The specific examples discussed below will focus on *intrinsic* higher-order topological phases with "strong" topology (which requires the preservation of point group crystalline symmetries, but is robust against breaking the lattice translation symmetry).

The first example involves a two-dimensional four-band model, originally proposed by Benalcazar, Bernevig, and Hughes (BBH) [44], that describes the low-energy physics of a (spinless) superconductor with a two-atom unit cell. The corresponding effective Hamiltonian is [44, 509]

$$
\begin{aligned}
H(k_x, k_y) &= \Gamma^0[m + t(2 - \cos k_x - \cos k_y)] + \Gamma^1 t \sin k_x + \Gamma^2 t \sin k_y \\
&+ \overline{\Gamma}^4 t'(\cos k_x - \cos k_y),
\end{aligned}
\tag{4.33}
$$

with $t' = t$ being real parameters and the Hermitian matrices Γ^j and $\overline{\Gamma}^4$ being defined in terms of two sets of Pauli matrices, τ_μ and σ_ν as: $\Gamma^0 = (\tau_2 \sigma_2 - \tau_1 \sigma_0)/\sqrt{2}$, $\Gamma^1 = -\tau_2 \sigma_3$, $\Gamma^2 = -\tau_2 \sigma_1$, and $\overline{\Gamma}^4 = (\tau_2 \sigma_2 + \tau_1 \sigma_0)/\sqrt{2}$. One can easily verify that these matrices are mutually anticommuting and square to one. The Hamiltonian in Eq. (4.33) is invariant under time–reversal $\mathcal{T}$, particle–hole conjugation $\mathcal{C}$, and chiral (sublattice) antisymmetry $\mathcal{S}$, with unitary operators $U_t = \tau_0 \sigma_0 \equiv 1$ and $U_c = U_s = \tau_3 \sigma_0$, respectively. In addition to these internal symmetries, the Hamiltonian satisfies two mirror symmetries, $\mathcal{M}_x$ and $\mathcal{M}_y$, with unitary operators $U_x = \tau_1 \sigma_3$ and $U_y = \tau_1 \sigma_1$, respectively,[7] and a fourfold rotation symmetry, C_4

$$
U_{C_4} H(k_x, k_y) U_{C_4}^{-1} = H(-k_y, k_x), \qquad U_{C_4} = \begin{pmatrix} 0 & \sigma_0 \\ -i\sigma_2 & 0 \end{pmatrix}.
\tag{4.34}
$$

Note that $(U_{C_4})^4 = -\tau_0 \sigma_0 \equiv -1$. Also note that the spectral gap of the Hamiltonian (4.33) vanishes at $\boldsymbol{k} = (0,0)$ for $m = 0$ and at $\boldsymbol{k} = (\pi, \pi)$ for $m = -4t$. The phase corresponding to $|m| \to \infty$ is manifestly topologically trivial.

To determine whether the gapless points represent topological quantum phase transitions (TQPTs), we first notice that the last term in Eq. (4.33) can be sent to zero, $t' \to 0$, without closing the spectral gap (for $m \neq 0, -4t$) and without violating the symmetries, i.e., without affecting the potential TQPTs. The simplified Hamiltonian has the "standard" Dirac form $H(\boldsymbol{k}) = \sum_{j=1}^{d} d_j(\boldsymbol{k}) \Gamma^j$, with $d_0 = m + t \sum_{j=1}^{d}(1 - \cos k_j)$, $d_j = t \sin k_j$ ($j = 1, \ldots, d$), and matrices Γ^j that mutually anticommute and square to the identity. Next, we investigate the existence of *mass terms* – perturbations proportional to Hermitian matrices that square to the identity, anticommute with the (simplified) Hamiltonian, and satisfy the relevant symmetry constraints. If no mass terms consistent with the internal symmetries exist, the gapless points are robust (and represent TQPTs) regardless of whether the spatial symmetries are enforced or not. In other words, the regime $-2dt < m < 0$

---

[7]The symmetry relations satisfied by the Hamiltonian are: $H(\boldsymbol{k}) = U_t H^*(-\boldsymbol{k}) U_t = -U_c H^*(-\boldsymbol{k}) U_c = -U_s H(\boldsymbol{k}) U_s$ and $H(k_x, k_y) = U_x H(-k_x, k_y) U_x = U_y H(k_x, -k_y) U_y$.

(where d is the spatial dimension) represents a "standard" AZ topological phase. If, on the other hand, there is a mass term that is consistent with the internal symmetries but breaks certain spatial symmetries, the regime $-2dt < m < 0$ corresponds to a topological crystalline phase protected by the relevant internal *and* spatial symmetries.

In the case of the BBH Hamiltonian, we identify two possible mass terms: $\delta H_M^{(1)} = \lambda_1 \overline{\Gamma}^4$ and $\delta H_M^{(2)} = \lambda_2 \tau_3 \sigma_0$. The term $\delta H_M^{(2)}$ is only compatible with $\mathcal{T}$ and with a twofold rotation, $C_2 = \mathcal{M}_x \mathcal{M}_y$. However, $\delta H_M^{(1)}$ is compatible with all symmetries, except the C_4 rotation (i.e., with $\mathcal{M}_x, \mathcal{M}_y, \mathcal{T}, \mathcal{C}$, and $\mathcal{S}$). Thus, if we do not enforce the fourfold rotation symmetry (i.e., if perturbations that break the rotation symmetry are allowed), $\delta H_M^{(1)}$ is an acceptable perturbation that opens finite spectral gaps at $m = 0$ and $m = -4t$; the corresponding system is topologically trivial for all m values. However, with rotation symmetry, the gapless points are robust (since $\delta H_M^{(1)}$ is not an acceptable perturbation) and the regime $-4t < m < 0$ corresponds to a topological crystalline phase. Thus, a system with $t' = 0$, i.e., without the $\overline{\Gamma}^4$ term in Eq. (4.33), is a *first order* topological crystalline superconductor with no gapless boundary states, since there is no possible (single) edge consistent with the C_4 rotation symmetry. On the other hand, a system with $t' \neq 0$ supports a *second order* topological crystalline superconductor with Majorana zero modes at the corners of a (globally) C_4-invariant boundary (e.g., a square boundary). We note that the corner modes are intrinsic – a consequence of the nontrivial topology of the bulk band structure – and, consequently, they are robust against symmetry-preserving perturbations of the boundary. Finally, to reinforce the importance of symmetry, we point out that breaking the particle–hole symmetry enables the removal of the corner states by applying local perturbations consistent with the remaining symmetries. However, one can show that, after the removal of the corner *states*, anomalous half-integer corner *charges* remain [44] – a signature of the system being in an obstructed atomic-limit phase. Again, these anomalous corner charges are intrinsic and cannot be eliminated by a C_4-compatible change of boundary termination.

The BBH model discussed above can be generalized to three dimensions. Following Ref. [449] we consider the effective low-energy Hamiltonian

$$
\begin{aligned}
H(\boldsymbol{k}) &= \Gamma^0 \left[m + t \sum_{j=1}^{3}(1 - \cos k_j) \right] + t \sum_{j=1}^{3} \Gamma^j \sin k_j \\
&+ \overline{\Gamma}^4 t' (\cos k_1 - \cos k_2),
\end{aligned}
\tag{4.35}
$$

where $\Gamma^0 = \tau_3 \sigma_0$, $\Gamma^j = \tau_1 \sigma_j$ $(j = 1, 2, 3)$, and $\overline{\Gamma}^4 = \tau_2 \sigma_0$. As before, the matrices mutually anticommute and square to one. Also, the "simplified" Hamiltonian corresponding to $t' = 0$ has the "standard" Dirac form mentioned above, with gapless points at $m = 0$ and $m = -6t$. In this case (i.e., for $t' = 0$), the Hamiltonian is invariant under time–reversal $\mathcal{T}$, with $U_t = \tau_0 \sigma_2$, and has a fourfold rotation symmetry C_4 (with respect to the z-axis), with $U_{C_4} = \tau_0 e^{-i\pi\sigma_3/4}$. The term proportional to $\overline{\Gamma}^4$ breaks both $\mathcal{T}$ and C_4, but is symmetric under the product $C_4\mathcal{T}$, so that the Hamiltonian in Eq. (4.35) satisfies the symmetry relation

$$(C_4\mathcal{T}) H(k_1, k_2, k_3) (C_4\mathcal{T})^{-1} = H(k_2, -k_1, -k_3). \tag{4.36}$$

Note that $[C_4, \mathcal{T}] = 0$ and we have $(C_4\mathcal{T})^4 = -\tau_0\sigma_0 \equiv -1$. The model has one mass term, $\delta H_M = \lambda\overline{\Gamma}^4$, which is compatible with the fourfold rotation symmetry but breaks time–reversal and the product $C_4\mathcal{T}$. Hence, for the $\mathcal{T}$–symmetric system with $t' = 0$ there is no mass term consistent with time–reversal symmetry and the regime $-6t < m < 0$ represents a "standard" first order topological insulator with gapless surface states, irrespective of the fourfold rotation symmetry. More specifically, the system has time–reversal symmetry with $\mathcal{T}^2 = -1$, hence it belongs to the AII symmetry class and the three dimensional topological insulating phase is characterized by a $\mathbb{Z}_2$ index (see Tables 3.1 and 3.3). With $t' \neq 0$, the points $m = 0$ and $m = -6t$ remain gapless (since the mass term is not compatible with the $C_4\mathcal{T}$ symmetry) and the phase corresponding to $-6t < m < 0$ is still topologically nontrivial. However, the system does not support gapless surface states, as they are gapped out on a generic surface in the absence of time–reversal symmetry, but supports chiral edge modes running along the crystal "hinges." These intrinsic gapless "hinge" modes characterize a second order topological crystalline insulator and are a manifestation of the topologically nontrivial bulk band structure. They cannot be removed by any change of the surface termination consistent with the $C_4\mathcal{T}$ symmetry.

We conclude this section with a comment regarding the classification of topological band structures in the presence of crystalline symmetries. The "standard" K-theory classification is based on bulk properties and does not consider the boundary signatures. Thus, for a given combination of internal and crystalline symmetries, the classifying group K includes first-order topological phases (with gapless states on boundaries of codimension one), higher-order topological phases (with gapless states on boundaries of codimension $n > 1$), and atomic-limit phases (with no protected boundary states, but possible anomalous boundary charges). A refined, boundary-resolved classification was proposed in Ref. [508]. In essence, the "addition" of band structures – e.g., taking the direct sum in Eq. (4.4) – does not reduce the number of boundary mass terms and, consequently, cannot lower the codimension associated with protected boundary modes. Formally, this implies that one can define a subgroup $K^{(n)}$ containing the elements of K that have intrinsic boundary signatures (i.e., protected gapless states) on boundaries of dimension *lower than* $d - n$. For example, the elements of the subgroup $K^{(1)}$ have intrinsic boundary signatures on boundaries of dimension $d - 2$ or lower, hence they do not include any first order topological insulator or superconductor. The refined classification is based on the subgroup sequence

$$K^{(d)} \subseteq K^{(d-1)} \subseteq \cdots \subseteq K^{(1)} \subseteq K. \qquad (4.37)$$

Thus, the bulk classification is complemented with a classification of anomalous boundary modes of codimension n described by the quotient

$$K_a^{(n)} = K^{(n)}/K^{(n+1)}. \qquad (4.38)$$

Note that $K^{(d)}$ can either be trivial or it may contain topological crystalline phases without protected boundary states (i.e., symmetry obstructed atomic insulators). The examples discussed in this subsection can be viewed as generators of topological classes in $K_a^{(1)}$ (i.e., second order topological phases).

4.2   GAPLESS TOPOLOGICAL PHASES

Gapped topological phases have been defined as equivalence classes of quantum states with respect to an equivalence relation involving the possibility of smoothly deforming the states into one another (while preserving certain symmetries) without closing the bulk gap. Essentially, the topological robustness of these states stems form the fact that perturbations, if they are not too large, do not qualitatively change the properties of a fully gapped system. Hence, the natural question: Is the presence of a bulk gap a requirement for nontrivial topology? In the absence of a gap, relatively small perturbations can significantly rearrange the low-energy degrees of freedom; hence, the possibility of having topological features in a gapless system is not obvious. Nonetheless, gapless topological phases can be defined and systems supporting such phases have already been studied experimentally, the most familiar example being the so-called Weyl semimetals. The topological features of a three-dimensional Weyl semimetal include protected gapless points in momentum space, topologically protected surface states (that have no equivalent in a purely two-dimensional system), and an electromagnetic response that depends only on the location of the gapless points in momentum space, being independent of other band structure details. Furthermore, the low-energy bulk excitations of the system – the so-called Weyl fermions – provide an example of the fruitful cross-pollination between high energy and condensed matter physics and join the list of quasiparticles representing condensed matter analogs of elementary particles – a list that also includes Dirac and Majorana fermions and axions, which are discussed throughout this book.

In this section, we provide a brief review of the rapidly growing area dedicated to the study of gapless topological phases, focusing on basic theoretical aspects. In Section 4.2.1, we consider three-dimensional Weyl semimetals and discuss the main ideas underlying the emergence of topological properties in these systems, while also presenting a few specific examples. In

Section 4.2.2, we generalize our discussion to other types of gapless topological systems and sketch the main ideas behind the topological classification of these phases. A discussion of the electromagnetic response of Weyl semimetals, including the relation to the so-called *chiral anomaly*, can be found in Chapter 7 (Section 7.3), as it fits naturally within the more general framework of *axion electrodynamics*. The reader interested in delving deeper into the study of gapless topological phases is referred to specialized review articles, e.g., [25, 90, 104, 238, 560], and the references therein.

4.2.1 Weyl semimetals in three-dimensional solids

The properties of weakly interacting crystalline materials depend critically on the filling factor ν – the number of valence electrons per unit cell per spin. If ν has noninteger values, the system is a metal characterized by the presence of gapless energy excitations near the Fermi surface – the surface in momentum space separating filled and empty single particle states and enclosing a "volume" that is directly proportional to the fractional part of ν (Luttinger's theorem). The low-energy physics in the vicinity of the Fermi surface controls the electronic properties of the metal. If, on the other hand, ν is an integer, the system is either an insulator(characterized by a finite energy gap) or a (gapless) semimetal. A so-called compensated semimetal can be obtained by smoothly deforming the Hamiltonian of an insulator so that the valence band partially raises above the Fermi energy creating one or more hole "islands," while the conduction band partially sinks below the Fermi level generating electron "lakes" that do not overlap with the hole "islands." Note that Luttinger's theorem is still satisfied (i.e., the total volume enclosed by the disjoint Fermi surfaces is zero, consistent with the integer value of ν) if we weight the volume of each pocket by the sign of the corresponding carriers. Hence, one can continuously connect a compensated semimetal to a band insulator without changing the filling or violating Luttinger's theorem and without closing the energy gap $E_c(\boldsymbol{k}) - E_v(\boldsymbol{k})$ between the valence and the conduction bands at any $\boldsymbol{k}$ in the Brillouin zone; the gapless character of a compensated semimetal can be viewed as an *essentially accidental* property.

A more interesting scenario for generating a semimetal involves the presence in the Brillouin zone of regions (e.g., points or lines) where the conduction and valence bands touch, $E_c(\boldsymbol{k}) - E_v(\boldsymbol{k}) = 0$. To determine the conditions for such band touchings, we focus on the low-energy subspace associated with the conduction and valence bands and describe the system in terms of an effective Hamiltonian $H(\boldsymbol{k})$. The dimension of the minimal low-energy effective Hamiltonian depends on the degeneracy of the two bands, which, in turn, is determined by symmetry. In the presence of spin rotation symmetry (e.g., in a system with no spin-orbit coupling and no Zeeman splitting) the bands are double degenerate. Double degenerate bands also arise in spin-orbit coupled systems in the presence of combined time-reversal ($\mathcal{T}$) and inversion symmetry ($\mathcal{I}$), $\widetilde{\mathcal{T}} = \mathcal{I}\mathcal{T}$. Indeed, since crystal momenta are invariant under $\widetilde{\mathcal{T}}$, the presence of this symmetry implies $\widetilde{\mathcal{T}} H(\boldsymbol{k}) \widetilde{\mathcal{T}}^{-1} = H(\boldsymbol{k})$; if $|u_{\alpha\boldsymbol{k}}\rangle$ is an eigenstate of $H(\boldsymbol{k})$ with eigenvalue $E_\alpha(\boldsymbol{k})$, $\widetilde{\mathcal{T}}|u_{\alpha\boldsymbol{k}}\rangle$ is also an eigenstate of $H(\boldsymbol{k})$ corresponding to the same eigenvalue. Furthermore, since $\widetilde{\mathcal{T}}$ is antiunitary and $\widetilde{\mathcal{T}}^2 = -1$, arguments similar to those used in the context of the Kramers theorem (see Section 3.2.1, page 73) lead to the conclusion that $|u_{\alpha\boldsymbol{k}}\rangle$ and $\widetilde{\mathcal{T}}|u_{\alpha\boldsymbol{k}}\rangle$ are orthogonal, i.e., the bands are double degenerate. The effective model describing a pair of double degenerate bands can be expressed in terms of 4×4 matrices, while the possible band touchings have fourfold degeneracy. We will address this case in Section 4.2.2.

Consider now a system with no spin rotation symmetry and no $\widetilde{\mathcal{T}}$ symmetry (which implies that time-reversal or inversion symmetry or both are broken). The corresponding valence and conduction bands are typically nondegenerate (except at the time-reversal invariant momenta in the presence of $\mathcal{T}$). The generic effective model for a pair of nondegenerate bands has the form

$$H(\boldsymbol{k}) = \epsilon_0(\boldsymbol{k})\, \sigma_0 + \sum_{j=1}^{3} \epsilon_j(\boldsymbol{k})\, \sigma_j, \qquad (4.39)$$

where σ_0 is the 2×2 identity matrix and σ_j are Pauli matrices. The corresponding energy bands are $E_\pm(\boldsymbol{k}) = \epsilon_0 \pm \sqrt{\epsilon_1^2 + \epsilon_2^2 + \epsilon_3^2}$, with an energy splitting $\Delta E(\boldsymbol{k}) = E_+ - E_- = 2\sqrt{\epsilon_1^2 + \epsilon_2^2 + \epsilon_3^2}$. Consequently, a band touching requires $\epsilon_1 = \epsilon_2 = \epsilon_3 = 0$. Let us assume that the functions $\epsilon_j(\boldsymbol{k})$ change sign throughout the Brillouin zone. Typically, the set of crystal momenta satisfying the condition $\epsilon_j(\boldsymbol{k}) = 0$ is a two-dimensional (2D) surface Σ_j^0. Furthermore, the intersection $\Sigma_1^0 \cap \Sigma_2^0$, if different from the empty set, is typically (i.e., in the absence of any fine tuning) a set of 1D lines, while the intersection $\Sigma_1^0 \cap \Sigma_2^0 \cap \Sigma_3^0$ is a set of points, $\{\boldsymbol{k}_0^{(1)}, \boldsymbol{k}_0^{(2)}, \dots\}$. In other words, in a system with nondegenerate bands the valence and conduction bands can touch (typically) at discrete points $\boldsymbol{k}_0^{(n)}$ in the Brillouin zone. Small perturbations of ϵ_j (i.e., small changes of the effective model parameters) will deform the surfaces Σ_j^0 and will (slightly) shift the nodes $\boldsymbol{k}_0^{(n)}$, but will not remove (i.e., gap) them. Consequently, if the set of nodes $\boldsymbol{k}_0^{(n)}$ is not empty, the system (with integer filling) is in a gapless *semimetal phase*.

To establish the connection to Weyl fermions, let us focus on a specific touching point, $\boldsymbol{k}_0$, and expand the Hamiltonian (4.39) about this point, $\boldsymbol{k} = \boldsymbol{k}_0 + \boldsymbol{q}$ (where $|\boldsymbol{q}|$ is small compared to the distance to the nearest node). We have

$$H(\boldsymbol{k}) = [\epsilon_0(\boldsymbol{k}_0) + \boldsymbol{v}_0 \cdot \boldsymbol{q}]\sigma_0 + \sum_{j=1}^{3} \boldsymbol{v}_j \cdot \boldsymbol{q}\, \sigma_j, \qquad (4.40)$$

where $\boldsymbol{v}_\mu = \boldsymbol{\nabla}_{\boldsymbol{k}} \epsilon_\mu(\boldsymbol{k})|_{\boldsymbol{k}=\boldsymbol{k}_0}$ are effective velocities. Choosing $\epsilon_0(\boldsymbol{k}_0) = 0$ and focusing on the special case $v_0 = 0$, $\boldsymbol{v}_j \cdot \boldsymbol{q} = \pm v q_j$ (with $v > 0$), we obtain the

Dirac, Weyl, and Majorana fermions

With the standard convention $c = 1 = \hbar$, the Dirac equation is

$$(i\gamma^\mu \partial_\mu - m)\Psi = 0, \tag{4.41}$$

where the 4×4 gamma matrices obey a so-called Clifford algebra. Assuming the Minkowski metric $\eta_{\mu\nu} = \mathrm{diag}(1, -1, -1, -1)$, the gamma matrices satisfy the anticommutation relation $\{\gamma^\mu, \gamma^\nu\} = 0$ for $\mu \neq \nu$ and $(\gamma^0)^2 = -(\gamma^j)^2 = \mathbf{1}$, with $j = 1, 2, 3$ and $\mathbf{1}$ designating the 4×4 unit matrix. Alternatively, we can write Eq. (4.41) in the "traditional" form $i\partial_t \Psi = H\Psi$, with the Dirac Hamiltonian $H = -i\alpha_j \partial_j + \beta m$ containing the matrices $\beta = \gamma^0$ and $\alpha_j = \gamma^0 \gamma^j$. Using two sets of Pauli matrices, σ_j and τ_j, the standard representation is $\alpha_j = \tau_x \otimes \sigma_j$ and $\beta = \tau_z \otimes \sigma_0$. In this representation, the free-particle solutions of the Dirac equation, $\Psi_{E\boldsymbol{p}}(\boldsymbol{r}, t) = u_E(\boldsymbol{p}) \exp[-i(Et - \boldsymbol{p} \cdot \boldsymbol{r})]$, are given by the eigenvectors $u_E(\boldsymbol{p})$ (corresponding to double degenerate eigenvalues $E = \pm\sqrt{m^2 + p^2}$) of the Dirac Hamiltonian

$$H_D(\boldsymbol{p}) = \begin{pmatrix} m\,\sigma_0 & \boldsymbol{p} \cdot \boldsymbol{\sigma} \\ \boldsymbol{p} \cdot \boldsymbol{\sigma} & -m\,\sigma_0 \end{pmatrix}. \tag{4.42}$$

In 1929, Hermann Weyl proposed a simplified version of the Dirac equation that describes *massless* spin-$\frac{1}{2}$ fermions with defined *chirality* (i.e., spin projection along the $\boldsymbol{p}$ direction) – so-called *Weyl fermions*. For $m = 0$, the Weyl (or chiral) representation $\alpha_j = \tau_z \otimes \sigma_j$, $\beta = -\tau_x \otimes \sigma_0$, which can be obtained from the standard representation through the unitary transformation $U_W = (\tau_0 + i\tau_y) \otimes \sigma_0 / \sqrt{2}$, results in a block-diagonal Hamiltonian, $H(\boldsymbol{p}) = \boldsymbol{p} \cdot \boldsymbol{\alpha}$, with solutions of the form $u_+ = (u_1, u_2, 0, 0)^T$ and $u_- = (0, 0, u_3, u_4)^T$ corresponding to positive chirality (i.e., right handedness) and negative chirality (i.e., left handedness), respectively. The nontrivial components of these spinors can be obtained as (two-component) eigenvectors of the Weyl Hamiltonian

$$H_W^\lambda(\boldsymbol{p}) = \lambda \boldsymbol{p} \cdot \boldsymbol{\sigma}, \tag{4.43}$$

with the chirality $\lambda = \pm 1$ corresponding to massless fermions that propagate parallel (or antiparallel) to their spin

In 1937, Ettore Majorana proposed a modification of the Dirac equation that can have a real (rather than complex) solution, which would describe a neutral particle that is its own antiparticle. Indeed, if one can find a purely imaginary representation of the gamma matrices, the solution of Eq. (4.41) can be a real spinor. This purely imaginary representation – the so-called Majorana representation – can be obtained from the standard (Dirac) basis via the unitary transformation $U_M = (\tau_z \otimes \sigma_0 + \tau_x \otimes \sigma_y)/\sqrt{2}$. We have

$$\gamma^0 = \tau_x \otimes \sigma_y, \quad \gamma^1 = i\tau_0 \otimes \sigma_z, \quad \gamma^2 = -i\tau_y \otimes \sigma_y, \quad \gamma^3 = -i\tau_0 \otimes \sigma_x. \tag{4.44}$$

In general, if a particle is described by the spinor Ψ, the *charge conjugate* spinor Ψ_c will describe its antiparticle. Explicitly, we have $\Psi_c = \mathcal{C}\Psi = \eta_c C \overline{\Psi}^T = -\eta_c \gamma^0 C \Psi^*$, where η_c is an arbitrary phase, $|\eta_c| = 1$. In all representations the charge conjugation matrix C transposes the gamma matrices, $C\gamma^\mu C^{-1} = -(\gamma^\mu)^T$, and satisfies the properties $C^{-1} = C^\dagger = C^T = -C$. In the Dirac representation, for example, we have $C = i\gamma^2\gamma_0 = -i\tau_x \otimes \sigma_y$ and $\gamma^0 = \tau_z \otimes \sigma_0$; the anti-unitary charge conjugation operator takes the form $\mathcal{C} = \eta_c \tau_y \otimes \sigma_y K$, where K is the complex conjugation. In the Majorana representation, we also have $C = -i\tau_x \otimes \sigma_y$, but γ^0 has a different form [see Eq. (4.44)] and, with the phase choice $\eta_c = -i$, the charge conjugation operator becomes $\mathcal{C} = K$. Hence, in the Majorana representation $\Psi_c = \Psi^*$ and a real solution of Eq. (4.41) describes a "truly neutral" particle that is its own antiparticle.

Weyl Hamiltonian (4.43) (see box on page 132). We will refer to the band touching at $\boldsymbol{q} = 0$ as a *Weyl node* and to the corresponding low-energy quasiparticles as *Weyl fermions*. Note that the chirality of the fermions described by Eq. (4.40) is $\lambda = \text{sign}(\boldsymbol{v}_1 \cdot \boldsymbol{v}_2 \times \boldsymbol{v}_3)$.

So far, we have established that 3D systems that do not have both TR and inversion symmetry are characterized by (typically) nondegenerate bands that can touch at discrete points in momentum space. If band touchings exist, they are robust against (small enough) perturbations and the low-energy physics in their vicinity is described by a generally anisotropic version of the Weyl equation. The topological origin of the stability of the Weyl points can be determined by recalling that a degeneracy point of a two-level system corresponds to a monopole of strength $1/2$ that generates a Berry flux $C2\pi = \pm 2\pi$ through any small surface (in parameter space, e.g., k space) surrounding the degeneracy point. Here, $C = \pm 1$ is the chirality (or topological charge) of the Weyl point. Indeed, Eq. (4.40) can be viewed as a special case of Eq. (1.34) from page 25. The stability of the degeneracy point is ensured by the conservation of the topological charge. In other words, the total Berry flux through a small surface surrounding a Weyl point is invariant against small perturbations, as it can only change if another monopole (i.e., Weyl point) goes through the surface (which requires a finite perturbation strength). Note, however, that the 3D system, which belongs to symmetry class A (if it breaks TR symmetry) or AII (with broken inversion symmetry), is characterized by a vanishing total Berry flux through the boundary of the Brillouin zone. This implies that the total chirality must vanish, i.e., the Weyl points occur in pairs with zero net charge (or opposite chirality). Hence, the Weyl semimetal phase corresponds to monopole–antimonopole pairs being separated in momentum space. As long as translation symmetry is preserved, a transition to a gapped phase can only occur if, by changing the Hamiltonian, the Weyl points of opposite chirality are brought together and annihilated pairwise. Furthermore, even in the presence of disorder the Weyl phase is protected if the disorder potential is sufficiently smooth and does not induce scattering between Weyl nodes with opposite chiralities.

To gain further insight, let us consider a simple four-band effective model of a Weyl semimetal given by the cubic-lattice Hamiltonian

$$
\begin{aligned}
H(\boldsymbol{k}) \;=\; & t(2 - \cos k_x - \cos k_y - \cos k_z)\tau_3\sigma_0 + \delta\,\tau_0\sigma_3 \\
& + \; t'(\sin k_x\,\tau_1\sigma_1 + \sin k_y\,\tau_1\sigma_2),
\end{aligned} \tag{4.45}
$$

where τ_i and σ_i are Pauli matrices and we assume, for simplicity, $t > 0$ and a lattice constant $a = 1$. Note that the Hamiltonian in Eq. (4.45) consists of two independent blocks, $H^{(\pm)}$, one $(H^{(+)})$ involving the elements $H_{\mu\nu}$ with $\mu,\nu \in \{1,4\}$ and the other $(H^{(-)})$ involving the elements with $\mu,\nu \in \{2,3\}$. The two blocks can be written in terms of the Pauli matrices s_i as

$$
H^{(\pm)} = t(2 - \cos k_x - \cos k_y - \cos k_z \pm \delta/t)s_3 + t'(\sin k_x\,s_1 \pm \sin k_y\,s_2). \tag{4.46}
$$

Note that each of these blocks provide a minimal model of an inversion symmetric Weyl semimetal [25, 90]. When $\delta = 0$, the Hamiltonian in Eq. (4.45) has time-reversal symmetry with $\mathcal{T} = \tau_0\sigma_2 K$ and inversion symmetry $\mathcal{I} = \tau_0\sigma_3$ and the bulk gap vanishes at $\boldsymbol{k}_\pm = (0,0,\pm\pi/2)$. The corresponding system is a *Dirac semimetal*. When one of the symmetries is broken, each of the Dirac nodes can be separated into a pair of Weyl nodes with opposite chiralities. Specifically, when $\delta \neq 0$ (with $|\delta| < t$) the bulk spectrum has four Weyl points located at $\boldsymbol{k}_0 = (0,0,\pm k_0^{(\pm)})$, with $\cos k_0^{(\pm)} = \pm\delta/t$. The location of the Weyl points in the Brillouin zone corresponding to $\delta \approx 0.25t$ is illustrated in Figure 4.8(a). The low-energy physics in the vicinity of the Weyl points, $\boldsymbol{k} = \boldsymbol{k}_0 + \boldsymbol{q}$, is described by contributions to the Hamiltonian that are linear in $\boldsymbol{q}$. In terms of the block Hamiltonians from Eq. (4.46) we have four Weyl-type contributions, $H_\pm^{(\pm)}(\boldsymbol{q}) \approx \sum_j [v_\pm^{(\pm)}]_j q_j s_j$, with $v_\pm^{(\pm)} = [t', (\pm)t', \pm t\sin k_0^{(\pm)}]$. Note that this corresponds to having velocities with components $[\boldsymbol{v}_j]_i = [v_\pm^{(\pm)}]_j\delta_{ij}$ in Eq. (4.40). The chirality of the Weyl fermions described by the linearized Hamiltonian is given by the sign of the product of these velocity components (see Figure 4.8).

The Weyl points shown in Figure 4.8(a) can be viewed as sources and sinks of Berry flux, with a total flux of 2π flowing from a node with chirality $+1$ into the node with chirality -1. Consequently, the Berry flux through a constant-k_z plane separating a pair of nodes, e.g., the plane $k_z = -\pi/2$, is 2π, which corresponds to a nonzero Chern number $C = +1$. Note that the Berry flux through the plane $k_z = -\pi/2$ is equal to the flux through the closed surface formed by the planes $k_z = -\pi/2$ and $k_z = -\pi$ and the lateral boundary of the Brillouin zone (BZ) between them, since there is no contribution from the BZ boundary. This closed surface can be smoothly deformed into a small sphere surrounding the Weyl node with the lowest k_z value, which has chirality $+1$. Alternatively, the Berry flux through the plane $k_z = -\pi/2$ can be expressed as minus the flux through the closed surface formed with the plane $k_z = \pi$

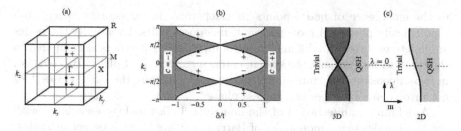

FIGURE 4.8 (a) Location in the Brillouin zone of the four Weyl points characterizing the Hamiltonian in Eq. (4.45) for $\delta \approx 0.25t$. The signs indicate the
chiralities of the Weyl points. (b) Dependence of the Weyl points positions
along the k_z axis on the TR-breaking parameter δ. When $|\delta| = t$, nodes with
opposite chiralities come together (pairwise) at $k_z = 0$ and $k_z = \pi$ (mod 2π)
and annihilate each other; for $|\delta| > t$, the system is gapped, representing a
weak 3D Chern insulator with Chern number $C = \pm 1$. The shaded area indicates the k_z values characterized by a nonzero Berry flux, $2\pi C$, through a
constant-k_z plane. (c) Schematic phase diagram for a system with TR symmetry and broken inversion symmetry. In 2D, the quantum spin Hall (QSH)
and trivial insulator phases are separated by a phase transition (driven by the
parameter m) characterized by a vanishing bulk gap (black line). In 3D, if the
system is not inversion symmetric (i.e., if $\lambda \neq 0$), the two gapped phases are
separated by a gapless (Weyl semimetal) phase (dark shaded area).

and the corresponding lateral BZ boundary. Again, the flux -2π through this
closed surface (or any other surface obtained by smoothly deforming it without
crossing a Weyl point) is consistent with the net chirality of the three Weyl
points surrounded by it. Of course, the flux through a constant-k_z plane that
separates the two pairs of nodes, e.g., $k_z = 0$, is zero.

The dependence of the position along k_z of the Weyl nodes on the TR-
breaking parameter δ is illustrated in Figure 4.8(b). The nodes associated
with $H^{(+)}$ emerge at the BZ boundary when $\delta/t = -1$ and, upon increasing
δ, move toward each other and annihilate at the zone center when $\delta/t = 1$.
The nodes associated with $H^{(-)}$ have a similar evolution when δ/t decreases
from 1 to -1. For $3t > |\delta| > t$ the system is gapped and the Berry flux
through every constant-k_z plane is $2\pi C$, with a Chern number $C = +1$ for
$\delta > t$ and $C = -1$ for $\delta < -t$. The corresponding gapped phases are weak
3D Chern insulators obtained by stacking 2D Chern insulator layers (also see
Figure 4.1). Thus the Weyl semimetal can be viewed as an intermediate state
between two topologically distinct insulating phases. This property is more
general. For example, it has been shown [376] that in 3D TR-invariant systems
with no inversion symmetry the trivial and quantum spin Hall (QSH) phases
are separated by an intermediate gapless phase. This scenario is schematically
illustrated in Figure 4.8(c). Starting in the trivial phase in the absence of
inversion symmetry (i.e., for $\lambda \neq 0$) and varying a certain parameter m leads

to the emergence of nodal points that separate (in momentum space) into topologically protected monopole–anti-monopole pairs. Eventually the nodes annihilate pairwise at a different value of m, changing partners as compared to the creation process, and the system enters a gapped QSH phase. By contrast, in two-dimensions the transition between the trivial and the QSH phases does not involve any intermediate gapless phase.

A defining manifestation of the nontrivial bulk topology associated with the Weyl nodes being monopoles of Berry curvature is the presence of gapless surface states. Since the bulk itself is gapless, defining surface states requires translation invariance. Then, at any energy E one can unambiguously define surface states labeled by crystal momenta $\tilde{\boldsymbol{k}} = (k_x, k_y)$ (within the 2D surface Brillouin zone) that do not label any bulk state $\boldsymbol{k} = (k_x, k_y, k_z)$ with energy $E_{\boldsymbol{k}} = E$. Assuming, for example, a pair of Weyl nodes at the Fermi energy ($E_F = 0$) with surface momenta $\pm \tilde{\boldsymbol{k}}_0$, one can define surface states with energy $E = E_F$ at any $\tilde{\boldsymbol{k}}$ different from $\pm \tilde{\boldsymbol{k}}_0$. The actual surface states that emerge in a Weyl semimetal form a *Fermi arc* that connects the two nodes, as illustrated in Figure 4.9. To determine the topological nature of the surface states, let us consider a plane in the Brillouin zone separating the two nodes. For concreteness, let us assume that the Weyl nodes are located at $\pm \boldsymbol{k}_0$ with $\boldsymbol{k}_0 = (k_0, k_0, 0)$. The Berry flux through the plane $k_x + k_y = K$, with $|K| < 2k_0$ is 2π, which implies that the 2D Hamiltonian $H(k_x, k_y, k_z)|_{k_x + k_y = K}$ describes a family of Chern insulators with Chern number $C = 1$. A surface perpendicular to the z direction represents a boundary along which the 2D Chern insulators will have chiral edge modes dispersing linearly near the Fermi energy, $\epsilon_{k_x + k_y}(k_x - k_y) \approx \epsilon^0_{k_x + k_y} + v_{k_x + k_y}(k_x - k_y)$, with ϵ^0 vanishing near the Weyl points, i.e., $\epsilon^0_{\pm 2k_0} = 0$. At $E = E_F = 0$, this family of boundary modes is labeled by crystal momenta $\tilde{\boldsymbol{k}} = (k_x, k_y)$ that form a Fermi arc connecting the Weyl nodes $\pm \tilde{\boldsymbol{k}}_0$. At finite energy, the projection onto the surface BZ of the bulk Fermi surface consists of filled disks enclosing the Weyl nodes, while the surface states are defined along arcs connecting the disks, as illustrated schematically in Figure 4.9. Note that the surface states disperse along the $x - y$ direction and inherit the chiral nature of the 2D Chern insulator edge states. Also note that, in a slab geometry, combining the Fermi arcs associated with the states on opposite surfaces would result in a closed Fermi line. Of course, the dispersion of the states hosted by the two surfaces is characterized by velocities having opposite signs – a property ultimately reflecting the presence of the nonzero chiral charges associated with the Weyl nodes. Experimentally, the presence of surface Fermi arcs was first established in the Weyl semimetal TaAs using photoemission spectroscopy [346, 559].

4.2.2 Topological semimetals and nodal superconductors

In the previous section, we have discussed key aspects associated with nontrivial topology in gapless systems by focusing on three-dimensional Weyl semimetals. Here, we sketch the basic ideas underlying the classification of

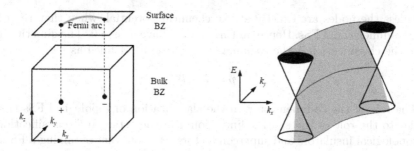

FIGURE 4.9 *Left*: Surface Fermi arc connecting the projections $\pm\tilde{k}_0$ of a pair of Weyl nodes onto the surface Brillouin zone. *Right*: Schematic representation of a surface state sheet (light gray) connecting two bulk state cones. Note that the sheet is tangent to the cones, as the surface states merge into the bulk.

gapless topological phases and list some of the systems that have attracted attention in recent years.

Each Weyl node and any (small) Fermi surface (FS) surrounding it are characterized by a well-defined topological charge – the chirality C equal to the Berry flux (in units of 2π) through a closed surface containing the FS.[8] We emphasize that the closed surface should contain a single FS. In general, a topologically nontrivial FS characterizing a lattice system is always accompanied by a "partner" with opposite topological charge, which is a consequence of the fermion doubling theorem [387]. Hence, since the total topological charge within the Brillouin zone (BZ) vanishes, the topological invariants have to be defined on submanifolds of the BZ, e.g., a closed surface containing a single FS or a plane separating a pair of FSs with opposite topological charges. Furthermore, pairs of FSs can be gapped by perturbations that connect them, i.e., by *nesting*. A (single) Fermi surface is said to be topologically stable if it is robust against perturbations that are smooth in real space (i.e., "local" in momentum space, to prevent nesting) and weak-enough (so that sufficiently far away from the FS the bulk gap remains finite and distinct FSs do not merge). In essence, the classification of gapless topological systems is the classification of topologically stable FSs.

Similar to the classification of topological insulators and superconductors, the classification of stable FSs depends on the symmetry class of the Hamiltonian describing the system and on the dimensions of the BZ and the FS. More specifically, let us consider a d-dimensional system, i.e., a BZ with dimension d. The dimension of the FS is, in general, dependent on the Fermi energy. We define the "minimal" dimension of the FS, d_{FS}, as the dimension of the corresponding band crossing, which is independent of energy. For example, the 3D Weyl semimetals discussed in the previous section are generally characterized by two-dimensional Fermi surfaces surrounding the Weyl nodes,

---

[8]Nodes occurring at the chemical potential are considered zero-dimensional FSs.

unless the nodes are exactly at the chemical potential, when they represent zero-dimensional FSs. Hence, in this case we have $d_{FS} = 0$. The quantity used in the classification is the *codimension* p of the FS defined as

$$p = d - d_{FS}. \tag{4.47}$$

The role of the codimension p in the classification of topological FSs is similar to the role of the spatial dimension d in the "standard" classification of topological insulators and superconductors. We note that obtaining a FS with the minimal dimension, d_{FS}, may require smooth deformations of the energy bands (without changing their topology) as, for example, in the case of type-II Weyl semimetals (see below).

To discuss the role of symmetry, let us focus on nonspatial AZ symmetries. We distinguish two types of Fermi surfaces: (i) "FS1" are Fermi surfaces located at the high symmetry points (or along symmetry-invariant lines) in the BZ and are invariant (individually) under anti-unitary AZ symmetries (which map $k \leftrightarrow -k$); (ii) "FS2" are located away from the high symmetry points (or symmetry-invariant lines) and are pairwise related to each other by the anti-unitary AZ symmetries. A few examples of "FS1" and "FS2" Fermi surfaces with codimension $p = 2$ are shown in Figure 4.10. The full classification of stable Fermi surfaces in terms of the ten AZ symmetry classes and FS codimension is given by Table 3.3 (see page 86) – the familiar "Periodic Table" of topological insulators and superconductors – with a column labeled by d corresponding to $p = d - 1 \pmod 8$ for "FS1" and $p = d + 1 \pmod 8$ for "FS2" [104, 358, 473]. For example, the column $d = 2$ in Table 3.3 describes the classification of stable Fermi surfaces of type "FS1" with codimension $p = 1$ and type "FS2" with codimension $p = 3$. We note that $\mathbb{Z}_2$ topological numbers do not protect Fermi surfaces located away from the high-symmetry points (or lines). For example, let us consider a Weyl semimetal with time-reversal symmetry and broken inversion symmetry described by a Hamiltonian obtained by taking $\delta = 0$ in Eq. (4.45) and adding a term $\lambda \sin k_z \tau_3 \sigma_3$, which is TR invariant but breaks inversion symmetry. The system, which belongs to symmetry class AII, has Weyl nodes located at $k_0 = (0, 0, \pm k_0^{(\pm)})$, with $\cot k_0^{(\pm)} = \pm \lambda / t$ and $0 < k_0^{(\pm)} < \pi$.[9] The codimension of the corresponding "FS2" is $p = 3$ and, according to Table 3.3, this would imply a $\mathbb{Z}_2$ topological classification. However, the Weyl nodes are protected by a $\mathbb{Z}$ topological invariant, e.g., the Chern number associated with planes that are not TR invariant and separate a pair of Weyl nodes (e.g., the plane $k_z = \pi/2$); in general, the invariant can include nonzero (integer) contributions from filled bands.

---

[9]With inversion symmetry (and broken TR), a Weyl node at k_0 is always accompanied by a partner node at $-k_0$ having the same energy and opposite chirality. In this case, the minimum number of Weyl nodes is two [see, e.g., the model Hamiltonian in Eq. (4.46)]. With TR symmetry (and broken inversion), the nodes at k_0 and $-k_0$ have the same topological charge and the presence of an additional pair of nodes (with opposite chirality) is required. The minimum number of Weyl nodes in this case is four.

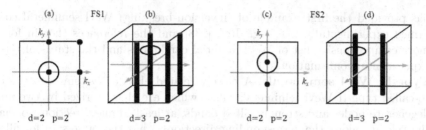

FIGURE 4.10 Examples of minimal Fermi surfaces (black points and lines) with codimension $p = 2$. Each "FS1" Fermi surface [panels (a) and (b)] is invariant under nonspatial symmetries, while the "FS2" Fermi surfaces [panels (c) and (d)] are pairwase related to one another by symmetry transformations that map $\boldsymbol{k} \leftrightarrow -\boldsymbol{k}$. The closed loops are contours on which the topological invariant (winding number) can be defined. For a three-dimensional system with zero-dimensional FSs (i.e., with $p = 3$) – see examples in Section 4.2.1 – the corresponding "contours" are closed surfaces surrounding the nodes.

In the presence of additional crystal symmetries, the classification of stable Fermi surfaces can be generalized along the lines discussed in Section 4.1.2. For example, in the presence of mirror symmetry, we first establish whether reflection commutes or anticommutes with the internal symmetries, i.e., we identify the values of $\eta_{t,c,s}$ in Eq. (4.8). Next, we determine how the FSs transform under reflection and internal symmetries. In general, we can have the following situations: (i) Each FS is invariant under internal and mirror symmetries. These FSs (denoted "FS1m") are located within mirror planes at high-symmetry points in the BZ. (ii) The FSs are pairwise related to each other by the AZ symmetries, but are invariant under reflection. These "FS2m" are located within mirror planes, but away from the high-symmetry points. (iii) The FSs are pairwise related to each other by both reflection and AZ symmetries. These FSs are located outside mirror planes and away from high-symmetry points. The topological classification of stable "FS1m" Fermi surfaces is given by Table 4.1 with the correspondence $d \to p = d - 1 \pmod 8$. Similarly, the classification of stable "FS2m" Fermi surfaces is given by Table 4.1 with the correspondence $d \to p = d + 1 \pmod 8$. A detailed discussion of the classification, including case (iii), can be found in Refs. [103] and [473].

We conclude this section with a list of gapless topological phases that have attracted attention in recent years, pointing out a few materials that were predicted to host these phases, some of them already confirmed experimentally. We note that a key role in identifying materials that potentially support topological phases is played by *ab initio* calculations [35].

Type-I Weyl semimetals. The systems discussed in Section 4.2.1 are examples of so-called type-I Weyl semimetals, which are characterized by two-fold degenerate nodes corresponding to crossings of a pair of nondegenerate bands having linear dispersion in all directions near the nodes. The TaAs family has provided the first example of (inversion-breaking) Weyl semimetal confirmed experimentally; see, e.g., Ref. [238] and the references therein for a more comprehensive list of Weyl material candidates and the status of their experimental confirmation.

Type-II Weyl semimetals. A closely related class of systems consists of so-called type-II Weyl semimetals [482], which are characterized by two-fold degenerate nodes and strongly tilted bands, at least in one direction, so that the velocity along the corresponding direction(s) has the same sign for all k values near the node, as illustrated in Figure 4.11. The result is that, even when a node is at the Fermi energy, each of the two bands has both positive and negative energy contributions forming electron and hole pockets, with the Weyl node emerging at the boundary between these pockets. We note that, when determining the minimal dimension, d_{FS}, for type-II Weyl FSs, we can assume that the tilt of the bands is smoothly reduced, so that the electron and hole pockets shrink and eventually disappear (i.e., we have $d_{FS} = 0$). Examples of type-II Weyl semimetals are the $Mo_x W_{1-x} Te_2$ family and the LaAlGe family [238].

Double-Weyl and triple-Weyl fermions. In the presence of additional crystal symmetries, the band dispersion near a Weyl node can be different from the standard linear dispersion. For example, systems with four-fold (six-fold) rotation axes can support Weyl nodes corresponding to crossings of a pair of singly-degenerate bands having quadratic (cubic) energy dispersion in the planes normal to the rotation axes (see Figure 4.11), while the dispersion along those axes is still linear. The corresponding low-energy excitations, called double-Weyl (triple-Weyl) fermions, have chiralities $C = \pm 2$ ($C = \pm 3$) and can be viewed as two (three) conventional Weyl fermions with the same chirality "sticking" together. Upon breaking the rotation symmetry, a double-Weyl (triple-Weyl) fermion splits into two (three) Weyl fermions with chirality $|C| = 1$. An proposed material candidate is $SrSi_2$ [256].

High-fold chiral fermions. In addition to two-fold degenerate Weyl nodes, nonmagnetic three-dimensional crystals allow band crossing involving three, four, six, or eight bands [74, 238]. A class of such high-fold crossings is characterized by low-energy excitations representing chiral fermions, a direct generalization of the two-fold Weyl fermions. In particular, it was shown that chiral crystal structures universally supports Weyl and high-fold chiral fermions near the high symmetry momenta [74, 238]. An n-fold crossings can be characterized by the chiral charges $(C_1, C_2, \ldots, C_{n-1})$ associated with the $n - 1$ inter-band gaps (see Figure 4.11). The RhSi family provides an example of material that hosts high-fold chiral fermions.

Nodal line semimetals. The band crossings characterizing the systems mentioned above are zero-dimensional (i.e., points in the BZ). However, higher dimensional crossings are possible. For example, the valence and conduction bands can cross along one-dimensional curves in momentum space [25, 89]. These crossings are protected by topological invariants given by winding numbers – integrals of the Berry connection along loops encircling the nodal lines

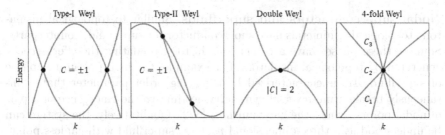

FIGURE 4.11 Schematic representation of band dispersions (along a representative direction in momentum space) associated with different types of Weyl fermions and their generalizations. From left to right: Crossings of non-degenerate valence and conduction bands giving rise to a pair of type-I Weyl nodes (black dots). Crossings of (strongly tilted) nondegenerate bands giving rise to a pair of type-II Weyl nodes. Quadratic dispersion in the vicinity of a double-Weyl node. Four-fold degenerate crossing with chiral charges C_1, C_2, C_3.

(see Figure 4.10). The boundary manifestation of the topologically nontrivial bulk structure are drumhead surface states connecting the bulk line nodes [89, 103], a two-dimensional generalization of the surface Fermi arcs that characterize Weyl semimetals. An example of material exhibiting drumhead surface states is PbTaSe$_2$ [238].

Dirac semimetals. If both inversion and TR symmetries are present, band crossings in three-dimensional materials are (at least) four-fold degenerate, which corresponds to a pair of Weyl nodes with opposite chiralities located at the same crystal momentum. This type of degeneracy has zero net Chern number and, consequently, is *not* topologically protected. However, in the presence of additional symmetries, these nodes can be stabilized as *symmetry-protected* degeneracies. The corresponding low-energy excitations are condensed matter realizations of Dirac fermions and the materials that host them are called Dirac semimetals. Dirac *nodes* can emerge, for example, in systems that are fine tuned to the critical point separating $\mathbb{Z}_2$ topological phases, where the bulk gap closes. Dirac semimetal *phases* can occur as a result of band inversion, when mixing of the inverted bands along certain symmetry lines is forbidden as a result of the bands having different symmetry properties. However, these Dirac points can be annihilated pairwise by continuously changing the system parameters (without affecting the symmetry) and, eventually, "un-inverting" the bands. A more robust mechanism involves band structures with space groups that require the presence of un-removable Dirac singularities. Systems having such singularities at or near the chemical potential are referred to as symmetry-enforced Dirac semimetals. A detailed discussion of these classes of Dirac semimetals, including aspects of their phenomenology (e.g., surface Fermi arcs) and materials considerations, can be found in Ref. [25].

Nodal superconductors and superfluids. Similar to topological insulators, topological semimetals have superconductor (or superfluid) counterparts. Some of these phases have a rather long history, predating the effervescence generated by topological insulators. For example, the high pressure A phase of superfluid ^{3}He is characterized by a pairing order parameter that spontaneously breaks time-reversal symmetry, leading to the emergence of a pair of nodal points (where the gap vanishes) [322]. Qualitatively, the system can be understood as a three-dimensional $p_x + ip_y$ superfluid with gapless points on the Fermi surface at $k_0 = (0, 0, \pm k_F)$. These points represent the superfluid analog of Weyl nodes and the corresponding low-energy quasiparticles are Weyl fermions [523]. The system supports Fermi arc surface states and displays the phenomenology associated with Weyl fermions (e.g., the chiral anomaly; see Section 7.3). More recent theoretical investigations have revealed that many nodal superconductors have nontrivial topology, including well known systems such as $d_{x^2-y^2}$ superconductors and heavy Fermion systems. While Weyl superconductors have features similar to those of ^{3}He-A, three-dimensional Dirac superconductors are characterized by protected fourfold degenerate nodal points and support surface Majorana arcs at zero energy [561]. Furthermore, by analogy with HOTIs (see Section 4.1.3), higher-order generalization of Dirac and Weyl superconductors featuring Majorana hinge modes have been proposed [569]. A review of some of the developments in the theoretical understanding of nodal topological superconductors can be found in Ref. [453].

4.3 FLOQUET TOPOLOGICAL INSULATORS

The theoretical study of topological condensed matter started with an explanation of the remarkable stability of quantum Hall plateaus in terms of topology and continued with the formulation of a few (rather exotic) models of topological quantum systems. An explosion of activity in the field was associated with the realization that, rather than being a simple (elegant) explanation, or a feature associated with a few exotic cases, topology is a paradigm that enables a deeper understanding of physical reality. People started to look for topology, and, behold, it was everywhere. Our discussion, so far, clearly illustrates the variety of topological quantum systems and the rather general conditions in which topological ideas are relevant. In this section, we further expand these conditions by considering nonequilibrium, driven systems. Our main purpose is to provide the reader with a taste of this interesting research direction by sketching the main ideas, without going into technical details. We will focus on two aspects – the basic formalism for describing periodically driven systems and the key ideas underlying the emerges of topology at the single-particle level – and we will briefly mention the main challenge (the heating problem in driven many-body systems) and some possible routes to overcoming this challenge and stabilizing the topological phases. Detailed discussions of topology in periodically-driven systems can be found in several recent review articles, for example Refs. [237], [395], and [436].

The description of a quantum system in terms of a time-independent Hamiltonian – the description that we have used so far in our discussion of topological quantum matter – is appropriate for isolated systems. More generally, open systems, i.e., systems coupled to an environment, can be described either in terms of a Hamiltonian corresponding to the "total" system (i.e., system of interest plus environment), or within a non-Hamiltonian framework, after integrating out the environment [83]. An interesting subclass of open systems – the so-called Floquet systems – can still be described by a (time-dependent) Hamiltonian, $H(t)$, the effect of the environment being incorporated as a time-periodic modulation of various coupling parameters. The time-dependent Schrödinger equation satisfied by the corresponding unitary time-evolution operator $U(t, t_0)$, which relates the state vectors at t_0 and t, is

$$i\hbar \frac{d}{dt} U(t, t_0) = H(t) \, U(t, t_0), \qquad (4.48)$$

with the initial condition $U(t_0, t_0) = \mathbb{I}$, where $\mathbb{I}$ is the identity operator. In addition, the time-evolution operator is unitary, $U^\dagger(t, t_0) = U^{-1}(t, t_0)$ and satisfies the group property $U(t_3, t_2) U(t_2, t_1) = U(t_3, t_1)$. Note that, as a consequence of these properties, we have $U^{-1}(t, t_0) = U(t_0, t)$. In Eq. (4.48) the Hamiltonian $H(t) \equiv H(t + T)$ is a periodic operator-valued function of time (with period T) corresponding to a specific Floquet "drive." As a direct consequence of this periodicity, we have $U(t + T, T) = U(t, 0)$, since the two operators satisfy the same equation and have the same initial condition. This implies that the full time evolution of the system can by obtained from $U(t, 0)$ with $0 \le t \le T$ using the group property and the identity

$$U(t + T, 0) = U(t, 0) U(T, 0), \qquad 0 \le t \le T. \qquad (4.49)$$

The one-cycle evolution operator $U(T) \equiv U(T, 0)$, which captures the stroboscopic dynamics of the system between discrete time values nT and $(n + 1)T$, with $n \in \mathbb{Z}$, can be written in terms of a Hermitian operator H_F as

$$U(T) = \exp\left(-\frac{i}{\hbar} H_F T\right). \qquad (4.50)$$

Using Eqs. (4.49) and (4.50) one can easily verify that the operator $\Phi(t) = U(t, 0) \exp(\frac{i}{\hbar} H_F t)$ is unitary and time-periodic (with period T). Consequently, the evolution operator of the periodically driven system can be decomposed in the form

$$U(t, 0) = \Phi(t) \exp\left(-\frac{i}{\hbar} H_F t\right), \qquad \Phi(t + T) = \Phi(t). \qquad (4.51)$$

Equation (4.51) is a formulation of Floquet's theorem [172]. The Hermitian operator H_F, called the Floquet Hamiltonian, characterizes the stroboscopic dynamics of the system [Eq. (4.50)], while the T-periodic unitary operator $\Phi(t)$ describes the "micromotion" of the system during a cycle.

To characterize the system, it is convenient to determine the eigenstates of $U(T)$ and the corresponding eigenvalues,

$$U(T)|\alpha\rangle = e^{-\frac{i}{\hbar}\epsilon_\alpha T}|\alpha\rangle, \tag{4.52}$$

where the so-called *quasienergy* ϵ_α, which is an eigenvalue of H_F, is a real quantity. While the eigenvalues of $U(T)$ are unique, the quasienergies are defined up to integer multiples of $2\pi\hbar/T = \hbar\omega$, where ω is the angular frequency of the drive. Formally, this nonuniqueness is related to the multi-valuedness of the complex logarithm that defines the Floquet Hamiltonian, $H_F = i\hbar/T \ln U(T)$, and expresses the fact that energy is not a conserved quantity in a periodically driven system, as it can exchange energy quanta $\hbar\omega$ (e.g., photons) with the environment. We define a special set of solutions of the time-dependent Schrödinger equation, called the *Floquet states*, as

$$|\psi_\alpha(t)\rangle = U(t,0)|\alpha\rangle, \qquad |\psi_\alpha(t+T)\rangle = e^{-\frac{i}{\hbar}\epsilon_\alpha T}|\psi_\alpha(t)\rangle. \tag{4.53}$$

Using Eq. (4.51) one can express a Floquet state in terms of a T-periodic state $|u_\alpha(t)\rangle = |u_\alpha(t+T)\rangle$ as $|\psi_\alpha(t)\rangle = |u_\alpha(t)\rangle \exp(-i\epsilon_\alpha t/\hbar)$, which is the analog of Bloch's theorem for time-periodic systems. A generic solution of the time-dependent Schrödinger equation can be written as a linear combination of Floquet states with *time-independent* coefficients, $|\psi(t)\rangle = \sum_\alpha a_\alpha|\psi_\alpha(t)\rangle$.

If we now consider a periodically-driven system with crystalline translation symmetry, the quantities introduced above can be defined as functions of the crystal momentum $\boldsymbol{k}$ taking values within the Brillouin zone. In particular, the time-independent Floquet Hamiltonian is given by a Hermitian operator $H_F(\boldsymbol{k})$ and the corresponding eigenvalues form quasienergy (Floquet–Bloch) bands, $\epsilon_\alpha(\boldsymbol{k})$. The basic idea behind the realization of topological band structures in periodically-driven systems is to apply the "standard machinery" for characterizing the topological states of equilibrium (noninteracting) Hamiltonians to the effective Floquet Hamiltonian, $H_F(\boldsymbol{k})$ [339]. In particular, by applying symmetry considerations similar to those discussed in the context of equilibrium systems one can obtain "periodic tables" that describe the topological classification of band structures associated with Floquet systems [395, 435]. Here, we will not discuss these developments. Instead, we point out two aspects that are specific to the physics of periodically-driven systems. First, the Floquet–Bloch bands characterizing a driven system can exhibit nontrivial topology even when the undriven system is topologically trivial [299, 339]. In other words, the drive induces a nontrivial band structure in an otherwise trivial system. Nonetheless, these topological band structures, which can be identified based on the properties of $H_F(\boldsymbol{k})$, have equilibrium counterparts [395]. A second, more interesting possibility is to have Floquet drives that lead to physical manifestations of nontrivial topology that can only occur in driven systems. In particular, while equilibrium systems are characterized by the presence of topologically protected boundary modes at interfaces between regions hosting bands with different topological indices (e.g., different values of the sum of Chern numbers associated with the occupied bands; see Section 3.4.2), the boundary modes emerging in Floquet systems are not necessarily governed by this type of bulk-boundary correspondence. This is because the Floquet Hamiltonian does not contain all the information regarding the topological properties of the system.

To better understand the source of the new physics in Floquet systems, let us first emphasize the difference between the energy spectrum of an equilibrium system, which is bounded from bellow and extends to arbitrarily high energy, and the quasienergy spectrum, which is periodic as a result of $\epsilon_\alpha(\boldsymbol{k})$ being defined up to integer multiples of $\hbar\omega$. As a result, the quasienergy is defined on a circle, similar to the crystal momentum being defined on a torus. Essentially, this analogy (as well as the analogy between Floquet's and Bloch's theorems mentioned above) stems from the fact that in both cases we have a continuous symmetry (spatial or time translation) that was reduces to a discrete symmetry, while the corresponding conserved quantity (momentum or energy) was "downgraded" to a quantity (crystal momentum or quasienergy) defined modulo contributions (reciprocal lattice vectors $\boldsymbol{K}$ or energy quanta $\hbar\omega$) determined by the periodicity of the system (primitive translation vectors or time period T).

To illustrate the consequences of the energy spectrum having a periodic structure, let us consider a 2D system in the semi-infinite strip geometry having two Floquet bands characterized by the Chern numbers C and $-C$, respectively, where $C = 0, \pm 1$. The corresponding quasienergy spectrum is represented schematically in Figure 4.12(a); its equilibrium counterpart is illustrated in Figure 3.3 (see page 93). The equilibrium system is characterized by one chiral edge mode crossing the bulk gap if $|C| = 1$ (i.e., if the system is topologically nontrivial; see Figure 3.3) and no edge mode that crosses the gap if $C = 0$ (trivial phase). The transition between the two phases is accompanied by the closing of the bulk gap (e.g., at $k_x = 0$). By contrast, the periodically-driven system can also undergo transitions associated with the closing of the bulk quasienergy spectrum at the edge of the energy "Brillouin zone," i.e. at $\epsilon = \pm\pi\hbar/T$ (near, e.g., $k_x = \pm\pi/2$ in Figure 4.12), without the bulk gap near $k_x = 0$ being affected. Thus, imagine starting from a trivial phase with $C = 0$ and no chiral edge modes (phase 1); close and reopen the gap near $k_x = 0$ ($\epsilon = 0$) and transition into a topological phase with $|C| = 1$ and one chiral mode crossing the $\epsilon = 0$ gap (phase 2); close and reopen the gap near $|k_x| = \pi/2$ ($|\epsilon| = \pi\hbar/T$) and transition into a phase with $C = 0$ and chiral modes crossing both gaps, at $\epsilon = 0$ (which was not affected) and $|\epsilon| = \pi\hbar/T$ (phase 3). While the bulk-boundary correspondence associated with phase 1 and phase 2 is the same as that characterizing equilibrium systems, the behavior associated with phase 3 has no equilibrium counterpart. Also, the possibility of having phase 3 (with chiral edge modes and trivial bulk

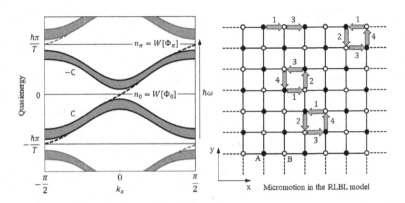

FIGURE 4.12 *Left*: Schematic representation of the quasienergy spectrum of a semi-infinite 2D system. The two Floquet bands have Chern numbers C and $-C$, respectively. Transitions between states with $C = 1$ and $C = 0$ are associated with the closing of the (bulk) quasienergy gap near $\epsilon_0 = 0$ or $\epsilon_\pi = \hbar\pi/T$. The number of chiral modes crossing a gap (n_0 or n_π) is determined by the corresponding winding number $W[\Phi_\epsilon]$, rather than the Chern number (see text). However, we have $C = W[\Phi_0] - W[\Phi_\pi]$. *Right*: Micromotion of a particle initialized on A (filled circles) and B (open circles) sites during one cycle in a semi-infinite system described by the RLBL model [437]. The numbers $1, \ldots, 4$ correspond to the four steps described in Eq. (4.54). Particles initialized on bulk sites (and B-type edge sites) return to the initial position after moving counterclockwise around a plaquette. Particle initialized on A-type edge sites move along the edge with velocity $2/T$.

band invariants) suggests that the description of the topological properties of the system in terms of an effective Hamiltonian (H_F) is incomplete.

To clarify this point, we consider a simple model introduced by Rudner, Lindner, Berg, and Levin (RLBL) [437]. The model is defined on a square lattice with two sub-lattices, A and B, and has time-periodically modulated nearest-neighbor hopping. For simplicity, we assume that the lattice constant is $a = 1$. Let $\boldsymbol{\delta}_x = (1, 0)$ and $\boldsymbol{\delta}_y = (0, 1)$ designate the vectors that couple A sites to B sites along the x and y directions, respectively, as shown in Figure 4.12(b). Also, let $t_{ij}^{AB}(t)$ be the hopping parameter along the bond coupling site $i = (i_x, i_y)$ on sub-lattice A to site $j = (j_x, j_y)$ on sub-lattice B. The periodic drive is defined by the following four-step protocol

$$t_{ij}^{AB}(t) = \begin{cases} J \text{ if } j = i + \boldsymbol{\delta}_x; & 0 \text{ otherwise,} & 0 < t < T/4, \\ J \text{ if } j = i - \boldsymbol{\delta}_y; & 0 \text{ otherwise,} & T/4 < t < T/2, \\ J \text{ if } j = i - \boldsymbol{\delta}_x; & 0 \text{ otherwise,} & T/2 < t < 3T/4, \\ J \text{ if } j = i + \boldsymbol{\delta}_y; & 0 \text{ otherwise,} & 3T/4 < t < T, \end{cases} \quad (4.54)$$

Choosing $J = 2\pi\hbar/T$ ensures that, during each $T/4$ time step, a particle initialized on a given site is transferred to a neighboring site (along the bond with nonzero hopping) with probability one. Over a full cycle, a particle initialized on a bulk site (or a B-type edge site) will move counterclockwise around a plaquette and return to the initial position, as illustrated in Figure 4.12(b). Thus, the stroboscopic dynamics of the bulk is trivial, with $U(T) = \mathbb{I}$, i.e., $H_F = 0$. The corresponding bulk quasienergy spectrum consists of a double-degenerate flat band, $\epsilon(\mathbf{k}) = 0$, characterized by a trivial Chern invariant. However, a particle initialized on a boundary A site moves two sites to the right over one cycle (Figure 4.12(b)), which implies the presence of a chiral edge mode with (group) velocity $+2/T$. We note that, in a (finite) strip geometry, a chiral mode with velocity $-2/T$ will propagate on the lower edge of the system.

The presence of the chiral edge modes reflects the topological nature of the system, which is not captured by the effective Hamiltonian description. In fact, the stroboscopic picture completely ignores the micromotion that takes place within each driving period, e.g., the counterclockwise movement of a particle around a plaquette. The information about this chiral movement is contained in the time-periodic operator $\Phi(t)$ from Eq. (4.51). Indeed, as shown in Ref. [437], the 2D system can be characterized by a topological winding number, $W[\Phi]$, defined in terms of the micromotion operator Φ as

$$W[\Phi] = \frac{1}{8\pi^2} \oint dt dk_x dk_y \, \mathrm{Tr} \left(\phi^\dagger \partial_t \phi [\phi^\dagger \partial_{k_x} \phi, \phi^\dagger \partial_{k_y} \phi] \right), \qquad (4.55)$$

where $[\,,\,]$ designates the commutator. For the RLBL model described above this number is $W = 1$, which coincides with the number of chiral edge modes. Moreover, if a Floquet system is characterized by a quasienergy gap around a quasienergy value ϵ, one can determine the (net) number n_ϵ of chiral modes that cross the gap by calculating the winding number $W[\Phi_\epsilon]$ corresponding to a time-evolution operator $U_\epsilon = \Phi_\epsilon$ (with $H_F = 0$) obtained by smoothly deforming the original operator $U(t, 0)$ so that all the quasienergy bands with $\epsilon_\alpha(\mathbf{k}) > \epsilon$ are "pushed" through the $\pi\hbar/T$ boundary of the energy "Brilouin zone" and, after re-appearing (due to periodicity) above the $-\pi\hbar/T$ boundary, are "squished" together with the bands that were already below the gap $(\epsilon_{\alpha'}(\mathbf{k}) < \epsilon)$ into a degenerate zero-quasienergy flat band. By construction, the (net) number of chiral edge modes that cross the (original) ϵ-gap is the same as the number of chiral edge modes of the system with time-evolution operator Φ_ϵ, which can be determined by calculating the winding number $W[\Phi_\epsilon]$ – see Figure 4.12(a). We note that the difference $W[\Phi_{\epsilon_1}] - W[\Phi_{\epsilon_2}]$ is equal to the sum of the Chern numbers of all Floquet bands that lie in between ϵ_1 and ϵ_2 [437], consistent with the intuitive picture sketched in Figure 4.12(a).

Our discussion of periodically-driven systems has focused, so far, on single-particle systems and the corresponding (possibly "topological") Floquet-Bloch band structure. However, real systems studied in the laboratory consist of many particles. Equilibrium noninteracting topological systems (e.g., topological insulators) are many-body systems obtained by "filling" with electrons a certain number of bands characterized by nontrivial topological invariants.

Weak interactions (which are always present) do not affect the topological properties of the system. By contrast, a weakly interacting Floquet system is expected to absorb energy from the environment and heat to infinite temperature, which maximizes the entropy. In this regime, which can be reached at long-enough times, all local operator expectation values become time independent and all signatures of nontrivial topology disappear.

Is there any way around this "heating catastrophe"? One possibility is to focus on short-times, before runaway heating occurs (see [436] and references therein). The short-time dynamics, which can reveal the opening of Floquet gaps, can be characterized using pump-probe measurements. We note that the heating rates can be suppressed in the high-frequency regime, when $\hbar\omega$ is large compared to the single-particle bandwidth and the strength of local interactions [436]. Such a regime can be practically realized in systems of cold atoms trapped in optical lattices. Systems that absorb energy very slowly can realize a so-called *prethermal* regime characterized by an equilibrium-like many-body state, which can exhibit topological phases over significant time periods [157]. Another possibility is to couple the system to a bath, which results (at long-enough times) in a steady state. For a Floquet system, the steady state will be depend on the microscopic details of the bath and of the system-bath coupling [306]. Establishing the conditions under which a dissipative coupling to a bath can stabilize topologically nontrivial states in a periodically-driven system and classifying the corresponding phases is an active research area [37, 461, 571].

Finally, an interesting possibility is to realize stable topological Floquet phases in *interacting* systems exhibiting many-body localization (MBL) [382]. A system in thermal equilibrium is completely characterized by a small number of intensive parameters (one parameter for each extensive conserved quantity), which means that the system has no memory of its (typically nonequilibrium) initial state. For an isolated quantum system, this property can be reconciled with unitary time evolution if the system satisfies the so-called *eigenstate thermalization hypothesis* (ETH). Consider a many-body system described by the Hamiltonian H. The thermal equilibrium properties are determined by the density operator $\rho^{eq}(T) = e^{-\beta H}/\text{Tr}[e^{-\beta H}]$, with $\beta = 1/k_B T$ (where T is now temperature). Assume that the completely isolated system is in a (many-body) state $|\psi_\alpha\rangle$, with $H|\psi_\alpha\rangle = E_\alpha|\psi_\alpha\rangle$. We define the temperature T_α through the relation $E_\alpha = \langle H \rangle_{T_\alpha} = \text{Tr}[\rho^{(eq)}(T_\alpha)H]$. Let us now divide the system into two subsystems, A and B; the subsystem A is described by the density operator $\rho_A^{(\alpha)} = \text{Tr}_B[|\psi_\alpha\rangle\langle\psi_\alpha|]$. The ETH asserts that in the thermodynamic limit subsystem A is at thermal equilibrium, i.e., $\rho_A^{(\alpha)} = \rho_A^{(eq)}(T_\alpha)$. We can intuitively understand this behavior by considering subsystem B as a reservoir coupled to A. Since the two subsystems are entangled, the information regarding the initial local properties of A is not locally available, being "distributed" throughout the entire system; subsystem A is thermalized. These considerations can be generalized to Floquet systems, leading to the conclusion that they can thermalize to infinite temperature, which corresponds to all pure states of any small subsystem being equiprobable. Consider now an interacting system with random couplings. Such a system exhibits many-body localization [382] and is characterized by a large set – $\mathcal{O}(N)$ – of spatially localized, mutually commuting ("one-bit") operators that commute with $U(T)$. As a result, the system effectively "breaks up" into local Rabi oscillators that do not transfer energy to each other, but can mutually affect their phases. The net energy absorbed during one cycle is zero and, consequently, the system does not heat up (breaking the Floquet analog of ETH). Furthermore, this property is robust, in the sense that weak perturbations to a drive that results in a Floquet-MBL regime will not affect that regime. Thus, a Floquet-MBL system is an excellent candidate for a system capable to support robust phases, including phases with nontrivial topological properties.

CHAPTER 5

Interacting Topological Phases

THE "TEN-FOLD WAY" UNDERLYING THE CLASSIFICATION OF NONIN-teracting topological insulators and superconductors maps the infrastructure of the "civilized" part of the topological world. With interactions, we step into the wild. The main challenges that stem from the presence of interactions can be organized into three categories. The first set of problems is generated by the lack of a general classification of interacting topological phases (i.e., the interacting analog of the "ten-fold way"), which could be used as the basis for a systematic understanding of these quantum states. Moreover, it is not even clear what type of mathematical structure should underlie such a classification. Nonetheless, significant progress has been made in the past few years involving partial classifications of certain types of interacting topological states based on tensor categories [512], group cohomology theory [98, 226], cobordism [276], and other "scary mathematical beasts."

The second type of challenge stems from the intrinsic difficulty of describing theoretically a many-body system with interactions, particularly a strongly correlated one. Topological quantum field theories, which may be describing the low-energy physics of some yet undiscovered physical systems that support topological phases, have been studied for decades. However, the long journey of topological quantum matter from the ivory tower of field theory to the lab should probably pass through some understanding of realistic topological models that could be realized in well-defined physical systems. Generating such models and understanding their physics are formidable tasks.

Finally, the experimental study of interacting topological phases is still in its infancy. The discovery of the fractional quantum Hall effect more than three decades ago stormed into the heart of condensed matter physics taking everybody by surprise. Many of the subsequent theoretical advances that define our current understanding of topological quantum phases were stimulated by this discovery. The experimental field, however, has since remained silent,

DOI: 10.1201/9781003226048-5

for the most part. One could only hope that the theoretical progress and the discovery of (noninteracting) topological insulators will eventually trigger some positive developments regarding the experimental study of interacting topological phases.

In this chapter, we will present a few highlights of the recent advances in understanding the topological quantum states of interacting many-body systems. Since this is a rapidly developing field and because any detailed description would involve a rather complex mathematical apparatus, we will only consider basic ideas and relatively simple examples. A more in-depth discussion can be found in specialized books and review articles, such as, for example [174, 440, 465, 514, 566].

5.1 TOPOLOGICAL PHASES: ORGANIZING PRINCIPLES

Broken symmetry and the associated long-range order form the foundation for classifying different phases of matter in traditional many-body physics. This paradigm fails in the case of topological quantum phases, which are distinct gapped ground states that have the same symmetry as the Hamiltonian of the many-body system and, consequently, cannot be distinguished from one another based on their symmetry. Yet intertwined with topology, symmetry still plays a central role as a guiding principle for organizing quantum matter, although in a dramatically different way. Within the new paradigm, quantum phases are distinguished based on their topological properties, but symmetry may be required to ensure the very existence of distinct topological properties. In other words, for some topological classes, two quantum states that are topologically distinct in the presence of a certain symmetry can be adiabatically deformed into one another in the absence of that symmetry. In this case, we say that the topological states are *symmetry protected*.

But symmetry protection is not an absolute requirement for having distinct topological phases. There is, in fact, another concept that plays a critical role in organizing topological quantum matter: quantum entanglement. In the presence of long-range entanglement, quantum states acquire so-called *intrinsic topological order*, which makes them topologically distinct even in the absence of a symmetry constraint. At the time of this writing, the most widely accepted perspective on topological quantum states of matter is based on the triplet entanglement – symmetry – topology. Below we provide a sketch of the "big picture" that emerges from this perspective.

5.1.1 Systems with no symmetry constraints

In Chapter 3, we have seen that many-body Hamiltonians can be organized into classes based on the dimensions of the underlying systems and on the symmetry constraints that they satisfy. We will call such a class of Hamiltonians having a given dimension and specified symmetry constraints an H-class. Within a given H-class, we define different quantum phases as the equivalence (sub)classes generated by a certain equivalence relation. Focusing on gapped phases, the equivalence relation is defined by the *gap-preserving continuous deformations*: we say that two Hamiltonians H_1 and H_2 are equivalent if they can be continuously deformed into each other (within the given H-class) without closing the gap that characterizes their excitations above the corresponding ground states.

Consider now the class of Hamiltonians describing d-dimensional systems without any symmetry constraint, i.e., the class of all (local) Hamiltonians of dimension d. How many distinct gapped phases are there within this H-class? At the time of this writing we do not have a complete answer to this question. However, it is known that the key element responsible for the existence of distinct quantum phases in this H-class is *quantum entanglement*. Note that these phases completely evade a classification based on Landau's symmetry breaking paradigm (since there is no symmetry to break).

Many-body entanglement. Entanglement is perhaps the most nonclassical (hence, counterintuitive) manifestation of quantum mechanics. It plays a central role in the context of EPR-like experiments, leading to correlations that violate Bell's inequalities. In Chapter 10, we will show that entanglement also plays a key role in quantum information, where it can be viewed as a new type of resource for information processing. Here, we focus on a few aspects regarding the characterization of many-body entanglement in quantum systems. We have already introduced the *entanglement entropy* [see Eq. (2.14)] as a measure of quantum entanglement in bipartite pure systems. We note that more complex situations involving multi-partite systems or mixed quantum states may require other entanglement measures (for a review, see [253, 409]).

A key observation is that entanglement is a property of a quantum state *relative* to a given set of subsystems [124]. Consider, for example, a system consisting of two fermions that occupy spatially separated (exponentially localized) orbitals centered around two sites. We can either consider the *particle partition* of the system [235] – in which case the subsystems are the two particles – or its *spatial partition* (the subsystems being the two sites). Relative to the spatial partitioning, the ground state is a direct product state with no entanglement. On the other hand, relative to the particle partitioning, this is the maximally entangled Bell state $[\phi_\alpha(1)\phi_\beta(2) - \phi_\alpha(2)\phi_\beta(1)]/\sqrt{2}$. Note also that particle partitioning entanglement is sensitive to quantum statistics, i.e., in the case of fermions, to the anti-symmetrization of the wave function. Finally, we point out that absence of entanglement relative to a real space partition can be understood intuitively as characterizing subsystems that contain within themselves all the information regarding their states. By contrast, in the presence of entaglement the information regarding the state of a subsystem is encoded within the whole system.

A second observation concerns the quantitative *measures* of entanglement. How can one characterize many-body entanglement and how can one distinguish among different types of entangled states? These are key questions because the presence of different types of entanglement is what differentiates (topological) quantum phases that otherwise cannot be distinguished by any local order parameter. It was conjectured [99] that all gapped quantum ground states described by the Landau symmetry breaking theory have *short-range* quantum entanglement. By contrast, quantum phases that are beyond the Landau paradigm, such as the fractional quantum Hall states, have deeply entangled ground states characterized by various *patterns of long-range entanglement*. But how can one formally distinguish between these types of entanglement? While a complete characterization of different types of entanglement and the classification of the corresponding quantum phases are objects of ongoing research, significant progress has already been made.

One approach [99] is based on the observation that two gapped ground states are in the same phase if (and only if) they are connected by a *local unitary evolution*.[1] In other words, local unitary transformations define an equivalence relation on a given H-class, the corresponding equivalence classes representing different quantum phases. Furthermore, with no symmetry condition imposed, local unitary transformations can "remove" local entanglement and evolve a short-range entangled state into a direct product state [99]. By contrast, states with long-range entanglement cannot be connected by a local unitary transformation to short-range entangled states. Consequently, in the absence of any symmetry constraint, the equivalence classes defined by local unitary transformations correspond to phases characterized by different patterns of long-range entanglement, in addition to one class containing all short-range entangled states.

Another approach is based on the so-called *topological entanglement entropy* introduced by Kitaev and Preskill [295]. In essence, if we consider the spatial partition of a two-dimensional system with ground state $|\Psi\rangle$ into two subsystems (A and B), the von Neumann entropy $S = -\text{Tr}_A[\rho_A \log \rho_A]$, where $\rho_A = \text{Tr}_B |\Psi\rangle\langle\Psi|$, has the form $S = \alpha L - \gamma + \mathcal{O}(L^{-1})$. Here, L is the length of the boundary (assumed to be large compared to the correlation length) and γ – the *topological entanglement entropy* – is a constant reflecting the topological properties of the entanglement that survive at arbitrarily long distances. In other words, a nonvanishing γ signals long-range entanglement. We note that the low-energy, long-wavelength observable properties of a many-body system with long-range entanglement are invariant under smooth deformations of space-time. Consequently, the low-energy effective theory of the system is a *topological quantum field theory* (TQFT). Remarkably, the properties of the (gapped) quasiparticles that emerge in such a system are part of the universal low-energy physics of the system and can be obtained within the framework of the TQFT (see Chapter 12 for more details).

Returning to the topological entanglement entropy, one can show [295] that it is related to the so-called *total quantum dimension* of the system.

---

[1] Note that a local unitary evolution is equivalent to a finite-depth local unitary quantum circuit [99] (see also Chapter 11).

Specifically, we have [295]

$$\gamma = \log \mathcal{D}, \tag{5.1}$$

where the total quantum dimension is given by

$$\mathcal{D} = \sqrt{\sum_a d_a^2}. \tag{5.2}$$

Here d_a (called the *quantum dimension* of type a quasiparticles) can be thought of as the asymptotic degeneracy per particle for a system containing N type a quasiparticles. For example, in a fractional quantum Hall system with filling factor $\nu = 1/q$ (where q is an odd integer) there are q distinct types of particles that are *Abelian anyons*. Abelian anyons always have $d_a = 1$, while non-Abelian particles have quantum dimension $d_a > 1$ (not necessarily an integer). Hence, the total quantum dimension for the $\nu = 1/q$ fractional quantum Hall system is $\mathcal{D} = \sqrt{q}$. More details and examples can be found in [384] and [399] (see also Chapter 12).

Intrinsic topological order. The presence of long-range entanglement leads to the emergence of a new type of quantum order dubbed *topological order* [532]. Quantum states characterized by different types of topological order cannot change into each other without a phase transition. One cannot differentiate topological orders using local order parameters, i.e., two quantum states characterized by different topological orders may have the same local properties. This situation is analogous to comparing a sphere and a torus: the two manifolds are locally identical and what differentiates them are their global topological properties, e.g., the fundamental groups, $\pi_1(S^2) = 1$ and $\pi_1(T^2) = \mathbb{Z} \times \mathbb{Z}$. Similarly, topological orders are characterized by certain *topological invariants*, which are universal quantum numbers that are independent of details regarding interactions, effective masses, perturbations, etc. It is worth noting that most of the theoretical work on topological order assumes zero temperature. The stability of topological order at nonzero temperature is an unsolved problem in condensed matter physics.

One topological invariant that can be used for characterizing topologically ordered states is the ground state degeneracy of a system defined on topologically nontrivial compact manifolds (e.g., on a torus). For example, a fractional quantum Hall state with $\nu = 1/q$ has q^g degenerate ground states on a Riemann surface of genus g [535]. Another invariant is the topological entanglement entropy [295]. As mentioned above, for a $\nu = 1/q$ fractional quantum Hall state we have $\gamma = \frac{1}{2} \log q$. A set of topological invariants of critical importance for quantum computation [291] characterizes the properties of the *anyonic excitations* that emerge in topologically ordered states. Note that these quasiparticles emerge in the *bulk* and have *fractional* quantum numbers (e.g., fractional charge). A brief summary of the key properties of anyonic excitations can be found in Chapter 12.

Basic classification and examples. In the absence of symmetry constraints, all short-range entangled (SRE) states belong to the same phase. On the other hand, there are many distinct phases of long-range entangled (LRE) quantum states, each characterized by a specific topological order. Different topological orders are characterized by distinct sets of topological invariants, such as ground state degeneracies on compact manifolds, non-Abelian geometric phases, and topological entanglement entropy. At the time of this writing, a general theory that would provide a complete classification of topologically ordered phases is not known. Also, various alternative approaches may not necessarily be equivalent.[2] Nonetheless, a lot is known about various specific cases, for example about the classification of Abelian fractional quantum Hall fluids [538]. Discussing these results in detail is beyond the scope of this introductory book. A schematic representation of possible gapped phases in the absence of symmetry, which summarizes the main ideas of this section, is shown in Figure 5.1(a). Some simple examples of quantum states with intrinsic topological order are briefly discussed in Section 5.2.

We conclude with a short list of known topologically ordered phases. *Superconductivity*, discovered experimentally in 1911, is a topologically ordered state characterized by $\mathbb{Z}_2$ topological order [234]. Note that, strictly speaking, superconductivity is described by a Ginzburg–Landau theory with *dynamical* $U(1)$ electromagnetic gauge field. It can be shown [234] that the corresponding effective low-energy theory is a *topological* $\mathbb{Z}_2$ gauge theory. However, superconductivity is often described as a state with *nondynamical* electromagnetic gauge field, which is a symmetry-breaking state with no topological order. The mean-field description of superconductivity that forms the basis for the classification of "topological" superconductors presented in Chapter 3 belongs to the second (i.e., symmetry-breaking) perspective. The *fractional quantum Hall* states provide the "standard" example of topologically ordered phases. Their discovery in 1982 [511] has stimulated significantly the theoretical progress in this area. The discovery of the non-Abelian $\nu = 5/2$ fractional quantum Hall state in 1987 [549] has provided the first example of a physical system that may support non-Abelian excitations, which could be used for topological quantum computation. Other examples of topologically ordered phases that were proposed theoretically are various types of *spin liquids* [22, 448], such as the chiral spin liquids [273, 536] and the $\mathbb{Z}_2$ spin liquids [428]. Kitaev's *toric code* [296], a two-dimensional spin lattice model with $\mathbb{Z}_2$ topological order, is also an example of topological quantum error correcting code.

---

[2]Integer quantum Hall fluids – the archetype of topological quantum matter – are examples of LRE states, if characterized as states which cannot be disentangled into a product state by a finite-depth local unitary quantum circuit; however, they do not possess topological order as measured by the topological entanglement entropy or the nontrivial topological ground state degeneracy. Also, as noninteracting systems of fermions, they share many features with other topological band insulators, which are short-range entangled, symmetry protected topological states. It is perhaps fitting to leave the categorization of integer quantum Hall states in limbo, as a reminder of our incomplete understanding of topological quantum matter.

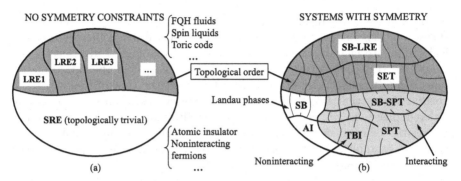

FIGURE 5.1 The "topological quantum world" (a) without symmetry constraints and (b) in the presence of symmetry. (a) All short-range entangled (SRE) states are topologically trivial and belong to the same phase. There are distinct long-range entangled (LRE) phases with (intrinsic) topological order. (b) Topologically trivial SRE states are either equivalent to the atomic insulator (AI) or belong to "standard" symmetry-breaking (SB) phases. Topologically nontrivial SRE phases either preserve the symmetry of the Hamiltonian – the symmetry protected topological (SPT) phases – or break it to a lower symmetry (SB-SPT). The noninteracting SPT phases are the topological (band) insulators and superconductors (TBI). LRE phases either preserve the symmetry of the Hamiltonian – the symmetry enriched topological (SET) phases – or break it (SB-LRE).

5.1.2 Systems with symmetry constraints

Consider d-dimensional systems described by many-body Hamiltonians that are invariant under some symmetry group $\mathcal{G}$. The corresponding H-class contains all Hamiltonians that satisfy this symmetry constraint. How many distinct gapped phases (i.e., equivalence sub-classes defined by $\mathcal{G}$-invariant gap-preserving continuous deformations) are there within this H-class?

In turns out that in the presence of symmetry the structure of the phase diagram becomes much richer [see Figure 5.1(b)]. The class of SRE states, which now contains different phases, has a nontrivial structure that depends on the specific group $\mathcal{G}$. First, there are the "standard" Landau symmetry-breaking phases (SB) corresponding to topologically trivial ground states that have a lower symmetry than the Hamiltonian. More specifically, for each SB phase the symmetry of the ground state is described by a specific subgroup $\mathcal{U}$ of $\mathcal{G}$.

Second, there are states that do not break any symmetry but have distinct topological properties that are *protected* by the symmetry group $\mathcal{G}$. We will call them *symmetry-protected topological* (SPT) phases. The (noninteracting) topological insulators and superconductors discussed in the previous chapter are examples of SPT states. In addition, there are interacting SPT phases, such as, for example, the Haldane phase of the spin-1 chain. The interacting

SPT phases, which can be viewed as a direct generalization of the noninteracting topological phases, occur in both fermionic and bosonic many-body systems. Their basic properties and some examples are discussed in Section 5.3. Note that SPT states do not possess (intrinsic) topological order and, for a given H-class, they all have the same symmetry. If the symmetry constraint is removed (i.e., if we allow for arbitrary deformations of the Hamiltonian), one can continuously deform an SPT state into the trivial band insulator. Finally, we note that symmetry protection can coexist with (partial) symmetry breaking. For example, there can be several topologically distinct phases with symmetry described by a (nontrivial) subgroup of $\mathcal{G}$. In Figure 5.1(b) we use the notation SB-SPT to designate these states.

While LRE states can belong to distinct phases even in the absence of any symmetry, imposing symmetry constraints results in a richer structure. On one hand, LRE can intertwine with symmetry breaking resulting in distinct phases with broken symmetry and long-range entanglement [denoted SB-LRE in Figure 5.1(b)]. On the other hand, in the presence of symmetry constraints there may be several distinct ground states that have the same symmetry as the Hamiltonian and cannot be continuously connected to one another (i.e., without a phase transition), but reduce to a single phase in the absence of symmetry. We will call them *symmetry-enriched topological* (SET) phases [158, 526, 533]. These are distinct phases of matter sharing the same topological order and symmetry group; upon breaking the symmetries the topological order is unaffected, but the distinctions among the SET phases disappear.

A key property of noninteracting fermion topological insulators (superconductors) is the presence of robust gapless states at the boundary between topologically distinct phases (e.g., between the system and vacuum). Since (interacting) SPT phases can be viewed as the minimal generalizations of noninteracting topological states, it is natural to ask about the fate of their boundary modes. We emphasize that, unlike LRE phases, SPT phases do not possess bulk topological order (i.e., no topological entanglement entropy, no ground state degeneracies, no bulk excitations with fractional quantum numbers, etc.). However, the boundary of a d-dimensional SPT phase is always nontrivial, in the sense that it cannot be realized in isolation, as a purely $(d-1)$-dimensional object. For example, the surface states of 3D topological insulators cannot be realized on a two-dimensional lattice. This impossibility stems from the fact that certain properties of the boundary would be ill-defined for a $(d-1)$-dimensional system existing in isolation. These properties are related to the so-called *quantum anomalies* known from work in quantum field theory [19] and can be used to identify and classify SPT phases [441, 440].

The properties of the boundary states can be summarized as follows [344]: i) The 0-dimensional boundary of a 1-dimensional (1D) SPT phase is always *gapless* (i.e., the system has zero-energy bound states localized near its ends). ii) The 1D boundary of a 2D SPT phase is either *gapless* or *spontaneously breaks the symmetry* that protects the phase. iii) The 2D surface of a 3D SPT phase is *gapless* or, if it has a gap, it either *spontaneously breaks the symmetry* or carries *intrinsic (boundary) topological order*. Note that the boundaries of noninteracting SPT phases protected by internal (nonspatial) symmetries are always gapless. Also note that the topologically ordered states that emerge at the surface of a 3D interacting SPT phase are LRE states that cannot be realized in a purely 2D system.

5.2 QUANTUM PHASES WITH TOPOLOGICAL ORDER

Topological order is a type of order that characterizes gapped quantum phases of matter, such as quantum spin liquids [273, 428, 536] and fractional quantum Hall fluids [319, 511]. Here, we define a gapped quantum phase as the equivalence class of ground states that can be continuously deformed into one another by varying the parameters of a (local) Hamiltonian in such a way as to maintain a nonzero energy gap for all excitations above the ground state. Topological order is, in some sense, the very essence of topological quantum matter: a type of order that is beyond the symmetry-breaking paradigm, cannot be captured by any local order parameter, and is insensitive to local perturbations. Note, however, that many topological phases (i.e., the SPT phases) do not possess topological order and can be viewed as distinct phases only in the presence of certain symmetries (i.e., they are not robust against perturbations that break those symmetries).

The essential ingredient of a topologically ordered state is the presence of long-range entanglement; distinct phases correspond to different "patterns" of long-range entanglement. At the time of this writing the general classification of topologically ordered phases is an open question, although interesting partial results have been obtained in recent years. We will refrain from discussing these results, which typically involve sophisticated mathematical formalisms. Instead, we focus on two paradigmatic examples of topologically ordered systems: the fractional quantum Hall liquids and the toric code. These examples will allow us to identify most of the key features that characterize topologically ordered phases, which can be summarized as follows:

- Robust *ground state degeneracy* for systems defined on (topologically nontrivial) compact manifolds (e.g., on a torus). Degeneracies are typically associated with symmetry; remarkably, in the presence of topological order all symmetries may be broken, yet the ground state has a degeneracy that only depends on the topology of space and cannot be lifted by any local perturbation.

- *Nonlocal excitations*. While "typical" quasiparticles are created by local operators, in a topologically ordered state excitations can only be created by infinite products of local operators. As a result, these excitations have intrinsically nonlocal properties, such as nontrivial statistics. The ground state degeneracy and the existence of nonlocal excitations are related properties stemming from the presence of long-range entanglement.

- The low-energy dynamics of systems defined on compact spaces depends only on the *topology* of the space and can be described in the framework of *topological field theory*. The theory completely captures the universal low-energy physics of the system, including the ground state degeneracy and the properties of the nonlocal excitations. All these properties reveal the existence of *long-range order* (stemming from long-range entanglement) without the presence of any *long-range correlations* for local operators.

- In the presence of global symmetries a topological phase can split into several *symmetry-enriched topological phases*. Moreover, the interplay between symmetry and topology leads to the emergence of nonlocal quasiparticle excitations that carry *fractionalized* quantum numbers of the global symmetry group (as compared to the corresponding numbers carried by local excitations). For example, the quasiholes in a $\nu = \frac{1}{m}$ Laughlin state carry electric charge $\frac{e}{m}$. The phases that support this type of excitations are often called *fractional topological phases*, e.g., fractional quantum Hall liquids, fractional Chern insulators, and fractional topological insulators.

5.2.1 Effective theory of Abelian fractional quantum Hall liquids

One possible approach to understanding topologically ordered systems is to construct an effective theory that captures the *universal* low-energy properties and provides labeling for (hence, classifies) different topological orders. A prominent example is the so-called *K-matrix* formulation [538] of two-dimensional topological phases with Abelian topological order, which is believed to provide a complete classification of all Abelian fractional quantum Hall (FQH) states [61, 425, 538]. In this section, we sketch the construction of the effective theory and point out some of its key implications. Details can be found in, for example, Refs. [327, 534, 538]. Systematic reviews of the FQH effect can found in many books, for example Refs. [129, 160, 563].

The FQH effect [511] is a physical phenomenon that occurs at low temperatures in two-dimensional electron systems in the presence of a high magnetic field. Its most striking manifestation is the emergence of precisely quantized Hall conductance plateaus at fractional values of e^2/h. A key step in understanding the FQH effect was the discovery by Laughlin of a many-body trial function for the ground states of FQH fluids with filling fraction $\nu = 1/m$ (where m is an odd integer) [319]. The Laughlin states support quasiparticle excitations that have fractional charge $q = e/m$ and satisfy anyonic statistics[3] with statistical angle $\theta = \pi/m$ [28]. Another key development involves the understanding of FQH states at filling fractions other than $\nu = 1/m$ using the so-called *hierarchical construction* proposed by Haldane [229] and further clarified by Halperin [232]. An alternative approach was proposed by Jain [262], in which a FQH liquid is viewed as an integer quantum Hall state of composite particles consisting of electrons having attached an even number of

---

[3]For details on the definition and properties of anyons see Chapter 12.

quantized vortices [243]. Note that different constructions may lead to different wave functions corresponding to the same filling fractions $\nu \neq 1/m$. Whether or not these states belong to the same universality class can be determined within the K-matrix formalism described below.

Effective theory of the Laughlin states. We start with the effective theory for the Laughlin state [535]. At filling fraction $\nu = 1/m$, the ground state of a two-dimensional interacting system consisting of N particles (bosons or fermions) moving in a strong perpendicular magnetic field is described by the Laughlin wave function [319]

$$\Psi(z_1, \ldots, z_N) = \prod_{i<j}(z_i - z_j)^m \exp\left(-\sum_i \frac{|z_i|^2}{4l_B^2}\right), \tag{5.3}$$

where $z_i = x_i + iy_i$ represents the coordinates of the ith particle, $l_B^2 = \hbar/eB$ is the magnetic length, and m is an even integer if the particles are bosons and an odd integer for fermions. Note that all particles are in the first Landau level and that there is an mth order zero for every pair of particles that approach each other, which indicates that the state (5.3) is highly correlated. The highly organized nature of the Laughlin state, which is not associated with any broken symmetry and cannot be captured by any local order parameter, reflects the topological order of the corresponding phase.

Consider now a small change δA_λ of the external electromagnetic field. Here, we use Greek letters to represent space-time indices; the quantities $A_0 = -\varphi$ and $(A_1, A_2) = \boldsymbol{A}$ represent the scalar and vector potentials, respectively. In response to the perturbation, the charge current changes by

$$(-e)\,\delta J^\mu = -\sigma_{xy}\epsilon^{\mu\nu\lambda}\partial_\nu\delta A_\lambda, \tag{5.4}$$

where $\sigma_{xy} = \frac{\nu e^2}{h}$ is the quantized Hall conductance and J^μ is the (conserved) particle number "current," with $J^0 = \sum_i \delta(\boldsymbol{r} - \boldsymbol{r}_i)$ and $\boldsymbol{J} \equiv (J^1, J^2) = \sum_i \boldsymbol{v}_i\delta(\boldsymbol{r} - \boldsymbol{r}_i)$ being the particle number density and current, respectively. In addition, we assumed that each particle carries electric charge $-e$ and that the uniform magnetic field applied to the system has $B_z = B > 0$. For simplicity, in this section, we will choose $\hbar = 1$ (i.e., $h = 2\pi$). The conservation of the particle number current, $\partial_\mu J^\mu = 0$, is automatically satisfied if we express J^μ in terms of a $U(1)$ gauge field a_μ,

$$J^\mu = \frac{1}{2\pi}\epsilon^{\mu\nu\lambda}\partial_\nu a_\lambda. \tag{5.5}$$

The dynamics of the gauge field will be given by some Lagrangian that generates Eq. (5.4) as an equation of motion. The effective Lagrangian (density) that satisfies this condition has the form

$$\mathcal{L} = \frac{m}{4\pi}\epsilon^{\mu\nu\lambda}a_\mu\partial_\nu a_\lambda - \frac{e}{2\pi}\epsilon^{\mu\nu\lambda}A_\mu\partial_\nu a_\lambda, \tag{5.6}$$

where A_μ represents the external electromagnetic field that couples to the charge current $(-e)J^\mu$. One can easily check that the Lagrange equations

$$\partial_\nu \left[\frac{\partial \mathcal{L}}{\partial(\partial_\nu a_\mu)} \right] = \frac{\partial \mathcal{L}}{\partial a_\mu} \tag{5.7}$$

produce the desired response to an external perturbation, i.e., Eq. (5.4). We note that, in the absence of a perturbation, the equation corresponding to $\mu = 0$ reads $mJ^0 = \frac{e}{2\pi}B$. Since $eB/2\pi = 1/2\pi l_B^2$ represents the number of degenerate (single particle) states per unit area, this equation tells us that the filling factor is $\nu = J^0 2\pi/eB = 1/m$.

We have constructed an effective theory for the quantum Hall state at filling fraction $\nu = 1/m$ that describes the linear response of the ground state to an external electromagnetic field. The first term in Eq. (5.6) is referred to as the *Chern–Simons* term and the effective theory defined by the Lagrangian (5.6) is a *Chern–Simons topological quantum field theory*[4] [101]. Note, however, that the effective theory is not complete without specifying the local particle excitations, i.e., the excitations corresponding to the underlying fermions or bosons.

Let us first consider an excitation that carries a (dimensionless) *gauge charge* l associated with a_μ, which corresponds to adding a source term $-la_0\delta(\mathbf{r} - \mathbf{r}_0)$. The Lagrange equation for $\mu = 0$ becomes

$$J^0 \equiv \frac{1}{2\pi}\epsilon^{ij}\partial_i a_j = \frac{e}{2\pi m}B + \frac{l}{m}\delta(\mathbf{r} - \mathbf{r}_0), \tag{5.8}$$

where $\epsilon^{ij} = \epsilon^{0ij}$. Equation (5.8) indicates that the particle density has increased due to the presence of the excitation by $\Delta J^0 = \frac{l}{m}\delta(\mathbf{r} - \mathbf{r}_0)$. The additional *electromagnetic* charge carried by the quasiparticle is

$$q_l = -e\frac{l}{m}. \tag{5.9}$$

Furthermore, we can define the a_μ-flux through a surface S as $\int_S d^2r\, b$, with $b = \epsilon^{ij}\partial_i a_j$. Hence, Eq. (5.8) tells us that, in addition to having a gauge charge l, the excitation carries a gauge flux $2\pi l/m$ (which means l/m a_μ-flux quanta, when expressed in units of $h = 2\pi$). This has profound implications for the exchange statistics of the quasiparticles. Taking an excitation that carries a_μ-charge l_1 around another excitation with a_μ-charge l_2 will induce a phase $\theta_{l_1l_2}$ that characterizes the *mutual statistics* of the two quasiparticles. Explicitly we have [534]

$$\theta_{l_1l_2} = 2\pi\frac{l_1l_2}{m}. \tag{5.10}$$

---

[4]A general discussion of the Chern–Simons theory can be found in Chapter 6. Its relation to the properties of anyonic excitations is mentioned in Chapter 12. More details can be found in books on quantum field theory, for example [174] and [534].

When $l_1 = l_2 = l$, interchanging the two (identical) excitations will induce half of the angle in Eq. (5.10) representing the *statistical angle* of the type-l quasiparticles,

$$\theta_l = \pi \frac{l^2}{m}. \tag{5.11}$$

Based on the above considerations, we can identify the fundamental particles that form the FQH fluid as the excitations that carry m units of a_μ-charge. Indeed, based on Eqs. (5.9) and (5.11), such a particle carries electromagnetic charge $q_m = -e$ and has a statistical angle $\theta_m = \pi m$, which corresponds to a fermion if m is odd and to a boson if m is even. This identification of the fundamental underlying particles completes the construction of the effective theory.

Remarkably, the identification of the fundamental particles with the excitations that carry m units of a_μ-charge imposes a constraint on the allowed excitations. First, note that a quasihole located at $\zeta = x + iy$ can be described by multiplying the ground state wave function (5.3) by a factor $\prod_i (\zeta - z_i)$. Taking a fundamental particle around this quasi-hole excitation results in a phase change of 2π. In other words, the fundamental particle and the quasi-hole excitation have *trivial* mutual statistics. In general, the single-valuedness of the wave function requires the allowed excitations to have trivial mutual statistics with the fundamental particles. Consequently, for an allowed excitation that carries a_μ-charge l, the angle θ_{ml} has to be a multiple of 2π. Using Eq. (5.10), we conclude that the a_μ-charge of an allowed excitation must be an *integer*. For example, the elementary quasihole excitation corresponds to $l = -1$, carries fractional electromagnetic charge $q_{-1} = +e/m$, and has fractional statistics $\theta_{-1} = \pi/m$, while the elementary quasiparticle excitation has a_μ-charge $l = 1$, carries electromagnetic charge $q_1 = -e/m$, and has statistics $\theta_1 = \pi/m$. These results reproduce the well-known properties of the quasiparticles in the Laughlin state [28]. The full effective theory of the Laughlin state with $\nu = 1/m$ in the presence of quasiparticles is given by the Lagrangian

$$\mathcal{L} = \frac{m}{4\pi} \epsilon^{\mu\nu\lambda} a_\mu \partial_\nu a_\lambda - \frac{e}{2\pi} \epsilon^{\mu\nu\lambda} A_\mu \partial_\nu a_\lambda - l a_\mu j^\mu, \tag{5.12}$$

where the quasiparticle current has $j^0 = \sum_i \delta(\boldsymbol{r} - \boldsymbol{r}_i)$ and $\boldsymbol{j} = \sum_i \boldsymbol{v}_i \delta(\boldsymbol{r} - \boldsymbol{r}_i)$.

Effective theory of the hierarchical states and general formulation.
Starting with a Laughlin state, we can construct a hierarchy of FQH fluids [229, 232]. Consider a system with filling fraction $\nu \gtrsim 1/m$, so that we create a certain number of elementary excitations with $l = 1$. Following Haldane's approach, we can view the gauge field a_μ, at the mean-field level, as creating a background "magnetic" field $b = \epsilon^{ij} \partial_i a_j$ for the quasiparticles, which behave like bosons in a uniform magnetic field. When the quasiparticle density satisfies $j^0 = (1/p) 2\pi/b$, where p is an even integer, the ground state of the bosons is again described by a Laughlin wave function. Thus, we can retrace the main steps from the previous section, starting with the introduction of a new $U(1)$

gauge field $\bar{a}_\mu$ and the representation of the (conserved) quasiparticle current in a form analog to Eq. (5.5),

$$j^\mu = \frac{1}{2\pi}\epsilon^{\mu\nu\lambda}\partial_\nu\bar{a}_\lambda. \tag{5.13}$$

The dynamics of the new gauge field will be given by a Chern–Simons term with coefficient p. The total effective theory, which includes the original particle condensate and the quasiparticle condensate, has the form

$$\mathcal{L} = \frac{m}{4\pi}\epsilon^{\mu\nu\lambda}a_\mu\partial_\nu a_\lambda - \frac{e}{2\pi}\epsilon^{\mu\nu\lambda}A_\mu\partial_\nu a_\lambda + \frac{p}{4\pi}\epsilon^{\mu\nu\lambda}\bar{a}_\mu\partial_\nu\bar{a}_\lambda - \frac{1}{2\pi}\epsilon^{\mu\nu\lambda}a_\mu\partial_\nu\bar{a}_\lambda. \tag{5.14}$$

The total filling factor, which can be determined from the Lagrange equations for a_μ and $\bar{a}_\mu$ corresponding to $\mu = 0$, is

$$\nu = \frac{1}{m - \frac{1}{p}}. \tag{5.15}$$

Thus, we have constructed an effective low-energy theory of a second-level hierarchical state [229]. We can write Eq. (5.14) in a more compact way using the notations $(a_\mu^1, a_\mu^2) = (a_\mu, \bar{a}_\mu)$, $t^T \equiv (t_1, t_2) = (1, 0)$ and introducing the integer matrix

$$K = \begin{pmatrix} m & -1 \\ -1 & p \end{pmatrix}. \tag{5.16}$$

With these notations, Eq. (5.14) becomes

$$\mathcal{L} = \frac{1}{4\pi}\epsilon^{\mu\nu\lambda}a_\mu^I K_{IJ}\,\partial_\nu a_\lambda^J - \frac{e}{2\pi}t_I\,\epsilon^{\mu\nu\lambda}A_\mu\partial_\nu a_\lambda^I, \tag{5.17}$$

where $I, J \in \{1, 2\}$ and summation over repeated indices is implied.

The generalization of Eq. (5.17) to the case of N gauge fields a_μ^I, $I = 1, 2, \ldots, N$, represents the low-energy effective theory of a generic Abelian quantum Hall state [538] with filling fraction

$$\nu = t^T K^{-1} t, \tag{5.18}$$

where $t^T = (t_1, t_2 \ldots, t_N)$. This so-called *K-matrix formulation* describes the bulk of a two-dimensional Abelian topological system in terms of an N-component $U(1)$ Chern–Simons theory that involves the $N \times N$ symmetric, nondegenerate integer matrix K and the N-component integer vector t (often called the *charge vector*). If the underlying particles are bosons, all diagonal elements of K are even integers, while in the case of fermions at least one diagonal entry is an odd integer.

The topological order of the FQH state is characterized, in part, by the degeneracy of the ground state for systems defined on topologically nontrivial compact manifolds. In the K-matrix formulation, the ground state degeneracy on a Riemann surface of genus g is given by $|\det K|^g$. Note that certain states with no topological order (i.e., SPT states) can be described in the K-matrix formalism by matrices with $|\det K| = 1$ [343]. In addition, the effective theory (5.17) describes the properties of the quasiparticle excitations. A generic quasiparticle is characterized by an integer vector l with components representing the integer gauge charge l_I carried by the excitation under each of the gauge fields a_μ^I. The quasiparticles can be introduced by adding to the Lagrangian (5.17) a term of the form $-l_I a_\mu^I j_l^\mu$, where j_l^μ is the current of type-l quasiparticles. The physical electric charge carried by each excitation is given by

$$q_l = -e\, l^T K^{-1} t, \tag{5.19}$$

This generalizes the result given by Eq. (5.9). Note that in the K-matrix formalism the Laughlin states correspond to $N = 1$, $K = m$ and $t = 1$. Furthermore, the mutual statistics corresponding to taking quasiparticle l_1 around quasiparticle l_2 is characterized by the phase

$$\theta_{l_1 l_2} = 2\pi l_1^T K^{-1} l_2. \tag{5.20}$$

Finally, the statistical angle θ_l of the type-l quasiparticle is given by an expression that generalizes Eq. (5.11),

$$\theta_l = \pi l^T K^{-1} l. \tag{5.21}$$

What are the "local" excitations corresponding to the fundamental particles in this generic K-matrix formalism? One can easily check that the quasiparticles with gauge charge $l = K\lambda$, where λ is an N-component integer vector, have electric charge $q_l = -e\,\lambda^T t$, which is an integer multiple of e. Hence, the quasiparticle with $l = K\lambda$ are local composites of the fundamental particles. Since at least one local excitation has to have the electric charge of the fundamental particle, the charge vector must satisfy the constraint

$$\gcd[t_1, t_2, \ldots, t_N] = 1, \tag{5.22}$$

where "gcd" denotes the greatest common divisor. Indeed, assuming that $t_I = 1$, the local excitation with $\lambda_I = 1$ and $\lambda_J = 0$ for all $J \neq I$ has charge $-e$. Furthermore, the statistical angle of a local excitation with $l = K\lambda$ is

$$\theta_l = \pi \lambda^T K \lambda = \pi \sum_I K_{II} \lambda_I^2 \quad (\text{mod } 2\pi). \tag{5.23}$$

If the underlying particles are bosons, the diagonal elements K_{II} must be even integers. If, on the other hand, the fundamental particles are fermions, we require that all local excitations with even charge $\lambda^T t$ (in units of $-e$) must be bosons, while those with odd charge must be fermions. This requirement imposes the constraint

$$K_{II} = t_I \quad (\text{mod } 2). \tag{5.24}$$

In the absence of additional symmetries (e.g., time reversal) the K-matrix and charge vector are unconstrained, except for the requirements discussed above.

Can two different K-matrices represent the same phase? The answer is yes. One can obtain an equivalent K-matrix through a redefinition of the gauge fields a_μ^I. However, it is important to emphasize that an Abelian topological phase is described by the Lagrangian (5.17) *together with* the quantization condition of the a_μ^I-charges. The transformations that keep the quantization condition unchanged are $N \times N$ integer matrices with unit determinant, i.e., elements of the group $SL(N, \mathbb{Z})$. Consequently, two effective theories described by (K_1, t_1) and (K_2, t_2), respectively, belong to the same equivalence class (i.e., describe the same FQH state) if there exists $W \in SL(N, \mathbb{Z})$ such that

$$K_2 = W K_1 W^T, \qquad t_2 = W t_1. \tag{5.25}$$

The corresponding gauge fields transform according to $a_\mu^I \to \sum_J W_{IJ} a_\mu^J$.

We close with a few important remarks. First, we note that FQH systems with boundaries always contain one-dimensional *gapless edge excitations*. These excitations should be distinguished from the *gapped* bulk excitations. The structure of the edge states reflects the bulk topological order and, consequently, there is a bulk-edge correspondence that allows us to analyze some of the properties of the topological system in terms of either bulk or edge degrees of freedom [174, 534]. For example, the chirality of the edge (i.e., the difference between the number of right and left moving modes) is given by the signature of the K-matrix (number of positive minus number of negative eigenvalues). In addition, the edge excitations provide a convenient way to probe the topological order of the bulk. Second, we note that non-Abelian FQH states (such as, for example, the $\nu = 5/2$ state) are *not* classified by K-matrices. On the other hand, the K-matrix formalism can be used to discuss topological phases in the absence of topological order (i.e., SPT phases) [343]. Finally, we note that two fractional excitations that differ by $\Delta l = K\lambda$ (i.e., by a local excitation vector) are equivalent. The number of distinct fractional quasi-particles is equal to the number of linearly independent, nonequivalent vectors l, which is given by the determinant of K. For example, the Laughlin state $\nu = 1/m$ supports m distinct quasiparticle excitations.

5.2.2 The toric code

The *toric code* is an exactly solvable spin-1/2 model on a two-dimensional lattice (typically a square lattice). The model, proposed by Alexei Kitaev [296], provides the canonical example of a topological phase. When defined on a surface of genus g (e.g., a torus[5]), the model has a ground state degeneracy of 4^g, which reveals the underlying topological order. More specifically, the toric code has $\mathbb{Z}_2$ topological order, a type of quantum order that was first studied in the context of spin liquids [428]. This highly entangled topological state supports emergent excitations that turn out to be Abelian anyons. In

---

[5]The torus is the natural geometry for a translation-invariant system, i.e., when imposing periodic boundary conditions. The "toric code" gets its name from this geometry.

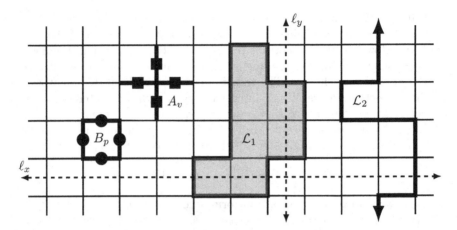

FIGURE 5.2 A piece of the square lattice on which the toric code is defined. Spin-1/2 degrees of freedom are associated with each *link* of the lattice. The star and plaquette operators A_v and B_p defined by Eq. (5.26) are products of σ_i^x operators (black squares) and σ_i^z operators (black circles), respectively. The "trivial" loop $\mathcal{L}_1$ is generated by acting with B_p for the gray plaquettes on the state $|11\ldots1\rangle$. For a system with periodic boundary conditions, the "nontrivial" loop $\mathcal{L}_2$ cycles the torus in the y direction. The large cycles ℓ_x and ℓ_y are independent, noncontractible loops on the torus. Note that $\mathcal{L}_1$ has "parity" $(\pi_x, \pi_y) = (1, 1)$, while $\mathcal{L}_2$ has $(\pi_x, \pi_y) = (-1, 1)$ (see main text).

addition, the toric code is the simplest example of a so-called *quantum double model* [291]. There is a direct relation between this type of model and *quantum error correction* (see Chapter 11). In this context, the toric code is an example of a so-called *stabilizer code* [92, 215].

Consider a set of spin-1/2 degrees of freedom defined on the *links* of a two-dimensional square lattice, as shown in Figure 5.2. The local operator acting on each link "i" is the spin operator $\boldsymbol{\sigma}_i = (\sigma_i^x, \sigma_i^y, \sigma_i^z)$. We introduce the vertex (or "star") operator A_v and the plaquette operator B_p as the products

$$A_v = \prod_{i \in v} \sigma_i^x, \qquad\qquad B_p = \prod_{i \in p} \sigma_i^z, \qquad (5.26)$$

where the star (or vertex) label "v" is associated with the four bonds that meet at a vertex and the plaquette label "p" corresponds to the bonds surrounding a plaquette (see Figure 5.2). The toric code is defined by the Hamiltonian

$$H_{tc} = -J \sum_v A_v - K \sum_p B_b, \qquad (5.27)$$

where J and K are positive coupling constants and the sums are over all sites v and plaquettes p. All star operators A_v commute with each other, as do the plaquette operators B_p. Moreover, since any given star and plaquette share either two edges or none, we also have $[A_v, B_p] = 0$, which means that all terms in the Hamiltonian commute. In addition, the operators A_v and B_p square to one, so that their eigenvalues are ± 1. Consequently, to minimize the energy of the system one should identify the many-body states that are eigenstates of A_v and B_p with eigenvalue $+1$ for every star and plaquette.

The ground states. Consider the σ^x basis, which contains all the states $|s\rangle = |s_1 s_2 \ldots s_N\rangle$ with the property

$$\sigma_j^x |s\rangle = s_j |s\rangle, \tag{5.28}$$

where $s_j = \pm 1$. From this basis, we select a subset $\mathcal{L}$ of states characterized by an *even* number of negative eigenvalues $s_j = -1$ on *every* star, i.e., states that satisfy the property $\prod_{j \in v} s_j = 1$ for every star v. Note that the links with $s_j = -1$ necessarily form closed loops (since an open string would imply the existence of stars with only one negative eigenvalue at the ends of the string). We will refer to the elements of $\mathcal{L}$ as "loop states."

Clearly, the loop states are eigenstates of the star operator with eigenvalue $+1$, i.e., $A_v |s\rangle = |s\rangle$ for every star v. On the other hand, all states $|s\rangle \notin \mathcal{L}$ have at least two sites v such that $A_v |s\rangle = -|s\rangle$. Consequently, the ground state should be constructed as a linear superposition of loop states,

$$|\Psi\rangle = \overset{(|s\rangle \in \mathcal{L})}{\sum_{s}} \alpha_s |s\rangle, \tag{5.29}$$

where the coefficients α_s have to be determined so that $|\Psi\rangle$ is also an eigenstate of the plaquette operators with eigenvalue $+1$.

Acting with σ_j^z on an eigenstate of σ_j^x will flip the spin, $s_j \to -s_j$. Consequently, since a plaquette operator B_p will flip two spins on each adjacent star, its action will transform a loop state into another loop state. Let us focus first on systems defined on the plane. We notice that in this case all the loops of a given loop state $|s\rangle$ can be viewed as the boundaries of a certain set of plaquettes (see Figure 5.2). We use the notation $\mathcal{P}_L(s)$ to designate this set. Based on these observations, we can write a generic loop state in the form

$$|s\rangle = \left(\overset{p \in \mathcal{P}_L(s)}{\prod_{p}} B_p \right) |11 \ldots 1\rangle \equiv B_{\mathcal{P}_L}(s)|11 \ldots 1\rangle, \tag{5.30}$$

where the operator $B_{\mathcal{P}_L}(s)$ is the "creation operator" for the loops of state $|s\rangle$.

Let us now consider the states of the form

$$|\psi(s)\rangle = \prod_p \frac{1}{\sqrt{2}}(1 + B_p)|s\rangle, \tag{5.31}$$

with $|s\rangle$ being an arbitrary loop state. Acting with products of plaquette operators on $|s\rangle$ will generate other elements of $\mathcal{L}$, hence $|\psi(s)\rangle$ is a superposition of loop states, i.e., it has a form given by Eq. (5.29). Moreover, since $B_p(1 + B_p) = B_p + 1$, the state $|\psi(s)\rangle$ is an eigenstate of the plaquette operator (with eigenvalue $+1$) for every plaquette p. We conclude that $|\psi(s)\rangle$ is a ground state of the toric code Hamiltonian (5.27). Note that the ground state (5.31) is a massive superposition of loop states and that it is highly entangled.

How many distinct ground states are there? For a system defined on the plane, we can write the state $|s\rangle$ in Eq. (5.31) in the form given by Eq. (5.30) and use the property $\prod_p(1 + B_p) B_{\mathcal{P}_L}(s) = \prod_p(1 + B_p)$, which holds for an arbitrary set $\mathcal{P}_L(s)$ because $B_p^2 = 1$. We conclude that in this case the ground state is *unique* and can be written in the form

$$|\Psi\rangle = \prod_p \frac{1}{\sqrt{2}}(1 + B_p)|11\ldots 1\rangle. \tag{5.32}$$

With periodic boundary condition, i.e., on the torus, we have to reconsider the arguments leading to Eq. (5.30). Indeed, a single nontrivial loop that winds in the x or y direction around the torus *cannot* be represented as the boundary of a set of plaquettes. The $B_{\mathcal{P}_L}$ operator can only create *pairs* of nontrivial loops, i.e., it cannot change the *parity* of the number of nontrivial loops. If ℓ_x and ℓ_y are two independent large cycles on the torus (see Figure 5.2), we describe this parity by defining the parameter $\pi_\mu(s) = (-1)^{M_{\ell_\mu}}$, where $\ell_\mu = \ell_x, \ell_y$ and M_{ℓ_μ} is the number of times the cycle ℓ_μ intersects a loop. Note that $\pi_\mu(s)$ is invariant to the action of plaquette operators. Consequently, we have four distinct classes of loop states characterized by $(\pi_x, \pi_y) = (1, 1)$, $(-1, 1)$, $(1, -1)$, and $(-1, -1)$, respectively.

The loop states given by Eq, (5.30) and the ground state (5.32) belong to the $(1, 1)$ class. There are three more ground states that belong to the other classes of loop states. We choose $|0\rangle_{11} \equiv |11\ldots 1\rangle$ as the representative of the $(1, 1)$ class. Similarly, we choose $|0\rangle_{-11}$ (the state with one loop along ℓ_x), $|0\rangle_{1-1}$ (the state with one loop along ℓ_y), and $|0\rangle_{-1-1}$ (the state with one loop along ℓ_x and one along ℓ_y) are representatives of the other three classes. With these notations, the four degenerate ground states of the toric code are

$$|\Psi\rangle_{\pi_x \pi_y} = \prod_p \frac{1}{\sqrt{2}}(1 + B_p)|0\rangle_{\pi_x \pi_y}. \tag{5.33}$$

We note that these degenerate states cannot be distinguished locally. Also note that for any local operator[6] $\mathcal{O}$ we have $_{\pi_x \pi_y}\langle \Psi | \mathcal{O} | \Psi \rangle_{\pi'_x \pi'_y} \propto \delta_{\pi_x \pi'_x} \delta_{\pi_y \pi'_y}$. As a result, the fourfold degeneracy of the toric code is robust against arbitrary perturbations (as long as they are below some threshold) and provides an invariant that characterizes the topological order of the system.

---

[6] A local operator is defined as a product of σ_j^μ operators over a finite cluster of links.

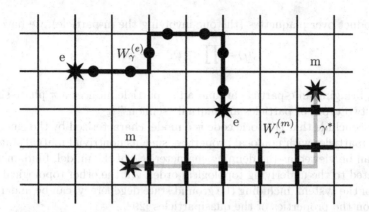

FIGURE 5.3 Anyons in the toric code. A pair of e-particles is created at the ends of a string $W_\gamma^{(e)}$ of σ^z operators, while a pair of m-particles is created by applying σ^x operators along a string γ^* defined on the dual lattice.

The excitations. The lowest energy excitations of the toric code can be generated by choosing one star or one plaquette to be negative. We refer to these excitations as "electric" (e) and "magnetic" (m) quasiparticles, respectively. The energy cost for creating an e-particle is $2J$, while for an m-particle the excitation energy is $2K$. For simplicity, we restrict our attention to the planar system. Interestingly, the quasiparticles can only be created in pairs. For example, applying σ_j^z to the ground state creates two e-particles, since the link j is shared by two stars. Applying σ^z along a string of links will create two e-particles at the ends of the string (see Figure 5.3). Similarly, a pair of m-particles can be generated by applying σ^x operators along a certain string defined on the *dual lattice*, as indicated in Figure 5.3. The wave functions that describe these states can be obtained by acting on the ground state (5.32) with the following path operators

$$W_\gamma^{(e)} = \prod_{j \in \gamma} \sigma_j^z, \qquad W_{\gamma^*}^{(m)} = \prod_{j \in \gamma^*} \sigma_j^x, \qquad (5.34)$$

where γ is a path on the lattice between sites v_1 and v_2, and γ^* is a path on the dual lattice between plaquettes p_1 and p_2 (see Figure 5.3).

The e- and m-particles have trivial bosonic self-statistics. For example, taking an e-particle around another e-particle does not change the wave function of the system. However, their mutual statistics is nontrivial. Consider, for example, a state $|\psi_i\rangle$ with an m-particle at the origin and an e-particle far away. We can move the e-particle from one site to the next by acting with σ_z on the link connecting the two sites. Completing a closed loop $\mathcal{C}$ around the origin corresponds to acting with $W_{\mathcal{C}}^{(e)}$ on the initial state. But, repeating the arguments that led to Eq. (5.30), we have $\prod_{j \in \mathcal{C}} \sigma_j^z = \prod_{p \in \mathcal{A}} B_p$, where $\mathcal{A}$ is the area bounded by $\mathcal{C}$. Since, there is exactly one negative contribution to the product over plaquettes (the one involving the m-particle), we have

$$|\psi_f\rangle = \prod_{j \in \mathcal{C}} \sigma_j^z |\psi_i\rangle = -|\psi_i\rangle. \qquad (5.35)$$

Hence, bringing an e-particle around an m-particle induces a π phase shift (i. e., the two types of particles are mutual "semions").

We conclude that the toric code is a model characterized by the emergence of quasiparticles with nonlocal properties, such as nontrivial mutual statistics. This can be viewed as the defining characteristic of the model, fundamentally connected to the underlying topological order. All the other topological properties of the system, including the ground state degeneracy, can be understood based on the properties of the quasiparticles [294].

5.3 SYMMETRY PROTECTED TOPOLOGICAL QUANTUM SATES

Symmetry protected topological (SPT) phases [225] are short-range entangled (SRE) gapped states whose topological properties rely on the presence of symmetries. SPT states do not possess intrinsic (bulk) topological order (i.e., they have no ground state degeneracies, topological entanglement entropy, exotic bulk excitations, etc.), but may have nontrivial surface states. With symmetry constraints, they represent distinct phases separated from trivial states (and from one another) by quantum phase transitions, but once the symmetry constraints are removed, SPT phases are continuously deformable to a trivial state (e.g., the atomic insulator).

SPT phases can be viewed as the minimal generalization of the concept of free fermion topological insulator (TI) to the interacting regime. For conciseness, we will refer to the free fermion TIs and superconductors as *topological band insulators* (TBIs). Clearly TBIs are special cases of SPT phases. Adding interactions generates qualitatively new features and changes the classification of topological phases. In essence, interactions can i) cause some (noninteracting) phases to merge, thus reducing the free fermion classification for certain symmetry classes, ii) allow for new topological phases to exist that necessarily require the presence of interactions, and iii) allow for the appearance of new types of boundary physics. In addition, interacting topological phases emerge in both fermionic *and* bosonic (or spin) systems. Below, we consider a few examples that illustrate these generic features, focusing on one-dimensional systems. We note that the general classification of SPT phases in dimensions higher than one is still an outstanding open problem, despite the significant progress made in recent years. Surveys of the progress in understanding and classifying SPT phases can be found in Refs. [440, 465, 514].

5.3.1 SPT phases in one dimension

In one (spatial) dimension (1D), there is no topological order, i.e., all gapped phases are SRE phases [519]. Consequently, without symmetry constraints all gapped states belong to the trivial phase, while in the presence of symmetry the gapped 1D ground states are either symmetry-breaking (topologically trivial) phases or SPT phases (with or without symmetry-breaking). The boundary states of 1D SPT phases are always gapless. Examples of 1D SPT states include noninteracting phases of fermions, such as the p-wave superconductor (class D) or the time-reversal symmetric 1D superconductor (class BDI), as well as interacting states, such as the Haldane phase of spin-1 Heisenberg antiferromagnetic chains.

The Haldane phase of spin-1 quantum spin chains. In 1983 Haldane suggested that the 1D antiferromagnetic Heisenberg spin chain has gapless (bulk) excitations for half-odd integer spins, while for integer spins the ground state is separated from all (bulk) excited states by a finite gap [230]. It turns out that the gapped states of integer spin chains can be topologically nontrivial. The spin-1 chain provides one of the canonical examples of SPT phase. In the Haldane phase, the ground state has the remarkable property that, although the system consists of spin-1 spins, it supports a spin-half moment at each end of the wire. Note that this fractional moment is robust against perturbations that do not break the symmetries protecting the SPT state (e.g., time-reversal) and cannot be "built" using a finite number of spin-1 spins. We emphasize, however, that this type of fractional excitation occurs only at the *boundary* of the system and has to be distinguished from *bulk* quasiparticles with fractional quantum numbers, which emerge in *topologically ordered* states (i.e., LRE states). Furthermore, for an *open* spin chain there are four degenerate ground states – the same number as one would expect for a pair of spin-$\frac{1}{2}$ particles – but, when defined on a *compact* manifold (i.e., with periodic boundary conditions), the system has a unique ground state. Again, this property is to be distinguished from the ground state degeneracy of topologically ordered phases defined on *compact* manifolds.

Consider the simple model of a spin-1 chain with nearest-neighbor exchange interactions and on-site single-ion anisotropy described by the Hamiltonian

$$H = J \sum_{\langle i,j \rangle} \boldsymbol{S}_i \cdot \boldsymbol{S}_j + D \sum_i (S_i^z)^2. \tag{5.36}$$

One can show [363] that for $-0.2 \lesssim D/J \lesssim 1$ the system is in the Haldane phase with a gapped ground state characterized by short-range antiferromagnetic spin correlations, $\langle S_0^\alpha S_n^\beta \rangle \propto (-1)^n \delta_{\alpha\beta}\, n^{-1/2} e^{-n\xi}$, where the correlation length ξ is of the order of a few lattice constants. The lowest energy (bulk) excitations form a massive magnon triplet.

The properties of Hamiltonian (5.36) and of related models have been investigated using a variety of theoretical tools and some of the predictions have been tested experimentally on quasi-one-dimensional $S = 1$ quantum magnets (see, for example, [363, 539] and references therein). We note that, in the large-S limit, the Heisenberg chain model can be mapped to a continuum field theory, the nonlinear sigma model (NLσM) with a topological term

[7, 174]. This mapping provides a basic understanding of the spin-1 chain, including its topological properties.

Further insight into the nature of the Haldane phase can be obtained from the ground state of a related spin model proposed by Affleck, Kennedy, Lieb, and Tasaki – the so-called AKLT model [8]. Let $\mathcal{P}_2(S_1 + S_2)$ denote the projection operator onto the subspace spanned by the states of $S_1 + S_2$ with total spin $S = 2$. We can express the projector in terms of spin operators as

$$\mathcal{P}_2(S_1 + S_2) = \frac{1}{2} S_1 \cdot S_2 + \frac{1}{6}(S_1 \cdot S_2)^2 + \frac{1}{3}, \qquad (5.37)$$

where we took into account that $S_1 \cdot S_2$ acting on states with $S = 0$, $S = 1$, and $S = 2$ has eigenvalues (in units of $\hbar$) -2, -1, and 1, respectively. The AKLT model is defined by the Hamiltonian

$$H = 2J \sum_i \mathcal{P}_2(S_i + S_{i+1}) = J \sum_i \left[S_i \cdot S_{i+1} + \frac{1}{3}(S_i \cdot S_{i+1})^2 + \frac{2}{3} \right], \qquad (5.38)$$

where J is a positive constant. Clearly, a state that would only contain combinations with total spin $S = 0$ and $S = 1$ for each pair of neighboring sites would be the exact ground state of the AKLT Hamiltonian (5.38).

To construct such a state, the main idea [8] is to represent the spin-1 at each site in terms of two $1/2$ spins, then put two of the $1/2$ spins associated with each bond into a singlet state. Specifically, we have $S_i = \mathcal{P}_1(\tilde{S}_{2i-1} + \tilde{S}_{2i})$, where $\tilde{S}_j$ are spin-$1/2$ operators and $\mathcal{P}_1$ is the projector onto spin-1 states. At each site, the standard spin-1 basis can be written in the form

$$|-\rangle = |\downarrow\downarrow\rangle, \qquad |0\rangle = \frac{1}{\sqrt{2}} \left[|\uparrow\downarrow\rangle + |\downarrow\uparrow\rangle \right], \qquad |+\rangle = |\uparrow\uparrow\rangle, \qquad (5.39)$$

where $|\uparrow\rangle$ and $|\downarrow\rangle$ are spin-$1/2$ eigenstates. Next, we construct the ground state of the AKLT Hamiltonian by imposing the condition that the spins $1/2$ from adjacent sites labeled by $2i$ and $2i + 1$ form a singlet. Since two of the four $1/2$ spins associated with sites i and $i+1$ form a singlet, the spins $2i - 1$, $2i$, $2i + 1$, and $2i + 2$ will be in a mixture of states with total spin 0 and 1. Acting with $\mathcal{P}_2(S_i + S_{i+1})$ on this state will give zero, which means that it is indeed the ground state of Hamiltonian (5.38). The AKLT ground state can be represented diagrammatically as in Figure 5.4. We note that similar ground states can be constructed for any integer spin $S > 1$; the resulting states are called *valence bond solid* (VBS) states.

To write down the AKLT state in terms of the original spin states, it is convenient to introduce the so-called *matrix product state* (MPS) representation of 1D ground states [100, 161, 412]. In general, a quantum state $|\Psi\rangle$ of a 1D lattice system (with periodic boundary conditions) can be represented as

$$|\Psi\rangle = \sum_{s_1, s_2, \ldots} \text{Tr}\left[A^{s_1} A^{s_2} \ldots \right] |s_1 s_2 \ldots\rangle, \qquad (5.40)$$

$\bullet = \alpha_\uparrow |\uparrow\rangle + \alpha_\downarrow |\downarrow\rangle$
α
spin-$\frac{1}{2}$ state

$\bullet\!\!-\!\!\bullet = \frac{1}{\sqrt{2}}(|\uparrow\downarrow\rangle - |\downarrow\uparrow\rangle)$
singlet pair

$\bigcirc = \mathcal{P}$
projection onto $S = 1$

FIGURE 5.4 Schematic representation of the VBS wave function representing the ground state of the $S = 1$ AKLT model (5.38). Each spin-1 degree of freedom (shaded ellipse) is represented by a pair of $S = 1/2$ spins (black circles). Pairs of $1/2$ spins from neighboring sites form singlets, i.e., valence bonds (thick lines). In an open chain, unpaired $S = 1/2$ degrees of freedom are present at each boundary.

where $|s_1 s_2 \ldots\rangle$ is a many-body basis state representing the direct product of spin states $|s_i\rangle$ associated with each site and A^{s_i} are m-dimensional matrices. We note that the ground state of any gapped 1D Hamiltonian can be represented as a MPS [518]. Furthermore, the MPS representation can be used to distinguish SPT phases and provides a complete classification of these phases in $d = 1$ dimensions [100, 412, 456]. For the AKLT model, we have $|s_i\rangle \to |-\rangle, |0\rangle, |+\rangle$ and $m = 2$. The corresponding A^{s_i} matrices are

$$A^- = -\frac{1}{\sqrt{6}}\sigma^-, \qquad A^0 = -\frac{1}{\sqrt{3}}\sigma_z, \qquad A^+ = -\frac{1}{\sqrt{6}}\sigma^+, \qquad (5.41)$$

where $\sigma^\pm = \sigma^x \pm i\sigma^y$ and σ^α represent Pauli matrices. One can easily check that for any two matrices A^{s_1} and A^{s_2} the state $A^{s_1} A^{s_2} |s_1 s_2\rangle$ does not contain any component with total spin $S = 2$.

While the trace in Eq. (5.40) corresponds to periodic boundary conditions, the four degenerate ground states of a (long) open chain are given by the four matrix elements of the product $A^{s_1} A^{s_2} \ldots A^{s_N}$ acting on the state $|s_1 s_2 \ldots s_N\rangle$. These degenerate ground states can be understood intuitively based on the schematic representation shown in Figure 5.4 as corresponding to the four boundary states associated with the unpaired $S = 1/2$ spins at the ends of the chain. In the thermodynamic limit ($N \to \infty$), these boundary modes become gapless. Also, using the MPS representation (5.40), one can show that the spin correlation function of the AKLT model decays exponentially. However, there are quantitative differences between the correlation functions that characterize the AKLT model and the spin-1 Haldane chain. Nonetheless, the two models belong to the same topological phase and can be adiabatically connected to one another.

Class BDI Majorana chain: The collapse of the noninteracting $\mathbb{Z}$ classification. To illustrate how the classification of topological band insulators and superconductors can be changed by interactions, we briefly discuss the collapse of the $\mathbb{Z}$ classification down to $\mathbb{Z}_8$ for the time reversal invariant Majorana chain (BDI symmetry class), which was first investigated by Fidkowski and Kitaev [169, 170]. Let us first consider the so-called *Kitaev chain* model [298], a lattice model for 1D topological superconductors involving nearest-neighbor hopping and pairing of spinless fermions (see Chapter 6). The model Hamiltonian can be written in terms of two real parameters u and v as $H_K[c^\dagger; c] = u H_1[c^\dagger; c] + v H_2[c^\dagger; c]$, where

$$
H_1[c^\dagger; c] = \frac{1}{2} \sum_i \left[-\left(c_i^\dagger c_{i+1} + c_{i+1}^\dagger c_i \right) + c_i^\dagger c_{i+1}^\dagger + c_{i+1} c_i \right].
$$

$$
H_2[c^\dagger; c] = \sum_i \left(c_i^\dagger c_i - \frac{1}{2} \right). \tag{5.42}
$$

Here, $c_i^\dagger$ and c_i are the creation and annihilation operators for spinless fermions on site i. One can easily verify that H_K has particle-hole symmetry with $\mathcal{C}^2 = +1$ and time-reversal symmetry for spinless particles ($\mathcal{T}^2 = 1$), hence it belongs to the BDI symmetry class (see Table 3.1). As we will show explicitly in the next chapter, when $|u| < |v|$ the system is topologically trivial, while for $|u| > |v|$ it is in a topological superconducting phase. The two phases are separated by a topological quantum phase transition at $|u| = |v|$, when the energy spectrum becomes gapless.

Next, consider N_f parallel Kitaev chains, each described by a Hamiltonian $H_K[(c^\alpha)^\dagger; c^\alpha]$, with c_i^α representing the fermion (annihilation) operator for chain α site i. In the absence of interaction, the topological phase corresponding to $|u| > |v|$ will be characterized by the winding number $\nu = N_f$. However, when $N_f = 0$ (mod 8) one can adiabatically connect the noninteracting $\nu = N_f$ topological phase to the topologically trivial phase by adding interactions [169, 170]. In the presence of interaction, the system is described by the Hamiltonian

$$
H = \sum_{\alpha=1}^{N_f} \left(u H_1[(c^\alpha)^\dagger; c^\alpha] + v H_2[(c^\alpha)^\dagger; c^\alpha] \right) + w H_3, \tag{5.43}
$$

where H_3 describes the interaction and can be written as a sum of terms each containing four fermion operators. For $N_f = 8$ and considering interactions that couple the same site i from different chains, i.e., for $H_3 = \sum_i W[(c_i^1)^\dagger, \ldots, (c_i^8)^\dagger; c_i^1, \ldots, c_i^8]$, one can identify an interaction W such that the energy spectrum remains gapped as the parameters u and v are varied, as long as $w \neq 0$ [169, 170]. The key simplification that allows an exact solution to this interacting problem is that in the *Majorana representation* (see Chapter 6 for details) the noninteracting system corresponding to $u = 0$, $v \neq 0$ (trivial phase) and $u \neq 0$, $v = 0$ (topological phase) becomes a collection of *disconnected dimers*. The interaction W does not couple the dimers along the chains, so the problem reduces to a finite size interacting problem (involving 8 dimers from different chains) that one can easily tackle.

The adiabatic evolution of the Hamiltonian (5.43) takes place in the parameter space (u, v, w). Starting from a noninteracting topological state with $|u| > |v|$, one can reach the trivial state $(0, v, 0)$ along the path $(u, v, 0) \xrightarrow{1} (u, 0, 0) \xrightarrow{2} (u, 0, w) \xrightarrow{3} (0, 0, w) \xrightarrow{4} (0, v, w) \xrightarrow{5} (0, v, 0)$. Note that during steps $2 - 5$, which involve the presence of interactions, the system is a collection of dimers and can be explicitly shown to be gapped. We conclude that, in the presence of interactions, all ground states of a system of 8 Kitaev chains can be adiabatically connected to the trivial state, hence they belong to a single phase. The arguments sketched above can be generalized to an arbitrary number N_f of parallel chains, which can be shown to possess $\nu = N_f \,(\mathrm{mod}\ 8) + 1$ distinct phases. For example, one can show that 9 copies of the Kitaev chain coupled by quartic interactions have two distinct states, similar to a single Kitaev chain. As a consequence, the noninteracting $\mathbb{Z}$ classification of the topological phases for 1D systems in the BDI symmetry class reduces to a $\mathbb{Z}_8$ classification in the presence of interactions.

5.3.2 SPT phases in two and three dimensions

The 1D examples discussed in the previous section illustrate some of the generic features of SPT phases in the presence of interactions, more specifically the possibility that these phases emerge in systems of both fermions and bosons (or spins) and the interaction-induced collapse of certain free fermion classifications. There are many higher dimensional examples that illustrate these properties. In addition, interactions can generate new topological phases that are not possible in noninteracting systems and induce new types of surface physics, including topologically ordered states that can only exist on the surface of an interacting 3D SPT phase. Below, we briefly mention a few results on the effect of interactions in two and three dimensions.

K-matrix classification of bosonic SPT phases in two dimensions. One approach to classifying and describing topological phases of interacting bosons in 2D is based on the K-matrix formulation [343], a theory originally formulated in the context of Abelian fractional quantum Hall fluids (see Section 5.2.1). To describe SPT phases (i.e., topological phases without intrinsic topological order), K has to be a *unimodular* matrix, i.e., $|\det K| = 1$. This condition ensures that the ground state is unique when the system is defined on topologically nontrivial compact manifolds (see Section 5.2.1). Furthermore, to describe bosons, all diagonal elements of K have to be *even integers*. Additional conditions incorporate information about the *chirality* of the state to be described and the *symmetries* of the system [343].

Consider first the chirality of the edge, i.e., the difference between the number of right and left moving modes. As mentioned in Section 5.2.1, this quantity is given by the signature of the K-matrix, i.e., number of positive minus number of negative eigenvalues $n_+ - n_-$. For example, a maximally chiral phase characterized by edge states propagating in a single direction would be described by a matrix K with all eigenvalues having the same sign. Note that the lowest dimension of a K matrix that describes maximally chiral states for bosons (without topological order) is 8 [343]. Without additional symmetry requirements, this leads to an integer classification of the chiral bosonic phases, which are characterized by an integer multiple of eight (i.e., $8n$, with $n \in \mathbb{Z}$) chiral edge modes and can be distinguished by the quantized value of the *thermal Hall conductance*, $\kappa_{xy}/T = 8n(\pi^2 k_B^2/3h)$. A nonchiral state, on the other hand, is described by a K-matrix having the same number of positive and negative eigenvalues, $n_+ = n_-$. It was shown [343] that a large class of nonchiral bosonic SPT phases can be described by the very simple K-matrix

$$K = \begin{pmatrix} 0 & 1 \\ 1 & 0 \end{pmatrix}, \tag{5.44}$$

which satisfies the necessary requirements, i.e., $|\det K| = 1$, even integers on the diagonal, and $n_+ = n_-$.

The Lagrangian of the low-energy effective theory, $\mathcal{L} = \frac{1}{4\pi}\epsilon^{\mu\nu\lambda}a_\mu^I K_{IJ}\partial_\nu a_\lambda^J - a_\mu^I j_I^\mu$, should be invariant under symmetry transformations. One can show [343] that a unitary symmetry, for example, can be implemented by a set $(W, \delta\phi_I)$, where $W \in SL(N, \mathbb{Z})$ is an integer matrix that leaves the K matrix invariant, $W^T K W = K$ and $\delta\phi_I$ are phase shifts (defined modulo 2π) associated with a global $U(1)$ phase transformation on the quasiparticle operator. These quantities satisfy certain constraints imposed by the condition that physical excitations (which are bosonic quasiparticles) must transform trivially under the identity element of the symmetry group.

The formalism leads to a classification of 2D SPT phases of bosons in the presence of various internal symmetries (e.g., time-reversal, $U(1)$ charge conservation, etc.). For example, the presence of $U(1)$ symmetry, leads to an integer classification; different phases can be distinguished by quantized values of the Hall conductance, which for bosonic systems with no topological order are *even integer* multiples of the universal conductance, $\sigma_{xy} = 2n(q^2/h)$.

Fermionic SPT phases in three dimensions. The above considerations are meant to provide a flavor of the type of problems one has to deal with when trying to understand (interacting) SPT phases and to give a sense of the richness and complexity of this field. In the same note, we end this section with a brief remark on some results regarding 3D interacting SPT phases. For a survey of recent progress in this area see Ref. [465].

Table 5.1 shows the classification of 3D fermionic SPT phases in the presence of interactions for five symmetry classes. The symmetry groups[7] are listed explicitly in the first column; $U(1)$ typically indicates the presence of charge conservation, $SU(2)$ corresponds to spin rotation symmetry, while $\mathbb{Z}_2^T$ denotes

---

[7]Time-reversal is anti-unitary and does not commute with the $U(1)$ phase rotations corresponding to charge conservation. Consequently, the symmetry group of TBIs (class AII), for example, is $U(1) \rtimes \mathbb{Z}_2^T$ (not $U(1) \times \mathbb{Z}_2^T$), where $\rtimes$ denotes the *semi-direct product*.

TABLE 5.1 Classification of 3D fermionic SPT phases for five symmetry classes. The second column shows the collapse of the free-fermion classification in the presence of interactions. The third column contains new (interacting) phases that correspond to bosonic composites of fermions. The suspected full classification is given in the fourth column.

Symmetry		Class	Interactions	Bosonic SPT	Complete
$U(1) \rtimes \mathbb{Z}_2^T$	$(\mathcal{T}^2 = +1)$	AI	$0 \to 0$	$\mathbb{Z}_2$	$\mathbb{Z}_2$
$U(1) \rtimes \mathbb{Z}_2^T$	$(\mathcal{T}^2 = -1)$	AII	$\mathbb{Z}_2 \to \mathbb{Z}_2$	$\mathbb{Z}_2^2$	$\mathbb{Z}_2^3$
$U(1) \times \mathbb{Z}_2^T$		AIII	$\mathbb{Z} \to \mathbb{Z}_8$	$\mathbb{Z}_2$	$\mathbb{Z}_8 \times \mathbb{Z}_2$
$\mathbb{Z}_2^T$	$(\mathcal{T}^2 = -1)$	DIII	$\mathbb{Z} \to \mathbb{Z}_{16}$	0	$\mathbb{Z}_{16}$
$SU(2) \times \mathbb{Z}_2^T$	$(\mathcal{T}^2 = +1)$	CI	$\mathbb{Z} \to \mathbb{Z}_4$	$\mathbb{Z}_2$	$\mathbb{Z}_4 \times \mathbb{Z}_2$

time-reversal (or TR-like) symmetry. The presence of interactions has three main consequences [168, 362, 524]: (1) All free fermion $\mathbb{Z}$ classifications (see also Table 3.3) are reduced by interactions to smaller classifications. This collapse of the integer classifications is illustrated in the second column of Table 5.1. Note that the 3D TBIs (class AII) are stable to weak electron-electron interactions [419]. (2) New phases emerge that are equivalent to bosonic SPT phases and cannot be adiabatically connected to the free-fermion topological states (column three of Table 5.1). Since SPT phases have a group structure [514], the "new" and "old" phases can be combined, resulting in the full classification of 3D fermionic SPT phases shown in column four. (3) The surface of a 3D interacting SPT phase may have intrinsic topological order [168, 362, 520, 524]. If it does, the surface is fully gapped and preserves the symmetry, but in a manner that is forbidden in strictly two-dimensional systems. For example, the surface of a 3D TR-invariant TBI can be gapped by strong interactions, which results in a symmetry-preserving surface with *non-Abelian* topological order. By contrast, the surface of a topological superconductor with spin $SU(2)$ and time-reversal symmetries (class CI) is always gapless, as long as the symmetries are preserved [524].

CHAPTER 6

Theories of Topological Quantum Matter

W HAT THEORETICAL APPROACHES ARE APPROPRIATE
for describing topological quantum matter? The answer depends on
the type of topological phase we want to study, e.g., noninteracting or interacting, and on the type of information we are interested in, e.g., generic
properties or system-specific properties that are relevant to a particular experiment or type of experiment. Noninteracting topological phases can be
described within *topological band theory* using the basic tools of the band theory of solids. However, the complexity of the model, i.e., how much detail has
to be included, depends on whether we are only interested in the topological
properties of the system or, in addition, we need to consider certain nontopological features. For example, predicting whether or not a specific material is a topological insulator requires detailed knowledge of its band structure,
which can be acquired using density functional theory (DFT) methods. Similar treatments may be required to account for certain experimental features
involving both topological and nontopological contributions. On the other
hand, to understand the classification of topological insulators, it is sufficient
to consider "bare bone" effective models, such as continuum Dirac models or
simple tight-binding models. Topologically, these descriptions with extremely
different levels of complexity are equivalent whenever one can smoothly connect the corresponding Hamiltonians (up to the addition/subtraction of trivial
bands) without closing the bulk gap. However, the effective models provide
only a crude description of experimentally relevant properties such as energy
gaps, band dispersion, and real space properties of boundary states. In some
sense, the situation is analogous to the characterization of a given mathematical object within topology, which only deals with properties that are robust
against deformations, and geometry, which accounts for properties described
by distances, angles, and shapes.

DOI: 10.1201/9781003226048-6

In the low-energy long-wavelength limit the properties of the system can be described using *topological field theories*. These theories are similar to their nontopological counterparts but involve certain topological terms that account for the universal topological properties of the system. Field theoretical approaches can be used for both noninteracting systems (as an alternative to topological band theory) and interacting topological phases. Of course, interacting lattice models of various degrees of complexity can also be used to investigate topological order and other aspects of topological phases in the presence of correlations. In general, problems involving interacting topological phases are extremely difficult. Currently, this class of problems represents one of the main theoretical battlegrounds for expanding the known topological realm into unexplored territory. In this chapter, we discuss a few examples of theoretical approaches to topological quantum matter focusing on simple models that capture the essential (topological) properties of the system.

6.1 TOPOLOGICAL BAND THEORY: CONTINUUM DIRAC MODELS

Noninteracting topological phases (i.e., topological insulators and superconductors) can be described within band theory. However, instead of only focusing on the band dispersion, as was "traditionally" the case in band theory, one has to also consider the properties of the eigenvectors, which ultimately determine the topological properties of the system. Since smooth deformations of the Hamiltonian that preserve the gap, plus additions/subtractions of trivial bands, do not affect the topological invariants, one can focus on the low-energy states (as measured relative to the Fermi energy) near the minimum of the bulk gap. The dispersion of the corresponding bands can be conveniently modified away from the minimum gap, while additional higher energy bands can be ignored altogether. In the continuum limit, the system will be described by a Dirac-type model with a minimum number of components that depends on symmetry class of the Hamiltonian. Note that the corresponding momentum space is the d-dimensional sphere[1] $\mathbb{S}^d$, rather than the Brillouin torus $\mathbb{T}^d$. Below, we discuss a few specific examples.

6.1.1 Graphene and Dirac fermions

Graphene is a two-dimensional (2D) form of carbon consisting of a honeycomb lattice with one atom at each vertex. The study of graphene has received a lot of attention in recent years [94, 197, 394, 570] due, in part, to the remarkable properties of this material, including its interesting electronic band structure characterized by two distinct Dirac points where the conduction and valence bands meet. Near these points, the electronic spectrum resembles the

---

[1]We map the momentum space $\mathbb{R}^d$ onto the punctured sphere $\mathbb{S}^d - \{NP\}$, then we *compactify* it by associating the point "at infinity" with the north pole (NP).

dispersion of massless relativistic particles. Graphene itself is not a topological insulator. We discuss it here because, on the one hand, it provides a paradigm for the emergence of quasiparticles with Dirac-like dispersion in solid state systems and, on the other hand, the simple two-band model of graphene represents the backbone for the Haldane model [231] of a Chern insulator and the Kane–Mele model [274] of a 2D quantum spin Hall insulator.

The simplest theoretical description of graphene is based on a two-band model corresponding to the p_z orbitals occupying the two inequivalent sites in the unit cell of the honeycomb lattice (see Figure 6.1). Choosing the translation vectors $\boldsymbol{a}_1 = a(\sqrt{3}, 0)$ and $\boldsymbol{a}_2 = \frac{a}{2}(\sqrt{3}, 3)$, we have the reciprocal lattice vectors $\boldsymbol{b}_1 = \frac{2\pi}{3a}(\sqrt{3}, -1)$ and $\boldsymbol{b}_2 = \frac{4\pi}{3a}(0, 1)$. The Brillouin zone can be taken as the hexagon with vertices at points $\boldsymbol{K}$ and $\boldsymbol{K}'$, where

$$\boldsymbol{K} = \frac{4\pi}{3a}\left(\frac{1}{\sqrt{3}}, 0\right), \qquad \boldsymbol{K}' = \frac{2\pi}{3a}\left(\frac{1}{\sqrt{3}}, 1\right), \tag{6.1}$$

up to translations by reciprocal lattice vectors (see Figure 6.1). In a tight-binding approximation with nearest-neighbor hopping, the Bloch Hamiltonian takes the form

$$\mathcal{H}(\boldsymbol{k}) = \boldsymbol{h}(\boldsymbol{k}) \cdot \boldsymbol{\sigma}, \tag{6.2}$$

where $\boldsymbol{\sigma} = (\sigma_x, \sigma_y, \sigma_z)$ are 2×2 Pauli matrices. Note that here we ignore the spin degree of freedom. Including spin results in the Bloch Hamiltonian being represented by a block-diagonal 4×4 matrix with each spin block given by Eq. (6.2). The vector $\boldsymbol{h}(\boldsymbol{k})$ has in-plane components given by

$$h_x(\boldsymbol{k}) + ih_y(\boldsymbol{k}) = -t\left(1 + e^{i\boldsymbol{k}\cdot\boldsymbol{a}_1} + e^{i\boldsymbol{k}\cdot\boldsymbol{a}_2}\right), \tag{6.3}$$

where $-t$ is the nearest-neighbor hopping matrix element. The corresponding spectrum is shown in Figure 6.1. In the presence of inversion symmetry ($\mathcal{I}$) and time-reversal symmetry ($\mathcal{T}$) the out-of-plane component vanishes, $h_z(\boldsymbol{k}) = 0$. Indeed, for spinless particles the time-reversal (TR) operator is represented by the complex conjugation operator[2] and for a TR invariant system we have $\mathcal{T}\mathcal{H}(\boldsymbol{k})\mathcal{T}^{-1} = \mathcal{H}^*(\boldsymbol{k}) = \mathcal{H}(-\boldsymbol{k})$. In particular, $h_z(\boldsymbol{k}) = h_z(-\boldsymbol{k})$. On the other hand, inversion symmetry changes the sign of both position and momentum (hence, unlike $\mathcal{T}$, leaves the commutator $[x, p_x]$ unchanged) and maps onto each other sites belonging to different sub-lattices, $\mathcal{I}c_{i,\alpha}\mathcal{I}^{-1} = (\sigma_x)_{\alpha\alpha'}c_{-i,\alpha'}$, where $\alpha \in \{A, B\}$ represents the sub-lattice index. For the Hamiltonian (6.2) we have

$$\mathcal{I}\mathcal{H}(\boldsymbol{k})\mathcal{I}^{-1} = \sigma_x\mathcal{H}(-\boldsymbol{k})\sigma_x = \mathcal{H}(\boldsymbol{k}). \tag{6.4}$$

For the z-component we have $h_z(\boldsymbol{k}) = -h_z(-\boldsymbol{k})$, which combined with the condition imposed by TR symmetry results in $h_z(\boldsymbol{k})$ being identically zero.

Next, we notice that $\boldsymbol{h}(\boldsymbol{k})$ actually vanishes at $\boldsymbol{K}$ and $\boldsymbol{K}'$. These Dirac points are locally stable against any (small) perturbation that preserves TR

---

[2]See Chapter 3 (Section 3.2.1) for details.

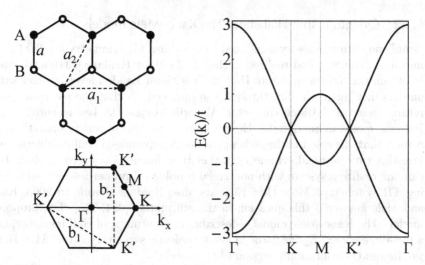

FIGURE 6.1 Honeycomb lattice, Brillouin zone, and the energy spectrum of graphene.

and inversion symmetries. In other words, any small perturbation $\delta\mathcal{H} = \epsilon(\boldsymbol{k}) + \boldsymbol{d}(\boldsymbol{k}) \cdot \boldsymbol{\sigma}$ with $d_z(\boldsymbol{k}) = 0$ will only shift the position of the Dirac points in momentum space, but will not open a gap. If, in addition, the perturbation preserves the C_3 rotation symmetry of the Hamiltonian, the points $\boldsymbol{K}$ and $\boldsymbol{K}'$ are stable Dirac nodes. In the vicinity of a Dirac node, i.e., for small $\boldsymbol{q} = \boldsymbol{k} - \boldsymbol{K}$, we have $\boldsymbol{h}(\boldsymbol{K} + \boldsymbol{q}) \approx \hbar v_F \boldsymbol{q} = \hbar v_F(q_x, q_y, 0)$, where $v_F = \frac{3}{2}ta/\hbar$. Similarly, we have $\boldsymbol{h}(\boldsymbol{K}' + \boldsymbol{q}) \approx \hbar v_F(-q_x, q_y, 0)$. Hence, the low-energy physics in the vicinity of $\boldsymbol{K}$ is described by the effective 2D massless Dirac Hamiltonian

$$\mathcal{H}(\boldsymbol{q}) = \hbar v_F \boldsymbol{q} \cdot \boldsymbol{\sigma}, \tag{6.5}$$

while a similar description with $(q_x, q_y) \longrightarrow (-q_x, q_y)$ holds in the vicinity of $\boldsymbol{K}'$. We note that breaking the inversion symmetry allows $h_z(\boldsymbol{k})$ to be nonzero, which results in the opening of an energy gap. The effective Dirac Hamiltonian (6.5) becomes massive,

$$\mathcal{H}(\boldsymbol{q}) = \hbar v_F \boldsymbol{q} \cdot \boldsymbol{\sigma} + m\sigma_z, \tag{6.6}$$

where $m = h_z(\boldsymbol{K})$, and the dispersion $E(\boldsymbol{q}) = \pm\sqrt{|\hbar v_F \boldsymbol{q}|^2 + m^2}$ is characterized by an energy gap $2|m|$. Since TR symmetry requires $h_z(\boldsymbol{K}') = h_z(-\boldsymbol{K}) = h_z(\boldsymbol{K})$, the Dirac quasiparticles near $\boldsymbol{K}'$ will have a mass with the same magnitude and sign, $m' = h_z(\boldsymbol{K}') = m$. On the other hand, breaking TR symmetry (instead of inversion) induces mass terms that have opposite signs at $\boldsymbol{K}$ and $\boldsymbol{K}'$, $m' = -m$. This condition is a consequence of Eq. (6.4), in other words a property required by inversion symmetry.

6.1.2 Quantum spin Hall state: The Kane–Mele model

Graphene with a mass term induced by breaking TR symmetry is a 2D Chern insulator. This was first realized in the late 1980s by Haldane [231], who aimed to obtain the integer quantum Hall effect without Landau levels, in fact without any net magnetic flux through the unit cell, so that the electron states retain their usual Bloch character. We will discuss the key features of the Haldane model in Section 6.2. Here, we note that this historical breakthrough reveals that the essential ingredient for getting quantized Hall conductance is breaking time-reversal symmetry, rather than having a nonzero net flux. But can one realize a system with nontrivial topological properties *without* breaking TR symmetry? More that 15 years after Haldane's work, in 2005, Kane and Mele answered this question in the affirmative [274, 275]. The proposed model – the Kane–Mele model – describes a quantum spin Hall insulator, i.e., a two-dimensional $\mathbb{Z}_2$ insulator in the symplectic symmetry class (AII). Here, we discuss the continuum version of this model.

In the low-energy, long-wavelength limit, the Haldane Hamiltonian can be written [54, 469] in the form

$$\mathcal{H} = -ihv_F(\sigma_x\tau_z\partial_x + \sigma_y\partial_y) + (M + \tau_z\lambda)\sigma_z, \qquad (6.7)$$

where the Pauli matrices σ_i are associated with the sub-lattice degree of freedom, while the matrix τ_z acts in the space of the Dirac points $\boldsymbol{K}$ and $\boldsymbol{K'}$. To simplify notation, here and below we systematically omit the corresponding unit matrices σ_0 and τ_0. In Eq. (6.7) we have used the space representation by taking $\boldsymbol{q} \to -i\boldsymbol{\partial}$. The mass term M, which corresponds to a staggered sub-lattice potential with values $+M$ for sub-lattice A and $-M$ for sub-lattice B, breaks inversion symmetry, while the λ term breaks TR symmetry and can be generated by a nonuniform magnetic field with zero flux through the unit cell. One can show (see Section 6.2) that the first Chern number of a system described by Eq. (6.7) is $\nu = \frac{1}{2}[\text{sign}(M+\lambda)-\text{sign}(M-\lambda)]$, hence for $|\lambda| > |M|$ the system is in a topologically nontrivial phase with $\nu = \pm 1$.

The Haldane Hamiltonian (6.7) describes spinless fermions. To obtain a nontrivial topological phase without breaking TR symmetry, one has to explicitly include spin, which results in an effective Hamiltonian represented by an 8×8 matrix. The basic idea of Kane and Mele was to introduce a λ-type mass term that has opposite signs for opposite spin orientations. This would be equivalent to having two copies of the Haldane model, one for each spin, with opposite orientations of the nonuniform magnetic field. Physically, such a mass term can be generated by a mirror-symmetric spin-orbit coupling,

$$\mathcal{H}_{SO} = \lambda_{SO}\sigma_z\tau_z s_z, \qquad (6.8)$$

where the Pauli matrix s_z is associated with the spin degree of freedom. To demonstrate that H_{SO} is a TR invariant term, we note that the representation of the TR operator for spin-$\frac{1}{2}$ fermions is (see Chapter 3) $\mathcal{T} = -is_y K$, were K is complex conjugation. In addition, under time-reversal $\boldsymbol{k} \to -\boldsymbol{k}$, hence

$\boldsymbol{K} + \boldsymbol{q} \to -\boldsymbol{K} - \boldsymbol{q} = \boldsymbol{K}' - \boldsymbol{q}$. In other words, the point $\boldsymbol{q}$ from the $\boldsymbol{K}$ sector is mapped onto point $-\boldsymbol{q}$ from the $\boldsymbol{K}'$ sector. In our matrix representation, switching the $\boldsymbol{K}$ and $\boldsymbol{K}'$ sectors is done by τ_x, so the TR operator can be represented in the form

$$\mathcal{T} = -i\tau_x s_y K, \tag{6.9}$$

where the unit matrix σ_0 corresponding to the sub-lattice degree of freedom was omitted for simplicity. It is straightforward to verify that $\mathcal{T} H_{SO} \mathcal{T}^{-1} = H_{SO}$, i.e., the spin-orbit coupling gap-opening term is TR invariant. In a system with broken mirror symmetry – e.g., in the presence of an applied or substrate-induced electric field in the z-direction – a Rashba type [58, 424] term, $(\boldsymbol{s} \times \boldsymbol{k}) \cdot \hat{\boldsymbol{z}}$ is also allowed by TR symmetry, $\mathcal{H}_R = \lambda_R(\sigma_x \tau_z s_y - \sigma_y s_x)$. Putting together all these TR invariant contributions, we obtain the following continuum version of the Kane–Mele model

$$\mathcal{H} = -i\hbar v_F(\sigma_x \tau_z \partial_x + \sigma_y \partial_y) + \lambda_{SO}\sigma_z \tau_z s_z + \lambda_R(\sigma_x \tau_z s_y - \sigma_y s_x) + M\sigma_z. \tag{6.10}$$

Solving the Schrödinger equation corresponding to Eq. (6.10) with fixed boundary conditions gives a complete low-energy description of the 2D system, including possible edge states. The Kane–Mele Hamiltonian for a translation-invariant system (i.e., a system with no boundaries), which describes the low-energy physics of the bulk, can be obtained from Eq. (6.10) through the substitution $-i\boldsymbol{\partial} \to \boldsymbol{k}$. We emphasize that these continuum models are low-energy effective models that approximately describe the electronic properties of a lattice system within a certain energy window near the Fermi level. To better appreciate this point, let us also consider the lattice version of the Kane–Mele model,

$$H = -t\sum_{\langle i,j \rangle} c_i^\dagger c_j + \frac{i\lambda_{SO}}{2\sqrt{3}}\sum_{\langle\langle i,j \rangle\rangle} \nu_{ij} c_i^\dagger s_z c_j + \frac{i\lambda_R}{2\sqrt{3}}\sum_{\langle i,j \rangle} c_i^\dagger (\boldsymbol{s} \times \boldsymbol{d}_{ij})_z c_j + M\sum_i \epsilon_i c_i^\dagger c_i, \tag{6.11}$$

where $c_i^\dagger = \{c_{i,\uparrow}^\dagger, c_{i,\downarrow}^\dagger\}$, while $\langle\dots\rangle$ and $\langle\langle\dots\rangle\rangle$ designate nearest-neighbor and next-nearest-neighbor pairs, respectively. The first term represents the nearest-neighbor hopping on a graphene lattice with an energy spectrum as shown in Figure 6.1. The next term, which involves second neighbor spin-dependent hopping, is a mirror symmetric spin-orbit interaction. Here, $\nu_{ij} = \frac{2}{\sqrt{3}}(\boldsymbol{d}_{kj} \times \boldsymbol{d}_{ik})_z = \pm 1$, where $\boldsymbol{d}_{kj}$ and $\boldsymbol{d}_{ik}$ are unit vectors along the two bonds $(j \to k$ and $k \to i$, respectively) connecting the next-nearest-neighbors j and i through the intermediate site k. At the low-energy points $\boldsymbol{K}$ and $\boldsymbol{K}'$ this term reduces to $\mathcal{H}_{SO}$ given by Eq. (6.8) and opens a gap $\Delta = 2|\lambda_{SO}|$ in the spectrum. The inversion-symmetry-breaking M-term, with $\epsilon_i = +1$ for A-type sites and $\epsilon_i = -1$ for B sites, also opens a gap $\Delta = 2|M|$. By contrast, the Rashba term (by itself) does not open a full gap in the spectrum. For example, at $\boldsymbol{K}$ the Rashba term becomes $\mathcal{H}_R = \lambda_R(\sigma_x s_y - \sigma_y s_x)$ and the low-energy spectrum is (approximately) given by the four eigenvalues $-\lambda_R \pm \sqrt{|\hbar v_F q|^2 + \lambda_R^2}$, $\lambda_R \pm \sqrt{|\hbar v_F q|^2 + \lambda_R^2}$, where $\boldsymbol{q} = \boldsymbol{k} - \boldsymbol{K}$ and v_F is

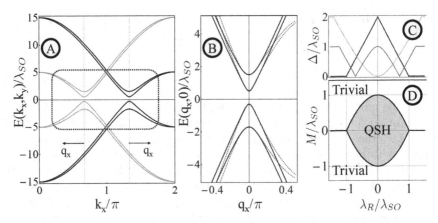

FIGURE 6.2 (A) Energy dispersion of the lattice Kane–Mele model as a function of k_x for $k_y = 0$ (black lines) and $k_y = \frac{2\pi}{3a}$ (gray lines). The model parameters (in units of λ_{SO}) are $t = 5$, $M = 0.5$, and $\lambda_R = 0.25$ and the unit length is $\sqrt{3}a = 1$. (B) Comparison between the low-energy spectra obtained using the continuum model (6.10) (full lines) and the corresponding lattice version (dashed lines). (C) Dependence of the bulk gap Δ on the strength of the Rashba term for $M = 0$ (black), $M = 0.5\lambda_{SO}$ (gray), and $M = \lambda_{SO}$ (dashed gray). (D) Phase diagram of the Kane–Mele model.

the Fermi velocity of graphene. The numerical factors appearing in Eq. (6.11) were chosen so that for $\boldsymbol{k} = \boldsymbol{K}$ and $\boldsymbol{k} = \boldsymbol{K}'$ each term of the lattice model reduces to the corresponding term of the continuum model (6.10).

The dispersion of the lattice Hamiltonian (6.11) along k_x cuts through $\boldsymbol{K}$ and $\boldsymbol{K}'$ is shown in Figure 6.2(A). For $|t| \gg \max\{|\lambda_{SO}|, |\lambda_R|, |M|\}$ the low-energy sector (i.e., the portion of the spectrum inside the dashed rectangle) is well approximated by the continuum model (6.10), as shown in Figure 6.2(B). Note that for $\lambda_R = 0$ one can block-diagonalize the Hamiltonians (6.10) and (6.11) into spin-dependent blocks corresponding to two copies of the Haldane model. The continuum Hamiltonian (6.10), for example, can be split into two blocks $\mathcal{H}_s$ given by Eq. (6.7) with $\lambda = \pm\lambda_{SO}$ for spin $\uparrow$ ($s = +$) and spin $\downarrow$ ($s = -$), respectively. The energy gap $\Delta = 2||\lambda_{SO}| - |M||$ is dominated by M in the inversion-symmetry-breaking dominated phase, vanishes when $|\lambda_{SO}| = |M|$, and is dominated by λ_{SO} in the spin-orbit-coupling dominated phase. We note that for $\lambda_R = 0$ we can define spin-dependent Chern numbers ν_s associated with the spin $\uparrow$ and spin $\downarrow$ sectors described by $\mathcal{H}_s$ with $s = +$ and $s = -$, respectively. For $|\lambda_{SO}| < |M|$ we have $\nu_+ = \nu_- = 0$, while for $|\lambda_{SO}| > |M|$, the spin-dependent Chern numbers are nonzero, $\nu_s = \text{sign}(s\lambda_{SO})$, signaling the presence of two counterpropagating, spin-polarized edge modes. In this case, the system is in a topological phase with vanishing (charge) Hall conductance (since the total Chern number is $\nu = \nu_+ + \nu_- = 0$) and quantized spin-Hall conductance $\nu_+ - \nu_- = \pm 2$ (in units of e^2/h).

For $\lambda_R \neq 0$, the S_z symmetry is broken and the two spin orientations get mixed together. One can no longer talk about two separate sectors, one with spin $\uparrow$ and one with spin $\downarrow$, characterized by spin-dependent Chern numbers and the spin conductance is not quantized. However, the topological phase and the associated low-energy edge states survive in the presence of finite Rashba coupling, as long as the bulk gap does not close. The topological invariant that characterizes this phase is the $\mathbb{Z}_2$ invariant introduced by Kane and Mele [275] (see Section 3.4.3). To calculate the $\mathbb{Z}_2$ invariant one needs to describe the system over the entire Brillouin zone, i.e., to use the lattice version of the model. Nonetheless, knowing that M (by itself) generates a trivial insulator, while λ_{SO} (by itself) generates a QSH insulating state, we can determine the full phase diagram using the continuum version of the Kane–Mele model. Basically, upon continuously varying the model parameters, the system remains in the same phase as long as the bulk gap does not close. Quantitatively, the eigenvalues of Hamiltonian (6.10) corresponding to $q = 0$ are $\lambda_{SO} \pm M$ and $-\lambda_{SO} \pm \sqrt{M^2 + 4\lambda_R^2}$, each value being double degenerate. Consequently, in the spin-orbit-coupling dominated phase the system is characterized by an energy gap

$$\Delta = 2|\lambda_{SO}| - |M| - \sqrt{M^2 + 4\lambda_R^2}. \tag{6.12}$$

The dependence of the gap on the model parameters is illustrated in Figure 6.2(C). The energy gap vanishes at the topological phase transition between the topological (spin-orbit-coupling dominated) phase and the trivial (inversion-symmetry-breaking dominated) phase. The corresponding phase boundary is given by the equation $|M/\lambda_{SO}| = (\lambda_R/\lambda_{SO})^2 - 1$ and is shown in Figure 6.2(D).

We close this section with a comment on the bulk-boundary correspondence. Figure 6.3 shows the energy bands for the lattice Hamiltonian (6.11) on a quasi one-dimensional strip with zigzag edges (i.e., cut along the a_1 direction, see Figure 6.1). The left panel is for a set of parameters corresponding to the trivial phase, while the right panel is calculated in the topological regime. We notice the presence of edge states in both regimes. However, it is only in the topological (QSH) phase that these edge states are gapless, as they cross the gap connecting the valence and conduction bands. There are two pairs of counterpropagating edge modes, one for each edge. These modes cross at a time-reversal (TR) invariant point ($k_x = \pi$) and the crossings are protected by TR symmetry. Consequently, for a *single* pair of counterpropagating modes (per edge) one can never open a gap as long as i) TR symmetry is preserved and ii) the bulk gap remains finite. The situation is completely different for a system with *two* pairs of counterpropagating modes on the same edge. By analogy with the Kane–Mele construction, one can realize such a system starting with two copies of a Chern insulator that has two chiral edge modes (propagating in a given direction for the first copy and in the opposite direction for the second). In this case, there will be intersection points away from the TR invariant momenta. These intersections are not protected

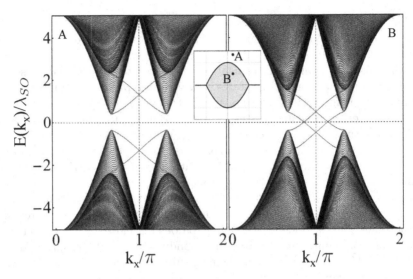

FIGURE 6.3 Energy bands for the lattice Kane–Mele model on a quasi one-dimensional strip with zigzag edges. The model parameters (in units of λ_{SO}) are $t = 5$, $\lambda_R = 0.25$, and $M = 1.35$ (panel A) or $M = 0.5$ (panel B). The location of these parameters in the phase diagram is shown in the inset. We take $\sqrt{3}a = 1$. In the QSH phase (panel B) two pairs of edge states (one on each edge) cross the gap and connect the valence and conduction bands.

by TR symmetry and the corresponding modes can scatter between one another, thus opening a gap in the spectrum. We emphasize that this is not the case in Figure 6.3(B) because the intersection points at $k_x \neq \pi$ (i.e., at $E(k_x) = 0$) are between modes located on *opposite edges* (hence, all scattering matrix elements vanish). The above considerations can be generalized to an arbitrary number of edge pairs: an odd number of pairs (per edge) ensures the protection of at least one pair of edge modes. This reveals the topological nature of the underlying bulk state. By contrast, an even number of edge pairs can always be gapped. At the bulk level, this even-odd property is captured by the topological invariant, which is a $\mathbb{Z}_2$ integer.

6.1.3 Three-dimensional four-component Dirac Hamiltonian

In this section, we construct an example of continuum 3D massive Dirac Hamiltonian and briefly discuss its symmetries and topological properties. The minimal number of components of the Dirac Hamiltonian for symmetry classes AII, DIII, and AIII is four, while for classes CI and CII this number is eight [454]. Note that these are the five symmetry classes that contain nontrivial topological phases in 3D. Here, we discuss the four-component case.

Let us consider the following four-component Dirac Hamiltonian [454]

$$\mathcal{H}(\boldsymbol{k}) = \tau_x \sigma_\mu k_\mu + m\tau_z = \begin{pmatrix} m & \boldsymbol{k} \cdot \boldsymbol{\sigma} \\ \boldsymbol{k} \cdot \boldsymbol{\sigma} & -m \end{pmatrix}, \qquad (6.13)$$

where $\mu = x, y, z$ and τ_μ, σ_μ are two sets of Pauli matrices. As usual, we omit the unit matrices, τ_0 and σ_0, and use the notation $\tau_\mu \otimes \sigma_\nu \equiv \tau_\mu \sigma_\nu$. The real-space correspondent of Eq. (6.13), which can be used, for example, to calculate the boundary modes, is obtained through the substitution $k_\mu \to -i\partial_\mu$. The energy spectrum of $\mathcal{H}(\boldsymbol{k})$ is given by $E(\boldsymbol{k}) = \pm\sqrt{k^2 + m^2}$ (twofold degenerate) and corresponds to a double degenerate gapped Dirac cone.

What is the symmetry class of Hamiltonian (6.13)? To answer this question, we first notice that it satisfies the following relations

$$i\sigma_y \mathcal{H}^*(\boldsymbol{k})(-i\sigma_y) = +\mathcal{H}(-\boldsymbol{k}), \qquad i\sigma_y(i\sigma_y)^* = -1, \qquad (6.14a)$$
$$\tau_y \sigma_y \mathcal{H}^*(\boldsymbol{k})\tau_y \sigma_y = -\mathcal{H}(-\boldsymbol{k}), \qquad \tau_y \sigma_y(\tau_y \sigma_y)^* = +1. \qquad (6.14b)$$

Comparison with Eqs. (3.14) leads to the interpretation of these relations as representing time-reversal symmetry for half-integer spin (with $U_t = i\sigma_y$) and (triplet) particle-hole symmetry (with $U_c = \tau_y \sigma_y$), respectively. Since the Hamiltonian has both time-reversal (TR) and particle-hole (PH) symmetries, it also has chiral (sub-lattice) symmetry corresponding to the product $S = \mathcal{T} \cdot \mathcal{C}$. This symmetry can be represented by τ_y, which anti-commutes with the Dirac Hamiltonian, $\tau_y \mathcal{H} \tau_y = -\mathcal{H}$. Consequently, according to the symmetry classification of single-particle Hamiltonians (see Table 3.1), the Dirac Hamiltonian (6.13) is a member of class DIII. We note that DIII is one of the Bogoliubov–de Gennes (BdG) classes, so we can interpret Eq. (6.13) as describing the dynamics of quasiparticles in a superconductor (or a superfluid). The generic form of a BdG Hamiltonian in momentum space is

$$\mathcal{H}(\boldsymbol{k}) = \begin{pmatrix} h(\boldsymbol{k}) & \Delta(\boldsymbol{k}) \\ \Delta^\dagger(\boldsymbol{k}) & -h^T(-\boldsymbol{k}) \end{pmatrix}, \qquad (6.15)$$

where $h = h^\dagger$ is a Hermitian matrix. Also, as a result of Fermi statistics, the off-diagonal block has the property $\Delta^T(-\boldsymbol{k}) = -\Delta(\boldsymbol{k})$, which implies $\Delta^\dagger(\boldsymbol{k}) = -\Delta^*(-\boldsymbol{k})$. We can bring the Dirac Hamiltonian (6.13) into the canonical form (6.15) through a unitary transformation $\mathcal{H} \to U^{-1}\mathcal{H}U$, with $U = \text{diag}(\sigma_0, -i\sigma_y)$, which yields

$$\mathcal{H}(\boldsymbol{k}) = \begin{pmatrix} m & \boldsymbol{k} \cdot \boldsymbol{\sigma}(i\sigma_y) \\ (-i\sigma_y)\boldsymbol{k} \cdot \boldsymbol{\sigma} & -m \end{pmatrix}. \qquad (6.16)$$

One can easily check that $\Delta(\boldsymbol{k}) = \boldsymbol{k} \cdot \boldsymbol{\sigma}(i\sigma_y)$ satisfies the condition imposed by Fermi statistics and that PH symmetry is now represented by $U_c = \tau_x$.

When identifying the symmetry class to which the Dirac Hamiltonian belongs, we have to keep in mind that the continuum model (6.13) represents a low-energy approximation of the lattice Hamiltonian describing the physical system and that the symmetry of this lattice Hamiltonian may be lower than the symmetry of (6.13). For example, the lattice Hamiltonian could have TR symmetry only, while the PH symmetry is just an approximate symmetry of the low-energy sector. In this case, the Dirac Hamiltonian (6.13) can be viewed as describing a topological insulator in class AII.

Similarly, the lattice Hamiltonian could have chiral symmetry only (TR and PH being approximate symmetries of the low-energy sector). Then, the 3D Dirac Hamiltonian (6.13) should be interpreted as describing an insulator in class AIII. Note that, in the presence of chiral symmetry, we can bring the Hamiltonian into block off-diagonal form by a rotation $(\tau_x, \tau_y, \tau_z) \rightarrow (\tau_x, -\tau_z, \tau_y)$. The 3D Dirac Hamiltonian (6.13) becomes

$$\mathcal{H}(\boldsymbol{k}) = \begin{pmatrix} 0 & \boldsymbol{k} \cdot \boldsymbol{\sigma} - im \\ \boldsymbol{k} \cdot \boldsymbol{\sigma} + im & 0 \end{pmatrix}. \tag{6.17}$$

The topological properties of the model can be determined from the eigenfunctions of the Hamiltonian (6.17). We have

$$u_{1,3}(\boldsymbol{k}) = \frac{1}{\sqrt{2}\lambda} \begin{bmatrix} \mp k_- \\ \pm(im + k_z) \\ 0 \\ \lambda \end{bmatrix}, \quad u_{2,4}(\boldsymbol{k}) = \frac{1}{\sqrt{2}\lambda} \begin{bmatrix} \pm(im - k_z) \\ \mp k_+ \\ \lambda \\ 0 \end{bmatrix}, \tag{6.18}$$

where $k_\pm = k_x \pm k_y$ and $\lambda(\boldsymbol{k}) = \sqrt{k^2 + m^2}$. In Eq. (6.18) the upper signs correspond to eigenfunctions with negative energy $E_1 = E_2 = -\lambda$, whereas the lower signs are for eigenfunctions with positive energy $E_3 = E_4 = +\lambda$. The projector (Q matrix) $Q(\boldsymbol{k}) = 1 - 2P(\boldsymbol{k})$, with $P(\boldsymbol{k}) = \sum_{i=1}^{2} |u_i\rangle\langle u_i|$, takes the off-diagonal form

$$Q(\boldsymbol{k}) = \begin{pmatrix} 0 & q(\boldsymbol{k}) \\ q^\dagger(\boldsymbol{k}) & 0 \end{pmatrix}, \quad \text{with } q(\boldsymbol{k}) = \frac{1}{\lambda(\boldsymbol{k})}(\boldsymbol{k} \cdot \boldsymbol{\sigma} - im). \tag{6.19}$$

We note that the class DIII massive Dirac Hamiltonian (6.16) can also be brought into block off-diagonal form, since it has chiral symmetry (see Chapter 3). The corresponding Q matrix is given by $q(\boldsymbol{k}) = \frac{1}{\lambda(\boldsymbol{k})}(\boldsymbol{k} \cdot \boldsymbol{\sigma} - im)(i\sigma_y)$, which satisfies $q^T(-\boldsymbol{k}) = -q(\boldsymbol{k})$, similar to the condition satisfied by $\Delta(\boldsymbol{k})$. A block off-diagonal Q matrix allows us to calculate the *winding number* ν associated with the map $\boldsymbol{k} \rightarrow q(\boldsymbol{k})$ from the momentum space (which for a 3D continuum model is the sphere S^3) to the space of projectors. For the projector (6.19), this space is the unitary group $U(2)$ and $\mathcal{H}(\boldsymbol{k})$ corresponds to a class AIII Hamiltonian.[3] Using Eqs. (3.38) and (3.39) we get

$$\nu[q] = \int \frac{d^3 k}{24\pi^2} \, \epsilon^{\mu\nu\rho} \, \text{Tr}[(q^{-1}\partial_\mu q)(q^{-1}\partial_\nu q)(q^{-1}\partial_\rho q)] = \frac{1}{2}\frac{m}{|m|}. \tag{6.20}$$

---

[3] For details on the Q matrix see Section 3.3. The winding number is defined in Section 3.4.2.

This rather puzzling result – a half integer winding number – is a common occurrence when using continuum descriptions and represents a consequence of the fact that Dirac models do not capture the correct structure of the wave functions at high energy [54, 231]. A continuum Dirac fermion spectrum does not have a finite bandwidth. In reality, the high energy dispersion of the bands is not linear, since they have to bend in order to accommodate a finite bandwidth. This results in additional contributions to the winding number, which becomes an integer. The same considerations apply to the Chern number of a continuum Dirac model. The conclusion is that one cannot fully determine the topological number of a filled band by analyzing only the low-energy part of the band, which can be described using a continuum Dirac model. To determine the topological invariant one has to use a lattice model. However, the *variation* of the topological number corresponding to the closing and reopening the gap *is* determined by the low-energy sector and can be calculated using a continuum description. Consider, for example, the transition from a phase with $m > 0$ to a phase with $m < 0$ through the critical point $m = 0$. If we already know that $m > 0$ corresponds to a trivial insulator (i.e., $\nu = 0$), Eq. (6.20) tells us that the Dirac Hamiltonian (6.17) with $m < 0$ describes a topological phase with $\nu = -1$. Another example – the calculation of the first Chern number for the Haldane model – is discussed in the next section.

6.2 TOPOLOGICAL BAND THEORY: TIGHT-BINDING MODELS

Continuum models are effective low-energy theories that capture the long wavelength physics in the vicinity of critical points associated with topological quantum phase transitions (where the bulk gap vanishes). To reveal the electronic band structure of a solid over the entire Brillouin zone for arbitrary values of the control parameters (e.g., external fields) requires a lattice description of the system. Knowing the band structure properties for the whole Brillouin zone allows us to calculate the relevant topological quantum numbers and identify the topological phase of the system. The simplest lattice models can be constructed in the tight-binding approximation, which corresponds to the limit of small overlap between orbitals associated with neighboring atoms in a lattice. Below, we briefly review a few basic tight-binding models of topological insulators.

6.2.1 Haldane model

The first model of a topological insulator was introduced by Haldane in 1988 [231]. The model consists of spinless fermions on a honeycomb lattice in the presence of a nonuniform (periodic) magnetic field with zero flux per unit cell, $\Phi = 6(\Phi_a + \Phi_b) + \Phi_c = 0$ (see Figure 6.4). This condition ensures that the full (nonmagnetic) translation symmetry of the lattice is preserved, while TR symmetry is broken. In fact, broken TR symmetry is the essential ingredient for having nonzero Hall conductance. The magnetic field does not affect

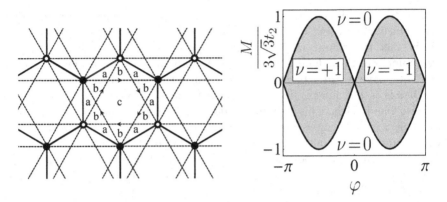

FIGURE 6.4 (Left) Haldane's model of a Chern insulator on the honeycomb lattice. The thick (full) and thin (dashed) lines designate nearest-neighbor and next-nearest-neighbor hoppings, respectively. Arrows correspond to hopping terms $t_2 e^{i\varphi}$. (Right) The phase diagram of the Haldane model.

the nearest-neighbor (n.n.) hopping t_1, but the next-nearest-neighbor (n.n.n.) hopping t_2 acquires a phase $e^{\pm i\varphi}$, with $\varphi = 2\pi(2\Phi_a + \Phi_b)/\Phi_0$, where $\Phi_0 = h/e$ is the flux quantum. The hopping directions corresponding to $e^{+i\varphi}$ are indicated by arrows in Figure 6.4. An on-site energy term with values $+M$ for A sites and $-M$ for B sites is included to break the inversion symmetry of the system. Explicitly, the Haldane model Hamiltonian can be written in the form

$$H = t_1 \sum_{\langle i,j \rangle} c_i^\dagger c_j + t_2 \sum_{\langle\langle ij \rangle\rangle} e^{-i\nu_{ij}\varphi} c_i^\dagger c_j + M \sum_i \epsilon_i c_i^\dagger c_i, \qquad (6.21)$$

where $c_i^\dagger$ (c_i) are creation (annihilation) operators for spinless fermions and all other notations are the same as in Eq. (6.11). After Fourier transforming Eq. (6.21), the Haldane Hamiltonian can be expressed in the two-component representation $(c_{\mathbf{k}A}^\dagger, c_{\mathbf{k}B}^\dagger)$ associated with sub-lattices A and B as

$$\mathcal{H}(\mathbf{k}) = \varepsilon(\mathbf{k}) + \mathbf{d}(\mathbf{k}) \cdot \boldsymbol{\sigma}, \qquad (6.22)$$

where σ_i are Pauli matrices corresponding to the sub-lattice degree of freedom. Choosing the lattice vectors $\mathbf{a}_1$ and $\mathbf{a}_2$ as in Figure 6.1, we have

$$
\begin{aligned}
\varepsilon(\mathbf{k}) &= 2t_2 \cos(\varphi)[\cos(\mathbf{k} \cdot \mathbf{a}_1) + \cos(\mathbf{k} \cdot \mathbf{a}_2) + \cos(\mathbf{k} \cdot (\mathbf{a}_1 - \mathbf{a}_2))], \\
d_1(\mathbf{k}) &= t_1[1 + \cos(\mathbf{k} \cdot \mathbf{a}_1) + \cos(\mathbf{k} \cdot \mathbf{a}_2)], \\
d_2(\mathbf{k}) &= -t_1[\sin(\mathbf{k} \cdot \mathbf{a}_1) + \sin(\mathbf{k} \cdot \mathbf{a}_2)], \\
d_3(\mathbf{k}) &= M + 2t_2 \sin(\varphi)[\sin(\mathbf{k} \cdot \mathbf{a}_1) - \sin(\mathbf{k} \cdot \mathbf{a}_2) - \sin(\mathbf{k} \cdot (\mathbf{a}_1 - \mathbf{a}_2))].
\end{aligned}
\qquad (6.23)
$$

Note that TR symmetry (for spinless fermions), $\mathcal{H}^*(\mathbf{k}) = \mathcal{H}(-\mathbf{k})$, requires $d_3(\mathbf{k}) = d_3(-\mathbf{k})$, which is satisfied only for $\varphi = 0$ or $\varphi = \pi$. Also, inversion with respect to the center of a hexagonal cell (represented by σ_x) demands

$d_3(\mathbf{k}) = -d_3(-\mathbf{k})$, which is satisfied only for $M = 0$. Hence, in general, the Hamiltonian (6.22) breaks both TR and inversion symmetries. The two energy bands can be easily determined by diagonalizing the 2×2 matrix in Eq. (6.22) and we have $E_{\pm k} = \varepsilon(\mathbf{k}) \pm d(\mathbf{k})$, where $d(\mathbf{k}) = \sqrt{\mathbf{d}(\mathbf{k}) \cdot \mathbf{d}(\mathbf{k})}$. The corresponding eigenstates are

$$|\mathbf{k}, +\rangle = \begin{pmatrix} \cos\frac{\theta}{2} e^{-i\phi} \\ \sin\frac{\theta}{2} \end{pmatrix}, \qquad |\mathbf{k}, -\rangle = \begin{pmatrix} \sin\frac{\theta}{2} e^{-i\phi} \\ -\cos\frac{\theta}{2} \end{pmatrix}, \qquad (6.24)$$

with $\theta = \arccos\frac{d_z(\mathbf{k})}{d(\mathbf{k})}$ and $\phi = \arctan\frac{d_y(\mathbf{k})}{d_x(\mathbf{k})}$. Assuming $|t_2| < |t_1|/3$, which is consistent with the physical intuition than the n.n.n. hopping should be much smaller that the n.n. hopping, the energy spectrum is either gapped, or the two bands touch at $\mathbf{K}$ and $\mathbf{K}'$. A linear expansion around these points provides the following low-energy effective Hamiltonians

$$\mathcal{H}(\mathbf{K} + \delta\mathbf{k}) = -3t_2\cos\varphi + \frac{3}{2}\frac{t_1 a}{\hbar}(\delta k_2 \sigma_x - \delta k_1 \sigma_y) + (M - 3\sqrt{3}t_2)\sin\varphi)\sigma_z,$$

$$\mathcal{H}(\mathbf{K}' + \delta\mathbf{k}) = -3t_2\cos\varphi - \frac{3}{2}\frac{t_1 a}{\hbar}(\delta k_2 \sigma_x + \delta k_1 \sigma_y) + (M + 3\sqrt{3}t_2)\sin\varphi)\sigma_z,$$

where the components δk_1 and δk_2 of $\delta\mathbf{k}$ are measured with respect to reference frame rotated (clockwise) by an angle $\pi/6$ with respect to the axes in Figure 6.1 (i.e., $\delta k_1 = \frac{\sqrt{3}}{2}\delta k_x - \frac{1}{2}\delta k_y$, etc.). Note that these equations are equivalent (up to an energy shift $\Delta E = -3t_2\cos\varphi$) with the continuum Haldane Hamiltonian given by Eq. (6.7), with $v_F = \frac{3}{2}t_1 a/\hbar$ and $\lambda = -3\sqrt{3}t_2\sin\varphi$.

To determine the phase diagram of the Haldane model, we calculate the corresponding Chern number. First, let us consider the low-energy contributions due to the Berry curvature of the occupied band in the vicinity of $\mathbf{K}$ and $\mathbf{K}'$. These contributions are captured by the continuum model. We note that the Berry curvature of a generic Dirac Hamiltonian of the form $\mathcal{H}(\mathbf{k}) = k_a \Lambda_{ab}\sigma_b + M\sigma_3$, with $a, b = 1, 2$, is $F_{xy}(\mathbf{k}) = \Omega_z(\mathbf{k}) = \frac{1}{2d^3}M\text{Det}(\Lambda)$, where $d^2 = k_a k_b \Lambda_{ac}\Lambda_{bc} + M^2$. Integrating over $\mathbf{k} \in \mathbb{R}^2$ we obtain the Chern number for the generic Dirac Hamiltonian as,

$$\nu = \frac{1}{2}\text{sign}(M)\,\text{sign}[\text{Det}(\Lambda)]. \qquad (6.25)$$

Note that ν is a half-integer because we ignored contributions from the Berry curvature at high energy, as discussed in the previous section. For the (continuum) Haldane model, we have two contributions of the form (6.25), one from $\mathbf{K}$ and the other from $\mathbf{K}'$, and the total Chern number becomes

$$\nu_H = \frac{1}{2}[\text{sign}(M + \lambda) - \text{sign}(M - \lambda)], \qquad (6.26)$$

with $\lambda = -3\sqrt{3}t_2\sin\varphi$. Note that ν_H is an integer, which suggests that in this particular case the high-energy contribution to the Chern number vanishes. Indeed, in the limit $|M| \to \infty$ we have $\nu_H = 0$, which is consistent with the wave function being localized on either A-type or B-type sites, i.e., the system being topologically trivial. In general, ν_H is zero for $|M| > 3\sqrt{3}|t_2 \sin \varphi|$ (trivial insulator) and ± 1 otherwise (Chern insulator). To confirm this result, we can use the solution (6.24) of the lattice model to calculate the Berry connection and the corresponding curvature over the entire Brillouin zone. The Chern number, obtained by integrating the Berry curvature over the BZ, is identical with the outcome of the continuum model calculation. The phase diagram of the Haldane model is shown in Figure 6.4.

6.2.2 Mercury telluride quantum wells: The BHZ model

The model that led to the discovery of the first 2D TI material [282] was proposed by Bernevig, Hughes, and Zhang (BHZ) in 2006 [55] as an effective four-band model for a CdTe/HgTe/CdTe quantum well. The system undergoes a band inversion at the Γ point as a function of the well width. In this structure, the electronic states closest to the Fermi level contain s and p orbitals. However, the HgTe well is characterized by an inverted band progression (as compared to CdTe and other "normal" semiconductors, e.g., GaAs) with the s-type Γ_6 band lying below the p-type Γ_8 band. To capture the low-energy physics, one can construct a minimal tight-binding Hamiltonian with a four-state basis [55]. Also, since the quantum well is a quasi-2D system and HgTe has a cubic (zinc-blende) crystal structure, it is natural to define the effective tight-binding model on a 2D square lattice with translation vectors $\boldsymbol{a}_1 = a\hat{\boldsymbol{x}}$ and $\boldsymbol{a}_2 = a\hat{\boldsymbol{y}}$. The corresponding Hamiltonian is

$$H_{\text{BHZ}} = \sum_{i,\alpha,\sigma} \epsilon_\alpha c_{i\alpha\sigma}^\dagger c_{i\alpha\sigma} + \sum_{i,j} \sum_{\alpha,\beta} \sum_\sigma t_\sigma^{\alpha\beta}(i,j) \, c_{i\alpha\sigma}^\dagger c_{j\beta\sigma}, \qquad (6.27)$$

where i and j are n.n. sites on the square lattice, $\alpha, \beta \in \{s, p\}$ are band indices, and $\sigma = \pm$ is the z component of the spin. In the orbital basis, the hopping matrix has the form

$$t_\sigma(i,j) = \begin{pmatrix} -t_{ss} & -t_{sp} e^{i\sigma\theta_{ij}} \\ -t_{sp} e^{-i\sigma\theta_{ij}} & t_{pp} \end{pmatrix}, \qquad (6.28)$$

where $\theta_{ij} \in \{0, \frac{\pi}{2}, \pi, \frac{3\pi}{2}\}$ is the angle between the x axis and the n.n. bond $j \to i$. To study the topological properties of the BHZ model it is convenient to Fourier transform the lattice Hamiltonian (6.27). We define the Dirac matrices

$$\Gamma^1 = \sigma_x \otimes s_x, \qquad \Gamma^2 = \sigma_x \otimes s_y, \qquad \Gamma^3 = \sigma_x \otimes s_z,$$
$$\Gamma^4 = \sigma_y \otimes s_0, \qquad \Gamma^5 = \sigma_z \otimes s_0, \qquad (6.29)$$

where σ_i and s_i are Pauli matrices associated with the orbital and spin degrees of freedom, respectively. Using these matrices, we can write the Fourier transform of the BHZ Hamiltonian as

$$\mathcal{H}(\boldsymbol{k}) = d_0(\boldsymbol{k}) + \sum_{a=1}^{5} d_a(\boldsymbol{k})\Gamma^a. \qquad (6.30)$$

The coefficients in Eq. (6.30) are $d_1(\boldsymbol{k}) = d_2(\boldsymbol{k}) = 0$ and

$$
\begin{aligned}
d_0(\boldsymbol{k}) &= \frac{1}{2}(\epsilon_s + \epsilon_p) - (t_{ss} - t_{pp})(\cos \boldsymbol{k} \cdot \boldsymbol{a}_1 + \cos \boldsymbol{k} \cdot \boldsymbol{a}_2), \\
d_3(\boldsymbol{k}) &= 2t_{sp} \sin \boldsymbol{k} \cdot \boldsymbol{a}_2, \\
d_4(\boldsymbol{k}) &= 2t_{sp} \sin \boldsymbol{k} \cdot \boldsymbol{a}_1, \\
d_5(\boldsymbol{k}) &= \frac{1}{2}(\epsilon_s - \epsilon_p) - (t_{ss} + t_{pp})(\cos \boldsymbol{k} \cdot \boldsymbol{a}_1 + \cos \boldsymbol{k} \cdot \boldsymbol{a}_2).
\end{aligned}
\tag{6.31}
$$

The eigenvalues of the Hamiltonian (6.30) are $E(\boldsymbol{k}) = d_0(\boldsymbol{k}) \pm \sqrt{\sum_a [d_a(\boldsymbol{k})]^2}$, both double degenerate.

The BHZ model has both TR symmetry and inversion symmetry. The $\mathbb{Z}_2$ invariant for this type of topological insulator is determined, within the framework developed by Fu and Kane [186] (see Section 3.4.3), by the product of the parities of the Kramers pairs at the TR invariant momenta (TRIM),

$$
(-1)^\nu = \prod_{i=1}^{4} \prod_{n=1}^{N} \xi_{2n}(\Lambda_i),
\tag{6.32}
$$

where $\xi_{2n} = \xi_{2n-1}$ is the parity of the nth Kramers pair, $2N$ is the total number of *occupied* bands, and Λ_i are the TRIM. Specifically, for the Kramers partners $|u_{2n-1,\Lambda_i}\rangle$ and $|u_{2n,\Lambda_i}\rangle$ we have $\Pi|u_{\alpha,\Lambda_i}\rangle = \xi_{2n}|u_{\alpha,\Lambda_i}\rangle$, where Π is the parity operator and $\alpha = 2n - 1, 2n$. We note that in our Γ representation the TR operator is constructed as $\mathcal{T} = -i\sigma_0 \otimes s_y K$, with K being the complex conjugation. On the other hand, since the s orbital is parity even and the p orbital is parity odd, the inversion operator is represented as $\Pi = \sigma_z \otimes s_0 = \Gamma^5$. Consequently, we have $\mathcal{T}\Gamma^a\mathcal{T}^{-1} = -\Gamma^a$ and $\Pi\Gamma^a\Pi^{-1} = -\Gamma^a$ for $a = 1, 2, 3, 4$, while $\mathcal{T}\Gamma^5\mathcal{T}^{-1} = +\Gamma^5$ and $\Pi\Gamma^5\Pi^{-1} = +\Gamma^5$. At a TRIM, the Hamiltonian satisfies the condition $\Pi\mathcal{H}(\Lambda_i)\Pi^{-1} = \mathcal{H}(\Lambda_i)$, so it must have the form

$$
\mathcal{H}(\Lambda_i) = d_0(\Lambda_i) + d_5(\Lambda_i)\Gamma^5.
\tag{6.33}
$$

The Hamiltonian in Eq. (6.33) is diagonal and its eigenstates are the basis states of the tight-binding model. These eigenstates form two Kramers pairs: the pair $|s,\uparrow\rangle$, $|s,\downarrow\rangle$ with even parity and energy $E_+(\Lambda_i) = d_0(\Lambda_i) + d_5(\Lambda_i)$ and the pair $|p_x + ip_y,\uparrow\rangle$, $|p_x - ip_y,\downarrow\rangle$ with odd parity and energy $E_-(\Lambda_i) = d_0(\Lambda_i) - d_5(\Lambda_i)$. If $d_5(\Lambda_i) > 0$, we have $E_+ > E_-$ and the odd-parity p band is occupied. On the other hand, if $d_5(\Lambda_i) < 0$, we have $E_+ < E_-$ and the even-parity s band is occupied. Finally, using the expression of $d_5(\boldsymbol{k})$ from Eq. (6.31) and taking into account that $(\boldsymbol{k} \cdot \boldsymbol{a}_1, \boldsymbol{k} \cdot \boldsymbol{a}_2)$ takes the values $(0, 0)$, $(0, \pi)$, $(\pi, 0)$, and (π, π) at the four TRIM, we arrive at the following conclusions:

1. If $\epsilon_s - \epsilon_p > 4(t_{ss} - t_{pp})$ we have $E_+(\Lambda_i) > E_-(\Lambda_i)$ for all Λ_i and the $\mathbb{Z}_2$ invariant given by Eq. (6.32) is $\nu = 0$, i.e., the system is a trivial insulator. Note that in this case the p-bands lie below the s-bands, which is the typical situation for a band insulator/semiconductor.

2. If $\epsilon_s - \epsilon_p < 4(t_{ss} - t_{pp})$ we have $E_+(\Lambda_i) < E_-(\Lambda_i)$ at $\Lambda_i = (0,0)$ (the Γ point of the 2D BZ) but $E_+(\Lambda_i) > E_-(\Lambda_i)$ for the other three TRIM. The $\mathbb{Z}_2$ invariant is $\nu = 1$, which means that the system is a topological insulator. Note that in this case the band order flips at the Γ point, i.e., the top of the valence band has s character, while the bottom of the conduction band has p character.

6.2.3 p-Wave superconductors in one and two dimensions

Superconductivity is an essentially interacting quantum phenomenon characterized by the vanishing of static electrical resistivity and the emergence of perfect diamagnetism (the Meissner effect) at low temperatures. In essence, the phenomenon is driven by an attractive effective interaction between electrons near the Fermi surface. This leads to the formation of bound states with energy lower than the energy of two free electrons, which produces an instability of the normal state (i.e., the Fermi sea). In conventional superconductors the attractive effective interaction between electrons is phonon-mediated, i.e., it is a consequence of the electron-phonon interaction. The basic theory of superconductivity was formulated by Bardeen, Cooper, and Schrieffer (BCS) in 1957 [36]. In the past few decades the focus has been on understanding the microscopic pairing mechanism in unconventional superconductors, such as, for example, the high temperature cuprates [38, 283], and on studying quasiparticle phenomena in various types of superconductors. Remarkably, there is a close analogy between the *mean-field* description of quasiparticles in a superconductor (or a superfluid) and the Hamiltonian of a noninteracting insulator. Below, we introduce the mean-field formulation of the quasiparticle physics in superconductors and discuss a few simple examples of tight-binding models.

The Bogoliubov–de Gennes formalism. We introduce the mean-field quasiparticle formalism by considering, for concreteness, the case of s-wave superconductors. We start with the many-body Hamiltonian of an interacting electron system

$$H = \sum_{\mathbf{k},\sigma} \xi_{\mathbf{k}} \, c_{\mathbf{k}\sigma}^{\dagger} c_{\mathbf{k}\sigma} + V_{\text{int}}, \qquad (6.34)$$

where V_{int} is an effective electron-electron interaction and $\xi(\mathbf{k}) = \epsilon(\mathbf{k}) - \mu$, with μ being the chemical potential and $\epsilon(\mathbf{k})$ the single-particle dispersion, e.g., $\epsilon(\mathbf{k}) = \hbar^2 k^2 / 2m$ for free electrons. We assume that the electrons interact attractively within a small energy window ω_D in the vicinity of the Fermi surface, which leads to an instability of the normal metal due to scattering between electrons in single-particle states $|\mathbf{k}, \uparrow\rangle$ and $|-\mathbf{k}, \downarrow\rangle$. Specifically, we have

$$V_{\text{int}} = \frac{1}{N} \sum_{\mathbf{k},\mathbf{k}'} V_{\mathbf{k}\mathbf{k}'} c_{\mathbf{k}\uparrow}^{\dagger} c_{-\mathbf{k}\downarrow}^{\dagger} c_{-\mathbf{k}'\downarrow} c_{\mathbf{k}'\uparrow}, \qquad (6.35)$$

where N is the number of (discrete) k values and $V_{kk'} = -V_0 < 0$ for $|\xi_k|, |\xi_{k'}| < \omega_D$, while $V_{kk'} = 0$ otherwise. Within the *mean-field approximation*, the product of operators in V_{int} is replaced by

$$\langle c_{k\uparrow}^\dagger c_{-k\downarrow}^\dagger \rangle c_{-k'\downarrow} c_{k'\uparrow} + c_{k\uparrow}^\dagger c_{-k\downarrow}^\dagger \langle c_{-k'\downarrow} c_{k'\uparrow} \rangle - \langle c_{k\uparrow}^\dagger c_{-k\downarrow}^\dagger \rangle \langle c_{-k'\downarrow} c_{k'\uparrow} \rangle, \qquad (6.36)$$

where $\langle \dots \rangle$ designates the thermodynamic average. This approximation leads to the following mean-field BCS Hamiltonian

$$H_{\mathrm{BCS}} = E_{\mathrm{BCS}} + \sum_{k,\sigma} \xi_k\, c_{k\sigma}^\dagger c_{k\sigma} + \sum_k [\Delta_k c_{k\uparrow}^\dagger c_{-k\downarrow}^\dagger + \Delta_k^* c_{-k'\downarrow} c_{k'\uparrow}], \qquad (6.37)$$

where $\Delta_k = \frac{1}{N}\sum_k V_{kk'}\langle c_{-k'\downarrow} c_{k'\uparrow} \rangle$ represents the superconducting order parameter and E_{BCS} is the energy of a pure condensate, which, in the limit $T \to 0$, is the BCS ground state energy. The last two terms in Eq. (6.37), which describe the dynamics of the quasiparticles in the superconducting state, can be diagonalized by defining a new set of fermionic operators constructed as linear combinations of electron creation and annihilation operators,

$$\gamma_{k\uparrow} = u_k^* c_{k\uparrow} + v_k c_{-k\downarrow}^\dagger, \qquad (6.38a)$$

$$\gamma_{-k\downarrow}^\dagger = u_k c_{-k\downarrow}^\dagger - v_k^* c_{k\uparrow}, \qquad (6.38b)$$

with $|u_k|^2 = \frac{1}{2}(1 + \xi_k/E_k)$, $|v_k|^2 = \frac{1}{2}(1 - \xi_k/E_k)$, and $u_k v_k = \Delta_k/(2E_k)$, where $E_k = \sqrt{\xi_k^2 + |\Delta_k|^2}$. Note that s-wave pairing ($\Delta_{-k} = \Delta_k$) implies $u_{-k} = u_k$ and $v_{-k} = v_k$. Using this mapping, called the Bogoliubov (or Bogoliubov–Valatin) transformation, the BCS Hamiltonian takes the simple form $H_{\mathrm{BCS}} = E_{\mathrm{BCS}} + \sum_{k,\sigma} E_k \gamma_{k\sigma}^\dagger \gamma_{k\sigma}$, where $E_k = \sqrt{\xi_k^2 + |\Delta_k|^2}$ is the quasiparticle excitation energy (see Figure 6.5). Note that the presence of a quasiparticle increases the energy of the system by at least $|\Delta_{k_F}|$ over the energy of a pure condensate, E_{BCS}. Also note that deep inside the Fermi sea $|u_k| \approx 0$ and $|v_k| \approx 1$, i.e., the Bogoliubov quasiparticles are hole-like, far above E_F $|u_k| \approx 1$ and $|v_k| \approx 0$, i.e., the quasiparticles are particle-like, while in the vicinity of the Fermi surface the Bogoliubov quasiparticles are combinations of holes and electrons with comparable amplitudes ($|u_k| = |v_k|$ exactly at E_F).

To describe the quasiparticle dynamics in a nonhomogeneous system and to make explicit the analogy with topological insulators, it is convenient to define a first-quantized Hamiltonian for the single-particle excitations. First, we note that, using the fermion anticommutation relations, we can write the second term in Eq. (6.37) as $\frac{1}{2}\sum_{k,\sigma}[\xi_k\, c_{k\sigma}^\dagger c_{k\sigma} - \xi_{-k}\, c_{k\sigma} c_{k\sigma}^\dagger + \xi_k]$. Also we have $\Delta_k c_{k\uparrow}^\dagger c_{-k\downarrow}^\dagger = \frac{1}{2}[\Delta_k c_{k\uparrow}^\dagger c_{-k\downarrow}^\dagger - \Delta_k c_{-k\downarrow}^\dagger c_{k\uparrow}^\dagger]$ and a similar relation for $\Delta_k^* c_{-k'\downarrow} c_{k'\uparrow}$. Second, we introduce the four-component spinor $\Psi_k^\dagger = (c_{k\uparrow}^\dagger, c_{k\downarrow}^\dagger, c_{-k\uparrow}, c_{-k\downarrow})$. Note that by doing this we *artificially double* the number of degrees of freedom of the system. Consider, for example, the many-body ground state of the normal system ($\Delta_k = 0$): each occupied (empty)

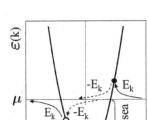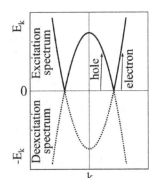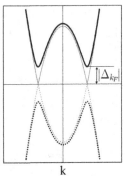

FIGURE 6.5 (Left) Single-particle band dispersion for a noninteracting electron system. The many-body ground state corresponds to a full Fermi sea. An excitation of energy $E_k = |\xi_k|$ is created by adding an electron with $\xi_k = \epsilon(k) - \mu > 0$ into an empty single-particle state or removing one electron from an occupied state with energy $\xi_k < 0$ (i.e., creating a hole). Deexcitation processes are symbolized by dashed arrows. (Center) Excitation and deexcitation spectra for a normal metal, $E_k = |\xi_k|$. (Right) Excitation (deexcitation) spectrum for an s-wave superconductor, $E_k = \sqrt{\xi_k^2 + \Delta_k^2}$ (black lines).

electron state with $\xi_k < 0$ ($\xi_k > 0$) is also represented as an empty (occupied) hole state with energy $\xi_k > 0$ ($\xi_k < 0$). The energies $E_k = |\xi_k|$ of the empty states (both electron-type and hole-type) give us the excitation spectrum (see Figure 6.5). Using this so-called Nambu spinor representation, we can write the BCS mean-field Hamiltonian in the form

$$H_{\text{BCS}} = \frac{1}{2} \sum_k \Psi_k^\dagger \mathcal{H}_{\text{BdG}}(k) \Psi_k + \text{const.}, \qquad (6.39)$$

where $\mathcal{H}_{\text{BdG}}$ is a first-quantized Hamiltonian called the Bogoliubov–de Gennes (BdG) Hamiltonian. For nonmagnetic superconductors, the Bloch (i.e., k space) BdG Hamiltonian has the form

$$\mathcal{H}_{\text{BdG}}(k) = \begin{pmatrix} \xi_k & 0 & 0 & \Delta_k \\ 0 & \xi_k & -\Delta_{-k} & 0 \\ 0 & -\Delta_{-k}^* & -\xi_{-k} & 0 \\ \Delta_k^* & 0 & 0 & -\xi_{-k} \end{pmatrix}. \qquad (6.40)$$

The eigenvalues of the BdG Hamiltonian (6.40) are $E_k = \pm\sqrt{\xi_k^2 + |\Delta_k|^2}$ (each being double degenerate) and correspond to the quasiparticle excitation (deexcitation) energies shown in Figure 6.5. We emphasize that the BdG description given by Eq. (6.40) is *intrinsically redundant*. Indeed, the positive energy eigenfunctions $\phi_{k\uparrow}^{(+)} = (u_k, 0, 0, v_k^*)^T$ and $\phi_{k\downarrow}^{(+)} = (0, u_k, -v_k^*, 0)^T$ describe quasiparticles corresponding to the creation operators $\Psi_k^\dagger \cdot \phi_{k\uparrow}^{(+)} = \gamma_{k\uparrow}^\dagger$ and $\Psi_k^\dagger \cdot \phi_{k\downarrow}^{(+)} = \gamma_{k\downarrow}^\dagger$, respectively. On the other hand, the negative energy solutions $\phi_{k\uparrow}^{(-)} = (-v_k, 0, 0, u_k^*)^T$ and $\phi_{k\downarrow}^{(-)} = (0, v_k, u_k^*, 0)^T$ correspond to the associated annihilation operators $\Psi_k^\dagger \cdot \phi_{k\uparrow}^{(-)} = \gamma_{-k\downarrow}$ and $\Psi_k^\dagger \cdot \phi_{k\downarrow}^{(-)} = \gamma_{-k\uparrow}$, respectively, i.e., they do not describe independent excitations.

As a final remark on the BdG formalism, we note that the Hamiltonian (6.40) can be written in the form

$$\mathcal{H}_{\text{BdG}}(k) = \xi_k \tau_z \otimes s_0 - \text{Re}[\Delta]\, \tau_y \otimes s_y - \text{Im}[\Delta]\, \tau_x \otimes s_y, \qquad (6.41)$$

where τ_i and s_i are Pauli matrices associated the particle-hole and spin degrees of freedom, respectively, and we have used the properties $\xi_{-k} = \xi_k$, $\Delta_{-k} = \Delta_k$. From Eq. (6.41) it is evident that the BdG Hamiltonian has particle-hole symmetry, $U_c \mathcal{H}^*(k) U_c^{-1} = -\mathcal{H}(-k)$, with $U_c = \tau_x \otimes s_0$. The fact that ξ_k and Δ_k are even functions of momentum plays no role in this result, i.e., particle-hole symmetry is a property of the BdG Hamiltonian (6.40) with no additional conditions.

One-dimensional p-wave superconductor. Consider a 1D superconducting system consisting of spinless fermions. At the mean-field level, pairing is described by $\frac{1}{2}\sum_k (\Delta_k c_k^\dagger c_{-k}^\dagger + \Delta_k^* c_{-k} c_k)$, where the pairing potential must have odd parity, $\Delta_{-k} = -\Delta_k$ (since even parity contributions from k and $-k$ cancel each other). In the basis $\Psi_k^\dagger = (c_k^\dagger, c_{-k})$, the BdG Hamiltonian has the form $\mathcal{H}_{\text{BdG}}(k) = \xi_k \tau_z + \text{Re}[\Delta]\, \tau_x - \text{Im}[\Delta]\, \tau_y$, where $\xi_k = \epsilon_k - \mu$ is the 1D non-interacting band dispersion. For simplicity, we assume that Δ is either real or purely imaginary, which we can always realize via a global phase rotation.

To construct a lattice model for this 1D superconductor, we consider spinless fermions with creation operators $c_i^\dagger$ hopping between the nearest-neighbor sites of a 1D wire with hopping amplitude $-t$ ($t > 0$) in the presence of a nearest-neighbor pairing potential Δ. Explicitly, we have

$$H_K = \sum_i \left[-t \left(c_i^\dagger c_{i+1} + c_{i+1}^\dagger c_i \right) - \mu c_i^\dagger c_i + \Delta \left(c_i^\dagger c_{i+1}^\dagger + c_{i+1} c_i \right) \right]. \quad (6.42)$$

We assume, for simplicity, that Δ is real. The Hamiltonian in Eq. (6.42) is the so-called *Kitaev model* for a 1D p-wave superconductor [298]. For a homogeneous system, we can write the Hamiltonian H_K, after performing a (lattice) Fourier transform and neglecting an energy constant, as $H_K = \frac{1}{2}\sum_k \Psi_k^\dagger \mathcal{H}_{\text{BdG}}(k)\Psi_k$, where the BdG Hamiltonian is

$$\mathcal{H}_{\text{BdG}}(k) = \begin{pmatrix} -2t\cos k - \mu & -2i\Delta\sin k \\ 2i\Delta\sin k & 2t\cos k + \mu \end{pmatrix}. \quad (6.43)$$

Using our "standard" notation, this becomes $\mathcal{H}_{\text{BdG}}(k) = \xi_k \tau_z + \Delta_k \tau_y$, with $\xi_k = -2t\cos k - \mu$ and $\Delta_k = 2\Delta\sin k$. The BdG Hamiltonian (6.43) has two energy bands, $E_\pm(k) = \pm E_k$ with $E_k = \sqrt{\xi_k^2 + \Delta_k^2}$.

To identify the topological phases of this 1D system, we first determine the symmetry of the BdG Hamiltonian. As a result of the particle-hole redundancy of the BdG representation, the Hamiltonian (6.43) has particle-hole symmetry represented by $U_c = \tau_x$. Note that $C^2 = U_c U_c^* = +1$. In addition, the model has time-reversal symmetry for spinless particles, $\mathcal{T} \mathcal{H}_{\mathrm{BdG}}(k) \mathcal{T}^{-1} = \mathcal{H}_{\mathrm{BdG}}(-k)$, with $\mathcal{T} = K$ (the complex conjugation). Note that $\mathcal{T}^2 = +1$. According to Table 3.1, the BdG Hamiltonian (6.43) can be viewed as a representative of the BDI symmetry class. Consequently, we can identify the distinct phases of this 1D (i.e., odd dimension) system by calculating the corresponding winding number. The BdG Hamiltonian (6.43) has chiral symmetry (represented by the product $\mathcal{S} = \mathcal{T} \cdot \mathcal{C}$) and can be brought into block off-diagonal form by a rotation $(\tau_x, \tau_y, \tau_z) \rightarrow (-\tau_z, \tau_y, \tau_x)$. The corresponding projector (Q matrix) is also block off-diagonal with $q(k) = (\xi_k - i\Delta_k)/E_k$. We calculate the winding number using Eqs. (3.38) and (3.39). We have

$$\nu = \frac{i}{2\pi} \int_{-\pi}^{\pi} dk \, q^{-1}(k) \frac{d}{dk} q(k) = \frac{-1}{2\pi} \int_{-\pi}^{\pi} dk \, \frac{\Delta(2t + \mu \cos k)}{(\mu + 2t \cos k)^2 + \Delta^2 \sin^2 k}. \quad (6.44)$$

Evaluating the integral we obtain $\nu = 0$ for $|\mu| > 2t$ and $\nu = -\mathrm{sign}(\Delta)$ for $|\mu| < 2t$, i.e., the system has two different phases. In the so-called *strong coupling* regime, which is characterized by the chemical potential being outside the band ($|\mu| > 2t$), the system is topologically trivial ($\nu = 0$) and, in the absence of pairing, it is a gapped band insulator. Note that in this regime superconductivity is characterized by the formation of "molecule-like" fermion pairs and does not emerge as a weak-coupling instability of the Fermi surface. On the other hand, in the *weak coupling regime*, which corresponds to μ being inside the band ($|\mu| < 2t$), the system is a BCS-like superconductor in a topologically nontrivial phase ($\nu = \pm 1$).

The spinless fermion model (6.42) can be viewed as a simplified model for a 1D spin-polarized electron system. The source of spin polarization can be, for example, an applied magnetic field. As a result, time-reversal (TR) symmetry is not an intrinsic property of the physical system, but rather a feature of our simplified model. In other words, more detailed models of the 1D p-wave superconductor are not TR invariant. From this perspective, we can ignore the "accidental" TR symmetry of Hamiltonian (6.43) and view it as a representative of symmetry class D. Table 3.3 tells us that in this case the system modeled by H_{BdG} is a one-dimensional $\mathbb{Z}_2$ superconductor. To distinguish the topological and trivial phases, we can calculate the so-called *Majorana number*, a $\mathbb{Z}_2$ invariant introduced by Kitaev [298].

To define the $\mathbb{Z}_2$ topological invariant and to reveal the consequence of the bulk-boundary correspondence – the emergence of unpaired Majorana bound states at the ends of the chain – we address the physics of the p-wave superconducting chain from a different perspective by introducing the so-called *Majorana representation*. The concept of *Majorana fermion* [352] will be discussed in more detail in Chapter 8. Here, we pursue a formal approach and simply replace each *complex* fermion (represented by the creation and annihilation operators $c_i^\dagger$ and c_i) by two *real* fermions, γ_{2i-1} and γ_{2i}, defined by the relations

$$\gamma_{2i-1} = c_i^\dagger + c_i, \qquad\qquad \gamma_{2i} = i(c_i^\dagger - c_i). \qquad (6.45)$$

The operators γ_j representing these real fermions, which are called *Majorana fermions*, have the property $\gamma_j^\dagger = \gamma_j$ and satisfy $\{\gamma_i, \gamma_j\} = 2\delta_{ij}$. In other words, each Majorana fermion is its own antiparticle and two distinct fermions are represented by operators that anticommute. Using the Majorana representation, the Hamiltonian (6.42) becomes (up to an energy constant)

$$H_K = \frac{i}{2} \sum_j \left[-\mu\gamma_{2j-1}\gamma_{2j} + (t - \Delta)\gamma_{2j}\gamma_{2j+1} - (t + \Delta)\gamma_{2j-1}\gamma_{2j+2} \right]. \qquad (6.46)$$

Note that this Hamiltonian has the form $H_K = \frac{i}{2} \sum_{l,m} \gamma_l A_{lm}\gamma_m$, where A is a real, anti-symmetric matrix. For a translation-invariant system, we have $A_{(2i+\alpha)(2j+\beta)} = B_{\alpha\beta}(i - j)$, where $\alpha, \beta = 1, 2$ and i, j label the lattice sites. After performing a Fourier transform, we obtain

$$\tilde{B}(k) = \begin{pmatrix} 0 & \xi_k + i\Delta_k \\ -\xi_k + i\Delta_k & 0 \end{pmatrix}, \qquad (6.47)$$

with $\xi_k = -2t\cos k - \mu$ and $\Delta_k = 2\Delta \sin k$. The matrix $\tilde{B}(k)$ satisfies $\tilde{B}^\dagger(k) = \tilde{B}^T(-k) = -\tilde{B}(k)$. The "Majorana number" $\mathcal{M}$ (also called the Pfaffian $\mathbb{Z}_2$ invariant) introduced by Kitaev can be expressed in terms of the Majorana representation matrix $\tilde{B}(k)$ as [298]

$$\mathcal{M} = \mathrm{sign}\left\{ \mathrm{Pf}\left[\tilde{B}(0) \right] \mathrm{Pf}\left[\tilde{B}(\pi) \right] \right\}, \qquad (6.48)$$

where $\mathrm{Pf}[\dots]$ denotes the Pfaffian of an anti-symmetric matrix.[4] For the matrix in Eq. (6.47) we have $\mathrm{Pf}\left[\tilde{B}(0) \right] = -(2t + \mu)$ and $\mathrm{Pf}\left[\tilde{B}(\pi) \right] = 2t - \mu$. Hence, the Majorana number takes the values $\mathcal{M} = +1$ in the trivial (strong coupling) phase and $\mathcal{M} = -1$ in the topological (weak coupling) phase.

Equation (6.48) allows us to calculate $\mathcal{M}$ for a closed chain. In the case of open boundary conditions, $\mathcal{M} = -1$, which corresponds to a nontrivial topological superconductor, signals the presence of an isolated Majorana bound state at each of the ends of the wire. To gain some intuition, we consider the Hamiltonian (6.46) on a finite lattice, $1 \le j \le N$, in two special limits.

1. Trivial phase with $\Delta = t = 0$ and $\mu < 0$. The Hamiltonian reduces to $H_K = -\mu\frac{i}{2}\sum_j \gamma_{2j-1}\gamma_{2j}$, which couples the Majorana operators on each site j but leaves the operators defined on neighboring sites uncoupled. The ground state corresponds to the eigenvalue $+1$ of the electron parity operator $P_j = -i\gamma_{2j-1}\gamma_{2j} = 1 - 2c_j^\dagger c_j$, which means that the ground state is the vacuum of c_j fermions. Adding a spinless fermion costs a finite energy $|\mu|$ and there are no low-energy states at the ends of the chain.

---

[4]For a definition of the Pfaffian see Eq. (3.42) on page 97.

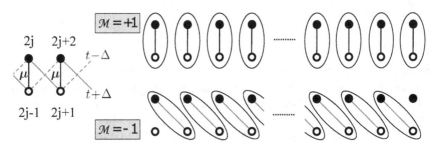

FIGURE 6.6 Schematic representation of the lattice p-wave Hamiltonian (6.46). The empty and filled circles represent the Majorana fermions γ_{2j-1} and γ_{2j}, respectively, corresponding to each physical site j. When $t = \Delta = 0$ and $\mu \neq 0$ the Majoranas "pair up" on the same site and the system is topologically trivial ($\mathcal{M} = +1$). In the limit $\mu = 0$ and $\Delta = t > 0$, which corresponds to the topological phase with $\mathcal{M} = -1$, the Hamiltonian couples Majoranas from adjacent sites leaving two "unpaired" Majoranas, γ_1 and γ_{2N}.

2. Topological phase with $-\Delta = t > 0$ and $\mu = 0$. The Hamiltonian reduces to $H_K = it \sum_j \gamma_{2j}\gamma_{2j+1}$, which only couples Majorana operators defined on adjacent physical sites. The ends of the chain support two "unpaired" Majorana modes that have zero-energy, since γ_1 and γ_{2N} are explicitly absent from the Hamiltonian. Defining the new fermion operators $a_j = \frac{1}{2}(\gamma_{2j+1} + i\gamma_{2j})$ and $f = \frac{1}{2}(\gamma_1 + i\gamma_{2N})$, we have $H_K = t \sum_{j=1}^{N-1} a_j^\dagger a_j$. The "bulk" is still gapped, since adding an a_j fermion implies an energy cost t. However, adding the highly nonlocal fermion f costs no energy, which results in a two-fold ground state degeneracy. Indeed, if $|0\rangle$ is a ground state that satisfies $f|0\rangle = 0$ (no f fermion), then $|1\rangle = f^\dagger|0\rangle$ is also a ground state, but with opposite fermion parity. By contrast, the ground state of a trivial gapped superconductor is nondegenerate and has even parity (since all fermions form Cooper pairs).

As one tunes the system away from these special points, the physics revealed by each of the two cases (which are schematically represented in Figure 6.6) survives as long as the (bulk) gap is preserved. We conclude that the boundary modes of a 1D topological p-wave superconductor are zero-energy Majorana bound states localized at the ends of the wire.

Two-dimensional chiral p-wave superconductor. Many aspects of the 1D superconductor physics discussed above can be generalized to higher dimensions. The simplest example is a 2D chiral p-wave superconductor described by a lattice model similar to the Hamiltonian in Eq. (6.42),

$$H = -\mu \sum_i c_i^\dagger c_i + \sum_{i,\delta} \left[-t \left(c_i^\dagger c_{i+\delta} + h.c. \right) + \Delta \left(\lambda_\delta c_i^\dagger c_{i+\delta}^\dagger + h.c. \right) \right], \quad (6.49)$$

where $i = (i_x, i_y)$ are sites in a square lattice of lattice constant $a = 1$,

$\delta = (\pm 1, 0), (0, \pm 1)$ are nearest-neighbor displacements, and $\lambda_{(\pm 1,0)} = 1$ while $\lambda_{(0,\pm 1)} = i$. We assume that Δ is real and $t > 0$. Performing a Fourier transform leads to a BdG Hamiltonian of the form

$$\mathcal{H}_{\text{BdG}}(\boldsymbol{k}) = \begin{pmatrix} \xi_{\boldsymbol{k}} & -i\Delta_{\boldsymbol{k}} \\ i\Delta_{\boldsymbol{k}}^* & -\xi_{\boldsymbol{k}} \end{pmatrix}, \qquad (6.50)$$

with $\xi_{\boldsymbol{k}} = -2t(\cos k_x + \cos k_y) - \mu$ and $\Delta_{\boldsymbol{k}} = 2\Delta(\sin k_x + i \sin k_y)$. Note that $\mathcal{H}_{\text{BdG}}$ has particle-hole symmetry (with $U_c = \tau_x$) but, unlike the 1D case in Eq. (6.43), it has no time-reversal symmetry because $\Delta_{\boldsymbol{k}}$ is intrinsically complex. Hence, $\mathcal{H}_{\text{BdG}}$ is a 2D representative of symmetry class D and, according to Table 3.3, it describes a $\mathbb{Z}$ topological superconductor. Calculating the Chern number reveals that the strong-coupling regime ($|\mu| > 4t$) corresponds to the trivial phase, while the weak-coupling regime ($|\mu| < 4t$) supports two topological superconducting phases (with $\mu < 0$ and $\mu > 0$, respectively) characterized by opposite values of the Chern number, $C = \pm 1$.

By analogy with the 2D Chern insulator, we expect the emergence of low-energy chiral edge modes when an open boundary system described by Hamiltonian (6.49) is in a topologically nontrivial phase. Remarkably, these edge states are not only chiral (i.e., for a given edge they propagate only in one direction), but they also satisfy the Majorana condition, hence they are *chiral Majorana edge modes*. To study these edge modes, we solve the lattice model (6.49) on a strip consisting of N_y infinite parallel chains (i.e., $-\infty < i_x < \infty$ and $1 \leq i_y \leq N_y$). Note that we can still perform a Fourier transform along the x direction. The corresponding BdG Hamiltonian $\mathcal{H}_{\text{BdG}}(k_x)$ is a $2N_y \times 2N_y$ matrix defined in the representation given by the spinor $\Psi_{k_x}^\dagger = (c_{k_x 1}^\dagger, \ldots, c_{k_x N_y}^\dagger, c_{-k_x 1}, \ldots, c_{-k_x N_y}) = (\psi_{k_x}^\dagger, \psi_{-k_x})$, where $c_{k_x j_y} = \sum_{j_x} e^{ik_x j_x} c_{(j_x j_y)}$. The energy spectrum of $\mathcal{H}_{\text{BdG}}(k_x)$ for a strip with $N_y = 50$, $\Delta = 0.45t$, and $\mu = -2.5t$ is shown in Figure 6.7(a). As expected in the weak-coupling regime ($|\mu| < 4t$), the superconducting gap is crossed by gapless modes, a feature that signals the nontrivial topological nature of the system.

From the numerical solution we can also determine the properties of the states $\chi^\alpha(k_x)$ associated with these gapless modes, where $\chi^\alpha(k_x) = (u_{k_x 1}^\alpha, \ldots, u_{k_x N_y}^\alpha, v_{k_x 1}^\alpha, \ldots, v_{k_x N_y}^\alpha) = (\mathcal{U}_{k_x}^\alpha, \mathcal{V}_{k_x}^\alpha)$ with $\alpha = L$ ($\alpha = R$) for the left (right) moving mode. We find the following generic features: i) The states are localized near the boundaries of the system, i.e., the amplitudes $|u_{k_x j_y}^{R(L)}|$ and $|v_{k_x j_y}^{R(L)}|$ decay with an exponential envelope away from the top (bottom) edge of the strip. In other words, the two gapless modes are chiral edge modes; see Figure 6.7(b). ii) The particle and hole components of the wave function, $u_{k_x j_y}^\alpha$ and $v_{k_x j_y}^\alpha$, differ only by a constant phase factor. With a convenient choice of the overall phase, we have $\chi^\alpha(k_x) = (e^{i\theta_\alpha} \mathcal{U}_{k_x}^\alpha, e^{-i\theta_\alpha} \mathcal{U}_{k_x}^\alpha)$, with $\mathcal{U}_{k_x}^\alpha$ being a real vector. iii) The states at k_x and $-k_x$ are identical, $\chi^\alpha(k_x) = \chi^\alpha(-k_x)$, which means that only edge modes with energy $E_k \geq 0$ are physically distinct (see Figure 6.7). In terms of edge mode operators, $(\gamma_{k_x}^\alpha)^\dagger = \chi^\alpha(k_x) \cdot \Psi_{k_x}^\dagger$,

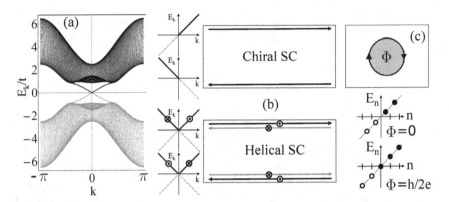

FIGURE 6.7 (a) Energy spectrum for a chiral p-wave superconductor described by the lattice Hamiltonian (6.49) on an infinite strip of width $N_y = 50$. The model parameters are $\Delta = 0.45t$ and $\mu = -2.5t$. The states with $E_k \geq 0$ (black lines) represent independent physical states. (b) Comparison between the edge states of a chiral p-wave superconductor (chiral Majorana or *Majorana–Weyl* modes) and those of a helical superconductor (spin-filtered counterpropagating Majorana modes). (c) Energy spectrum versus angular momentum (in units of $\hbar$) for a trivial disk surrounded by a topological $p + ip$ superconductor. Threading a flux $\Phi = h/2e$ through the disk leads to states with integer angular momentum and a spectrum that includes a zero-energy Majorana mode.

these properties imply $(\gamma^\alpha_{-k_x})^\dagger = e^{i\theta_\alpha}\mathcal{U}^\alpha_{k_x} \cdot \psi^\dagger_{-k_x} + e^{-i\theta_\alpha}\mathcal{U}^\alpha_{k_x} \cdot \psi_{k_x} = \gamma^\alpha_{k_x}$. The corresponding real-space operators

$$\gamma^\alpha_{j_x} = \sum_{k_x} e^{ik_x j_x}\gamma^\alpha_{k_x} = \sum_{k_x} e^{ik_x j_x}(\gamma^\alpha_{-k_x})^\dagger = \sum_{k_x} e^{-ik_x j_x}(\gamma^\alpha_{k_x})^\dagger = (\gamma^\alpha_{j_x})^\dagger, \quad (6.51)$$

satisfy explicitly the Majorana condition. We conclude that the topological phase of a 2D chiral p-wave superconductor is characterized by the emergence of chiral Majorana modes localized near the boundaries of the system. The modes with positive energy are physically distinct, while the negative energy states reflect the redundancy of the BdG formalism.

The chiral Majorana edge modes are gapless when the boundary between the topological and trivial regions is (infinitely) long. For finite boundaries, on the other hand, the spectrum is discrete. For example, the modes propagating around a small (trivial) disk embedded into a large topological region are gapped and are characterized by half-integer values of the angular momentum, as shown in Figure 6.7(c). Threading a magnetic flux $h/2e$ through the disk changes the boundary conditions and shifts the angular momentum to integer values [15]. As a result, a zero-energy Majorana mode becomes localized at the disk boundary. Physically, this situation can be realized when a magnetic flux penetrates the bulk of a (type II) topological superconductor. The core of the resulting vortex, which consists of normal metal, becomes gapped due to the small size of the normal region and can be adiabatically connected to a trivial insulator. Hence, in a spinless $p + ip$ 2D superconductor, each $h/2e$ vortex threading a topological region binds a localized zero-energy Majorana mode.

As a final remark, we mention that one can generate models for time-reversal invariant superconductors (in symmetry class DIII) following the strategy leading from the Haldane model of a Chern insulator to the Kane–Mele model for the quantum spin Hall (QSH) insulator. We have seen (Section 6.1) that a QSH insulator can be viewed as two copies of the Chern insulator, one for each spin and having opposite orientations of the TR breaking field. Similarly, a TR invariant superconductor (also called *helical superconductor*) can be constructed using two copies of the chiral superconductor having opposite chiralities: a $p_x + ip_y$ superconductor made of spin-up electrons and a $p_x - ip_y$ superconductor with spin-down electrons. The topological and trivial phases are distinguished by the two values of a $\mathbb{Z}_2$ topological invariant. In systems with boundaries, the topological phase is characterized by the presence of robust counterpropagating, spin-polarized edge modes. A comparison between the edge modes in chiral and helical superconductors is shown schematically in Figure 6.7(b).

6.3 TOPOLOGICAL FIELD THEORY

Quantum field theory is the native language of topological quantum matter. Field theory is the very birthplace of the fundamental ideas about the nature of topological quantum phases, which were developed long before topological insulators reached the radar of mainstream condensed matter physics. These ideas – the result of a fruitful relation between physics and topology – revealed the important role that global (rather than local) properties play in certain quantum systems. A *topological quantum field theory* (TQFT) is a field theory that is insensitive to the metric of space-time and is characterized by correlation functions that are topological invariants. Essentially, a many-body condensed matter system is in a topological quantum phase if its low-energy, long wavelength effective theory is a TQFT. The role of TQFTs in describing topological phases of matter is, basically, similar to the role of Landau–Ginzburg field theories in describing symmetry-breaking phases of matter.

Having pointed out the significance of this theoretical tool, we should also mention that this is an area of considerable technical complexity. For the purpose of this introductory book, it would be probably meaningless to attempt to summarize it. Instead, we will just mention a few aspects concerning a specific type of TQFT – the Chern–Simons quantum field theories. Our goal here is to provide some perspective for the field-theoretic description of the Abelian fractional quantum Hall states in Chapter 5 (see Section 5.2.1 starting on page 159), the discussion of the so-called *axion electrodynamics* characterizing the electromagnetic response of topological insulators and Weyl semimetals in Chapter 7, and the discussion of anyonic properties in Chapter 12 (Section 12.2 on page 379). In addition, we would like to show that field theory represents a convenient tool for describing not only interacting topological phases, but also noninteracting topological band insulators (like those discussed in the previous section). For a systematic introduction to quantum field theoretical methods in condensed matter physics the reader is referred to the books by E. Frandkin [174], Altland and Simons [17], and X.-G. Wen [534].

Chern–Simons quantum field theories

Chern–Simons theories are examples of Schwarz-type TQFTs [460]. We note that a second general class of TQFTs includes the so-called Witten-type TQFTs, e.g., the topological Yang–Mills theory in four dimensions [552]. In a Schwarz-type theory, the correlation functions of the system can be written as path integrals of *metric independent* action functionals. For example, the Abelian 2D *BF theory* – another example of Schwarz-type TQFT – is described by an action of the form

$$S_{BF} = \frac{k}{2\pi} \int d^3x \, \epsilon^{\mu\nu\lambda} a_\mu \partial_\nu b_\lambda, \tag{6.52}$$

where $\mu, \nu, \lambda \in \{0, 1, 2\}$ are space-time indices, a and b are $U(1)$ gauge fields, and k is a parameter that describes the braiding statistics of the quasiparticles. Note that the action does not contain a metric tensor (unlike, e.g., the action corresponding to an electromagnetic field), but the Levi–Civita symbol. Since $\epsilon^{\mu\nu\lambda}$ is manifestly invariant under coordinate transformations, the action (6.52) defines a TQFT. We note that the gauge fields a and b behave differently under time-reversal, $(a_0, a_1, a_2) \rightarrow (a_0, -a_1, -a_2)$ and $(b_0, b_1, b_2) \rightarrow (-b_0, b_1, b_2)$. Consequently, the BF theory (6.52) is time-reversal symmetric and could be regarded as an effective low-energy description of TR-invariant topological insulators [106]

A Chern–Simons (CS) theory in $(2+1)$ dimensions is defined by the action

$$S_{CS}[a] = \frac{k}{4\pi} \int_M d^3x \, \epsilon^{\mu\nu\rho} \, \text{Tr} \left(a_\mu \partial_\nu a_\rho + i \frac{2}{3} a_\mu a_\nu a_\rho \right), \tag{6.53}$$

where the gauge field a_μ is an element of a certain Lie group G, which can be Abelian (i.e., commutative, for example $U(1)$) or non-Abelian (e.g., $SU(2)$). In the Abelian case, the second term in the action vanishes. The trace in Eq. (6.53) is over the so-called *fundamental representation* of the gauge group G. If $g \in G$ is a generic element of the group, it can be expressed in terms of n generators T^a as $g = \exp(i\lambda^a T^a)$, where λ^a are real parameters. The generators T^a satisfy the normalization condition $\text{Tr}(T^a T^b) = \frac{1}{2}\delta_{ab}$ and the *Lie algebra* commutation relations

$$[T^a, T^b] = i f^{abc} T^c, \tag{6.54}$$

where the parameters f^{abc} are the *structure constants* of the Lie algebra.[5] For example, in the case of the group $G = SU(2)$, the generators $T^a = J^a$ are the components of the total angular momentum, which satisfy the algebra $[J^a, J^b] = i\epsilon^{abc} J^c$. In the Abelian case, $f^{abc} = 0$. Note that the gauge field can be expressed in terms of generators as $a_\mu(x) = a_\mu^a(x) T^a$.

The action (6.53) is manifestly metric-independent. To describe a physical system, it has to generate *gauge invariant* expectation values for all the observables, when the theory is defined on a compact manifold M, e.g., on the $\mathbb{S}^3$ sphere or the $\mathbb{T}^3$ torus. On a manifold with a boundary, the gauge invariance requirement demands the presence of an additional term, which describes the gapless boundary physics. In the Abelian case, a gauge transformation corresponds to $a_\mu \to a_\mu + \partial_\mu \omega$ and establishing the gauge invariance of S_{CS} is straightforward. For non-Abelian fields, the gauge transformation reads

$$a_\mu \to U^\dagger a_\mu U - i U^\dagger \partial_\mu U, \tag{6.55}$$

with $U(x) \in G$. Under this gauge transformation, the action acquires two additional terms: a total derivative, which vanishes for a system defined on a compact manifold, and a term $2\pi k\, w[U]$, with $w(U)$ being the *winding number* of the gauge transformation,

$$w[U] = \frac{1}{24\pi^2} \int_M d^3x\, \epsilon^{\mu\nu\rho} \operatorname{Tr}\left(U^\dagger \partial_\mu U U^\dagger \partial_\nu U U^\dagger \partial_\rho U\right). \tag{6.56}$$

The winding number is an integer that counts how many times the gauge transformation $U(x)$ covers the whole group G as x covers the manifold M (once). Note that $w[U]$ is a unique characteristic of the gauge transformation U and cannot be changed by local continuous coordinate transformations (i.e., $w[U]$ is a topological invariant).

What are the physical observables in a Chern–Simons theory? In accordance with the general requirements of the theory, the operators representing these observables have to be gauge-invariant and metric-independent. The so-called *Wilson loop* – which is obtained from the holonomy of the gauge field a_μ around a given loop C – satisfies these conditions. The Wilson loop is defined by the trace of the path-ordered exponential of the gauge field

$$W_C = \operatorname{Tr}\left[\mathcal{P} \exp\left(i \oint_C dx^\mu a_\mu\right)\right], \tag{6.57}$$

---

[5]The Lie algebra can be represented by sets of Hermitian matrices satisfying the normalization condition and Eq. (6.54). Given G, many matrix representations having different dimensions are possible. The Pauli matrices, for example, are a two-dimensional representation of the generators of $SU(2)$. For $SU(N)$, the N-dimensional representation is called *fundamental*. In general, a representation of a group assigns to every element $g \in G$ a matrix $\rho(g)$ in such a way as to preserve the multiplication structure of the group, i.e., $g = g_1 g_2 \to \rho(g) = \rho(g_1)\rho(g_2)$. Note that a matrix $\rho(g)$ of dimension d can be viewed as a linear transformation acting on a d-dimensional Hilbert space. For $SU(2)$, the eigenstates $|j, m_j\rangle$ of the total angular momentum operator J^2 and J_3 provide a basis for the Hilbert space associated with the $d = 2j + 1$ dimensional representation, where j is a nonnegative integer of half integer.

where $\mathcal{P}$ is the path-ordering operator. Physically, the action of the Wilson loop operator corresponds to the creation and subsequent annihilation of a pair of elementary particle-antiparticle excitations of the quantum field. The expectation value of $W_\mathcal{C}$ can be calculated as a path integral of the form $\int \mathcal{D}a \, W_\mathcal{C} \exp(iS_{CS}[a])$, which, of course, has to be gauge invariant. Since under a gauge transformation U the CS action acquires an extra contribution $2\pi k \, w[U]$, with $w[U]$ integer, gauge invariance requires k to be an integer, i.e., $e^{2\pi i k \, w[U]} = 1$. We conclude that the coupling k of a non-Abelian Chern–Simons theory (also known as the *level* of the theory) satisfies the quantization condition $k \in \mathbb{Z}$. Note that such a condition does not exist in the Abelian case.

A specific example of CS field theory is the effective theory of Abelian fractional quantum Hall liquids discussed in Chapter 5 (Section 5.2.1, page 159). The relation between CS theories and the properties of Abelian and non-Abelian anyons is briefly mentioned in Chapter 12 (see Section 12.2, page 379). Here, we would like to point out that CS theories (and, more generally, TQFTs) can also describe the response of a topological quantum system to an external gauge field. More specifically, assume that a 2D Chern insulator is coupled to an external electromagnetic field and that we integrate out the degrees of freedom associated with the 2D quantum system. The response of the system is described by the CS field theory

$$S_{\text{eff}} = \frac{C_1}{4\pi} \frac{e^2}{\hbar} \int d^3x \, \epsilon^{\mu\nu\rho} A_\mu \partial_\nu A_\rho, \qquad (6.58)$$

where A_μ is the external field.[6] The response equations are $j^\mu = \delta S_{\text{eff}}/\delta A_\mu = C_1(e^2/h)\epsilon^{\mu\nu\rho} \partial_\nu A_\rho$, where j^μ is the particle number current density of the 2D system. In particular, the spatial components are $j_i = C_1(e^2/h)\epsilon^{ij}E_j$, which is the quantum Hall response with Hall conductance $\sigma_H = C_1(e^2/h)$. The parameter C_1 is the topological invariant of the Chern insulator, which for a band insulator is the sum of the first Chern numbers [see Eq. (3.37)] corresponding to the occupied bands. More generally, the topological invariant can be expressed in terms of the single-particle Green's function $G(k) = G(\boldsymbol{k}, \omega)$ of an interacting system as [528]

$$C_1 = \frac{\pi}{3} \int \frac{d^3k}{(2\pi)^3} \text{Tr} \left[\epsilon^{\mu\nu\rho} G\partial_\mu G^{-1} G\partial_\nu G^{-1} G\partial_\rho G^{-1} \right], \qquad (6.59)$$

where the trace is over band indices.

These considerations can be generalized to higher odd space-time dimensions [421]. For example, in (4+1) dimensions the effective theory that describes the low-energy physics acquires a topological CS-type term of the

---

[6] At this point we should emphasize that, within the field theory approach, it is common to use units corresponding to, e.g., $\hbar = c = e = 1$, a practice inherited from high energy physics. Supplemented with different choices for the electromagnetic units, this generates a rather annoying variety of essentially equivalent expressions, with frequent (and not unexpected, considering the circumstances) erroneous factors of c, 2π, etc.

form

$$S_{\text{eff}} = \frac{C_2}{24\pi^2} \frac{e^3}{\hbar^2} \int d^5x \, \epsilon^{\mu\nu\rho\sigma\tau} A_\mu \partial_\nu A_\rho \partial_\sigma A_\tau. \tag{6.60}$$

The system described by this action represents the 4D generalization of the quantum Hall effect. Note that, unlike the 2D case, this theory describes the *nonlinear* response to an external field, $j_\mu \propto \epsilon^{\mu\nu\rho\sigma\tau} \partial_\nu A_\rho \partial_\sigma A_\tau$. In general, the coupling constant C_2 can be written in terms of the single-particle Green's function of the system [528], while in the noninteracting limit it reduces to the second Chern number.

A significant difference between the 2D action in Eq. (6.58) and the 4D action in Eq. (6.60) is given by their behavior under time-reversal (TR). Since under TR we have $(A_0, A_1, A_2) \to (A_0, -A_1, -A_2)$, the 2D term breaks time-reversal, while the 4D action is explicitly TR invariant. In fact, Eq. (6.60) describes the "fundamental" TR invariant topological insulator from which all lower-dimensional topological insulators (i.e., the familiar 3D and 2D TIs) can be derived systematically by a procedure called *dimensional reduction* [54, 419], which is sketched in Chapter 7. Thus, analyzing the response of the system to an external electromagnetic field provides an alternative way to establishing the $\mathbb{Z}_2$ classification of topological insulators (see Chapter 7), in addition to arguments involving the bulk or surface properties of the system, which were discussed in Chapters 3 and 4.

Axion Electrodynamics in Topological Quantum Matter

W HAT SIGNATURES OF NONTRIVIAL TOPOLOGY CAN be observed in the electromagnetic response of topological quantum matter? So far, we have characterized topological systems based on the bulk and surface properties of microscopic or low-energy effective models. In this chapter, we focus on the observable response of the system to applied electromagnetic fields, going beyond the the well-known quantized Hall effect discussed in Section 3.4.1. In particular, we will consider the so-called topological magneto-electric effect arising from a topological term in the effective action that has the same form as the action describing the coupling between a photon and a hypothetical elementary particle called *axion* [258]. The axion – named by Frank Wilczek after a homonymous laundry detergent [545] – was proposed as a possible solution to the so-called strong charge-parity problem in quantum chromodynamics [258]. A candidate for dark matter, the axion joins the list of yet undetected elementary particles with analogs emerging as low-energy quasiparticles in condensed matter topological quantum systems, a list that also includes the Majorana and Weyl fermions discussed earlier in this book. This brief summary of axion electrodynamics in topological quantum systems includes a discussion of the quantized magneto-electric effect in three-dimensional topological insulators with time-reversal symmetry and axion insulators with "effective" time-reversal symmetry (Section 7.1), the emergence of dynamical axion fields in topological magnetic insulators (Section 7.2), the topological electromagnetic response of Weyl semimetals (Section 7.3), and the response of topological superconductors to temperature gradients and mechanical rotations (Section 7.4). More details can be found

DOI: 10.1201/9781003226048-7

in specialized review articles and the references therein, for example Refs. [26, 386, 463].

7.1 QUANTIZED MAGNETO-ELECTRIC EFFECT IN TOPOLOGICAL INSULATORS AND AXION INSULATORS

We start by sketching the derivation of the so-called θ term that characterizes the effective field theory of a time-reversal (TR) invariant three dimensional (3D) topological insulator in the presence of an applied electromagnetic field using the *dimensional reduction* procedure [54, 419]. In (4+1) dimensions, the electromagnetic response is described by the Chern–Simons field theory in Eq. (6.60) (see Section 6.3). Let us consider field configurations characterized by a constant A_4 component and $A_\mu(\boldsymbol{r}, t) = A_\mu(x_1, x_2, x_3, x_0)$, with $\mu = 0, 1, 2, 3$, independent of the "extra dimension" x_4, which is compactified into a small circle. The effective field theory in (3+1) dimensions is obtained by performing the integral over x_4 in Eq. (6.60). Note that the integrand becomes $\epsilon^{\mu\nu\rho\sigma\tau} A_\mu \partial_\nu A_\rho \partial_\sigma A_\tau = 3A_4 \, \epsilon^{4\mu\nu\rho\sigma} \, \partial_\mu A_\nu \partial_\rho A_\sigma$. Explicitly, we obtain

$$S_\theta = \frac{\theta}{8\pi^2} \frac{e^2}{\hbar} \int dt \, d^3r \, \epsilon^{\mu\nu\rho\sigma} \, \partial_\mu A_\nu \partial_\rho A_\sigma = \frac{\theta}{2\pi} \frac{e^2}{h} \int dt \, d^3r \, \boldsymbol{E} \cdot \boldsymbol{B}, \qquad (7.1)$$

where $\theta = (e/\hbar)C_2 \oint dx_4 \, A_4 = C_2\Phi$, with Φ being the magnetic flux that threads the compactified circle (in units of $\hbar/e$). The topological term given by Eq. (7.1) generates the so-called *axion electrodynamics* [543]. Note that the factor e^2/h is proportional to the fine-structure constant, $e^2/h = 2\epsilon_0 c\alpha$, with $\alpha = e^2/(4\pi\epsilon_0\hbar c) \approx 1/137$. Also note that the physics generated by this term is invariant under a shift of the axion angle θ by 2π. This property can be understood in terms of invariance with respect to adding a magnetic flux quantum $\Phi_0 = h/e$ that threads the compactified circle (since taking $C_2 = 1$ implies $\theta = \Phi + \Phi_0$, with $\Phi_0 = 2\pi$ in units of $\hbar/e$). On the other hand, under TR we have $\theta \to -\theta$ (a property inherited from the behavior of A_4 under TR). Note that $\boldsymbol{E} \cdot \boldsymbol{B}$ is also odd under time-reversal. Consequently, TR invariance is consistent with either $\theta = 0$ (mod 2π) or $\theta = \pi$ (mod 2π), i.e., with only two values of θ (mod 2π). We therefore conclude that in 3D there are two different classes of TR invariant insulators: the topologically trivial insulator with $\theta = 0$ and the topologically nontrivial insulator characterized by $\theta = \pi$. The arguments sketched above represent a field-theoretical route for deriving the $\mathbb{Z}_2$ classification of 3D TIs in symmetry class AII. Of course, the parameter θ/π represents the $\mathbb{Z}_2$ topological invariant. For a noninteracting system, θ can be expressed in terms of the Berry connection $\mathcal{A}_j^{\alpha\beta} = i\langle u_\alpha(\boldsymbol{k})|\partial_j|u_\beta(\boldsymbol{k})\rangle$, where $|u_\alpha(\boldsymbol{k})\rangle$ is the cell-periodic Bloch function and $\partial_j = \partial/\partial k_j$. Explicitly, we have [159, 419]

$$\theta = \frac{1}{2\pi} \int_{BZ} d^3k \, \epsilon^{ijk} \, \text{Tr} \left[\mathcal{A}_i \partial_j \mathcal{A}_k - i\frac{2}{3}\mathcal{A}_i \mathcal{A}_j \mathcal{A}_k \right], \qquad (7.2)$$

where the trace is over occupied bands. Equation (7.2) expresses the $\mathbb{Z}_2$ topological invariant as the integral of the *Chern–Simons 3-form* over the 3D momentum space. We point out that, as an alternative to the dimensional reduction method sketched here, the θ term given by Eq. (7.1) can be obtained directly starting with a microscopic model of the 3D topological insulator in the presence of an electromagnetic field and "integrating out" the fermionic degrees of freedom [463].

The constraint imposed by TR symmetry on the possible values of θ (i.e., θ is an even multiple of π in trivial insulators or an odd multiple of π in topological insulators) also holds in systems with broken TR symmetry in the presence of an "effective" TR symmetry consisting of a combination of time-reversal and a lattice translation (which characterizes a class of 3D antiferromagnetic insulators called *axion insulators*), as well as in systems with inversion symmetry (since $\boldsymbol{E} \cdot \boldsymbol{B}$ is odd under spatial inversion). On the other hand, if both time-reversal ("true" or "effective") and inversion symmetries are broken, the axion angle can take arbitrary values. Moreover, θ can even acquire spatial and time dependence becoming an *axion field*, $\theta(\boldsymbol{r}, t)$. Thus, in general, axion electrodynamics is generated by a topological contribution to the Lagrangian density of the form

$$\mathcal{L}_\theta = \frac{\theta(\boldsymbol{r}, t)}{2\pi} \frac{e^2}{h} \boldsymbol{E} \cdot \boldsymbol{B}, \tag{7.3}$$

which is added to the "standard" Maxwell Lagrangian. As a consequence, the Gauss and Ampère laws are modified taking the form

$$\boldsymbol{\nabla} \cdot \boldsymbol{E} = \frac{\rho}{\epsilon_0} - 2c\alpha \boldsymbol{\nabla} \left(\frac{\theta}{2\pi} \right) \cdot \boldsymbol{B}, \tag{7.4}$$

$$\boldsymbol{\nabla} \times \boldsymbol{B} = \mu_0 \boldsymbol{j} + \frac{1}{c^2} \frac{\partial \boldsymbol{E}}{\partial t} + \frac{2\alpha}{c} \left[\boldsymbol{B} \frac{\partial}{\partial t} \left(\frac{\theta}{2\pi} \right) + \boldsymbol{\nabla} \left(\frac{\theta}{2\pi} \right) \times \boldsymbol{E} \right]. \tag{7.5}$$

We note that, alternatively, one can use the ordinary Maxwell equations and incorporate the θ contributions into the constitutive relations,

$$\boldsymbol{D} = \epsilon_0 \boldsymbol{E} + \boldsymbol{P} + \alpha \frac{\theta}{\pi} \sqrt{\frac{\epsilon_0}{\mu_0}} \boldsymbol{B}, \tag{7.6}$$

$$\boldsymbol{H} = \frac{1}{\mu_0} \boldsymbol{B} - \boldsymbol{M} - \alpha \frac{\theta}{\pi} \sqrt{\frac{\epsilon_0}{\mu_0}} \boldsymbol{E}. \tag{7.7}$$

The θ-dependent terms in Eqs. (7.6) and (7.7) can be viewed as contributions to the polarization ($\overline{\boldsymbol{P}} \propto \theta \boldsymbol{B}$) and magnetization ($\overline{\boldsymbol{M}} \propto \theta \boldsymbol{E}$), respectively, generated by a topological magneto-electric effect. The corresponding contribution to the linear magneto-electric polarizability is

$$\overline{\alpha}_{ij} = \left. \frac{\partial \overline{M}_j}{\partial E_i} \right|_{\boldsymbol{B}=0} = \left. \frac{\partial \overline{P}_i}{\partial B_j} \right|_{\boldsymbol{E}=0} = \alpha \frac{\theta}{\pi} \sqrt{\frac{\epsilon_0}{\mu_0}} \delta_{ij}. \tag{7.8}$$

We point out that in materials with inversion and time-reversal symmetry the "standard" (i.e., orbital, spin, and ionic) contributions to the magneto-electric polarizability vanish. However, strong topological insulators (with $\theta = \pi$) are characterized by a topological contribution $\bar{\alpha}$ that is an order of magnitude larger than the polarizability of typical magneto-electric materials (e.g., Cr_2O_3).

Next, we consider several interesting physical phenomena generated by the θ term (7.1) in systems with TR or inversion symmetry [419, 421]. Note that, since θ is a constant (with θ/π taking integer values), it does not generate bulk contributions to Eqs. (7.4) and (7.5), but produces surface effects.

Surface half-quantized anomalous Hall effect. A direct consequence of having a quantized axion angle in the bulk is the emergence of a half-quantized Hall effect on the surface of a strong TI, if TR symmetry is broken on the surface (hence, the surface states are gapped). To illustrate this phenomenon, we start with an effective, Dirac-like effective Hamiltonian describing the surface states of a topological insulator

$$H_{\text{surf}}(\boldsymbol{k}) = v_F(k_y\sigma_x - k_x\sigma_y), \tag{7.9}$$

where v_F is the Fermi velocity of the surface modes and σ_j are Pauli matrices. The Hamiltonian in Eq. (7.9) is a 2×2 matrix with eigenvalues $E_\pm(\boldsymbol{k}) = \pm v_F\sqrt{k_x^2 + k_y^2}$ corresponding to a gapless Dirac cone in the (k_x, k_y, E) space. Let us now assume that TR symmetry in broken on the surface due to exchange interaction between electrons and magnetic impurities or between electrons and a ferromagnetic layer deposited on the surface. The homogeneous component of the exchange Hamiltonian, which has the form $H_{\text{exch}} = \boldsymbol{m} \cdot \boldsymbol{\sigma}$, constitutes a perturbation to the surface Hamiltonian that generates a gapped spectrum,

$$E_\pm(\boldsymbol{k}) = \pm\sqrt{(v_F k_x + m_y)^2 + (v_F k_y - m_x)^2 + m_z^2}. \tag{7.10}$$

Note that the gap is generated by the m_z component, while the in-plane components of $\boldsymbol{m}$ only shift the position of the Dirac cone. For simplicity, we consider the case $m_x = m_y = 0$. The corresponding effective surface Hamiltonian becomes $H(\boldsymbol{k}) = \boldsymbol{d}(\boldsymbol{k}) \cdot \boldsymbol{\sigma}$, with $\boldsymbol{d}(\boldsymbol{k}) = (v_F k_y, -v_F k_x, m_z)$. The surface Hall conductance is proportional to the winding number of the unit vector $\hat{\boldsymbol{d}} = \boldsymbol{d}/|\boldsymbol{d}|$, i.e., the area on the unit sphere covered by $\hat{\boldsymbol{d}}$ as $\boldsymbol{k}$ varies within the 2D Brillouin zone. Explicitly, we have [523]

$$\sigma_H = -\frac{e^2}{h}\frac{1}{4\pi}\int_{BZ} d^2k\,\hat{\boldsymbol{d}} \cdot \left(\frac{\partial\hat{\boldsymbol{d}}}{\partial k_x} \times \frac{\partial\hat{\boldsymbol{d}}}{\partial k_y}\right). \tag{7.11}$$

The integral in Eq. (7.11), which is the area covered by $\hat{\boldsymbol{d}}$, can be easily evaluated by noting that at $\boldsymbol{k} = 0$ the unit vector $\hat{\boldsymbol{d}} = (0, 0, \text{sgn}(m_z))$ points toward the north (or south) pole of the unit sphere, while for $|\boldsymbol{k}| \gg |m_z|/v_F$ the unit vector lies in equatorial plane. Hence, $\hat{\boldsymbol{d}}(\boldsymbol{k})$ covers half of the unit sphere and we have

$$\sigma_H = -\text{sgn}(m_z)\frac{e^2}{2h}. \tag{7.12}$$

Note that the Hall conductance is finite even in the limit $m_z \to 0$ (i.e., for an infinitesimally small gap) and has a jump at $m_z = 0$; the direction of the Hall current depends on the sign of m_z, i.e., on the direction of the impurity-generated (or proximity-induced) magnetization. Also note that the experimental observation of this phenomenon in a ferromagnetically ordered topological insulator thin film [96] involves contributions (corresponding to the same sign of m_z) from both top and bottom surfaces, which leads to an observed Hall conductance $\sigma_H = \pm e^2/h$.

We emphasize that the half-quantized anomalous Hall effect is a property of the boundary of a 3D topological insulator that cannot be realized in a purely 2D system, which has a Hall conductance that is an integer multiple of e^2/h. An analogy between the emergence of the half-quantized Hall effect on the boundary of a topological insulator and the emergence of half-integer "end charges" in a symmetry obstructed 1D insulator (see Section 4.1.1) can be found in Ref. [26]. We also note that, although the derivation of σ_H sketched above is based on an idealized effective model of the surface, one can show that the result is robust against disorder (generated, e.g., by magnetic impurities) and is not affected by contributions from (nonlinear) large-momentum corrections to the Hamiltonian [421].

Topological magneto-electric effect. In general, the magneto-electric effect involves a magnetization induced by an electric field, or, alternatively, a charge polarization induced by a magnetic field. The topological magneto-electric effect emerges in strong topological insulators with a fully gapped surface, which can be realized, for example, by cladding a cylindrical TI with a ferromagnetic film with magnetization pointing outward (see Figure 7.1). The effect can be viewed as a direct consequence of the surface half-quantized anomalous Hall effect. Indeed, applying an electric field parallel to the axis of the cylinder generates an anomalous Hall current on the surface,

$$\boldsymbol{j}_H = \sigma_H\,\hat{n} \times \boldsymbol{E} = -\frac{e^2}{2h}\hat{n} \times \boldsymbol{E}, \tag{7.13}$$

where $\hat{n}$ is the unit vector normal to the (lateral) surface of the TI and we assumed $\text{sgn}(m_z) = +1$ (i.e., induced magnetization pointing outward). Note that $\boldsymbol{j}_H$ represents a surface (rather than bulk) current density. The Hall current induces a magnetization [see also Figure 7.1(a)]

$$\boldsymbol{M} = \frac{e^2}{2h}\boldsymbol{E}, \tag{7.14}$$

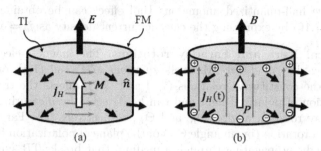

(a)   (b)

FIGURE 7.1 Schematic representation of a proposed realization of the topo-logical magneto-electric effect in a cylindrical 3D topological insulator (TI) wrapped by a ferromagnetic film (FM) with magnetization orthogonal to the interface [419]. The lateral surface of the TI is gapped by the proximity-coupled ferromagnet and exhibits a half-quantized anomalous Hall effect. (a) Applying an electric field E parallel to the axis of the cylinder generates an anomalous surface Hall current j_H, which induces a magnetization M par-allel to the applied electric field. (b) Applying a magnetic field B induces a (transient) electric field that generates an anomalous surface Hall current j_H, which results in a finite electric polarization P.

which can be easily obtained using Ampere's law. On the other hand, ap-plying a magnetic field B parallel to the axis of the cylinder induces, ac-cording to Faraday's law, $\nabla \times E = -\partial B/\partial t$, an azimuthal electric field $E = R/2(\partial B/\partial t)\hat{\varphi}$, where R is the radius of the cylinder and $\hat{\varphi}$ is the unit vector in the azimuthal direction. In turn, the induced electric field generates a Hall current (flowing parallel to the axis of the cylinder), $I_H = 2\pi R\, j_H = (\pi R^2)e^2/2h\, \partial B/\partial t$. This current can be viewed as a polarization current $\partial P/\partial t = I_H/(\pi R^2)$ corresponding to an electric polarization [see also Figure 7.1(b)]

$$P = \frac{e^2}{2h}B. \qquad (7.15)$$

Thus, the surface half-quantized anomalous Hall effect is clearly related to a (bulk) magneto-electric effect. Furthermore, since $\alpha\sqrt{\epsilon_0/\mu_0} = e^2/2h$, Eqs. (7.14) and (7.15) correspond to the topological terms $\overline{M}$ and $\overline{P}$ in Eqs. (7.7) and (7.6), respectively (with $\theta = \pi$). We point out that satisfying Eqs. (7.14) and (7.15) requires the presence of a θ term of the form given by Eq. (7.1) in the action. For a constant axion field (e.g., $\theta = \pi$), the corresponding Lagrangian density is a total derivative, since $\epsilon^{\mu\nu\rho\sigma}\,\partial_\mu A_\nu\,\partial_\rho A_\sigma = \epsilon^{\mu\nu\rho\sigma}\,\partial_\mu(A_\nu\,\partial_\rho A_\sigma)$. Consequently, after integrating over x_μ in Eq. (7.1), we can define the surface action

$$S_\theta = \frac{\theta}{8\pi^2}\frac{e^2}{\hbar}\int dt\, d^2r\, \epsilon^{\nu\rho\sigma}\, A_\nu\partial_\rho A_\sigma. \qquad (7.16)$$

The surface half-quantized anomalous Hall effect can be obtained directly from Eq. (7.16) by expressing the electric current density as $j_\nu = \delta S_\theta / \delta A_\nu$.

Topological Kerr and Faraday rotations. The magneto-electric effect captured by the "modified" Maxwell equations (7.4) and (7.5), or, alternatively, by the constitutive relations (7.6) and (7.7), affects the transmission and reflection of polarized light and can be detected by measuring the Kerr and Faraday rotation angles, Θ_K and Θ_F, respectively. The Faraday effect consists of a rotation (by an angle Θ_F) of the plane of polarization for linearly polarized light propagating through a medium that breaks TR symmetry. A similar rotation (by an angle Θ_K) occurs when polarized light is reflected by a TR symmetry breaking surface – the so-called magneto-optical Kerr effect. For a (semi-infinite) strong TI with broken TR symmetry at the interface with a trivial insulator (e.g., vacuum), the rotation angles can be determined by solving the modified Maxwell equations. We obtain [348, 419]

$$\tan \Theta_F = \frac{\theta}{\pi} \frac{\alpha}{\sqrt{\epsilon_1/\mu_1} + \sqrt{\epsilon_2/\mu_2}}, \tag{7.17}$$

$$\tan \Theta_K = \frac{\theta}{\pi} \frac{2\alpha\sqrt{\epsilon_1/\mu_1}}{\epsilon_2/\mu_2 - \epsilon_1/\mu_1 + \alpha^2\theta^2/\pi^2}, \tag{7.18}$$

where ϵ_1 (μ_1) and ϵ_2 (μ_2) are the dielectric constants (relative magnetic permeabilities) of the trivial insulator and topological insulator, respectively. In general, the rotation angles depend on the optical properties of the materials and include additional (nontopological) contributions. However, measuring both angles (at reflectivity minima) in a geometry involving a slab of topological insulator with vacuum on one side and a substrate on the other provides the quantized coefficient $\alpha\theta/\pi$, which is given by the combination [348]

$$\frac{\cot \Theta_K + \cot \Theta_F}{1 + \cot^2 \Theta_F} = \alpha\frac{\theta}{\pi}. \tag{7.19}$$

We close this section with a few remarks on the so-called *axion insulators*. While TR invariant topological insulators are characterized by a topological magneto-electric effect with a quantized axion angle $\theta = \pi$ (mod 2π), the value of θ (generally) becomes arbitrary in systems with broken TR symmetry. However, in the presence of a combination of time-reversal and lattice translation corresponding to an "effective" TR symmetry, the quantization of θ is preserved. Systems characterized by an "effective" TR symmetry are 3D antiferromagnetic insulators, such as the $MnBi_2Te_4$ family of layered van der Waals compounds [463]. For concreteness, let us assume that the system has two spins per unit cell (oriented antiferromagnetically) and that a translation by half a unit cell along the direction of a primitive translation vector (say a_3) results in flipping the spin orientations (see Figure 7.2). The corresponding translation operator can be represented as [368]

$$T_{\frac{1}{2}}(\boldsymbol{k}) = e^{-\frac{i}{2}\boldsymbol{k}\cdot\boldsymbol{a}_3} \begin{pmatrix} 0 & 1 \\ 1 & 0 \end{pmatrix}, \tag{7.20}$$

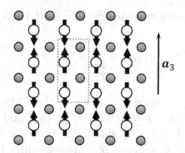

FIGURE 7.2 Schematic representation of the lattice structure of an antiferromagnetic topological insulator with "effective" time-reversal symmetry, $S = \mathcal{T} T_{\frac{1}{2}}$. The dashed line corresponds to a (magnetic) unit cell.

where $\mathbb{1}$ is the identity matrix on half of the unit cell. Note that $T_{\frac{1}{2}}^2 = e^{-ik \cdot a_3}$ (where the unit matrix on the unit cell is omitted, for simplicity) represents a translation by the primitive vector a_3. The translation operator commutes with the time-reversal operator, which (for spin-$\frac{1}{2}$ systems) can be represented as $\mathcal{T} = -i\sigma_y K$, with K being the complex conjugation (see Section 3.2.1). In terms of representations, we have $\mathcal{T} T_{\frac{1}{2}}(k) = T_{\frac{1}{2}}(-k) \mathcal{T}$. The combination $S = \mathcal{T} T_{\frac{1}{2}}$ is an antiunitary operator represented by $S_k = \mathcal{T} T_{\frac{1}{2}}(k)$. An atiferromagnetic system that is invariant under the symmetry operation S (i.e., translation by half unit cell followed by a spin flip) is characterized by a Bloch Hamiltonian satisfying

$$S_k H(k) S_k^{-1} = H(-k), \tag{7.21}$$

which is formally similar to the condition satisfied by TR invariant systems (see Section 3.2.1). However, while $\mathcal{T}^2 = -1$, the operator S^2, which is represented by $S_{-k} S_k = -e^{-ik \cdot a_3}$, is the minus identity operator up to a translation by a unit cell. Nonetheless, on the Brillouin zone plane $k_3 = 0$ the symmetry properties of the system under the S operation are formally identical to the properties of a TR invariant system, which leads to a $\mathbb{Z}_2$ topological classification [368]. In particular, the antiferromagnetic topological insulators with S symmetry (also called axion insulators) are characterized by a quantized axion angle, $\theta = \pi$ (mod 2π), and exhibit key aspects of the axion phenomenology discussed in this section.

7.2 DYNAMICAL AXION FIELDS IN TOPOLOGICAL MAGNETIC INSULATORS

In systems that break both inversion symmetry and time-reversal symmetry (real and "effective"), the axion angle can take arbitrary values. To illustrate this point, let us consider a specific effective model given by the Hamiltonian

$$H(\boldsymbol{k}) = \sum_{j=1}^{5} d_j(\boldsymbol{k})\Gamma^j, \tag{7.22}$$

where Γ^j are 4×4 Dirac matrices satisfying the Clifford algebra $\{\Gamma^i, \Gamma^j\} = 2\delta_{ij}$, $d_j(-\boldsymbol{k}) = -d_j(\boldsymbol{k})$ for $j = 1, 2, 3$, $d_4 = m_0$, and $d_5 = m_5$. The Dirac matrices Γ^j with $j = 1, 2, 3$ are odd under time-reversal and inversion, Γ^4 is invariant under these symmetry operations, while $\Gamma^5 = \Gamma^1\Gamma^2\Gamma^3\Gamma^4$. Note that a system characterized by $m_5 = 0$ is invariant under both time-reversal and inversion and corresponds to a trivial insulator when $m_0 > 0$ and a topological insulator for $m_0 < 0$. A nonzero m_5 mass can be induced by coupling the electrons to a staggered (Néel) field in a system with long-range antiferromagnetic order. The corresponding axion field has the form [463]

$$\theta = \frac{\pi}{2}[1 - \operatorname{sgn}(m_0)] - \tan^{-1}\left(\frac{m_5}{m_0}\right). \tag{7.23}$$

The first term in Eq. (7.23) is zero for a trivial system ($m_0 > 0$) and π for a topologically nontrivial insulator ($m_0 < 0$), while the second term characterizes the deviation from the quantized values. Also note that, in the presence of a nonzero m_5 mass, the surface states of a topologically nontrivial system are gapped regardless of the surface orientation, the surface gap value being m_5. We conclude that in magnetic insulators the presence of a mass term $m_5\Gamma^5$ that breaks both time-reversal and inversion symmetries leads to arbitrary values of θ, the deviation from the quantized values being proportional to the staggered magnetization.

In the above considerations we assumed a static staggered magnetization, which, in turn, generates a static axion field. However, in general the antiferromagnetic system supports spin-wave excitations, which induce a dynamical component of the axion field. Thus, the axion field can be decomposed in a static contribution θ_0 (characterizing the ground state) and a dynamical component $\delta\theta(\boldsymbol{r}, t)$, referred to as the *dynamical axion field*: $\theta(\boldsymbol{r}, t) = \theta_0 + \delta\theta(\boldsymbol{r}, t)$. Considering, for example, the antiferromagnetic phase of Bi_2Se_3 doped with magnetic impurities and assuming a Néel field oriented in the z direction, we have $\delta\theta \propto \delta m_5$. After calculating the action of the antiferromagnetic insulator in the presence of an external electromagnetic field, we obtain [334, 463]

$$\begin{aligned}
S &= g^2 J \int dt d^3r \left[(\partial_t \delta\theta)^2 - (v_i \partial_i \delta\theta)^2 - m^2 \delta\theta^2 \right] \\
&+ \frac{e^2}{2\pi h} \int dt d^3r \, (\theta_0 + \delta\theta) \boldsymbol{E} \cdot \boldsymbol{B},
\end{aligned} \tag{7.24}$$

where $g = m_0$ and J, v, and m are the stiffness, velocity, and mass of the $\delta M_z(r,t)$ spin-wave mode. Explicitly, we have

$$J = \int_{BZ} \frac{d^3k}{(2\pi)^3} \frac{\sum_{j=1}^{4} d_j^2}{16|d|^5}, \tag{7.25}$$

$$m^2 = \frac{m_5^2}{J} \int_{BZ} \frac{d^3k}{(2\pi)^3} \frac{1}{4|d|^3}. \tag{7.26}$$

The first term in Eq. (7.24) describes a massive axion, while the second term represents the topological coupling between the axion and the external electromagnetic field.

What are the possible observable consequences of realizing a dynamical axion field in a condensed matter system? Assume, for example, that the axion is coupled to polarized light in the presence of a uniform, static magnetic field B_0 parallel to the electric field E of the photon. The corresponding coupling is linear in E [see Eq. (7.24), with $B = B_0$]. The linear coupling of a collective mode (e.g., an optical phonon or a magnon) to photons leads to the emergence of hybridized propagating modes called *polaritons*. Thus, in an antiferromagnet, the linear coupling of the axion mode to light (in the presence of a uniform magnetic field) generates a new type of polariton called the *axionic polariton* [334]. The equations of motion for the coupled axion and electromagnetic field can be obtained from the action in Eq. (7.24), after adding the field contribution, $S_{\text{field}} = \frac{1}{2} \int dt d^3r \, (\epsilon E^2 - 1/\mu B^2)$. Neglecting the dispersion of the axion (which is much weaker that that of the photon), we obtain the following dispersion of the axionic polariton [334]

$$\omega_\pm^2(k) = \frac{1}{2} \left(\tilde{c}^2 k^2 + m^2 + b^2 \right) \pm \frac{1}{2} \sqrt{\left(\tilde{c}^2 k^2 + m^2 + b^2 \right)^2 - 4\tilde{c}^2 k^2 m^2}, \tag{7.27}$$

where $\tilde{c}$ is the speed of light in the material and $b^2 = \alpha^2 B_0^2 / 8\pi^3 \epsilon g^2 J$. Note that ω_- has a maximum equal to m at large k values, while the mode ω_+ has a minimum at $k = 0$, $\omega_+(0) = \sqrt{m^2 + b^2}$. The gap between m and $\sqrt{m^2 + b^2}$ is controlled by the applied magnetic field, B_0. Also note the similarity between the dispersion relation in Eq. (7.27) and the optical phonon polariton, with the significant difference that the coupling between the axion and the electric field is *tunable*, being determined by the external magnetic field. Experimentally, one can detect the gap of the axionic polariton spectrum using a setup consisting of polarized light incident perpendicular to the surface of the sample and a static magnetic field parallel to the electric field of light. A significant increase of the reflectivity should be observed when the frequency of the incident light is within the gap. The key point is that the frequency band within which this increased reflectivity is observed is controlled by the magnitude of B_0.

Another measurable consequence of having a dynamical axion field involves a (bulk) electric current response. The electric current density induced by external electromagnetic fields can be obtained from the modified Maxwell equations as

$$j(r,t) = \frac{e^2}{2\pi h} \left[\partial_t \theta(r,t) B + \nabla \theta(r,t) \times E \right]. \tag{7.28}$$

The component of this current generated by the applied magnetic field is associated with the so-called *chiral magnetic effect*, first studied in nuclear physics. The electric field-induced component is the result of an anomalous Hall effect. Note that, unlike the surface half-quantized anomalous Hall effect discussed in Section 7.1, the current in Eq. (7.28) is a bulk current and can be obtained directly from the constitutive relation (7.6) as a polarization current, $\partial \overline{P}/\partial_t = e^2/(2\pi h)\partial_t\theta\, B$. Similarly, the magnetic field component can be obtained as a magnetization current from Eq. (7.7), $\nabla \times \overline{M} = e^2/(2\pi h)\nabla\theta \times E$. Hence, an alternating current generated by a magnetic field (i.e., the dynamical chiral magnetic effect) is predicted to emerge due to the time dependence of the antiferromagnetic order parameter (e.g., in a so-called antiferromagnetic resonance state) [462]. Similarly, an anomalous Hall effect is expected to arise as a result of spatial variations of the order parameter (e.g., variations associated with antiferromagnetic domain walls) [462].

We conclude this section by mentioning a proposal [353] for using topological antiferromagnetic insulators with dynamical axion fields to detect axion dark matter with mass ranging from 0.7 to 3.5 meV. In essence, the axion particles are expected to couple to the axionic polariton emerging inside an antiferromagnetic insulator. At the boundary, the axionic polaritons convert to propagating photons (with frequencies in the THz regime), which are detected using low-noise methods. The conversion process is significantly enhanced when the frequency of the (dark) axion is equal to the axionic polariton frequency. Note that the resonant frequency can be scanned by varying the applied magnetic field B_0, which provides sensitivity to axion dark matter in a parameter regime currently inaccessible to other detection methods. If successful, such an experiment would represent a fitting way of closing the "conceptual circle" that connects particle physics – where ideas regarding the topological properties of matter were first formulated – and condensed matter physics – the area that brought topological quantum matter in the laboratory.

7.3 TOPOLOGICAL ELECTROMAGNETIC RESPONSE OF WEYL SEMIMETALS

The nontrivial response of Weyl semimetals can be viewed as a consequence of the so-called *chiral anomaly*. The chiral anomaly, which was originally discovered by Adler, Bell and Jackiw in the context of particle physics [6, 43], refers to the anomalous nonconservation of chiral charge for massless relativistic particles, which break classical conservation requirements based on symmetry. For concreteness, let us consider a system of massless fermions described by the Dirac Hamiltonian $H(\boldsymbol{k}) = \gamma^0\gamma^j k_j$, where γ^μ (with $\mu = \overline{0,3}$) are Dirac matrices satisfying the anticommutation relations $\{\gamma^\mu, \gamma^\nu\} = 2\eta_{\mu\nu}\mathbf{1}$, with the metric $\eta_{\mu\nu} = \text{diag}(1, -1, -1, -1)$. If ψ is a Dirac spinor, we can define the ordinary (charge) current as the spatial component of the four vector $j^\mu = \psi^\dagger \gamma^0 \gamma^\mu \psi$ and the conservation of electric charge takes the form of the continuity equation $\partial_\mu j^\mu = 0$. However, one would expect an additional conservation law: the conservation of chiral charge. Indeed, let us consider the matrix $\gamma^5 = i\gamma^0 \gamma^1 \gamma^2 \gamma^3$. The Dirac field ψ can be decomposed into components with well defined chirality (i.e., left-handed and right-handed components) as

$$\psi_L = \frac{1}{2}(1 - \gamma^5)\psi, \qquad \psi_R = \frac{1}{2}(1 + \gamma^5)\psi. \tag{7.29}$$

Note that ψ_L and ψ_R are eigenspinors of γ^5 with eigenvalues -1 and $+1$, respectively. The density of chiral charge and the corresponding current can be expressed in terms of the components of the four vector $j_5^\mu = \psi^\dagger \gamma^0 \gamma^\mu \gamma^5 \psi$. Since γ^5 anticommutes with the Dirac matrices γ^μ and, consequently, commutes with the Hamiltonian, $[H, \gamma^5] = 0$, we would expect chiral charge conservation: $\partial_\mu j_5^\mu = 0$. This additional conservation holds if one couples the massless fermions (described within a first quantization framework) to a classical (nonquantized) electromagnetic field. However, in the case of the corresponding quantum field theory, the currents, which contain products of Dirac operators at short distances, require a consistent definition, i.e., some regularization. In the case of masless Dirac fermions coupled to an electromagnetic field, the regularization cannot simultaneously preserve both gauge invariance (i.e., charge conservation) and chiral symmetry (i.e., chiral charge conservation). Given the physical implications of gauge invariance, this implies the breaking of chiral symmetry. We note that, although the anomaly arises from the regularization of the short distance singularities in the quantum field theory, it is a finite and universal term, which suggests that is has a topological nature.

To build some intuition, let us first consider a $(1 + 1)$-dimensional theory of masless Dirac fermions with charge $e > 0$ coupled to a background gauge field A^μ. The corresponding Lagrangian density (in natural units) is

$$\mathcal{L} = \bar{\psi} i\gamma^\mu \partial_\mu \psi - e\bar{\psi}\gamma^\mu \psi A_\mu, \tag{7.30}$$

where $\bar{\psi} = \psi^\dagger \gamma^0$, the 2×2 Dirac matrices are $\gamma^0 = \sigma_1$ and $\gamma^1 = i\sigma_2$, and we define $\gamma^5 = -\gamma^0\gamma^1 = \sigma_3$. In this chiral basis, the components of the Dirac spinor correspond to left and right movers, $\psi = (\psi_L, \psi_R)^T$. The number densities of left and right movers are $n_L = \psi_L^\dagger \psi_L$ and $n_R = \psi_R^\dagger \psi_R$, respectively, while the corresponding gauge current, $\boldsymbol{j} = (j^0, j^1)$, and chiral current, $\boldsymbol{j}_5 = (j_5^0, j_5^1)$, can be written as

$$j^0 = n_R + n_L = -j_5^1, \qquad j^1 = n_R - n_L = j_5^0. \tag{7.31}$$

In the temporal gauge, $A_0 = 0$, $-A_1 = A^1 \equiv A_x$, the Dirac equation takes the form

$$i\partial_t \psi_R = -(i\partial_x + eA_x)\psi_R, \qquad i\partial_t \psi_L = (i\partial_x + eA_x)\psi_L. \tag{7.32}$$

When $A_x = 0$ (no background field), the plane wave solutions of Eq. (7.32) with momentum p have energies $\varepsilon = +p$ (for right movers) and $\varepsilon = -p$ (for left movers). If we assume a system of length L (with periodic boundary conditions), p takes discrete values, $p = (2\pi/L)n_p$, with $n_p \in \mathbb{Z}$. The ground state corresponds to a filled Dirac sea, with all negative energy states being occupied, while the states with $\varepsilon > 0$ are empty (see Figure 7.3). Assume now that we turn on a uniform electric field $E = -\partial_t A_x > 0$. The charged fermions will accelerate and their momentum will change in time. Based on the semiclassical equation of motion, we have $dp/dt = eE$. Each time this change in momentum becomes an integer multiple of $2\pi/L$, an additional right moving particle is added to the system and a left moving particle is removed (i.e., a left moving antiparticle is added), as illustrated in Figure 7.3. Thus, the density of right/left movers will change in time as

$$\frac{dn_{R/L}}{dt} = \pm \frac{e}{2\pi} E \rightarrow \pm \frac{e}{h} E, \tag{7.33}$$

where in the rightmost expression we have restored Planck's constant. This implies that the electric (gauge) charge is conserved, e.g., $n_R + n_L = 0$, while the chiral charge increases (if, as assumed, $eE > 0$). In covariant notation, the nonconservation of the chiral (axial) charge takes the form

$$\partial_\mu j_5^\mu = \frac{e}{h} \eta^{\mu\nu} F_{\mu\nu}, \tag{7.34}$$

where $\eta^{\mu\nu} = \text{diag}(1, -1)$ and $F_{\mu\nu} = \partial_\mu A_\nu - \partial_\nu A_\mu$.

These simple arguments obscure the connection between the chiral (or axial) anomaly given by Eq. (7.34) and the UV regularization of the quantum field theory corresponding to the Lagrangian density in Eq. (7.30). However, our (implicit) assumption that the creation of right moving particles does not generate any right moving antiparticle (i.e., the Dirac see is infinitely deep), but results in an equal number of left moving antiparticles, corresponds to a gauge-invariant regularization that conserves the electric (gauge) charge. The anomaly occurs because no gauge-invariant regularization conserves the axial charge. Furthermore, Eq. (7.34) tells us that the anomaly is proportional to a total divergence,[1] which reveals its topological nature.

Consider now a 3D Weyl semimetal having a pair of isotropic nodes with opposite chirality coupled to a uniform magnetic field $\mathbf{B} = B\hat{z}$. The low-energy states are solutions of the Weyl equation $[i\hbar\partial_t \mp v(\mathbf{p} - e\mathbf{A}) \cdot \boldsymbol{\sigma}]\psi_{R/L} = 0$, where R (L) corresponds to chirality $C = +1$ $(C = -1)$. Using the gauge $\mathbf{A} = (0, Bx, 0)$, one can look for solutions of the form $\psi_{R/L} = [i\hbar\partial_t \pm v(\mathbf{p} - e\mathbf{A}) \cdot \boldsymbol{\sigma}]\Phi$, with Φ being an eigenstate of p_y and p_z. The problem reduces to finding the solutions of a harmonic oscillator problem and the corresponding eigenvalues

---

[1]We have $\eta^{\mu\nu} F_{\mu\nu} = 2\partial_\mu(\eta^{\mu\nu} A_\nu)$. The integral of this quantity is proportional to the *topological charge* of the gauge field, i.e., the number of instantons – topologically nontrivial classical solutions describing the background field.

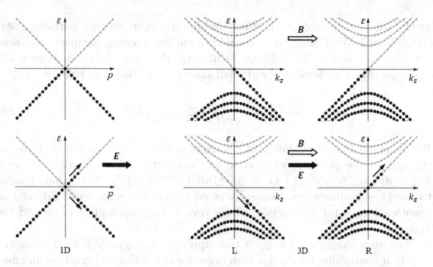

FIGURE 7.3 Schematic illustration of the chiral anomaly in 1D (left two panels) and 3D (right four panels). The black points represent occupied states, while the small gray dots represent empty states. *Left*: The top panel corresponds to a filled Dirac sea, with all negative energy states being occupied. Upon applying a uniform electric field (bottom panel), the Fermi momentum changes as indicated by the arrows, which results in the creation of right moving particles and left moving antiparticles. *Right*: The top panels show the Landau level spectra of Weyl fermions near two nodes with opposite chirality, $C = -1$ (L) and $C = +1$ (R), in the presence of a uniform magnetic field $\boldsymbol{B} = B\hat{\boldsymbol{z}}$. Note the presence of chiral zero modes [$n = 0$ in Eq. (7.35)]. Applying a uniform electric field parallel to $\boldsymbol{B}$ (bottom panels) results in the creation of right moving particles and left moving antiparticles. For simplicity, the particles are assumed to carry positive charge, $e > 0$; otherwise, $\boldsymbol{E}$ should have opposite orientation. Note that each Landau level (hence, each dot) has a degeneracy $\nu_B = L_x L_y / 2\pi \ell_B^2$.

are the Landau levels

$$\varepsilon_{nk_z} = \operatorname{sgn}(n) \frac{\hbar v}{\ell_B} \left(2|n| + \ell_B^2 k_z^2 \right)^{\frac{1}{2}}, \qquad n = \pm 1, \pm 2, \pm 3, \ldots,$$

$$\varepsilon_{0k_z} = C \hbar v k_z, \tag{7.35}$$

where $\ell_B = \sqrt{\hbar/eB}$. Note that each Landau level is degenerate with respect to k_y, which can take $\nu_B = L_x L_y / 2\pi \ell_B^2$ different values. Also note that the zero modes ($n = 0$) are chiral and have a linear dispersion, similar to the 1D case discussed above, as illustrated in Figure 7.3 (right panels). If we apply a uniform electric field $\boldsymbol{E}$ parallel to $\boldsymbol{B}$, the particles will accelerate and k_z will change in time. The occupation of the completely filled or completely empty $n \neq 0$ levels will not change, but for the zero mode the same considerations as in the 1D case apply, the only difference being that, upon changing k_z by

$2\pi/L_z$ we add ν_B (rather than one) additional right moving particles (near the $C = +1$ Weyl node) and ν_B additional left moving antiparticles (near the $C = -1$ node). This change is illustrated schematically in Figure 7.3. Consequently, the density of right/left movers will change in time as

$$\frac{dn_{R/L}}{dt} = \pm\frac{1}{4\pi^2\ell_B^2}eE = \pm\frac{e^2}{h^2}EB. \tag{7.36}$$

We conclude that the density of electrons in the vicinity of the Weyl nodes changes in the presence of electric and magnetic fields according to [388] $dn_{R/L}/dt = \pm(e^2/h^2)\boldsymbol{E}\cdot\boldsymbol{B} = \pm(e^2/8h^2)\eta^{\mu\nu\alpha\beta}F_{\mu\nu}F_{\alpha\beta}$. Note that, similar to the $(1+1)$-dimensional case discussed above, the right hand side of this equation is a total divergence, which reveals the topological nature of the anomaly.

The above arguments suggest that applying a magnetic field B along the direction connecting two nodes with opposite chiralities and creating an effective chemical potential difference $\delta\mu$ between them would result in a current $j = e^2/h^2 B\Delta\mu$. Of course, since there is no applied voltage, j cannot be an equilibrium dc current. However, it can be realized as a nonequilibrium current, e.g., by applying an oscillating electric field (parallel to the magnetic field) with a frequency higher than the inter-node relaxation rate. The generation of an electric current along an external magnetic field due to a chirality imbalance is a manifestation of the *chiral magnetic effect* [288].

Another simple manifestation of nontrivial topology in Weyl semimetals, similar to the properties of topological insulators discussed in Section 7.1, is the anomalous Hall effect. Since breaking time-reversal is required, let us assume, for simplicity, a magnetic Weyl semimetal with a pair of Weyl nodes separated by the crystal wave vector $\boldsymbol{q} = 2b\hat{\boldsymbol{z}}$ pointing from the $C = +1$ node to the $C = -1$ node (see Figure 7.4). As discussed in Section 4.2.1, each k_x–k_y plane with $-b < k_z < b$ can be viewed as a 2D Chern insulator supporting a chiral edge mode with a contribution e^2/h to the Hall conductance, as illustrated in Figure 7.4. When the chemical potential is at the Weyl nodes, the total number of edge modes, $q/(2\pi/L_z)$, is proportional to the separation between the nodes and the (Fermi arc) surface states generate an anomalous Hall effect with

$$\sigma_{xy} = \frac{e^2}{2\pi h}2b. \tag{7.37}$$

The dependence on the chemical potential is expected to be weak while the bulk Fermi surfaces surrounding the nodes remain separate (see, e.g., Figure 4.9 on page 137). Note that, since $q = 2b$ is defined modulo reciprocal lattice vectors, σ_{xy} is determined up to an integer multiple of the quantized conductance, which, physically, implies that nonzero contributions from filled bands are possible. Also note that in crystals with cubic symmetry the anomalous Hall effect vanishes due to the absence of a preferred axis, unless the symmetry is broken (e.g., by applying uniaxial strain).

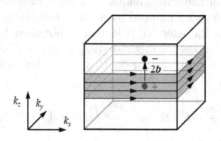

FIGURE 7.4 Anomalous Hall effect generated by the chiral Fermi arc surface states of a Weyl semimetal. The magnitude of the effect is proportional to the separation in momentum space between the Weyl nodes, $2b$. Note that each k_x–k_y plane with $-b < k_z < b$ can be viewed as a 2D Chern insulator that supports a chiral edge mode.

What is the effective field theory that formally describes these observable phenomena? For concreteness let us assume a Weyl semimetal having a pair of nodes separated by a vector $2b$ in momentum space and shifted in energy by $2b_0$. The low-energy physics of the system is described by the Lagrangian density

$$\mathcal{L} = \bar{\psi}\, i\gamma^\mu \partial_\mu \psi - e\, \bar{\psi}\gamma^\mu \psi A_\mu - e\, \bar{\psi}\gamma^\mu \gamma^5 \psi b_\mu. \tag{7.38}$$

For simplicity, we have absorbed the (possibly anisotropic) Fermi velocity in the definition of momentum and we have used natural units. Note that the last term in Eq. (7.38) can be written as $j_5^\mu b_\mu$, i.e., it describes the linear coupling of the chiral current to the four vector b_μ, which is similar to the coupling of the electric current to the ordinary gauge field, $j^\mu A_\mu$. Also note that the Lagrangian density possesses a chiral symmetry, $\psi \to \exp(-i\gamma^5\theta)\psi$, suggesting the separate conservation of the numbers of left and right movers. In particular, a gauge transformation with $\theta(\boldsymbol{r}, t) = 2\boldsymbol{b}\cdot\boldsymbol{r} - 2b_0 t$ eliminates the term $j_5^\mu b_\mu$ from Eq. (7.38), which would suggest that this system is equivalent to one having the Weyl nodes at the same momentum and energy, i.e., a system that does not possess the special transport properties mentioned above. Note, however, that in a quantum field theory the quantities of interest are correlation functions that can be obtained using, for example, the path integral formalism. The path integrals involve a weight, given by an exponential of the action obtained from the Lagrangian density (7.38), and a measure, $D\psi^\dagger D\psi$. The subtlety is that, while the action possesses chiral symmetry, the measure is modified by a chiral transformation. Note that $\psi^\dagger$ and ψ are not complex conjugate quantities, but independent variables in the fermion path integral. In other words, the chiral transformation $\psi \to \exp(-i\gamma^5\theta)\psi$ is a symmetry of the action, $S = \int dt\, d^3r\, \mathcal{L}$, but not a symmetry of the measure, $D\psi^\dagger D\psi$, hence not a symmetry of the generating functional $\mathcal{Z} = \int D\psi^\dagger D\psi \exp(iS/\hbar)$. This is the formal source of the chiral anomaly. Taking into account the change in the path integral measure induced by a chiral transformation gives rise to an additional term in the effective field theory obtained by integrating out the fermions. Using Fujikawa's method [192], one can show [575] that this additional term is a θ term similar to that characterizing the response of topological insulators,

$$S_\theta = \frac{e^2}{h} \int dt\, d^3r\, \frac{\theta(\boldsymbol{r},t)}{2\pi} \boldsymbol{E} \cdot \boldsymbol{B} = -\frac{e^2}{2\pi h} \int dt\, d^3r\, \partial_\mu \theta\, \eta^{\mu\nu\alpha\beta} A_\nu \partial_\alpha A_\beta, \quad (7.39)$$

where the rightmost expression was obtained after expressing the fields in terms of the gauge potential A_μ and integrating by parts. The "axion" field $\theta(\boldsymbol{r},t)$ has the form

$$\theta(\boldsymbol{r},t) = 2\boldsymbol{b} \cdot \boldsymbol{r} - 2b_0 t. \quad (7.40)$$

Thus, the direct consequence of the chiral anomaly is the emergence of a topological θ term in the effective field theory describing a Weyl semimetal. In turn, the observable transport properties of the system can be derived from the θ term. Indeed, varying the effective action with respect to A_ν gives a four current of the form $j_\nu = (e^2/2\pi h) b_\mu\, \eta^{\mu\nu\alpha\beta} \partial_\alpha A_\beta$. The corresponding charge and current densities are

$$j^0 = \frac{e^2}{2\pi h} 2\boldsymbol{b} \cdot \boldsymbol{B}, \quad (7.41)$$

$$\boldsymbol{j} = \frac{e^2}{2\pi h} (2\boldsymbol{b} \times \boldsymbol{E} - 2b_0 \boldsymbol{B}). \quad (7.42)$$

The second term in Eq. (7.42) describes the chiral magnetic effect discussed above, i.e., the generation of an electric current along the external magnetic field due to an energy difference between the nodes $\Delta\mu = 2b_0$. We emphasize that this is not an equilibrium current and that Eq. (7.42) has to be supplemented by an equation describing the relaxation of the chiral charge. The first term in Eq. (7.42) describes the anomalous Hall effect and corresponds to the Hall conductance given by Eq. (7.37). Finally, we point out that these effects are closely related to the topological magneto-electric effect discussed in Section 7.1, which characterizes TR-invariant topological insulators. Indeed, writing the last term in Eq. (7.42) as a polarization current, $\boldsymbol{j} = \partial_t \boldsymbol{P}$, and the energy shift as $2b_0 = -\partial_t \theta$ we obtain $\boldsymbol{P} = (e^2/2\pi h)\theta\, \boldsymbol{B}$, which reduces to Eq. (7.15) for $\theta = \pi$ (i.e., in the presence of time-reversal symmetry). Similarly, the first term in Eq. (7.42) can be written as a magnetization current, $\boldsymbol{j} = \nabla \times \boldsymbol{M}$; with $2\boldsymbol{b} = \nabla\theta$ in the presence of a constant electric field this gives $\boldsymbol{M} = (e^2/2\pi h)\theta\, \boldsymbol{E}$, which reduces to Eq. (7.14) for $\theta = \pi$. Note, however, that, unlike the related quantity in TR-invariant topological insulators, the θ field characterizing Weyl semimetals has time and bulk position dependence (similar to the dynamical field discussed in Section 7.2) and, consequently, leads to observable effects in the bulk of the material.

7.4 AXION "GRAVITOELECTROMAGNETISM" IN TOPOLOGICAL SUPERCONDUCTORS

In the previous three sections, we have discussed the response of three-dimensional topological insulators and Weyl semimetals to applied electromagnetic fields. A key property underlying the emerging physics is charge conservation, which characterizes these systems. Hence, the natural question: Is there a similar type of response associated with topological superconductors, in which charge and spin are not conserved? In particular, the gapless boundary states of topological superconductors are charge neutral Majorana modes; a quantum Hall measurement, or a similar type of electric-transport probe, cannot be used to reveal the topological nature of the system. However, energy and angular momentum are still conserved, which suggests that studying the responses of topological superconductors and superfluids to temperature gradients and rotations of the system is a meaningful enterprise. In this section, we briefly discuss these responses, which can be viewed as the thermodynamic analog of the axion electromagnetic response of topological insulators and Weyl semimetals. The results will be formulated in the language of *gravitoelectromagnetism* – a term that refers to a set of formal analogies between Maxwell's equations and an approximation to the Einstein field equations for general relativity [5]. An alternative way to incorporate topological superconductors in our discussion of axion electrodynamics (which we will not pursue here) is to consider their actual electromagnetic response. We simply point out that an external electromagnetic field couples to the superconducting phase fluctuations and, for a topological superconductor, this coupling is predicted to have the same form as the coupling of axions with an Abelian gauge field [420].

Let us now consider the response of a 3D topological superconductor to a temperature gradient and a mechanical rotation. For simplicity, the sample is assumed to have a cylindrical geometry, with the surface modes being gapped by a magnetization pointing outward (generated, e.g., by magnetic impurities doped near the surface), similar to the setup illustrated in Figure 7.1. Since energy is conserved, the topological nature of the system is reflected by its thermal transport properties, in particular by the thermal Hall conductivity. Specifically, we have $j_E = \kappa_H \hat{n} \times \nabla T$, where j_E is the energy current density, ∇T is the temperature gradient, and $\hat{n}$ is the unit vector perpendicular to the surface. Note that in a superconductor j_E coincides with the thermal current, since there is no additional charge current contribution. For boundary Majorana modes of mass m, the thermal Hall conductivity is [393, 426]

$$\kappa_H = \text{sgn}(m)\frac{\pi^2}{6}\frac{k_B^2}{2h}T. \qquad (7.43)$$

The temperature gradient can be expressed in terms of a "gravitoelectric field" E_g as $E_g = -\nabla T/T$, while the the angular velocity $\Omega = \Omega \hat{z}$ of the system rotating around the z axis corresponds to the "gravitomagnetic field"

$B_g = \frac{2}{v}\Omega$, where v is the Fermi velocity of the system [393]. Furthermore, by analogy with electromagnetism we can define the *energy magnetization* M_E as the density of dipole moment of the energy current. In terms of the spatio-temporal components of the energy-momentum tensor, $T^{\mu\nu}$, we have [393] $j_E^\alpha = vT^{\alpha 0}$; the energy magnetization (antisymmetric) tensor is given by the finite temperature average $M_E^{\alpha\beta} = \frac{1}{2}\langle x^\alpha T^{\beta 0} - x^\beta T^{\alpha 0}\rangle$. Note that this definition is similar to that of the angular momentum, $L^{\alpha\beta} = \frac{1}{v}\langle x^\alpha T^{\beta 0} - x^\beta T^{\alpha 0}\rangle$, hence we have (in terms of axial vectors) $M_E = \frac{v}{2}L$. The variation of the free energy density of a rotating system can be written as

$$dF = -SdT - L \cdot d\Omega = -SdT - M_E \cdot dB_g. \tag{7.44}$$

Thus, a generalization of the Streda formula [494] to the thermal Hall conductivity leads to [393]

$$\kappa_H = v\left(\frac{\partial M_E}{\partial T}\right)_{B_g} = v\left(\frac{\partial S}{\partial B_g}\right)_T = \frac{v^2}{2}\left(\frac{\partial L}{\partial T}\right)_\Omega = \frac{v^2}{2}\left(\frac{\partial S}{\partial \Omega}\right)_T, \tag{7.45}$$

where all vectors are oriented along the z direction.

To illustrate the consequences of Eq. (7.45), we consider a cylindrical 3D topological superconductor, as described above, having a temperature gradient in the (negative) z direction (i.e., a "gravitoelectric field" E_g parallel to $\hat{z}$). As a result, a Hall energy current $j_E = -\kappa_H \partial_z T\hat\varphi$ will flow on the surface, carrying total momentum $P_\varphi = (2\pi r\ell)j_E/v^2$, where r is the radius of the cylinder and ℓ its length. The corresponding induced angular momentum per unit volume is

$$L_z = \frac{rP_\varphi}{\pi r^2\ell} = -\frac{2\kappa_H}{v^2}\partial_z T. \tag{7.46}$$

In the language introduced above, this equation can be written as $M_E = \frac{T\kappa_H}{v}E_g$. On the other hand, if the system is rotated with angular velocity $\Omega = \Omega\hat{z}$ (in the absence of a temperature gradient), Eq. (7.45) implies the emergence of an entropy (or thermal energy) density accumulation on the top and bottom surfaces of the system as a result of the angular velocity varying from Ω (inside the system) to zero (outside). Specifically, we have

$$\Delta Q(z) = \frac{2T\Omega}{v^2}\kappa_H\left[\delta(z) - \delta(z - \ell)\right]. \tag{7.47}$$

This equation can be conveniently expressed by introducing the *thermal polarization* P_E defined (again, by analogy with electromagnetism) in terms of excess thermal energy density: $\Delta Q = -\nabla \cdot P_E$. Thus, Eq. (7.47) implies the presence of a uniform (bulk) thermal polarization $P_E = \frac{T\kappa_H}{v}B_g$. At this point, the analogy with the θ contributions to the constitutive relations (7.6) and (7.7) is fully transparent. In particular, the analog of the linear magneto-electric polarizability given by Eq. (7.8) is

$$\left.\frac{\partial M_E^j}{\partial E_g^i}\right|_{B_g=0} = \left.\frac{\partial P_E^i}{\partial B_g^j}\right|_{E_g=0} = \frac{T\kappa_H}{v}\delta_{ij}, \tag{7.48}$$

with $\frac{T\kappa_H}{v} \leftrightarrow \frac{e^2}{2h}$. Finally, we note that the angular momentum can be obtained from an internal energy functional U_θ as $-\delta U_\theta / \delta \Omega$. The corresponding functional is

$$U_\theta = -\int d^3r \, \frac{2\kappa_H}{v^2} \boldsymbol{\nabla} T \cdot \boldsymbol{\Omega} = \frac{k_B^2 T^2}{24\hbar v} \int d^3r \, \theta_g \boldsymbol{E}_g \cdot \boldsymbol{B}_g, \qquad (7.49)$$

where $\theta_g = \pi$. In the derivation of this results we implicitly assumed a single boundary Majorana mode. We note that class DIII three-dimensional topological superconductors (with a $\mathbb{Z}$ classification) have an integer number of surface Majorana modes. Assuming N of these modes having uniform (same sign) mass gaps, each giving rise to a surface half-integer *thermal* Hall effect with κ_H given by Eq. (7.43), we have $\theta_g = N\pi$ in Eq. (7.49).

Equation (7.49) represents the analog of the axion magneto-electric coupling in Eq. (7.1). The consequence of this coupling is a nontrivial correlation between the response of a topological superconductor (or superfluid) to a thermal gradient and its response to a mechanical rotation, a correlation formally similar to the magneto-electric effect in topological insulators. As a final note, we point out that in a $(p+is)$-wave superconductor θ_g can be written in terms of the relative phase between the two superconducting gaps [472]. Since the fluctuations of the relative phase (the so-called Legget modes) can be position and time dependent, this represents a "gravitational" analog of the dynamical axion field discussed in Section 7.2). The corresponding field theory is described by the action [472]

$$S_\theta^g = \frac{k_B^2 T^2}{24\hbar v} \int dt \, d^3r \, \theta_g(\boldsymbol{r}, t) \boldsymbol{E}_g \cdot \boldsymbol{B}_g, \qquad (7.50)$$

which is the "gravitational" analog of the second term in Eq. (7.24). A direct consequence of having a dynamical "gravitational" axion field is the emergence of a bulk thermal current response, similar to the electric response in Eq. (7.28). Specifically, we have

$$\boldsymbol{j}_E(\boldsymbol{r}, t) = \frac{k_B^2 T^2}{12\hbar v} \left[\partial_t \theta_g(\boldsymbol{r}, t) \boldsymbol{B}_g + \boldsymbol{\nabla} \theta_g(\boldsymbol{r}, t) \times \boldsymbol{E}_g \right]. \qquad (7.51)$$

The first term, representing the *chiral gravitomagnetic effect*, corresponds to a thermal (bulk) current induced by a mechanical rotation, while the second term, representing the anomalous thermal Hall effect, corresponds to a heat current induced by a temperature gradient and flowing perpendicular to its direction.

CHAPTER 8

Majorana Zero Modes in Solid-State Heterostructures

D EEP IN THE HEART OF THE TOPOLOGICAL WORLD LIVES THE MAJO-
rana fermion. There are several reasons the Majorana quasiparticle occupies a focal point within topological quantum matter. Of course, conceptually, it is a classical example of boundary mode that reveals the topological
properties of the bulk, in this case a topological superconductor; an example
that captures extremely well the very character of topological matter: simple
yet subtle, hiding in plain view. But what enables the Majorana to play an
important symbolic role is its conspicuous position at the crossroad that defines the new paradigm underlying the study of topological quantum matter.
A child of particle physics, the Majorana was reborn in condensed matter
physics and grew under the guidance of quantum computation. Lost before it
was even found, it is fervently searched for in solid state and cold atom systems because it carries the promise to revolutionize the manner in which we
process information and understand the foundations of the quantum world.
Whether or not the Majorana will ever fulfill this promise is a minor detail;
its main role is that of a legend, to give hope and move things forward. Practically, the search for zero-energy Majorana modes will play an important role
in achieving the key condition for getting experimental access to topological
quantum matter and quantum computation: an unprecedented level of control
over quantum systems. At the time of this writing (spring 2016; revised fall
2023), the full technological impact of these developments cannot be clearly
foreseen.

Below, we provide a brief overview of the rapidly developing field that focuses on the realization of Majorana zero modes in solid state systems. This
material should be viewed in connection with a number of closely related

DOI: 10.1201/9781003226048-8

topics presented in this book, in particular the simple models of topological superconductors discussed in Chapter 6, the realization of Majorana modes in cold atom systems (Chapter 9), the properties of non-Abelian anyons (Chapter 12), and the implementation of topological quantum computation with Majorana zero modes (Chapter 12). The interested reader can find more technical details in review articles, such as Refs. [15, 40, 131, 156, 323, 489].

8.1 THEORETICAL BACKGROUND

What is a *Majorana zero mode*? Is there any difference between a Majorana zero mode and a *Majorana fermion*? Is it possible to have *non-Abelian anyons* emerging in a symmetry-protected topological state, i.e., in a quantum phase that does not possess (intrinsic) topological order? Is there any connection between the concept of non-Abelian anyon and the particle hypothesized by Ettore Majorana in 1937? Before discussing specific schemes for the realization of Majorana zero modes in solid state systems, we briefly address these basic questions, as they touch upon some important issues regarding the terminology used in this field and, more importantly, address several fundamental aspects of topological quantum matter.

8.1.1 Majorana zero modes

Ettore Majorana's name is linked to concepts associated with three different strands that thread particle physics, condensed matter physics, and quantum computation. The key idea originated in relativistic quantum mechanics and was motivated by a critical analysis of the Dirac equation [144], which describes spin-$\frac{1}{2}$ fermions. A complex solution ψ of this equation is not an eigenstate of the charge conjugation operator $\mathcal{C}$ and the bi-spinors ψ and $\psi_c = \mathcal{C}\psi$ describe a particle (e.g., an electron) and its antiparticle (positron), respectively. Majorana discovered that real solutions of the Dirac equation with the property $\psi_c = \psi$ are possible (see box on page 132) and suggested that neutral fermions, such as neutrons and neutrinos, might be represented by such solutions [352]. A particle described by a spinor that satisfies the Majorana property – dubbed *Majorana fermion* – is identical with its antiparticle. We know that this is not the case for the neutron, but for the neutrino the jury is still out [32]. In addition, supersymmetric theories postulate the existence of Majorana supersymmetric partners associated with each bosonic particle [546]. Still awaiting experimental confirmation, the "original" Majorana fermion is a neutral elementary particle that is identical with its antiparticle and obeys (standard) Fermi–Dirac statistics.

The Majorana fermions of condensed matter physics are not elementary particles, but quasiparticle excitations emerging in certain types of many-body systems. In metals and semiconductors, the fermionic quasi-particles and their antiparticles, the holes, are always charged and, therefore, distinct. In superconductors, on the other hand, charge conservation is violated due to the presence of a Cooper-pair condensate and the quasiparticle excitations (the so-called *Bogoliubov quasiparticles*) become superpositions of electrons and holes. Thus, superconductors provide a natural environment for the emergence of Majorana fermions. In fact, we have already seen in Chapter 6 (Section 6.2.3) that the gapless boundary states of a topologically nontrivial *p*-wave superconductor are Majorana modes. Let γ_j denote the corresponding real space operator [see Eq. (6.51)]. The object created by $\gamma_j^\dagger$ is its own antiparticle in the sense that $\gamma_j^\dagger = \gamma_j$ and $\gamma_j^2 = 1$, as one can easily verify following the discussion leading to Eq. (6.51). In addition, γ_j are fermionic operators satisfying the anti-commutation rule $\{\gamma_i, \gamma_j^\dagger\} = 0$ for any $i \neq j$. Hence, in the context of condensed matter physics, one can view the Majorana fermions as boundary modes emerging in topological superconductors and being represented by second quantized operators that satisfy the Majorana condition

$$\{\gamma_i, \gamma_j\} = 2\delta_{ij}, \qquad \gamma_j^\dagger = \gamma_j. \tag{8.1}$$

We emphasize that these Majorana boundary modes are *propagating* modes that become gapless in the limit of an infinitely long boundary. If the wave vector k is a good quantum number (e.g., in the strip geometry, with $k = k_x$), the corresponding operator $\gamma_k = \sum_j e^{ikj}\gamma_j$ satisfies the k-space "Majorana condition" $\gamma_k^\dagger = \gamma_{-k}$, which expresses the intrinsic redundancy of the Bogoliubov–de Gennes (BdG) description. Indeed, consider the (time independent) BdG equation

$$\mathcal{H}_{\text{BdG}} \, \psi_n = E_n \psi_n, \tag{8.2}$$

where n is an integer that labels the quasiparticle energies and $\psi_n(r) = (u_{n\uparrow}, u_{n\downarrow}, v_{n\uparrow}, v_{n\downarrow})^T$ are 4-component spinors. As a consequence of particle-hole symmetry, we have $E_{-n} = -E_n$ and the corresponding spinors are not independent, so that (with a convenient choice of phases) we have $v_{-n\sigma}(r) = [u_{n\sigma}(r)]^*$. In the language of second quantization, the Bogoliubov quasiparticle described by $\psi_n(r)$ is created by the operator

$$\hat{\psi}_n^\dagger \equiv \sum_{r,\sigma} \left\{ u_{n\sigma}(r)\hat{c}_{r\sigma}^\dagger + [u_{-n\sigma}(r)]^* \hat{c}_{r\sigma} \right\} = \hat{\psi}_{-n}, \tag{8.3}$$

where $\hat{c}_{r\sigma}^\dagger$ and $\hat{c}_{r\sigma}$ are the electron creation and annihilation operators (corresponding to position r and spin σ), respectively. Based on Eq. (8.3), one can naturally establish a correspondence between Bogoliubov quasiparticles and the Majorana fermions described by the operators

$$\gamma_{n1} \equiv \hat{\psi}_n^\dagger + \hat{\psi}_n = \hat{\psi}_{-n}^\dagger + \hat{\psi}_{-n}, \qquad \gamma_{n2} \equiv i(\hat{\psi}_n^\dagger - \hat{\psi}_n) = -i(\hat{\psi}_{-n}^\dagger - \hat{\psi}_{-n}). \tag{8.4}$$

Hence, as a result of the equivalence modulo 2e of charge $+e$ and charge $-e$ excitations[1] and because only half of the degrees of freedom associated

---

[1] The Bogoliubov quasiparticle – Majorana fermion correspondence breaks down if the Coulomb interactions become significant and remove this equivalence modulo 2e.

with the BdG equation are independent (which can be understood in terms of a set of "particles" that are indistinguishable from their "antiparticles"), Majorana fermions appear rather naturally in the mean-field description of a *generic* superconductor [95]. The Majorana representation is useful when addressing physical phenomena such as, for example, the pair annihilation of Bogoliubov quasiparticles [39]. There is a close formal analogy between the propagating Majorana fermions that emerge in a superconductor and their (hypothesized) particle physics cousins [156]. Furthermore, given the excellent agreement of theoretical predictions based on the BdG formalism with a large body of experimental data, the existence of this type of Majorana fermion is well established.

There is, however, a related and much more remarkable phenomenon that emerges in a *topological* superconductor: the *Majorana zero mode* (MZM). In essence, MZMs are midgap excitations occurring at exactly zero energy that are localized in the vicinity of topological defects, such as vortices and domain walls. MZMs are represented by Majorana operators γ_j that, in addition to satisfying the condition (8.1), commute with the Hamiltonian,

$$[H, \gamma_j] = 0 \qquad (8.5)$$

The "composite" consisting of a MZM and the associated topological defect has nontrivial statistical properties and represents a *non-Abelian anyon*. In turn, non-Abelian anyons constitute the foundation of topological quantum computation. It is the property that MZMs give rise to this type of quantum objects that motivates the strong interest in the study of a variety of condensed matter systems predicted to support topological superconducting phases. We note that in the literature the term "Majorana fermion" is often used when referring to a "Majorana zero mode." We emphasize, however, that generically Majorana fermions do not have zero energy and do not exhibit non-Abelian statistics.

The basic properties of anyons and their relation to (topological) quantum computation will be discussed in more detail in Chapter 12. Here, we just point out a few key ideas, to better understand the nature of MZMs and their significance in the context of quantum computation.

- Anyons are (quasi)particles that occur in two-dimensional systems and have statistical properties that are neither fermionic nor bosonic. Exchanging a pair of anyons twice (say, counter-clockwise) does not leave the wave function invariant (unlike the double exchange of bosons and fermions). If the quantum state describing a system of $2N$ anyons at fixed positions is nondegenerate, any exchange of an anyon pair results in a nontrivial phase factor (i.e., a factor different from $e^{in\pi}$, with n integer) and the anyons are called *Abelian*, since the order of the exchanges is not important. If, on the other hand, the quantum state is degenerate, exchanging a pair of anyons corresponds to a rotation in the Hilbert space of degenerate states. In this case, the final state depends on the order of the exchanges and the anyons are called *non-Abelian*. *Topological quantum computation* is an approach to fault-tolerant quantum computation based on the nontrivial braiding of non-Abelian anyons.

- Anyons can occur either i) as finite energy excitations of an interacting system that is in a quantum phase with (intrinsic) topological order, or ii) as quasiparticles bounded by extrinsic defects, e.g., vortices, lattice dislocations, and domain walls, occurring in ordered systems that host a (symmetry protected) topological phase (e.g., topological superconducting state). The (bulk) excitations of a fractional quantum Hall fluid and the excitations of the toric code (see Chapter 5) are examples of anyons that are intrinsic topological quasiparticles. A defect supporting a Majorana zero mode is an example that illustrates the second mechanism for generating anyons. We emphasize that a (mean-field) topological superconductor is a symmetry-protected topological phase and, consequently, does not support bulk anyonic excitations. The non-Abelian statistics of the MZMs requires the presence of defects, which are not quantum excitations but semiclassical objects that rely on the winding of some global textures (see Figure 2.5 on page 56).

- MZMs localized near defects in a topological superconductor are examples of a particular type of non-Abelian anyons called *Ising anyons* (see Chapter 12). Ising anyons can also emerge as quasiparticles in a topologically ordered state, e.g., as bulk excitations in a $\nu = 5/2$ fractional quantum Hall fluid. We note that these intrinsic excitations have finite energy. By contrast, the defects (e.g., vortices) that bind MZMs involve an energy cost that grows logarithmically with the distance between them. We say that the defects are *confined*. In practice, creating a pair of defects (e.g., vortex-antivortex) involves tuning some external parameters and requires energy proportional to the system size.

- Loosely speaking, a pair of Majorana zero modes (γ_1, γ_2) forms an ordinary Dirac fermion,

$$c = \frac{1}{2}(\gamma_1 + i\gamma_2), \qquad c^\dagger = \frac{1}{2}(\gamma_1 - i\gamma_2). \tag{8.6}$$

In other words, a Majorana operator γ_i can be viewed as a *fractionalized* zero-mode representing "half" of an ordinary fermion. One cannot meaningfully say that γ_i is occupied or unoccupied. However, a pair of MZMs has two well-defined states labeled by the fermion parity $(-1)^n$, where $n = 0, 1$ are the eigenvalues of the occupation number operator $c^\dagger c = \frac{1}{2}(1 + i\gamma_1\gamma_2)$ (equivalently, the fermion parity is given by the eigenvalues ± 1 of the operator $i\gamma_2\gamma_1$). Note that the two states have the same energy. Again, loosely speaking, we can say that the pair of MZMs is unoccupied ($c^\dagger c = 0$) or occupied ($c^\dagger c = 1$) by one regular fermion. These two states (corresponding to even and odd fermion parity, respectively) can be used to encode a *qubit*. Remarkably, the spatial separation between the two MZMs can be arbitrarily large, so that the quantum information is stored in a highly nonlocal manner. This key property endows the qubit with (topological) protection against local perturbations.

- A system containing $2N$ Majorana zero modes has degenerate ground states labeled by the eigenvalues ± 1 of the operators $i\gamma_{2j-1}\gamma_{2j}$, where $j = 1, \ldots, N$. Assuming that the (overall) parity of the system (i.e., the product of all these eigenvalues) is fixed, there are 2^{N-1} distinct ground states that can be used to encode $N - 1$ qubits. Braiding the MZMs enables unitary transformations within the 2^{N-1} dimensional low-energy subspace spanned by the degenerate ground states. Since the unitary operations depend only on the topological class of the braid, the corresponding quantum computation is endowed with fault-tolerance.

8.1.2 "Synthetic" topological superconductors

Majorana zero modes are predicted to occur near defects in one- and two-dimensional (1D and 2D) *topological superconductors* [298, 426, 522]. According to the general classification of topological insulators and superconductors summarized in Chapter 3, such topological superconducting phases can occur in several symmetry classes (see Table 3.3). In Chapter 6, we have discussed two simple models of p-wave superconductors that have topological phases: the 1D Kitaev model given by Eq. (6.42) and the 2D chiral p-wave superconductor corresponding to the Hamiltonian (6.49). The MZMs occur as unpaired Majorana modes at the ends of the Kitaev chain or as zero-energy modes bound by vortices in the 2D topological superconductor (see Section 7.2.3). Both models involve *spinless fermions* (which implies a time-reversal operator $\mathcal{T} = K$ that squares to $+1$) and have *particle-hole symmetry* with $U_c = \tau_x$ (i.e., $\mathcal{C}^2 = +1$). Generically, the models belong to symmetry class D (see Table 3.1), which has topologically nontrivial phases in 1D and 2D. Note that the Hamiltonian (6.42) of the Kitaev model has an "accidental" time-reversal symmetry that places it in symmetry class BDI (which hosts 1D topological phases). These general considerations bring us to the key question: Are there actual physical systems that realize these topological phases?

Materials in which p-wave superconductivity emerges "intrinsically" through many-body interactions are rare. The best known candidates are the superfluid phases of ^{3}He [322, 521, 523] and the spin-triplet superconductor Sr_2RuO_4 [349, 351]. However, the experimental realization of half-quantum vortices capable of hosting MZMs represents a serious challenge in both superfluid ^{3}He-A thin films and Sr_2RuO_4 [366]. In addition, it is not clear how one can practically detect and manipulate Majorana zero modes in these systems. Furthermore, the realization of spin-polarized p-wave superfluidity in ultracold atom systems [500] (see also Chapter 9) poses serious challenges due to the extremely short lifetime of p-wave bound pairs [102]. These problems make the actual realization, detection, and manipulation of MZMs in "intrinsic" topological superconductors a daunting experimental task.

An appealing and, so far, very promising alternative is to engineer "synthetic" topological superconductors by combining ordinary superconductors with other materials, such as semiconductors and topological insulators. The pioneering proposals by Fu and Kane envisioned the realization of topological superconductivity using conventional s-wave superconductors in combination with topological insulators [187, 188]. What these proposals have shown is that one can rely on rather abundant and typically well studied materials, instead of having to deal with "exotic" superconductors. This was a real breakthrough. The next step was to replace the topological insulators (which are still experimentally quite challenging) with semiconductors (the most technologically friendly materials) [14, 345, 396, 446]. Another development involves a proposal [377] for the realization of MZMs in ferromagnetic atomic chains (e.g., Fe) on superconductor substrates (e.g., Pb). We will briefly discuss these schemes in the next section. Here, we focus on the basic idea on which these proposals are formulated.

Let us try to identify a physical system that realizes the Kitaev chain model. The necessary ingredients are i) spinless fermions and ii) p-wave superconductivity. Of course, the fermions in a real solid state system are electrons and they carry spin-1/2. Therefore, the first problem we need to address is how to "freeze out" half of the spin degrees of freedom, so that the system becomes *effectively spinless*. A straightforward approach is to fully spin-polarize the electrons and ensure that a *single* spin-split band is occupied. However, it would be difficult (if not impossible) to drive such a system into a superconducting state. An alternative approach is to exploit the property of strong spin-orbit coupling to lock the electron's spin to its momentum. The surface (or edge) states of topological insulators and the spin-split bands of semiconductors with strong spin-orbit coupling are ideal candidates.

Consider, for example, the low-energy spectrum of a 1D semiconductor wire near the bottom of the conduction band (see Figure 8.1). In the absence of spin-orbit coupling and Zeeman splitting the band is double degenerate, as shown in panel (a). Adding Rashba-type spin-orbit coupling [Figure 8.1(b)] removes this degeneracy, except at $k = 0$, where it is protected by the Kramers theorem. The remaining degeneracy can be removed by breaking time-reversal symmetry, e.g., using a Zeeman field [Figure 8.1(c)]. If the Fermi energy is sufficiently close to $\epsilon = 0$, only the lowest energy band is occupied and the system is effectively spinless, i.e., it has only one active fermionic species. Note that the combination of (strong) spin-orbit coupling and Zeeman splitting gives rise to a nontrivial spin texture in momentum space, as shown in Figure 8.1. In particular, the states from the lower energy band in panel (c) with wave vectors $+k$ and $-k$ have nonzero anti-parallel spin components (along the y direction); this is key for realizing proximity-induced pairing using an s-wave superconductor.

Obtaining the second ingredient (p-wave superconductivity) seems more challenging, since, as discussed above, one cannot rely on "intrinsic" p-wave superconductors. Fortunately, the task can be accomplished with ordinary

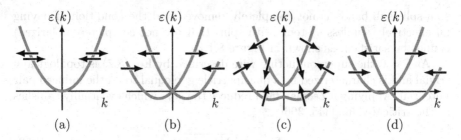

FIGURE 8.1 Energy spectrum for a 1D wire modeled by the Hamiltonian in Eq. (8.7) with $\Delta = 0$. (a) No spin-orbit coupling and no Zeeman field. (b) Nonzero spin-orbit coupling and $\Gamma = 0$. (c) Zeeman field oriented *perpendicular* to the spin-orbit coupling field ($\alpha_R \neq 0$, $\Gamma \neq 0$). (d) Zeeman field $\Gamma \sigma_y$ oriented *parallel* to the spin-orbit coupling field. The arrows show the spin orientation; by convention a horizontal arrow correspond to the spin parallel to the y axis.

s-wave superconductors. In essence, coupling the semiconductor (or topological insulator) wire to a long-range ordered superconductor results in the "spinless" electrons acquiring Cooper pairing through *superconducting proximity effect*. In addition, as a result of the nontrivial spin texture generated by the combination of spin-orbit coupling and Zeeman field, the induced superconductivity is a mixture of s-wave and p-wave components. Thus, a topological superconducting regime can be achieved in the 1D wire by tuning the chemical potential (to ensure single-band occupancy) and the Zeeman field.

We conclude that synthetic topological superconductors can be engineered by combining a) strong spin-orbit coupling, b) Zeeman splitting, and c) proximity-induced superconductivity. The first two ingredients provide an effectively spinless fermion system with a nontrivial spin texture, while the third provides (proximity-induced) Cooper pairing. More details on specific implementations and variations of this basic scheme are discussed in the next section. The essential components of a synthetic topological superconducting wire are captured by a simple effective model of the form

$$H_{\text{eff}} = \int \frac{dk}{2\pi} \left[\psi_k^\dagger \left(\frac{\hbar^2 k^2}{2m} - \mu + \alpha_R k \sigma_y + \Gamma \sigma_x \right) \psi_k + \Delta \psi_{k\uparrow} \psi_{k\downarrow} + \text{h.c.} \right], \quad (8.7)$$

where $\psi_k = (\psi_{k\uparrow}, \psi_{k\downarrow})^T$ is the electron annihilation operator, m is the effective mass, α_R is the Rashba spin-orbit coupling coefficient, Γ is the Zeeman field, and Δ is the induced pair potential. The chemical potential μ can be changed using external gates, while the Zeeman spin splitting can be modified by tuning an external magnetic field, $\Gamma = \frac{1}{2} g \mu_B B$, where g the effective Landé g-factor, B the applied magnetic field, and μ_B the Bohr magneton. We emphasize that, to be able to drive the system into a topological regime, it is essential that the magnetic field be perpendicular to the spin-orbit field. In Eq. (8.7), for example, the magnetic and spin-orbit fields are oriented along the x and y directions, respectively. If the two fields are parallel, the degeneracy of the spin-split sub-bands is not completely removed and the condition of having an effectively spinless system (with spins that are not completely polarized) cannot be satisfied, as shown in Figure 8.1(d).

At $\Gamma = 0$ the superconducting wire described by Eq. (8.7) is topologically trivial and is characterized by a quasiparticle gap equal to Δ, the induced pair potential. Applying a Zeeman field reduces the gap, which eventually vanishes at the critical value [345, 396]

$$\Gamma_c = \sqrt{|\Delta|^2 + \mu^2}.$$ (8.8)

The vanishing of the quasiparticle gap signals a topological quantum phase transition [426, 446]. For $\Gamma > \Gamma_c$ the gap opens again and the system enters the topological superconducting phase.

8.2 REALIZATION OF MAJORANA ZERO MODES: PRACTICAL SCHEMES

Various schemes for engineering topological superconductors by skilfully combining conventional s-wave superconductors with other materials have been proposed following the seminal work of Fu and Kane [187, 188]. Below, we touch upon some key features associated with two types of heterostructures that have received most of the experimental attention so far, focusing on semiconductor wires proximity coupled to conventional superconductors.

8.2.1 Semiconductor-superconductor hybrid structures

In 2012 the search for Majorana zero modes in solid state structures entered a new phase when, following earlier theoretical predictions [345, 396], an experimental group in Delft published evidence consistent with the existence of these modes in InSb nanowires proximity-coupled to a superconductor (NbTiN) [375]. More specifically, it was found that a zero-bias tunneling conductance peak (see below, Section 8.3.1) develops when the magnetic field applied parallel to the wire exceeds a certain value, consistent with the emergence of MZMs when the topological criterion $\Gamma > \Gamma_c$ is satisfied. Similar findings were subsequently reported by other groups [112, 128, 137, 171].

Figure 8.2(a) illustrates the basic structure of a semiconductor wire-based Majorana device: a semiconductor wire with strong spin-orbit coupling (e.g., InAs or InSb) is proximity-coupled to a conventional s-wave superconductor (e.g., Nb or Al) and placed in a magnetic field, typically oriented parallel to the wire. A normal metal lead may be added to probe the system by performing charge tunneling measurements. Back gates are used to generate potential barriers and control the chemical potential. Real wires are quasi one-dimensional systems that, typically, have several active confinement-induced bands. The bands are spin-split due to the spin-orbit coupling and applied Zeeman field, as shown in Figure 8.2(b). In general, the BdG spectrum of the system is characterized by a finite quasiparticle gap. If the chemical potential

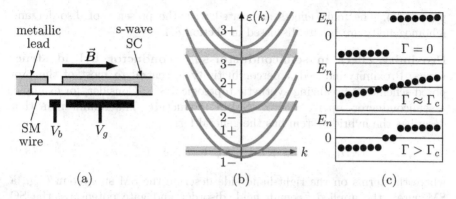

(a)   (b)   (c)

FIGURE 8.2 (a) Basic architecture of a semiconductor (SM) wire – superconductor (SC) device. The back gate V_g controls the chemical potential, while V_b creates a tunnel barrier. (b) Schematic structure of a multi-band energy spectrum. The energies of the $\alpha-$ ($\alpha+$) sub-bands decrease (increase) with increasing Zeeman field. When the chemical potential is tuned within the shaded regions, the system is topologically nontrivial. (c) Low-energy BdG spectrum of a finite wire showing the energies of the states ψ_n with $|n| \leq 7$. The quasiparticle gap vanishes at the critical Zeeman field Γ_c; two zero-energy modes emerge when $\Gamma > \Gamma_c$.

is tuned close to the bottom of a confinement-induced band, a pair of zero-energy Majorana modes emerge when the applied Zeeman field exceeds a certain critical value, as illustrated in Figure 8.2(c). The MZMs are localized at the two ends of the wire segment covered by the superconductor.

The low-energy physics of an *ideal* semiconductor (SM) wire proximity-coupled to a superconductor (SC) is described by the effective Hamiltonian (8.7). However, the real world happens to be slightly more complex. Quite generally, the physical systems that support topological phases are governed by an interplay of topological and nontopological features. For the synthetic topological SCs that host MZMs this is both a curse and a blessing: a curse because it opens paths for circumventing the topological protection and a blessing because it provides knobs for probing and controlling the system, thus facilitating its experimental study and the implementation of operations necessary for quantum computation. Below we discuss a few examples of "real world" theoretical problems that need to be addressed if we want to make a connection between the "ideal picture" and the physics of the structures realized and observed in the laboratory. We note that this is not an exhaustive list of experimentally relevant issues. Other important problems include understanding the role of different types of disorder, disentangling topologically trivial and nontrivial features, and incorporating details regarding the geometry of the heterostructure, specific material properties, and the presence of nonhomogeneous fields (e.g., applied magnetic fields and effective confining potentials). The main effects associated with the presence of disorder and inhomogeneity are briefly discussed in Section 8.4

Proximity effect in semiconductor-superconductor hybrid structures. Proximity-induced superconductivity is central to most of the proposed schemes for realizing synthetic topological SCs. Consider, for example, a semiconductor-superconductor (SM-SC) structure. The Hamiltonian that describes the hybrid system has the general form

$$H_{\text{tot}} = H_{\text{SM}} + H_{\text{Z}} + H_{\text{V}} + H_{\text{SC}} + H_{\text{SM-SC}}, \tag{8.9}$$

where the terms on the right-hand side describe the SM subsystem (e.g., a SM wire), the applied Zeeman field, disorder and gate potentials, the SC subsystems, and the coupling between the SM and the SC, respectively. To account for the effect of the SC on the SM subsystem, we need to model (at a certain level) the superconductor and the coupling term; then we can integrate out the SC degrees of freedom. For simplicity, we describe the system using a simple tight-binding Hamiltonian and incorporate superconductivity, at the mean field level, through a constant pairing amplitude Δ_0. Explicitly, we have

$$H_{\text{SC}} = -t_{sc} \sum_{\langle i,j \rangle, \sigma} a_{i\sigma}^\dagger a_{j\sigma} - \mu_{sc} \sum_{i,\sigma} a_{i\sigma}^\dagger a_{i\sigma} + \Delta_0 \sum_i (a_{i\uparrow}^\dagger a_{i\downarrow}^\dagger + a_{i\downarrow} a_{i\uparrow}),$$

$$H_{\text{SM-SC}} = -\sum_{m,\sigma} \sum_{\langle l_0, j_0 \rangle} \left(\lambda_{m\sigma} c_{l_0 m}^\dagger a_{j_0\sigma} + \lambda_{m\sigma}^* a_{j_0\sigma}^\dagger c_{l_0 m} \right), \tag{8.10}$$

where i and j label the lattice sites associated with the bulk SC, l_0 and j_0 designate lattice sites on the SM and SC sides of the SM-SC interface, respectively, $\langle \dots \rangle$ designates nearest-neighbors, $a_{i\sigma}^\dagger$ is the creation operator for a single-particle state with spin σ localized inside the SC, and $c_{lm}^\dagger$ creates a single-particle state inside the SM characterized by the set of quantum numbers $m = (\alpha, s)$ associated with orbital and spin degrees of freedom. The model parameters are the SC hopping (t_{sc}), the chemical potential of the SC (μ_{sc}), the pairing amplitude (Δ_0), and the hopping across the SM-SC interface ($\lambda_{m\sigma}$). If the SM is described by a single-band model (corresponding to the conduction band), $m = \pm 1$ labels the spin projection along the z-axis and we have $\lambda_{m\sigma} = \lambda \, \delta_{m\sigma}$. The low-energy physics of the SM is described by an effective action obtained by integrating out the SC degrees of freedom. In terms of Green functions, the coupling to the SC induces an interface self-energy contribution to the SM Green function [414, 489]. Neglecting nonlocal terms, the superconducting proximity effect is captured by the (local) self-energy [489]

$$\Sigma_{mm'}(l_0; \omega) = -\nu_F \sum_{s,s'} \lambda_{ms} \left[\frac{\omega + \Delta_0 \sigma_y \tau_y}{\sqrt{\Delta_0^2 - \omega^2}} + \zeta \tau_z \right]_{ss'} \lambda_{m's'}, \tag{8.11}$$

where ν_F is the surface density of states of the SC subsystem in the normal state (at the Fermi energy), σ_μ and τ_μ are Pauli matrices associated with the spin and Nambu spaces, respectively, and ζ is a proximity-induced shift of the chemical potential. For a single-band model, the superconducting proximity effect depends on the coupling constant $\tilde{\gamma} = \nu_F |\lambda|^2$, i.e., on the SC density of states and the transparency of the interface. In addition, the strength of the proximity effect is determined by the spatial profile of the wave functions corresponding to the low-energy states of the semiconductor. More specifically, assuming (for simplicity) that the characteristic amplitude of the electron wave functions at the SM-SC interface is ψ_0, the strength of the proximity effect is given by the *effective* SM-SC coupling

$$\gamma = \nu_F |\lambda|^2 |\psi_0|^2. \tag{8.12}$$

Note that, more generally, the wave functions associated with different confinement-induced bands have different amplitudes at the interface and, consequently, the effective SM-SC coupling is a matrix $\gamma_{\alpha\beta}$, where α and β are band indices [489]. If the inter-band energy separation is large compared to the SM-SC effective coupling, the relevant terms are the diagonal elements $\gamma_\alpha \equiv \gamma_{\alpha\alpha}$.

The main consequences of the superconducting proximity effect become transparent in the low-energy, weak coupling limit defined by $\omega \ll \Delta_0$ and $\gamma < \Delta_0$. In this limit, the self-energy can be considered within the static approximation $\sqrt{\Delta_0^2 - \omega^2} \approx \Delta_0$ and the SM Green function becomes [414, 489] $G^{-1} \approx Z^{-1}(\omega - \mathcal{H}_{\text{eff}})$, where $Z = (1 + \gamma/\Delta_0)^{-1}$ is the reduced quasiparticle weight and $\mathcal{H}_{\text{eff}}$ is the effective low-energy BdG Hamiltonian describing the SM sub-system. Explicitly, we have

$$\mathcal{H}_{\text{eff}} = \frac{Z}{2}[(\tau_z + 1)\mathcal{H}_0 + (\tau_z - 1)\mathcal{H}_0^T] + \Delta_{\text{ind}}\sigma_y\tau_y, \tag{8.13}$$

where $\mathcal{H}_0 = \mathcal{H}_{\text{SM}} + \mathcal{H}_Z + \mathcal{H}_V$ is the (first quantized) Hamiltonian for the semiconductor, which includes the Zeeman field and gate (and/or disorder) potential, and $\Delta_{\text{ind}} = \gamma\Delta_0/(\gamma + \Delta_0)$ is the *induced* pairing potential. Note that in the limit $\gamma \ll \Delta_0$ we have $\Delta_{\text{ind}} \approx \gamma$. We emphasize that in general (i.e., for arbitrary γ) the topological superconducting phase is realized when the Zeeman field exceeds a critical value determined by the effective SM-SC coupling (rather than the induced gap) [487],

$$\Gamma_c = \sqrt{\gamma^2 + \mu^2}. \tag{8.14}$$

This reduces to the expression given by Eq. (8.8) in the limit $\gamma \ll \Delta_0$, when the induced gap is approximately equal to the effective SM-SC coupling.

Based on this simplified analysis, we can formulate several observations regarding the superconducting proximity effect. (i) The effective Hamiltonian describing a proximity-coupled SM wire has, indeed, the general structure of the simple model given by Eq. (8.7), but only in the low-energy, weak-coupling limit. For energies comparable to the bulk SC pairing amplitude Δ_0 the explicit frequency-dependence of the self-energy has to be taken into account. (ii) The parameters of the Hamiltonian $\mathcal{H}_0$ describing the semiconductor (and the external fields) are renormalized by the coupling to the superconductor. The renormalization factor $Z < 1$ can be interpreted as the probability of finding an electron (which occupies a certain low-energy single-particle state) inside the semiconductor, rather than the SC. In the strong-coupling limit ($\gamma > \Delta_0$) the electrons "live" mostly inside the SC (i.e., $Z \ll 1$). (iii) In a multi-band system the SM-SC coupling $\gamma_{\alpha\beta}$ is a nondiagonal matrix, which results in an effective inter-band coupling via the superconductor. Consequently, the band structure of the effective Hamiltonian may differ significantly from that of the bare SM Hamiltonian. (iv) If the transparency of the SM-SC interface is spatially nonuniform, the coupling constant $\tilde{\gamma}$ becomes position dependent. In turn, this generates not only a position-dependent induced pair potential, but also a highly nontrivial renormalization of the low-energy physics. Both effects represent important sources of disorder.

Multi-band occupancy and phase diagram. In a real, quasi-1D wire several confinement-induced bands may be occupied. To understand the structure of the corresponding topological phase diagram, let us focus on the ideal case of an infinitely long wire. We assume that the spin-orbit field is oriented in the y-direction. For $\Gamma=0$, the BdG spectrum is gapped (see Figure 8.3) and the corresponding superconducting phase is topologically trivial. As mentioned before, driving the system into a topological superconducting state requires a Zeeman field oriented perpendicular to the spin-orbit field and exceeding a certain critical value. Note that applying an external magnetic field along the direction of the spin-orbit field (i.e., the y-direction) reduces the quasiparticle gap, which eventually vanishes at $\Gamma^* \sim \gamma$. However, the gap does not open again upon further increasing the Zeeman splitting so that for $\Gamma > \Gamma^*$ the system remains in a gapless state, as shown in Figure 8.3(c).

Consider now a Zeeman field oriented perpendicular to the spin-orbit field, i.e., along the x- or z-direction. In this case, the quasiparticle gap corresponding to any finite momentum is nonzero, $|E_\alpha(k)| > 0$ for $k \neq 0$, where α is the band index. However, the quasiparticle gap at $k = 0$ vanishes when the system undergoes a topological quantum phase transition; see Figure 8.3(b). If the Zeeman field and the effective SM-SC coupling are small compared to the inter-band spacing and the chemical potential is near the bottom of the confinement-induced band α, the topological criterion for the ideal (one-dimensional) wire holds, but the chemical potential has to be measured relative to the top-most occupied band α_m (also called the "Majorana band"). Explicitly, we have

$$\Gamma > \Gamma_c = \sqrt{\gamma_\alpha^2 + (\delta\mu_\alpha)^2}, \qquad \delta\mu_\alpha = \mu - \epsilon_\alpha(0), \qquad (8.15)$$

where $\epsilon_\alpha(k)$ is the energy of the α band in the absence of induced superconductivity and applied Zeeman field , i.e., when $\gamma_{\alpha\beta} = 0$ and $\Gamma = 0$. In essence,

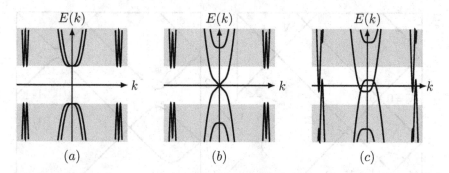

FIGURE 8.3 BdG spectra for an infinite quasi-1D wire with the chemical potential at the bottom of the second band [sub-bands 2+ and 2− in Figure 8.2(b)]. Shaded regions correspond to energies $|E(k)| > \Delta$. (a) No Zeeman field ($\Gamma = 0$); the qp gap is equal to the induced pair potential Δ. (b) Topological quantum phase transition ($\Gamma = \Gamma_c$ applied parallel to the wire); the top occupied band (i.e., sub-band 2−) becomes gapless at $k = 0$. The dispersion of the low-energy sub-bands (1+ and 1−) is weakly dependent on the applied Zeeman field. For $\Gamma > \Gamma_c$ the qp gap reopens and the SC phase becomes topologically nontrivial. (c) Zeeman field parallel to the spin-orbit field. The system becomes gapless for $\Gamma > \Gamma^*$.

for low-enough Zeeman fields, the pairs of spin-split sub-bands corresponding to low-energy occupied bands are "passive" – both sub-bands from each pair cross the chemical potential and are gapped by proximity effect – and the topological properties of the system are determined by the "active" Majorana band: if only one active spin-split sub-band is occupied, the system is effectively spinless and the superconducting phase is topologically nontrivial; otherwise, the wire is a trivial SC.

With increasing Zeeman field, i.e., when Γ is comparable to (or larger than) the inter-band spacing, spin-split sub-bands corresponding to different confinement-induced bands cross, as shown in Figure 8.4. At this point, it is worth noting that in a quasi-1D wire the Rashba spin-orbit coupling has both a *longitudinal* contribution (proportional to the wave vector k) and a *transverse* term that couples different confinement-induced bands. In a basis $\psi_{\alpha\sigma}(k)$, where α is the band index, the Rashba spin-orbit coupling has the generic form

$$[\mathcal{H}_{\text{SOI}}]_{\alpha\beta} = \delta_{\alpha\beta} \, \alpha_R \, k\sigma_y + iq_{\alpha\beta} \, \sigma_x, \tag{8.16}$$

where $q_{\alpha\beta} = -q_{\beta\alpha}$ are matrix elements that can be evaluated using specific models [489]. The existence of the transverse Rashba term impacts the topological phase diagram in two major ways.

First, it changes the symmetry of the Hamiltonian. Indeed, *without* the transverse Rashba coupling, the effective Hamiltonian has, in addition to particle-hole symmetry, an artificial "time-reversal" symmetry with $\mathcal{T} = K$, where K is the complex conjugation. More specifically, the effective

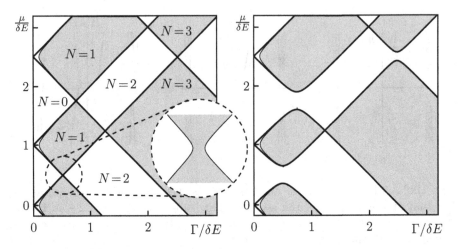

FIGURE 8.4 Topological phase diagram of a multi-band nanowire with Zeeman field oriented along the wire (left) and in the z-direction (right). Energies are in units of inter-band spacing δE. In the absence of transverse Rashba coupling ($q_{\alpha\beta} = 0$) the two diagrams become identical and the topological phases are indexed by the $\mathbb{Z}$ invariant N. When $q_{\alpha\beta} \neq 0$, the topological phases are labeled by a $\mathbb{Z}_2$ invariant that, for the left panel, is $N_2 = N$ (mod 2). The inset shows a change of the phase boundaries (corresponding to a Zeeman field oriented along the wire) in the presence of inter-band pairing, $\Delta_{12} \neq 0$.

Hamiltonian satisfies the condition $\mathcal{H}^*_{\text{eff}}(k) = \mathcal{H}_{\text{eff}}(-k)$. Consequently, the Hamiltonian belongs to the symmetry class BDI and its distinct 1D topological phases are classified by a $\mathbb{Z}$ topological invariant. Let α_+ and α_- denote the spin-split bands corresponding to the confinement-induced band α and let us assume that the energy of α_+ (α_-) increases (decreases) with Γ. Also, assume that for a given value of the Zeeman field $\Gamma > 0$ we have ν_- occupied α_- sub-bands and ν_+ occupied α_+ sub-bands. Then, the $\mathbb{Z}$ topological invariant of the corresponding phase is, basically, $N = \nu_- - \nu_+$, as shown in Figure 8.4. Furthermore, in a long (but finite) wire there will be N MZMs localized at each end, as a manifestation of the bulk-boundary correspondence.

In the presence of the transverse Rashba term the artificial time-reversal symmetry is broken and the Hamiltonian belongs to symmetry class D with distinct 1D topological phases labeled by a $\mathbb{Z}_2$ invariant. Sufficiently far from the phase boundaries, the topological invariant N is given by the *parity* of the sub-band occupation number, $N_2 = \nu_- + \nu_+$ (mod 2). An odd number of occupied sub-bands corresponds to a topological SC phase ($N_2 = 1$), while an even number corresponds to a trivial SC ($N_2 = 0$). In a long (but finite) wire, there will be exactly N_2 MZMs at each end. Note, however, that if the symmetry is weakly broken – e.g., when the relevant matrix elements $q_{\alpha\beta}$ are small – the wire will still host N low-energy modes localized near each end;

N_2 of these modes have exactly zero energy, while $N - N_2$ of them can be viewed as weakly energy-split Majorana modes.

The second way in which the transverse Rashba term impacts the topological phase diagram is to introduce a dependence of the sub-band energies $\epsilon_{\alpha_\pm}(0)$ on the orientation of the Zeeman field. More specifically, when the field is parallel to the wire the crossing of the sub-bands α_+ and β_- is not affected by the transverse Rashba coupling. By contrast, when Γ is oriented along the z direction the two sub-bands anticross if $q_{\alpha\beta} \neq 0$, as illustrated in the right panel of Figure 8.4.

Other important factors that impact the location of the phase boundaries are the orbital effect of the external magnetic field, the presence of disorder and nonhomogeneous confinement, the strength of the semiconductor-superconductor coupling, and the off-diagonal induced pairing. Strong SM-SC coupling (γ), for example, renormalizes the low-energy physics of the nanowire, including the effective value of the Zeeman field. With increasing γ, the phase boundary between the trivial ($N = 0$) and nontrivial ($N = 1$) phases is pushed to higher values of Γ, as suggested by Eq. (8.15). Also, off-diagonal induced pairing $\Delta_{\alpha\beta}$ (where $\alpha \neq \beta$ are band indices) emerges generically in a SM-SC structure with no particular spatial symmetry [see Eq. (8.11)]. For systems with a magnetic field oriented parallel to the wire, inter-band pairing leads to an expansion of the topological SC phase near "crossing points," as illustrated in the inset of Figure 8.4.

Finite size effects and topological protection. We defined the Majorana zero modes in terms of Majorana fermion operators that *commute* with the Hamiltonian. However, Eq. (8.5) corresponds to the idealized case when the MZMs are infinitely far away from one another. In real physical systems, there is always a finite characteristic length L associated with the separation between two MZMs. For example, in the case of a superconducting wire that supports MZMs at its ends, L is the length of the wire. More generally, consider a physical system supporting $2N$ modes localized near points r_j and described by Majorana operators γ_j. The *operational* definition of the Majorana zero modes replaces Eq. (8.5) with the requirement that the commutator of γ_j with the Hamiltonian be exponentially small,

$$[H, \gamma_j] \sim \exp\left(\frac{L}{\xi}\right), \tag{8.17}$$

where $L = \min_i(|r_j - r_i|)$ and ξ is a length scale associated with the Hamiltonian H that characterizes the "size" of the MZMs. The characteristic length ξ depends on microscopic parameters, such as the effective mass, the effective SM-SC coupling, the Zeeman field, and the chemical potential; it diverges at the topological quantum phase transition and has a minimum at values of the Zeeman field slightly higher than the critical field, $\Gamma \gtrsim \Gamma_c$. In the limit $L \gg \xi$, Eq. (8.17) ensures the quasi-degeneracy of the system of $2N$ MZMs and generates a 2^{N-1}-dimensional low-energy subspace. This property allows using the system as a platform for topological quantum computation.

The effective Hamiltonian obtained by projecting H onto the low-energy subspace has the form $H_{\text{eff}} = -i\sum_{j,k} \epsilon_{jk}\gamma_J\gamma_k$, where $\epsilon_{jk} = -\epsilon_{kj}$ are exponentially small energy splittings, $\epsilon_{jk} \propto e^{-|\mathbf{r}_j-\mathbf{r}_k|/\xi}$. If the distances between every two MZMs are much larger that the characteristic length scale ξ, the low-energy subspace becomes effectively degenerate and can be used for encoding quantum information nonlocally, as occupied or unoccupied states of the nonlocal Dirac fermion modes (8.6) constructed using pairs of MZMs. For this to work, it is essential that the global fermion parity of the system be preserved. This can be realized by having a finite quasiparticle gap above the 2^{N-1}-dimensional low-energy subspace. The existence of such a gap suppresses exponentially the rate of changing the fermion parity (e.g., due to thermal fluctuations).

Let us consider a simple 1D tight-binding model of a proximitized semiconductor wire. The effective Hamiltonian, including the superconducting proximity effect and the applied Zeeman field, has the form

$$H = -t_0 \sum_{\langle i,j \rangle} c_i^\dagger c_j - \mu \sum_i c_i^\dagger c_i + \frac{i\alpha_R}{2} \sum_{\langle i,j \rangle} \left(c_i^\dagger \sigma_y c_j - c_j^\dagger \sigma_y c_i \right)$$
$$+ \Gamma \sum_i c_i^\dagger \sigma_x c_i - \sum_i \left(\Delta\, c_{i\uparrow}^\dagger c_{i\downarrow}^\dagger + \Delta^* c_{i\downarrow} c_{i\uparrow} \right), \tag{8.18}$$

where σ_μ are Pauli matrices and $c_i^\dagger = (c_{i\uparrow}^\dagger, c_{i\downarrow}^\dagger)$ represents the creation operator in spinor form corresponding to lattice site i, where $1 \leq i \leq N$. The model parameters are the nearest-neighbor hopping t_0, the chemical potential μ, the Rashba spin-orbit coupling coefficient α_R, the Zeeman field Γ, and the induced pairing potential $\Delta = |\Delta|e^{i\phi}$.

The dependence of the low-energy BdG spectrum on the applied Zeeman field is shown in Figure 8.5. In the long wire limit (left panel), the quasicontinuous spectrum is characterized by a quasiparticle (qp) gap Δ at $\Gamma = 0$. Upon increasing the Zeeman field the qp gap decreases and eventually vanishes at the critical field $\Gamma_c = \sqrt{\mu^2 + \Delta^2}$ corresponding to the topological quantum phase transition. For $\Gamma > \Gamma_c$ the *bulk* qp gap reopens, but two localized midgap modes emerge – the MZMs localized at the ends of the wire. Additional in-gap states can occur in the presence of disorder (see Section 8.4).

In a short wire (right panel), the spectrum is discrete and the bulk qp gap does not completely close, signaling that we can distinguish two different phases (i.e., topologically trivial and nontrivial) only operationally (e.g., by defining a finite energy resolution δ_E and considering any state with $|E_n| < \delta_E$ as an *effectively* zero-energy state). The *effective* MZMs are not well separated spatially (i.e., $L \gtrsim \xi$) and have an energy splitting ϵ that oscillates as a function of the Zeeman field. The spatial profiles of the MZMs are shown in the inset. To be useful as a platform for topological quantum computation, the system has to be long-enough and clean enough so that $\epsilon < \delta_E \ll \Delta_{qp}$.

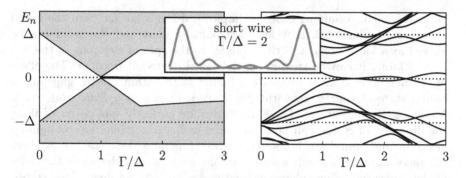

FIGURE 8.5 Dependence of the low-energy spectrum of Hamiltonian (8.18) on the Zeeman field. *Left*: Long wire. The spectrum is quasi-continuous and the bulk gap closes at $\Gamma_c = \Delta$ (for $\mu = 0$). Two MZMs emerge when $\Gamma > \Gamma_c$. *Right*: Short wire. The bulk gap does not close and the MZMs exhibit energy splitting oscillations. *Inset*: Spatial profiles of the Majorana wave functions.

8.2.2   Shiba chains

In addition to the significant progress toward the experimental demonstration of topological superconductivity and Majorana zero modes in semiconductor nanowire structures, several experimental studies [163, 289, 378, 452] have reported evidence consistent with the presence of MZMs in chains of magnetic atoms (e.g., Fe) on a conventional superconductor substrate (e.g., Pb). More specifically, using scanning tunneling spectroscopy (STS), it was found that zero-bias peaks qualitatively similar to the expected signature of MZMs develop near the ends of the atomic chains. If further validated as a viable platform for topological superconductivity, chains of magnetic atoms could provide a significant boost to the study of Majorana excitations by i) giving direct access to the Majorana end states using STS (which provides information on both the *spectral* and *real space* properties of the MZMs) and ii) opening the possibility to manipulate the adatoms using a scanning tunneling microscope (which is potentially useful for studying topological phase transitions by varying the distance between atoms and for realizing more complex structures that may enable braiding).

In essence, the emergence of MZMs in chains of magnetic atoms relies on the same basic recipe that we discussed in the context of semiconductor Majorana wires: realizing a 1D *helical* electron system that is proximity coupled to a conventional s-wave superconductor (SC). Below, we briefly summarize the key ideas underlying the theoretical understanding of this alternative platform for engineering topological superconductivity and MZMs in solid state systems [87, 107, 300, 333, 355, 377, 407].

Consider a single magnetic impurity in a conventional superconductor, which is a classical problem in superconductivity well studied since the 1960s

[439, 471, 564]. Assuming that the impurity d-levels are far from the Fermi energy of the host SC, they are electronically inert and the atom can be viewed as a local moment S that couples to the spin of electrons in the SC. If S is large, it can be treated as a classical degree of freedom. The effect of the magnetic atom is to create a bound state within the SC gap (i.e., a bound state of energy E_0 with $|E_0| < \Delta_0$), known as a *Shiba state* (or a Yu–Shiba–Rusinov state). The Shiba states are perfectly *spin polarized* along the direction of S and can be represented as linear combinations of spin-up electrons and spin-down holes. Their wave functions decay as $1/r$ at short distances and exponentially above a certain length scale (which is of the order of the superconducting coherence length $\xi_0 = \hbar v_F/\Delta_0$ for deep Shiba states, $|E_0| \ll \Delta_0$).

Next, we consider a chain of magnetic impurities. Depending on the distance between adjacent adatoms, we have two different regimes: i) the atomic wire limit (dense chain with large hopping between the d-levels of neighboring atoms; the resulting d-bands cross the Fermi energy of the SC substrate) and ii) the Shiba chain limit (dilute chain with small hopping; the d-band is electronically inert, but the Shiba states hybridize into a Shiba band).

Atomic wires. In this regime, the mechanism responsible for the emergence of topological superconductivity is very similar to the generic mechanism summarized in Section 8.1.2. Basically, hopping between neighboring magnetic adatoms leads to the formation of spin-split d-bands that cross the Fermi level of the SC at different hopping amplitudes. In a certain parameter range (i.e., when the distance between neighboring adatoms is within a certain range), only one band crosses the Fermi energy and the atomic chain becomes a 1D system of effectively spinless fermions. The superconducting correlations are provided by the SC substrate.

We note that for the atomic wire the Zeeman field is "built in," so it does not require an applied magnetic field. In addition, the Zeeman splitting is very large. This reduces the efficiency with which the spin-orbit coupling in the wire generates p-wave superconductivity by a factor approximately equal to the ratio between the spin-orbit coupling strength and the spin splitting. Nonetheless, one can still induce a significant p-wave gap if there is strong spin-orbit coupling in the SC substrate [110].

Another specific aspect of the atomic wire concerns the localization length ξ_M of the MZM. Naively, one expects it to be of the order of the superconducting coherence length, $\xi_M = \hbar v_F/\Delta$, where v_F is the Fermi velocity of the 1D electron systems and Δ the (induced) topological gap. Furthermore, v_F should be comparable to the Fermi velocity in typical metals (i.e., comparable to the Fermi velocity of the substrate) and $\Delta < \Delta_0$, which implies $\xi_M \gtrsim \xi_0$, where ξ_0 is the SC coherence length. This estimate is in sharp contrast with a recent experimental observation [378], which suggests that the localization length is of the order of the interatomic distance, i.e., orders of magnitude smaller that ξ_0.

A possible explanation of this observation involves the renormalization of the Fermi velocity in the 1D wire due to the superconducting proximity effect [403]. As discussed in Section 8.2.1, coupling to the SC induces a renormalization of the energy scale in the wire by a factor equal to the reduced quasiparticle weight $Z \approx \Delta_0/\gamma$, where γ is effective wire-superconductor coupling. For the atomic wire, $\gamma \sim 1\text{eV}$, while the SC gap is of the order of 10K. This generates a substantial renormalization of the Fermi velocity which, in turn, leads to a reduction of ξ_M that can be consistent with the experimental observation.

Shiba chains. If the broadening of the d-levels is small, they remain electronically inert and the low-energy physics is controlled by overlapping Shiba states forming a "Shiba band." To get some intuition, we consider a chain of deep Shiba states with energy E_0 and include only the hybridization between neighboring states, which is characterized by an amplitude t. We note that a more detailed model has to take into account the slow $1/r$ decay of the Shiba states at short distances, which implies that hybridization beyond nearest-neighbors may be relevant. This feature can be accounted for using a tight-binding model with long range hopping [407], but does not change qualitatively the conclusions based on the simplified analysis sketched below.

The key problem is understanding the emergence of induced pairing correlations in a chain of Shiba states. Recall that a Shiba state is spin-polarized along the direction of the impurity spin. Consequently, pairing should necessarily involve different sites. But how are different impurity spins oriented with respect to one another?

One possibility is that the spins, which interact via superconductor-mediated RKKY interactions, order ferromagnetically [87]. In this case, inducing pairing correlations requires spin-orbit coupling in the superconducting substrate, as in the atomic wire regime. A second scenario relies on the magnetic impurities forming a *spin helix* [78, 377, 407, 517]. In this case, two neighboring Shiba states are polarized along slightly different directions, making possible the proximity coupling to the singlet Cooper pairs of the host SC. An effective low-energy model that captures the hybridization of adjacent Shiba states (with amplitude t) and the emergence of pairing correlations (of strength Δ) has the form

$$H = E_0 \sum_j c_j^\dagger c_j - t \sum_j \left[c_{j+1}^\dagger c_j + c_j^\dagger c_{j+1} \right] + \Delta \sum_j \left[c_{j+1} c_j + c_j^\dagger c_{j+1}^\dagger \right]. \quad (8.19)$$

Remarkably, Eq. (8.19) is nothing but the Kitaev model (6.42) discussed in Section 6.2.3 (with $\mu \to -E_0$). Based on this observation and on the analysis of the Kitaev model in Section 6.2.3, the following basic picture emerges. For very dilute chains, the bandwidth of the Shiba band is $2t < |E_0|$, which corresponds to the *strong coupling* regime of the Kitaev chain; the system is topologically trivial. Reducing the distance between atoms enhances the hybridization of the Shiba states and, consequently, the bandwidth. When $2t = |E_0|$ the quasiparticle gap closes, then (i.e., for $2t > |E_0|$) it reopens; now,

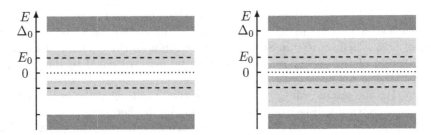

FIGURE 8.6 Excitation spectrum of a Shiba chain. *Left*: Dilute system. The atomic Shiba levels (of energy E_0) broaden into narrow bands that do not cross the chemical potential. *Right*: Dense chain. The broad Shiba bands overlap at the center of the gap, but pairing correlations reopen the SC gap. The corresponding topological SC phase hosts MZMs at the ends of the chain.

the system is in a topological p-wave superconducting state. This dependence on the distance between adatoms is illustrated schematically in Figure 8.6.

We note that the topological superconducting states realized in the two regimes discussed above (i.e., the atomic wire and Shiba chain limits) are adiabatically connected, i.e., they represent the same topological phase. This phase is expected to host MZMs at the ends of the atomic chain. There are indications [378, 403] that the localization length of the MZMs might be very short (of the order of the distance between adatoms). Direct access to the MZMs (by STS) and the possibility of manipulating the adatoms and engineering more complex structures are probably the most attractive features of this proposal.

8.3 EXPERIMENTAL DETECTION OF MAJORANA ZERO MODES

Being able to probe the Majorana zero modes is not only a requirement for their experimental investigation, but a key step toward their controlled manipulation, which is necessary for quantum computation. The properties that need to be demonstrated experimentally can, roughly, be divided into three categories: (i) MZMs are *zero-energy* bound states localized at topological defects in a topological SC, (ii) a pair of MZMs corresponds to a highly *non-local* Dirac fermion, and (iii) MZMs have *non-Abelian statistics*. Property (iii) would represent the ultimate validation of the predicted MZMs. However, probing these different properties involves measurements of increasing difficulty and complexity, so demonstrating the existence of a zero-energy bound state, although not a definitive demonstration of MZM, represents a natural first step. At the time of this writing (spring 2016; revised fall 2023) there is experimental evidence for (i), although certain aspects are rather controversial. Improving the experimental conditions (e.g., reducing disorder, enhancing the nanostructure design, etc.) appears to be a necessary condition for being able to corroborate the existing evidence with measurements of nonlocal properties

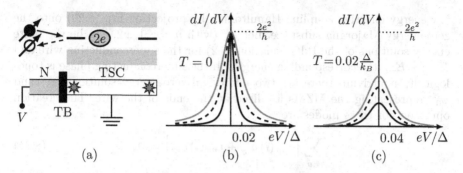

FIGURE 8.7 (a) Normal lead (N) coupled to a topological superconductor (TSC) with MZMs at its ends. When the bias voltage across the tunnel barrier (TB) is $V \approx 0$, an incoming Ψ_1 electron with specific spin orientation is reflected as a hole with the same spin, while $2e$ charge is transfered to the TSC. Electrons with opposite spin (Ψ_2) are totally reflected as electrons. (b) Differential conductance at zero temperature for effective coupling constant $\tilde{g}$ (black line), $2\tilde{g}$ (dashed) and $3\tilde{g}$ (gray). (c) Finite temperature dI/dV.

and, eventually, non-Abelian statistics. Below we review some basic types of measurement schemes that can be used to detect the presence of MZMs.

8.3.1 Tunneling spectroscopy

A Majorana zero-mode is a neutral object that does not couple to an external electric field. Also, the MZM does not carry spin and, consequently, does not couple to an external magnetic field. However, because of the charge non-conservation associated with the presence of the superconducting condensate, electrons tunneling into a topological superconductor *do couple* to a MZM. Thus, charge and spin tunneling provide powerful and conceptually simple tools for detecting Majorana zero-modes.

To understand the basic idea, let us consider the measuring scheme shown schematically in Figure 8.7. The system consists of a normal lead (N) coupled to a topological superconductor (TSC) with Majorana zero-energy end modes (e.g., a semiconductor wire proximity-coupled to a bulk s-wave SC). In practice, the tunnel barrier (TB) between the normal lead and the SC can be controlled using a gate potential. We model the system using an idealized tight-binding model on a 1D lattice, with sites $i \leq 0$ representing the normal metal and $i \geq 1$ the SC wire. Within this model, the N-SC coupling has the form

$$H_c = \sum_{\sigma} (\tilde{t}\, c_{0\sigma}^\dagger a_{1\sigma} + \tilde{t}^* a_{1\sigma}^\dagger c_{0\sigma}), \qquad (8.20)$$

where $\tilde{t}$ is the effective coupling parameter, $c_{i\sigma}$ (with $i \leq 0$) is the electron annihilation operator for the normal metal, and $a_{i\sigma}$ (with $i \geq 1$) the annihilation operator for the SC. To get further insight, it is convenient to define a low-energy effective coupling Hamiltonian by projecting Eq. (8.20) onto the zero-energy Majorana subspace. Let ψ_n (with $n = \pm 1, \pm 2, \dots$) be the finite energy solutions of the BdG equation (8.2) for the superconducting wire and $E_n = -E_{-n}$ the corresponding energies. If the superconducting phase is topologically nontrivial, there are two additional zero-energy solutions, ψ_0^L and ψ_0^R, representing the MZMs localized at the ends of the wire. The creation operators for these modes are

$$\hat{\psi}_n^\dagger = \sum_{i,\sigma} \left[u_{n\sigma}(i)\, a_{i\sigma}^\dagger + v_{n\sigma}(i)\, a_{i\sigma} \right] = \hat{\psi}_{-n},$$
$$\tag{8.21}$$

$$\gamma_{L(R)}^\dagger = \sum_{i,\sigma} \left[u_{0\sigma}^{L(R)}(i)\, a_{i\sigma}^\dagger + v_{n\sigma}^{L(R)}(i)\, a_{i\sigma} \right] = \gamma_{L(R)},$$

where we have used the property $v_{-n\sigma} = u_{n\sigma}^*$, which is a consequence of particle-hole symmetry and, to simplify notation, we omit the hats on the Majorana operators γ_L and γ_R. Note that for the Majorana modes the particle and hole components have equal amplitude, $|u_{0\sigma}^\alpha| = |v_{0\sigma}^\alpha|$, with $\alpha = L, R$. Choosing the overall phase so that $(v_{0\sigma}^\alpha)^* = u_{0\sigma}^\alpha$, the zero-energy solutions of a BdG Hamiltonian given by Eq. (8.18) can be written in the form

$$\psi_0^L = \left(u_{0\uparrow}^L e^{i\frac{\phi}{2}}, u_{0\downarrow}^L e^{i\frac{\phi}{2}}, u_{0\uparrow}^L e^{-i\frac{\phi}{2}}, u_{0\downarrow}^L e^{-i\frac{\phi}{2}} \right)^T,$$
$$\tag{8.22}$$

$$\psi_0^R = \left(iu_{0\uparrow}^R e^{i\frac{\phi}{2}}, iu_{0\downarrow}^R e^{i\frac{\phi}{2}}, -iu_{0\uparrow}^R e^{-i\frac{\phi}{2}}, -iu_{0\downarrow}^R e^{-i\frac{\phi}{2}} \right)^T,$$

where ϕ is the phase angle of the SC order parameter and $u_{0\sigma}^\alpha(i)$ are now real functions of the position along the wire. Note that using this notation, we need to change $u_{0\sigma}^R \rightarrow iu_{0\sigma}^R e^{i\frac{\phi}{2}}$, etc., in Eq. (8.21). Taking $\phi = 0$ and considering the quasiparticle operators from Eq. (8.21), we can express the original fermion operator in the form

$$a_{i\sigma} = u_{0\sigma}^L(i)\, \gamma_L + iu_{0\sigma}^R(i)\, \gamma_R + \sum_{n \geq 1} \left[u_{n\sigma}(i)\, \hat{\psi}_n + v_{n\sigma}^*(i)\, \hat{\psi}_n^\dagger \right].$$
$$\tag{8.23}$$

When projecting onto the low-energy subspace only the first two terms survive. In addition, $u_{0\sigma}^R(1) = 0$ (since that MZM is localized near the other end of the wire). Consequently, the effective coupling Hamiltonian becomes

$$H_c = \tilde{t}\gamma \left(u_\uparrow^* c_{0\uparrow}^\dagger + u_\downarrow^* c_{0\downarrow}^\dagger - u_\uparrow c_{0\uparrow} - u_\downarrow c_{0\downarrow} \right) = \tilde{g}\gamma \left[\Psi_1^\dagger(0) - \Psi_1(0) \right], \tag{8.24}$$

where we used the simplified notations $\gamma \equiv \gamma_L$ and $u_\sigma = u_{0\sigma}^L(1)$ and introduced the effective coupling constant $\tilde{g} = \tilde{t}(|u_\uparrow|^2 + |u_\downarrow|^2)$. The new fermion operators for the lead, $\Psi_{1(2)}(i)$, are defined through the unitary transformation $\Psi_1(i) = (u_\uparrow c_{i\uparrow} + u_\downarrow c_{i\downarrow})/(|u_\uparrow|^2 + |u_\downarrow|^2)$ and $\Psi_2(i) = (-u_\downarrow c_{i\uparrow} + u_\uparrow c_{i\downarrow})/(|u_\uparrow|^2 + |u_\downarrow|^2)$. Note that the MZM couples only to the Ψ_1 electrons, while the Ψ_2 electrons are completely decoupled from the SC wire in the low-energy limit. Regarding the two species of fermions as electrons with opposite spin orientation along the direction determined by the spinor $(u_\uparrow, u_\downarrow)^T/(|u_\uparrow|^2 + |u_\downarrow|^2)$, i.e., a direction dictated by the properties of the topological SC, we conclude that the coupling to the MZM is spin selective.

Based on this simple analysis, we can understand the tunneling of charge into a topological superconductor as a MZM-induced selective equal spin *Andreev reflection* [242]. As shown in Figure 8.7, an incoming Ψ_1 electron (with energy close to zero) is reflected as a hole with the *same* spin orientation, while $2e$ charge is transfered to the superconducting condensate. An incoming Ψ_2 electron, on the other hand, is reflected as a Ψ_2 electron and does not contribute to charge transport. We emphasize that this mechanism is different from ordinary Andreev processes [62], in which the reflected holes have opposite spin as compared to the incoming electrons and there is no spin selection.

The electrical conductance of a 1D wire with N-SC coupling given by Eq. (8.24) takes the form

$$G(V,T) = \frac{2e^2}{h} f(T, V; \tilde{g}, \Gamma, \mu), \qquad (8.25)$$

where T is temperature, V the bias voltage between the normal lead and the SC, and f a function that depends implicitly on the coupling strength $\tilde{g}$ and the properties of the SC phase (presence/absence of MZMs, amplitudes $u_{0\sigma}^L$, etc.), which are controlled by the Zeeman field Γ and the chemical potential μ.

Consider first the zero temperature case, $T = 0$. In the topologically trivial regime (e.g., $\Gamma = 0$), ordinary (double degenerate) Andreev processes are allowed, but they are strongly suppressed in the high barrier, low-coupling limit. The resulting low-bias conductance is nonuniversal, $0 \leq G(0,0) \leq 4e^2/h$. Next, we assume that the tunnel barrier height is set to a weak tunneling regime corresponding to $G(0,0) \approx 0$ for $\Gamma = 0$ and apply an external Zeeman field. The conductance remains practically zero as long as the system is in the trivial SC regime ($\Gamma < \Gamma_c$). By contrast, when $\Gamma > \Gamma_c$, the emergence of MZMs induces selective equal spin Andreev processes, with perfect (spin-selected) Andreev reflection (i.e., $f = 1$) in the limit $T \to 0$ and $V = 0$. Consequently, the presence of a MZM at the end of the wire is signaled by a zero-bias conductance peak with a quantized maximum height $G(0,0) = 2e^2/h$.

The emergence at $T = 0$ of a robust quantized zero-bias conductance peak (ZBCP) for $\Gamma > \Gamma_c$ represents a unique signature of the Majorana zero mode. While the weight of the ZBCP depends on the N-SC coupling strength (i.e., on the tunnel barrier potential), the height of the peak remains quantized, as illustrated in Figure 8.7. Furthermore, the ZBCP height remains quantized upon further increasing the Zeeman field or changing the chemical potential, as long as the system is in a topologically nontrivial SC phase. However, at finite temperature, the height of the ZBCP is no longer quantized [see Figure 8.7(c)], making it difficult to unambiguously identify the ZBCP as a signature of Majorana zero-modes, since other (nontopological) mechanisms such as disorder [340] and smooth confinement [284] can lead to similar (non-quantized) features.

Tunneling spectroscopy is an appealing method of detecting the presence of MZMs because, basically, it is relatively straightforward to implement. The observation of perfect Andreev reflection, which corresponds to a quantized ZBCP, and its robustness against variations of the control parameters (e.g., tunnel barrier potential, Γ, and μ), would constitute solid evidence for the presence of MZMs. The main problem with this type of measurement is that realizing the experimental conditions necessary for the observation of quantized conductance (e.g., low-enough temperature) is nontrivial, while non-quantized ZBCPs can be generated by alternative mechanisms. Nonetheless, this method can provide additional information if implemented using setups other than the simple one described above. For example, tunneling into a MZM from a spin-polarized lead is predicted to be strongly dependent on the spin polarization of the lead [242]. By contrast, tunneling into an ordinary low-energy fermionic state is not spin-selective (in general) and is expected to result in a weak dependence on the spin polarization. Perhaps even more interesting are the proposals for probing *nonlocal* features associated with MZMs. For example, observing the splitting oscillations generated by the overlap of two MZMs localized at the ends of a short wire [134] provides basic information about the spatial properties of these modes (e.g., their characteristic length scale). More "exotic" transport properties emerge when both MZMs are explicitly involved in the tunneling process (see Section 8.3.3).

8.3.2   Fractional Josephson effect

In the previous section, we have discussed signatures of MZMs that occur in out-of-equilibrium charge transport. Another useful probe of MZMs involves the equilibrium electrical current that flows between two superconductors in the absence of an applied voltage, more specifically the *supercurrent* generated by Cooper pairs tunneling across the weak link between the SCs, i.e., the *Josephson current* [506].

Consider first two normal SCs separated by a thin insulator (or a weak link) and characterized by a superconducting phase difference ϕ. Specifically, we assume that the pairing potentials for the left and right SCs are $\Delta_L = \Delta$ and $\Delta_R = e^{i\phi}\Delta$, respectively. The matrix element for single-electron tunneling across the junction is $\tilde{t}$. Note that the BdG Hamiltonian describing the junction depends on the phase difference (through the pairing term containing $e^{i\phi}\Delta$) and is 2π periodic in ϕ. The corresponding energy of the junction, which has the form $E(\phi) = -J\cos\phi$, is also periodic in ϕ with period 2π. The parameter J, called the Josephson coupling, is proportional to the square of the tunneling amplitude, $J \propto \tilde{t}^2$. The supercurrent I_J can be determined as the derivative

$$I_J = \frac{2e}{\hbar}\frac{dE(\phi)}{d\phi}, \tag{8.26}$$

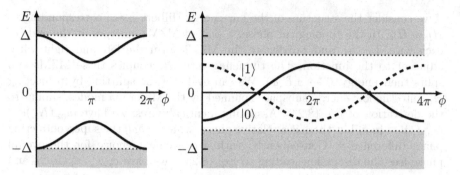

FIGURE 8.8 Energy as function of the phase difference across a Josephson junction. The shaded areas correspond to higher energy excited states. *Left*: Topologically trivial SCs ($\Gamma = 0$). *Right*: Topological SCs ($\Gamma > \Gamma_c$). The crossings at $\phi = \pi, 3\pi$ are protected by quasiparticle parity conservation. Transitions between the states $|0\rangle$ and $|1\rangle$ are possible if a fermion is transfered to/from stray quasiparticles or the "outer pair" of MZMs (see main text).

and, consequently, is periodic in ϕ with period 2π. The current is generated by Cooper pairs tunneling through the insulator and carrying charge $2e$.

Next, consider a junction between two topological superconductors. In this case, MZMs are generated on both sides of the Josephson junction. The key difference from the standard Josephson effect is that the presence of the MZMs enable the tunneling of *single electrons* carrying charge e. In turn, these processes give rise to the so-called *fractional* Josephson effect characterized by a 4π periodicity in ϕ. We emphasize that the doubling of the period is not a consequence of changing the Hamiltonian, but of modifying the physical state. For simplicity, let us focus on the 1D case. The full Hamiltonian is $H = H_L + H_R + H_c$, with the two superconductors being modeled by 1D tight-binding Hamiltonians given by Eq. (8.18) with $c_i^\dagger \rightarrow c_{ai}^\dagger = (c_{ai\uparrow}^\dagger, c_{ai\downarrow}^\dagger)$ representing the creation operator (in spinor form) for electrons in the left ($a = L$) and right ($a = R$) SCs and $\Delta \rightarrow \Delta_a$ representing the corresponding pair potentials. The single-electron tunneling across the barrier is described by the coupling term

$$H_c = -\tilde{t}\left(c_{LN_L}^\dagger c_{R1} + c_{R1}^\dagger c_{LN_L}\right), \tag{8.27}$$

where N_L (N_R) is the number of lattice sites of the left (right) SC wire and we assume $\tilde{t} > 0$. Since ϕ enters only through $\Delta_R = e^{i\phi}\Delta$, the total Hamiltonian is manifestly 2π periodic in the phase difference.

The low-energy modes can be obtained by simply diagonalizing the total Hamiltonian. In the topologically trivial regime (e.g., $\Gamma = 0$), $E(\phi)$ has period 2π, as shown in Figure 8.8. By contrast, in the topological regime ($\Gamma > \Gamma_c$) the period doubles (see Figure 8.8). We can better understand this feature if we consider the structure of the low-energy Hilbert space corresponding to $H_L + H_R$. In the topological phase, a pair of MZMs emerges at the ends of each wire. Let γ_L and γ_R denote the MZMs form the left and right wires adjacent to the junction. The tunneling term H_c couples these MZMs and splits their energy, $0 \to \pm E(\phi)$. We can evaluate the splitting by projecting H_c onto the low-energy subspace spanned by the Majorana modes, similar to the derivation of Eq. (8.24). Assuming identical wires, we have $|u_{0\sigma}^L(N_L)| = |u_{0\sigma}^R(1)| \equiv |u_\sigma|$, but there is a relative phase, which is *half* the superconducting phase difference ϕ. Consequently, with a convenient choice for the overall phase [see the discussion leading to Eq. (8.23)], we have $c_{L N_L \sigma} \to i u_\sigma \gamma_L$ and $c_{R 1 \sigma} \to u_\sigma e^{i\phi/2} \gamma_R$ (with u_σ real) and the *effective* low-energy Hamiltonian becomes

$$H_{\text{eff}} = i \gamma_L \gamma_R \Lambda \cos\left(\frac{\phi}{2}\right) = 2\Lambda\left(\hat{n} - 1/2\right) \cos\left(\frac{\phi}{2}\right), \qquad (8.28)$$

where $\Lambda = \tilde{t}(u_\uparrow^2 + u_\downarrow^2)$ and $\hat{n} = f^\dagger f$ is the occupation number of the fermionic state corresponding to the pair of MZMs adjacent to the junction, $f = (\gamma_L + i\,\gamma_R)/2$. Note that $[H_{\text{eff}}, \hat{n}] = 0$, which means that the fermion occupation number is conserved (as long as the low-energy subspace is protected by a finite gap) and can be used to label the states of the junction, $|n\rangle$.

Assuming that the system is in a state with fixed parity $i\gamma_L\gamma_R = \pm 1$ (i.e., $n = 0$ or $n = 1$), the single-electron component of the current across the junction is

$$I_J = \frac{2e}{\hbar} \frac{dE_{\text{eff}}}{d\phi} = \frac{4e\Lambda}{\hbar}(1/2 - n) \sin\left(\frac{\phi}{2}\right). \qquad (8.29)$$

The current in Eq. (8.29) exhibits 4π periodicity in ϕ as a result of Majorana-mediated single-electron tunneling (instead of the conventional Cooper-pair tunneling). Observing this phenomenon, known as the *fractional Josephson effect*, would provide strong evidence for the presence of MZMs. However, such an observation is not trivial. First, in addition to the 4π component (8.29), the Josephson current has a "regular" 2π-periodic component that, potentially, can be much larger. Hence, the current measurement has to be precise-enough to be able to disentangle the two components. Second, 4π periodicity in Eq. (8.29) relies on the conservation of the occupation number n (i.e., on the parity $i\gamma_L\gamma_R$ remaining fixed during the measurement). Fermions localized in low-energy states or thermally excited quasiparticles can modify this parity (thus reducing the measured periodicity to 2π). The (exponentially small) overlap of the MZMs localized at the junction with the MZMs at the "free" ends of the wires, γ_L' and γ_R', can also change this parity, even in the absence of stray quasiparticles. The total parity $\gamma_L \gamma_R \gamma_L' \gamma_R'$ is conserved, but a fermion can be transfered from the inner pair of MZMs to the outer pair. To circumvent this problem, the observation has to involve an AC measurement with a frequency high-enough compared to the splitting energy but low relative to the (bulk) quasiparticle gap. This is expected to lead to the observation of *Shapiro steps*[2] with missing odd conventional values.

8.3.3 Nonlocal transport

Another class of measurements that can be used to probe Majorana zero modes addresses their nonlocal nature, in essence the fact that two spatially separated MZMs correspond to one regular fermion. For example, coupling both ends of a grounded topological SC nanowire to quasi 1D metallic leads generates Majorana-mediated *crossed Andreev reflection* (also known as *Cooper pair splitting*). In essence, an electron injected at one end of the wire is accompanied by a hole emitted at the other end, the net result being the (nonlocal) injection of a Cooper pair into the topological SC wire [390]. Such processes generate a cross-correlation of the currents in the two leads, which could be detected in a current noise measurement [320, 390].

Nonlocal tunneling processes involving both Majoranas occur when the two modes are coupled [65, 184, 501]. In a two-lead configuration, one electron is injected into, say, the left end of the wire flipping the fermion parity. Subsequently, the electron escapes into the right lead and the fermion parity returns to its initial value. This process can be viewed as *Majorana-assisted electron transfer* (also referred to as electron "teleportation") and can be detected in a transport measurement. The coupling between the two Majoranas may be due to the overlap of the wave functions in a (relatively) short wire, or the finite charging energy of the (isolated) superconductor. In the second case the length of the wire is irrelevant, but the s-wave SC that induces the superconducting correlations in the wire has to be a very thin superconducting island. The coupling of the two Majoranas is characterized by an energy splitting δE_M (which is given by the overlap splitting ϵ or the charging energy E_C), while the coupling to the leads induces level broadening. Nonlocal tunneling requires the level splitting to be large compared to the level broadening.

Transport measurements, while relatively straightforward, raise some questions concerning possible ambiguities in the interpretation of the experimental results. For example, nonquantized zero-bias peaks (ZBPs) could also be generated by localized low-energy states in a topologically-trivial system, as mentioned above. Similarly, one can imagine scenarios in which nonlocal properties expected to occur in a system hosting MZMs are "simulated" by a trivial wire with (regular) bound states localized near its ends. However, we emphasize that corroborating several of these potentially "ambiguous" observations (e.g., the emergence of ZBPs above a critical Zeeman field, correlated splitting oscillations in short wires, Majorana-assisted electron transfer, etc.) and

---

[2]A Josephson junction subjected to an electromagnetic wave of frequency ω develops a DC voltage across the junction characterized by steps $V_n = n\frac{h}{2e}\omega$, $n = 0, 1, 2, \ldots$. In the presence of MZMs, the height of the steps is expected to double, $V_n = n\frac{h}{e}\omega$, as a result of single-electron tunneling. This corresponds to the "disappearance" of the odd conventional Shapiro steps.

confirming their expected dependence on the relevant parameters (e.g., external magnetic field, gate potentials, etc.) would practically *demonstrate* the realization of Majorana zero modes in solid state systems. Nonetheless, this is not the end of the journey. The most remarkable theoretical prediction about MZMs is that they have non-Abelian statistical properties; this requires direct experimental confirmation. In addition, if we want to make progress toward quantum computation with MZMs, we have to be able to control and take advantage of these properties. This would involve being able to braid and fuse MZMs in a controlled manner – tasks that pose significantly greater technical challenges than the experiments realized so far. Implementing, for example, interferometric schemes [11, 132, 220, 447] to detect the Majorana modes could be an important step in the right direction. In Chapter 12, we discuss the nature and origin of the non-Abelian properties of MZMs from the more general perspective of the physics of non-Abelian anyons and sketch how these properties could be used in the implementation of topological quantum computation.

8.4   EFFECTS OF DISORDER IN HYBRID MAJORANA NANOWIRES

More than a decade after the first reports of experimental observations consistent with the presence of MZMs, in the context of a significant growth of available data showing similar features, questions regarding the "true nature" of the low-energy modes responsible for the observed phenomenology still loom large. Essentially, the observed features do not really possess the expected robustness associated with topological states and emerge within rather small regions in parameter space, often requiring some sort of "fine tuning." This experimental situation was described as being affected by "confirmation bias" [133] and a "reproducibility crisis" [183]. The root cause of this situation is twofold. On the one hand, the theoretical prediction of MZMs preceded their experimental observation. In the context of topological quantum matter, this scenario is rather generic, with the notable exception of quantum Hall states. Theoretical predictions are typically based on idealized models and do not (initially) incorporate "real life details," such as the presence of inhomogeneity and disorder. On the other hand, disorder is ubiquitously present in solid state devices and, in the case of hybrid Majorana nanowires, it can generate effects that substantially complicates their experimental study: (i) Trivial low-energy states that mimic the (local) phenomenology of MZMs; (ii) A nontrivial interplay with finite size effects; (iii) The "migration" of the topological phase in parameter space and the eventual destruction of topological superconductivity. In this section, we briefly describe the main consequences of the presence of disorder in a topological superconductor, focusing on Majorana semiconductor-superconductor hybrid structures. We note that many of these disorder effects have close correspondents in other topological systems.

Consider first a 1D topological superconductor in the thermodynamic limit (i.e., an infinitely long system). It was shown that the presence of weak disorder in a SM-SC nanowire hosting a topological SC phase does not destroy the phase, even if interaction effects are included [341]. However, for a p-wave superconductor (i.e., the effective topological phase supported by the SM-SC structure), it was established that strong-enough disorder destroys the topological phase and causes a quantum phase transition to a trivial localized phase (i.e., an Anderson phase) [373]. The onset of this transition is associated with the localization length (which is reduced by disorder) becoming shorter than the SC coherence length (which increases with decreasing the topological gap that characterizes the clean system). Thus, roughly speaking, the topological SC phase is associated with a large gap-to-disorder ratio, while the trivial localized phase corresponds to small values of this ratio. In addition, the presence of disorder induces low-energy (sub-gap) localized states, creating an effective Griffiths phase [373].

This scenario becomes slightly more complicated in finite systems, particularly in the context of interpreting experimental signatures associated with low-energy states. In particular, in "short" wires (of size comparable to or shorter than the coherence length) the topological quantum phase transition (TQPT) between the trivial and topological SC phases becomes ill defined even in the absence of disorder, with the bulk gap never closing and pairs of overlapping Majorana modes exhibiting energy splitting oscillations (see Figure 8.5). Disorder induces low-energy states that may hybridize with the overlapping Majorana modes, which results in an interplay of finite size and disorder effects that cannot be captured by a simple picture based on the long-wire physics. The key elements characterizing the low-energy physics of a finite Majorana wire in the presence of disorder are briefly summarized below. Some of these elements are illustrated in Figure 8.9, which shows the dependence of the energy spectrum of a finite disordered wire on the applied Zeeman field for two values of the chemical potential. The bulk spectrum of the clean wire (shaded areas) is shown for comparison.

- The most conspicuous effect of disorder is the emergence of low-energy states within the the quasiparticle gap of the clean system, both above ($\Gamma > \Gamma_c^0$) and below ($\Gamma < \Gamma_c^0$) the critical field Γ_c^0 associated with the topological quantum phase transition of the clean system. The system still remains gapped at low Zeeman fields, below a certain (μ-dependent) value $\Gamma_{min}(\mu) \geq \Gamma_{min}(0)$, where $\Gamma_{min}(0)$ is the minimum Zeeman field associated with the TQPT of the clean system. Within a minimal effective model, we have $\Gamma_{min}(0) = |\Delta|$; if we explicitly incorporate the proximity effect, $\Gamma_{min}(0) = \gamma$. Hence, the energy scale associated with the effective SM-SC coupling, γ, (or with the induced gap, Δ, in the weak coupling limit) corresponds to the minimum Zeeman field associated with the emergence of zero-energy states, which can be easily determined experimentally. Note that in the presence of strong disorder we have $\Gamma_{min}(\mu) \approx \Gamma_{min}(0)$ over a wide range of chemical potential values, i.e., the zero energy states emerge above a magnetic field value that depends weakly on μ. For Zeeman field values above $\Gamma_{min}(\mu)$, the system becomes practically gapless due to the proliferation of disorder-induced sub-gap states.

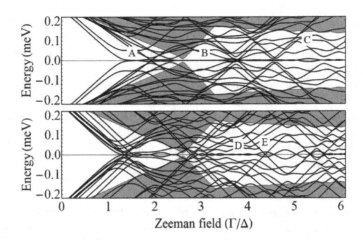

FIGURE 8.9 Low-energy spectrum as function of the applied Zeemann field for a disordered wire of length $L = 4$ μm, induced pair potential $\Delta = 0.25$ meV, and chemical potential $\mu = +2.4\Delta$ (top panel) and $\mu = -2.4\Delta$ (bottom). The shaded areas correspond to the bulk spectrum of the clean system; notice the (near) closing of the spectral gap at $\Gamma_c^0 = 2.6\Delta$, the critical field associated with the (clean) TQPT. *Top*: The lowest-energy mode in region A ($\Gamma \approx 1.6\Delta$) is an Andreev bound state localized near one end of the wire. Region B is characterized by the presence of a pair of MZMs; the Majorana modes can be traced from $\Gamma < \Gamma_c^0$ up to the largest Zeeman field value. The finite energy modes in region C are disorder-induced low-energy states with characteristic length scales comparable to L; they can hybridize with both Majorana modes "removing" them from zero energy. *Bottom*: The lowest energy mode (including region D) consist of trivial bound states pinned by disorder, typically involving a pair of partially separated Majorana bound states localized within a finite segment of the wire. The linearly dispersing modes (e.g., the mode in region E) are strongly localized ABSs pinned near the minima of the disorder potential.

- The (near) zero-energy modes that emerge immediately above $\Gamma_{\min}(\mu)$ are typically trivial Andreev bound states (ABSs) localized near the boundary of the system. For specific disorder realizations and control parameter values, these ABSs may consists of a pair of partially separated Majorana bound states (also called quasi-Majorana modes), which were first discussed in the context of smooth confinement [284]. The partially separated ABSs "stick" to zero energy over a finite range of Zeeman field values (see, e.g., region A in Figure 8.9) and mimic the (local) phenomenology of MZMs. We note that, in general, a (partially separated) ABS does not have a counterpart at the opposite end of the wire and does not generate end-to-end correlation in a charge tunneling experiment.

- One can qualitatively distinguish two types of disorder-induced sub-gap states. The first type consists of strongly localized ABSs bounded by the minima of the disorder potential. The characteristic length scale of these states (typically of the order $10^1 - 10^2$ nm) decreases with increasing the disorder strength. Their spectral signature is a (nearly) linear dispersion as function of the control parameters (e.g., Γ), as illustrated by the mode that crosses zero energy in region E in the lower panel of Figure 8.9. We note that, except in very short wires, these strongly localized ABSs can hybridize with at most one MZM and, consequently, do not affect the stability of the Majorana modes. The second type of disorder-induced low-energy states are characterized by large length scales (of the order $10^1 - 10^2$ μm) that can be comparable to the wire length. These modes remain gapped in the presence of weak disorder, but upon increasing the disorder strength the gap collapses. The finite energy modes in region C (top panel of Figure 8.9) are example of large length scale disorder-induced states. We emphasize that, although in a long-enough wire these states are localized (and, consequently, do not affect the stability of the Majorana modes), in a system with L of the order of one to several microns (the typical length of systems realized experimentally) they represent *effectively delocalized* states. These states contribute to the nonlocal transport through the wire and, when their energy vanishes, they mimic a TQPT, which is characterized by the presence of a truly delocalized zero-energy state. Furthermore, the large length scale disorder-induced states can hybridize with a pair of Majorana modes localized at the opposite ends of the system and destabilize them.

- The fate of the MZMs in the presence of disorder depends strongly on the control parameters (e.g., μ and Γ). For low values of the chemical potential (bottom panel in Figure 8.9), even moderate disorder can destroy the topological superconducting phase and cause a transition to a (trivial) localized phase. For example, the near-zero energy mode in region D corresponds to the presence of a trivial localized state consisting of a pair of (partially overlapping) Majorana bound states localized at the ends of a finite segment of the wire. Increasing the length of the wire will result in the emergence of multiple disconnected segments hosting such overlapping Majorana pairs. Increasing the disorder strength will result in a reduction of the characteristic length of these segments and will result in the enhancement of the energy splitting oscillations characterizing the lowest energy mode. There are no unpaired Majorana modes at the ends of the system, consistent with the topologically trivial nature of the underlying phase.

- For $\mu > 0$ (top panel in Figure 8.9) the topological phase is more stable against disorder. Moreover, for intermediate values of the disorder strength the topological phase extends beyond the phase boundary characterizing the clean system (i.e., inside the previously trivial region). For example, the Majorana mode from region B extends to the left of the (clean) critical point Γ_c^0. Nonetheless, the stability of MZMs is affected by disorder. In relatively short (experimental size) wires, effectively delocalized states (e.g., those from region C) can hybridize with both MZMs and gap them. We note that the maximum amplitude of the energy splitting oscillations of the system considered in Figure 8.9 would be less than 1.5 μeV in the absence of disorder (i.e., unobservable on the energy scale used in the figure); the energy splitting characterizing the lowest energy mode in Figure 8.9 is disorder-induced. Increasing the length of the system (to eliminate the possibility of having effectively delocalized states) will result in a stable zero energy mode. However, the corresponding MZMs may be slightly "pushed away" from the ends of the wire by the disorder potential. This scenario is also relevant in short systems. As a result, a charge tunneling experiment may show no evidence of ZBCP at any relevant finite temperature at one end of the wire (or both), generating a false negative signature of topological superconductivity. On the other hand, increasing the disorder strength eventually leads to the Majorana modes being bound by disorder within finite segments of the wire, as discussed for $\mu < 0$. In short wires, the size of such a segment can be comparable to L, leading to false positive signatures of topological superconductivity.

- In practice, the combined result of having false negatives (generated by disorder-induced deformations of the MZM wave function) and finite size effects is that systems with intermediate disorder and parameters consistent with a topological superconducting phase show clear signatures of MZMs only within small "islands" in the parameter space, with characteristic sizes smaller than Γ_{min} (i.e., smaller than γ or Δ). On the other hand, false positives characterizing a (relatively short) system with strong disorder and parameters consistent with a trivial localized phase also occur within small "islands" in the parameter space. The fundamental ambiguity of "Majorana" signatures emerging within small "islands" in the parameter space sits at the core of the controversy affecting the experimental investigation of semiconductor-superconductor hybrid wires. Possible paths for overcoming this challenge involve increasing the wire length, optimizing the systems parameters (e.g., the SM-SC coupling) and, most importantly, reducing the disorder strength. For example, enhancing the length of (potentially topological) systems with intermediate disorder should result in some increase of the characteristic size of "topological islands" (as a result of reducing the finite size effects). By contrast, enhancing the length of strongly disordered system (hosting a trivial localized phase) is expected to lead to the shrinking and eventual disappearance of these "islands." Ultimately, a significant reduction of the disorder strength is expected to lead to the emergence of large "topological islands," with characteristic sizes comparable to (or larger than) Γ_{min}. This would constitute strong evidence for the presence of MZMs and would create the conditions for meaningful fusion and braiding experiments.

CHAPTER 9

Topological Phases in Cold Atom Systems

ULTRACOLD ATOMIC GASES HAVE EMERGED AS AN IDEAL TESTGROUND for fundamental many-body physics. The "quantum revolution" triggered by the work of John Bell and the subsequent advances in the experimental control of few-body quantum systems has nowadays entered a new phase associated with the rise of quantum information science and the quest for control over macroscopic quantum systems. The exceptional level of control enabled by cold atom systems has placed them at the frontier of modern quantum physics as a promising route to realizing quantum simulators, engineering topological quantum matter, and emulating interacting gauge fields. By allowing the realization of tunable interactions and synthetic spin-orbit couplings, cold atom systems offer the possibility of engineering topological phases that are not supported by solid-state systems, including experimental realizations of ideal models. In this chapter, we briefly summarize some recent developments in this area and point out the basic physical ideas behind these advances. For a more detailed survey of the field the reader is referred to specialized books and review articles, for example Refs. [60, 210, 331, 332].

9.1 BRIEF HISTORICAL PERSPECTIVE

The experimental observation of Bose–Einstein condensation (BEC) in 1995 [20, 73, 136] marked a breakthrough moment in which the atomic, molecular, and optical (AMO) physics has reached the frontiers of condensed matter physics. This significant achievement was made possible by the earlier development of efficient methods to cool and trap atoms using laser light [108, 116, 406]. With the observation of BEC, the study of many-body systems took central stage in AMO physics. Nonetheless, the condensed matter community remained rather reserved, mainly because the BEC phenomenon can be well understood within an effective Ginzburg–Landau picture [203] in

DOI: 10.1201/9781003226048-9

which the coherent many-body state of the quantum system is described by a macroscopic wave function $\Psi(\boldsymbol{r}, t)$, a very familiar concept in superconductivity and superfluidity. A complete and quantitative description of the static and time-dependent properties of a BEC can be obtained by solving the so-called Gross–Pitaevskii equation [221, 408], a nonlinear Schrödinger-type equation for the macroscopic wave function,

$$i\hbar \frac{\partial \Psi(\boldsymbol{r}, t)}{\partial t} = \left(-\frac{\hbar^2}{2m} \nabla^2 + V(\boldsymbol{r}) + g|\Psi(\boldsymbol{r}, t)|^2 \right) \Psi(\boldsymbol{r}, t), \qquad (9.1)$$

where $V(r)$ is an external potential (e.g., the trapping potential) and g is a coupling constant describing the weak interaction between two particles. The elementary excitation of the BEC can be described by considering small fluctuations $\delta\Psi$ with respect to the equilibrium wave function, which leads to the well-known Bogoliubov theory of weakly interacting Bose gases.

The birth of quantum information was a significant driving force behind the unprecedented level of quantum engineering (i.e., preparation, manipulation, control, and detection of quantum systems) developed in AMO physics. For example, systems of cold trapped ions were investigated with the goal of realizing quantum gates, such as the Cirac–Zoller controlled-NOT gate [113], and building a scalable quantum computer [249, 450]. Perhaps more interestingly, cold atom systems can naturally be employed as *quantum simulators* [88], following the idea originally suggested by Feynman [166]. Trapped ions, for example, can be used to simulate interacting spin chains [451]. Since the computational power of classical computers is likely to be limited (at least in the foreseeable future) to simulations of about 50 qubits [422], using analog quantum simulators provides a solution for dealing with certain types of classically intractable problems and an appealing route toward a deeper understanding of (strongly) interacting quantum systems consisting of $\mathcal{O}(10^2)$ qubits (e.g., 100 spin-1/2 particles). We should think about an *analog* quantum simulator as a many-body quantum system that mimics a simple model (or family of models). In turn, the model is relevant to understanding a certain challenging problem in condensed matter (or in other areas, e.g., high-energy physics). The system has to satisfy certain requirements. More specifically, we assume that i) the initial state and its dynamics can be precisely controlled, ii) the relevant parameters can be manipulated, and iii) the readout of the important characteristics of the final state can be efficiently performed. However, compared to quantum computers, analog quantum simulators are predicted to have substantially lower constraints regarding the number of required qubits and the fidelities of quantum operations [88].

Ultracold atoms have become prime candidates for the realization of quantum simulators as a result of three major developments that have considerably enlarged the range of physics that is accessible to AMO systems: (i) the ability to tune the interaction strength using *Feshbach resonances* [122, 257], (ii) the possibility of generating strong periodic potentials through *optical lattices* [218], and (iii) the possibility of engineering laser-induced *gauge po-*

tentials [210], including non-Abelian fields that generate an effective *spin-orbit coupling* [193, 567] and *dynamical gauge fields* for simulating interacting gauge theories with cold atoms [540]. These advances opened wide possibilities for addressing interesting problems in condensed matter physics, quantum electrodynamics, or quantum chromodynamics using the cold atoms toolbox and have attracted the attention of different physics communities. In principle, studying cold atom systems could shed light on challenging problems, such as, for example, understanding the effects of strong correlations in many-body quantum systems using Hubbard-type [264] and spin models [147, 194], understanding disorder effects and the interplay between disorder and interactions [307, 455], realizing topologically nontrivial quantum phases [12, 53, 120, 211, 467, 488] and Majorana bound states [266, 500], and investigating various quantum chromodynamics phenomena (e.g., the confinement of dynamical quarks) [574]. The future will tell us how fruitful this path really is and whether or not using cold atom systems as quantum simulators can provide fundamentally new insight into some basic open questions in physics.

9.2 MANY-BODY PHYSICS WITH ULTRACOLD GASES: BASIC TOOLS

In this section, we briefly describe the basic tools for creating trapped cold atom systems. Here, we only sketch some key ideas; the interested reader can find the relevant details in specialized books and review articles (see, for example, [118] and [331] and references therein).

9.2.1 Cooling and trapping of neutral atoms

The molecules and atoms in a gas at room temperature have speeds on the order of the speed of sound. Cooling the gas down to, say, 4 K (the condensation temperature of helium) will reduce this thermal velocity to tens of meters per second. If the system is at equilibrium, cooling it further will produce the condensation of the gas. It may seem that studies of free atoms have to be done with fast moving particles. This, however, limits the observation time (and, consequently, the spectral resolution) and generates Doppler shifts and relativistic effects that cause the displacement and broadening of spectral lines – a major limitation for precision measurements and atomic clocks. Furthermore, observing any collective quantum phenomenon (e.g., BEC) requires low particle densities n (to prevent the transition to a solid phase) and thermal de Broglie wavelengths λ_{dB} of the order of the inter-atomic spacing, so that the waves of neighboring atoms overlap and interfere constructively. Realizing this *quantum degenerate* regime requires a *phase space density* $n\lambda_{dB}^3 > 2.6$. The path to realizing this condition involves slowing down an atomic beam, trapping the atoms, then cooling the system further.

Doppler cooling. When placed in a laser field, neutral atoms experience a velocity-dependent dissipative force generated by light scattering processes.

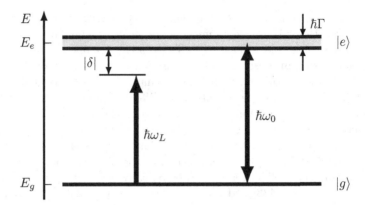

FIGURE 9.1 Schematic representation in terms of a two-level model of an atomic transition via red-detuned light.

In certain conditions, the scattered light carries away more energy than the amount of energy absorbed by the atoms. This leads to the so-called *Doppler cooling* scheme, which was proposed independently by Hänsch and Schawlow [241] and by Wineland and Dehmelt [550] in 1975, and to the concept of *optical molasses*.

Consider an atom that absorbs (or emits) a photon of momentum $\hbar k$. As a result of this process, the momentum p of the atoms shifts to $p \pm \hbar k$. The corresponding atomic recoil velocity is $\hbar k/m$, where m is the mass of the atom; in the case of cesium, for example, the recoil velocity is of the order of 3 mm/s. Since spontaneous emission occurs with the same probability in all directions, for an atom in a laser field the average change in atomic momentum (after many absorption-spontaneous emission cycles) is $\langle \delta p \rangle = \hbar k_L$, where k_L is the laser photon wavenumber. The resulting radiative force is $\hbar k_L$ times the rate of absorption-spontaneous emission cycles.

Next, consider an idealized two-level atom with ground state $|g\rangle$ and unstable excited state $|e\rangle$ of energy $\hbar \omega_0$ and lifetime Γ^{-1}, as shown in Figure 9.1. For an atom at rest, the mean scattering rate has a Lorentzian dependence on the frequency of the laser with a maximum at $\omega_L = \omega_0$ [118]. The key idea is to use the Doppler effect to make this scattering rate (hence, the corresponding radiative force) velocity dependent. For a given direction, the scheme involves two counter-propagating lasers having the same amplitude and the same, slightly *red detuned* frequency $\omega_L = \omega_0 + \delta$, with $\delta < 0$. As illustrated schematically in Figure 9.2, for an atom moving with velocity v, the counter-propagating laser beam (with wave vector $-k_L$ in the laboratory frame) is perceived with a frequency $\omega_L + v \cdot k_L$ closer to ω_0, while the co-propagating beam $(+k_L)$ is perceived with a frequency $\omega_L - v \cdot k_L$ that is further detuned from the atomic transition frequency. This results in more photons being absorbed from the counter-propagating wave than from the co-propagating wave and generates a friction-like force that, for slow-enough

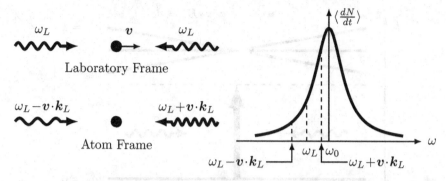

FIGURE 9.2 *Left*: Atom moving in the field created by two counter-propagating lasers with frequency ω_L and wave vectors $\boldsymbol{k}_L$ and $-\boldsymbol{k}_L$, respectively. In the reference frame of the atom the frequencies of the lasers are Doppler shifted in opposite directions. *Right*: The frequency dependence of the mean scattering rate $\langle dN/dt \rangle$. The imbalance between the radiation pressure forces of the two counter-propagating lasers results in a drag-like force.

atoms, has the form [118]

$$\boldsymbol{F} = -\alpha\boldsymbol{v}, \tag{9.2}$$

where α is a positive coefficient. Of course, to provide cooling along all directions, one has to use three (relatively orthogonal) pairs of counter-propagating lasers. Note that Eq. (9.2) describes the motion of a particle in a viscous medium, the solution corresponding to an exponential damping of the velocity toward $\boldsymbol{v} = 0$. The viscous medium created by the lasers is the so-called *optical molasses* [109].

Since Doppler cooling involves absorption and spontaneous emission processes, equilibrium corresponds to a Brownian motion of the atom (rather than a particle at rest). Consequently, there is a limit for the minimal temperature T_D that can be reached using this scheme. Explicit calculations using the two-level model give [214, 551]

$$k_B T_D = \frac{\hbar\Gamma}{2}. \tag{9.3}$$

For cesium atoms, for example, $T_D \approx 120\,\mu K$.

Magnetic and magneto-optical trapping. The methods used for trapping atoms can be roughly divided into three groups. The corresponding key elements that enable the trapping of neutral particles are (i) induced atomic dipole moments, (ii) nonhomogeneous magnetic fields, and (iii) a combination of radiation pressure and static fields. Below, we briefly summarize the main ideas behind the last two types of traps; dipole traps, which are based on induced atomic dipole moments, will be discussed later in the context of optical lattices. Note that, in general, wall-free confinement is necessary to prevent the atoms from sticking to the surface of the container.

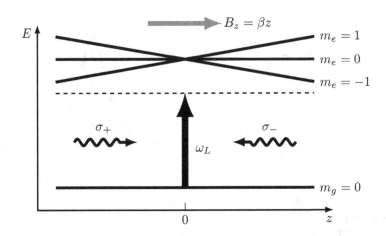

FIGURE 9.3 Magneto-optical trap. An atom with $j_g = 0$ and $j_e = 1$ is placed in a magnetic field $B_z = \beta z$ in the presence of two circularly polarized counter-propagating lasers. For $z > 0$ ($z < 0$) the σ_- (σ_+) beam, which drives the $\Delta m = -1$ ($\Delta m = +1$) transition, is Zeeman shifted toward resonance and pushes the atom back toward $z = 0$.

Once the atoms are cold enough (i.e., after Doppler cooling the system), they can be confined in a magnetic trap. The basic idea is to create a non-homogeneous magnetic field that has a local minimum and use neutral particles that have nonzero magnetic moment. The atoms with magnetic moments aligned opposite to the field (the so-called "low-field-seekers") will try to minimize their energy by occupying locations with lower field strengths. Consequently, the local minimum of the magnetic field will trap low-field-seeking atoms that do not have enough kinetic energy to escape.

Another trapping method is based on the idea of making the radiative force not only v-dependent but also position-dependent. Assume that the ground state has angular momentum $j_g = 0$, while the excited level has $j_e = 1$, and the corresponding states are $|e, m_e\rangle$, where $m_e = -1, 0, 1$ is the projection of the angular momentum along the z axis. The counter-propagating lasers are σ_+ and σ_- circularly polarized, respectively, and a position-dependent magnetic field $B_z = \beta z$ is applied along the quantization axis z, as shown in Figure 9.3. The σ_+ laser couples the ground states to the excited state $|e, 1\rangle$, while the σ_- laser couples it to $|e, -1\rangle$. In the presence of a magnetic field, the energies ϵ_{m_e} of the excited states are Zeeman shifted, so that ϵ_+ is closer to the resonance frequency for $z < 0$, while ϵ_- is closer to resonance when the atom is in the region $z > 0$. Consequently, the radiative force will acquire an additional, position-dependent term $\Delta F_z = -\kappa z$, which corresponds to an effective harmonic trapping potential.

Sub-Doppler cooling. Since 1988, systematic experimental investigations of atoms cooled in optical molasses have found temperatures well below the

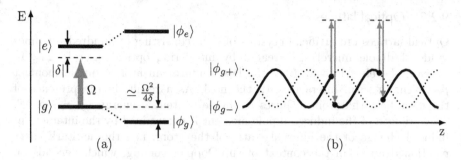

FIGURE 9.4 (a) ac-Stark shift induced by atom-light interaction. The energies of the "dressed states" $|\phi_g\rangle$ and $|\phi_e\rangle$ are shifted with respect to the energies of the corresponding bare states $|g\rangle$ and $|e\rangle$ by the light shifts (9.7). (b) Sisyphus cooling. An atom climbs the "hills" of the position dependent light shifts of a ground state sub-level (thus losing kinetic energy), but is optically pumped to the lower sub-level and has to climb again.

Doppler limit [236, 325, 470]. In addition, these low temperatures were observed at large laser detunings, $\delta \gg \Gamma$ (instead of the optimal value $\delta = \Gamma/2$ predicted for Doppler cooling). The mechanism responsible for these effects involves, in essence, the following components [117]: In the presence of a laser field, the (bare) atomic levels shift and mix; this is the so-called *ac-Stark effect* or *light shift* illustrated schematically in Figure 9.4(a). The light shifts depend on the laser polarization. Since two counter-propagating lasers can give rise to a (periodic) position-dependent polarization, the corresponding light shifts are equivalent to a periodic, state-selective potential. If, for example, the ground level has $j_g = 1/2$, the potential "hills" of $|g, -1/2\rangle$ will coincide spatially with the "valleys" of $|g, +1/2\rangle$ (and vice versa). Finally, a transition to the excited states is more likely when the atom is on top of a $|g, m\rangle$ hill; this is followed by a spontaneous emission that brings the atom into the $|g, -m\rangle$ valley, as shown in Figure 9.4(b). The net result is that the atoms keep climbing potential hills, but end up in potential valleys, thus losing kinetic energy. The mechanism is called *Sisyphus cooling* or *polarization gradient cooling* [117].

As a final note, we mention that the Sisyphus cooling limit is given by the height of the potential hills and is of the order of $10^2 E_R$, where $E_R = \hbar^2 k^2/2m$ is the recoil energy. In general, for multi-level atoms the fundamental laser cooling limit is characterized by the recoil energy. Schemes to overcome the recoil limit include velocity selective coherent population trapping [29], stimulated Raman cooling [278], and forced evaporative cooling [286]. To achieve Bose–Einstein condensation, for example, one can use laser cooling to "precool" the gas, which is then trapped in a magnetic trap; next, forced evaporative cooling is applied as a second cooling stage. This is realized by reducing the trap depth and allowing the most energetic atoms to escape.

9.2.2 Optical lattices

Optical lattices are artificial crystals of light consisting of hundreds of thousands of dipole microtraps created by interfering optical laser beams. In essence, the electric field E of the laser induces an atomic dipole moment p, which, in turn, interacts with the field. As a result, in the presence of the laser field the atom acquires a (dipole) potential energy $V_d = -p \cdot E$. The strength of the induced dipole moment is determined by the interaction-induced change of the internal states of the atom, i.e., the ac-Stark effect mentioned above in the context of sub-Doppler cooling, which depends on the laser frequency, light intensity, and polarization. Linear response gives $p = \alpha(\omega_L) E(r,t)$, where $\alpha(\omega_L)$ is the polarizability of the atom and ω_L the frequency of the laser. Since the center-of-mass motion of the atom is much slower than the inverse laser frequency, ω_L^{-1}, the effective potential will be given by the time-averaged laser intensity $I(r) \propto \langle E^2(r) \rangle_t = \frac{1}{2}|E(r)|^2$, and we have $V_d(r) = -\frac{1}{2}\alpha(\omega_L)|E(r)|^2$. Consequently, a spatially dependent intensity induces a potential energy that can be used to trap atoms. An optical lattice corresponds to a periodic potential energy landscape induced by the standing wave pattern generated by interfering laser beams.

To get better insight, let us consider an atom described by the two-level toy model introduced above in the presence of a classical electromagnetic field $E(r,t) = \mathcal{E}(r)e^{-i\omega_L t} + \mathcal{E}^*(r)e^{i\omega_L t}$. Since the atom does not have a permanent dipole moment, the dipole operator $\hat{d} = -e\sum_k r_k$, where r_k designates the position of the kth electron, is off-diagonal in the two-level basis:

$$\hat{d} = \mu|e\rangle\langle g| + \mu^*|g\rangle\langle e|, \tag{9.4}$$

where $\mu = \langle e|\hat{d}|g\rangle$. Consequently, the Hamiltonian $H = H_0 - \hat{d} \cdot E$ of the atom-laser field system can be written as

$$H = \hbar\omega_0|e\rangle\langle e| - (\mu|e\rangle\langle g| + \mu^*|g\rangle\langle e|) \cdot \left(\mathcal{E}(r)e^{-i\omega_L t} + \mathcal{E}^*(r)e^{i\omega_L t}\right), \tag{9.5}$$

In the limit of small detuning $|\delta| = |\omega_L - \omega_0| \ll \omega_0$, it is convenient to work in the rotating frame of the laser by performing the unitary transformation $U(t) = \exp\left(-i\omega_L t\, \hat{\sigma}_z/2\right)$, with $\hat{\sigma}_z = |e\rangle\langle e| - |g\rangle\langle g|$. Neglecting rapidly oscillating terms $\propto e^{\pm i(\omega_L + \omega_0)t}$, which generate effects that average out to zero over relevant timescales, we obtain the time-independent Hamiltonian

$$H' = -\frac{\hbar}{2}\delta\,\hat{\sigma}_z + \frac{\hbar}{2}\left[\Omega(r)|e\rangle\langle g| + \Omega^*(r)|g\rangle\langle e|\right], \tag{9.6}$$

where $\Omega(r) = -\frac{2}{\hbar}\mathcal{E}(r) \cdot \mu$ is the so-called *Rabi frequency*. When the detuning is large compared to the Rabi frequency, $|\delta| \gg |\Omega|$, the ac-Stark shifts given by second-order perturbation theory are

$$\Delta E_g = \hbar\frac{|\Omega(r)|^2}{4\delta}, \qquad \Delta E_e = -\hbar\frac{|\Omega(r)|^2}{4\delta}. \tag{9.7}$$

We conclude that the ac-Stark shifted energy of an atom (in the ground state) is, in general, position-dependent and corresponds to an optical potential $V_d(\mathbf{r}) = \hbar\frac{|\Omega(\mathbf{r})|^2}{4\delta}$. Note that, as expected, the potential is proportional to the laser intensity, $|\Omega(\mathbf{r})|^2 = (2/\hbar)^2|\langle e|\hat{d}_{\mathcal{E}}|g\rangle|^2|\mathcal{E}|^2 \propto |\mathbf{E}|^2$, where $\hat{d}_{\mathcal{E}}$ is the dipole operator in the direction of the field. Also, within this simplified model, the polarizability takes the form $\alpha(\omega_L) \approx -\frac{1}{\hbar}|\langle e|\hat{d}_{\mathcal{E}}|g\rangle|^2/(\omega_L - \omega_0)$. Since the atom experiences a force $\mathbf{F} = -\nabla V_d$, it will be attracted to the minima or to the maxima of the laser intensity for blue-detuned ($\omega_L > \omega_0$) or red-detuned ($\omega_L < \omega_0$) laser light, respectively. This property is often described in the literature by calling the atoms "weak field seekers" and "strong field seekers" for the cases of blue and red detuning, respectively.

The simplest periodic optical potential is obtained by overlapping two counter-propagating laser beams of equal frequency that create a standing wave interference pattern. The resulting trapping potential has the form

$$V_{latt}(x) = V_0 \sin^2(kx), \tag{9.8}$$

where $k = 2\pi/\lambda$ is the wave vector of the laser light and V_0 is the lattice depth (which is determined by the Rabi frequency and the detuning, as shown above). Equation (9.7) defines a one-dimensional (1D) lattice of parallel confining planes (or, more precisely, confining disks, if we take into account the profile of the laser) with a lattice constant $a = \lambda/2$. A two-dimensional lattice of confining potential tubes can be created using two orthogonal sets of counter propagating laser beams. Similarly, a 3D simple cubic array of confining centers is obtained by superimposing three orthogonal standing waves. Note that using laser beams with two different wavelengths gives rise to superlattice structures.

A large variety of optical lattices can be created by controlling the number and the geometry (i.e., relative angles) of the lasers, their frequencies, intensities and polarizations, as well as the type of atom to be loaded in the optical lattice. For example, one can create different effective optical potentials for different magnetic sublevels, thus obtaining *spin-dependent* optical lattices [265]. The interested reader can find more details in specialized books and review articles, for example [59, 224, 331]. The versatility of these periodic potential landscapes makes optical lattices a critical element in the effort to establish cold atom systems as effective quantum simulators for condensed matter systems. In this context, we note that, in contrast to real materials, optical lattices are free of disorder, structural defects, and lattice vibrations, which makes them perfect candidates for simulating idealized quantum models.

9.2.3   Feshbach resonances

The ability to control the particle-particle interaction strength is another key feature of cold atom systems. Feshbach resonances represent an essential tool that enables this type of control and the most direct way of reaching a strong-interaction regime. Below, we sketch the basic physics of the Feshbach resonance and make a few general remarks on the role of this phenomenon in cold atom physics. For a detailed discussion of Feshbach resonances in ultracold gases see, for example, the review article by Chin et al. [102].

Basic scattering theory. One of the best approaches to understanding the interaction between particles is to scatter them off each other. Consider first the collision of two *structureless* atoms of masses m_1 and m_2 that interact through a potential $V(\boldsymbol{R})$, where $\boldsymbol{R} = \boldsymbol{r}_2 - \boldsymbol{r}_1$ is the relative position vector. For simplicity, we consider the case of a potential with spherical symmetry, $V(\boldsymbol{R}) = V(R)$. The relative motion of the pair is described by the wave function $\psi_{\boldsymbol{k}}(\boldsymbol{R})$ and we assume that the system is prepared in a plane wave with relative momentum $\hbar\boldsymbol{k}$ and relative kinetic energy $E = \hbar^2 k^2/2\mu$, where $\mu = m_1 m_2/(m_1 + m_2)$ is the reduced mass. In the limit of large separation, the wave function is a superposition of the incident plane wave and a scattered spherical wave,

$$\psi_{\boldsymbol{k}}(\boldsymbol{R}) \simeq e^{i\boldsymbol{k}\cdot\boldsymbol{R}} + f_k(\theta)\frac{e^{ikR}}{R}, \tag{9.9}$$

where θ is the angle between the initial and final directions of relative motion and the function $f_k(\theta)$, called the *scattering amplitude*, provides information about the relative amplitude and phase of the scattered wave along a given direction. Below, to simplify the notation, we drop the k-dependence in $f_k(\theta)$. We note that the scattering amplitude is related to the *differential cross section*[1], $\frac{d\sigma}{d\Omega} = |f(\theta)|^2$. The scattering amplitude can be represented as a sum over partial waves,

$$f(\theta) = \sum_{\ell=0}^{\infty} (2\ell + 1) f_\ell(k) P_\ell(\cos\theta), \tag{9.10}$$

where P_ℓ are Legendre polynomials and $f_\ell(k)$ are *partial scattering amplitudes* that characterize the scattering in states with (relative) angular momentum $\ell = 0, 1, 2, \ldots$. We note that in the case of identical particles the wave function (9.9) has to be properly symmetrized (for bosons) or antisymmetrized (for fermions). In particular we have $e^{i\boldsymbol{k}\cdot\boldsymbol{R}} \rightarrow e^{i\boldsymbol{k}\cdot\boldsymbol{R}} \pm e^{-i\boldsymbol{k}\cdot\boldsymbol{R}}$ and $f(\theta) \rightarrow f(\theta) \pm f(\pi - \theta)$, consistent with the fact that one cannot say which one of the particles scatters in the direction θ and which one in the opposite direction (i.e., $\pi - \theta$). Consequently, for identical bosons or fermions only partial scattering amplitudes with even or odd values of ℓ are possible, respectively. Usually, at low temperatures (i.e., in the sub-millikelvin regime) the lowest angular momentum collisions dominate, more specifically s-wave collisions for ultracold bosons and p-wave collisions for fermions.

---

[1]The differential cross section represents the number of particles scattered into direction (θ, ϕ) per unit time per unit solid angle, divided by the incident flux.

It is convenient to express the partial scattering amplitudes in terms of the so-called *S-matrix* element $S_\ell = e^{2i\delta_\ell}$, which is given by the *scattering phase shift* δ_ℓ. Specifically, we have

$$f_\ell(k) = \frac{e^{i\delta_\ell}\sin\delta_\ell}{k} = \frac{1}{k\cot\delta_\ell - ik} = \frac{S_\ell - 1}{2ik}. \tag{9.11}$$

For *s*-wave scattering in the low-energy regime, which is relevant for ultracold bosonic systems, we have $k\cot\delta_0(k) \approx -\frac{1}{a}$, where a is the so-called *scattering length*. In other words, in this limit the two-body scattering problem is uniquely specified by a single parameter, the scattering length a, and we have

$$f_0(k) = -\frac{a}{1+ika}. \tag{9.12}$$

Note that the total scattering cross section is $\sigma \simeq 4\pi a^2$, i.e., the scattering length a characterizes the effective "size of target." Moreover, a can be either positive or negative, corresponding to an *effectively* repulsive or attractive interaction, respectively. One can show that Eq. (9.12) is the exact scattering amplitude for the *pseudopotential*

$$\widetilde{V}(\boldsymbol{R}) = \frac{2\pi\hbar^2 a}{\mu}\delta(\boldsymbol{R}). \tag{9.13}$$

Hence, the two-body interactions in ultracold gases can be described by the pseudopotential $\widetilde{V}(\boldsymbol{R})$ with a strength determined by the scattering length a, which is typically an experimentally determined parameter.

In general, the poles of the scattering amplitude in the upper complex k-plane are related to the *bound states* of the interaction potential $V(R)$. More specifically, the pole $k = i/a$, with $a > 0$, corresponds to the highest energy bound state with wave function $\propto \exp(-R/a)$ and energy $E_{10} = -\hbar^2/2\mu a^2$, where $0 > E_{1\ell} > \cdots > E_{N_\ell \ell}$ are the energies of the discrete set of N_ℓ bound states with relative angular momentum ℓ (see Figure 9.5). Note that the top bound state has to be within a certain energy range E_c below the continuum threshold, $|E_{10}| < E_c$; if such a bound state does not exist (e.g., if $E_{10} < -E_c$), the interaction will be *effectively* attractive with $a < 0$. Finally, if there is a zero-energy bound state, $E_{10} \lesssim 0$, the two-particle collision is characterized by a *divergent* scattering length, $a \to \infty$, i.e., an infinitely strong pseudopotential.

<u>Collision channels</u>. Real atoms have nontrivial internal structures, in particular nontrivial spin structures. For example, the total electronic angular momentum $\boldsymbol{J} = \boldsymbol{L} + \boldsymbol{S}$ may be coupled to the nuclear spin $\boldsymbol{I}$ to give the total angular momentum $\boldsymbol{F} = \boldsymbol{J} + \boldsymbol{I}$. In this case, the eigenstates of the atom are labeled (at zero magnetic field) by the quantum numbers f and m corresponding to the total angular momentum and its projection along the z axis, respectively. In general, the states of the atom are indexed by a composite label α and the corresponding energies are ϵ_α. The set of quantum numbers that characterize a *pair* of atoms in the limit of large separation,

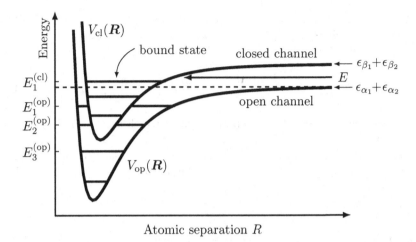

FIGURE 9.5 Two-channel model for a Feshbach resonance. The energy differ-
ence $\Delta\epsilon = (\epsilon_{\beta_1}+\epsilon_{\beta_2}) - (\epsilon_{\alpha_1}+\epsilon_{\alpha_2})$ between the closed and open channels can be
tuned by a magnetic field. When the energy of the top molecular bound state
supported by the closed channel approaches zero, $E_1^{(cl)} \to 0$, it couples resonantly to the open channel and, as a result, the scattering length for collisions
near zero energy, $E \to 0$, becomes very large.

e.g., $(\alpha_1; \alpha_2) = (f_1, m_1; f_2, m_2)$, defines a certain *scattering channel*. If the
energy of the system is $E > \epsilon_{\alpha_1} + \epsilon_{\alpha_2}$, the channel is open, otherwise (i.e., for
$E < \epsilon_{\alpha_1} + \epsilon_{\alpha_2}$) the channel is said to be closed (see Figure 9.5).

Consider now a simplified two-channel model corresponding to the states
$|\alpha_1; \alpha_2\rangle$ and $|\beta_1; \beta_2\rangle$. We assume that the atoms are prepared in the *open
channel* $(\alpha_1; \alpha_2)$ and that the second channel is closed, i.e., $\epsilon_{\alpha_1} + \epsilon_{\alpha_2} < E <
\epsilon_{\beta_1} + \epsilon_{\beta_2}$. Since, in general, the interaction potential $V(\mathbf{R})$ is spin-dependent,
the two channels are coupled. More specifically, the effective Hamiltonian that
describes the scattering process takes the form

$$H = \begin{pmatrix} -\frac{\hbar^2}{2\mu}\nabla^2 + V_{op}(\mathbf{R}) & W(\mathbf{R}) \\ W^\dagger(\mathbf{R}) & -\frac{\hbar^2}{2\mu}\nabla^2 + V_{cl}(\mathbf{R}) \end{pmatrix}, \qquad (9.14)$$

where $V_{op}(\mathbf{R}) = \langle\alpha_1; \alpha_2|V(\mathbf{R})|\alpha_1; \alpha_2\rangle$ and $V_{cl}(\mathbf{R}) = \langle\beta_1; \beta_2|V(\mathbf{R})|\beta_1; \beta_2\rangle$ are
the potentials of the open and closed channels, respectively, and $W(\mathbf{R}) =
\langle\alpha_1; \alpha_2|V(\mathbf{R})|\beta_1; \beta_2\rangle$ is the coupling between the two channels.

Feshbach resonances. In a two-channel model the scattering length has a
background component a_{bg} that is solely determined by the properties of the
open channel (i.e., the scattering length corresponding to the absence of any
coupling to the closed channel, $W(\mathbf{R}) = 0$) and a resonant component that
becomes very large when the energy of the system is nearly resonant with a
bound state of the closed channel potential V_{cl} (see Figure 9.5). If the magnetic

moments corresponding to the two channels are different, we can shift the closed-channel energy with respect to the open-channel energy by varying the magnetic field, thus changing the scattering length. The dependence of the s-wave scattering length on the magnetic field B near a magnetically tuned Feshbach resonance is described by the expression [367]

$$a(B) = a_{\text{bg}} \left(1 - \frac{\Delta_B}{B - B_0} \right), \tag{9.15}$$

where the parameter B_0 gives the *resonance position* and Δ_B characterizes the *resonance width*. Note that Δ_B can be positive or negative. At the magnetic field $B^* = B_0 + \Delta_B$, the scattering length crosses zero.

Feshbach resonances arise from the coupling of a discrete (bound) state to the continuum. This type of phenomenon was systematically investigated by Ugo Fano, in the context of atomic physics [162], and Herman Feshbach, in the context of nuclear physics [165]. Sometimes the phenomenon is referred to as a *Fano–Feshbach resonance*, although Fano's name is typically associated with the asymmetric line-shape of such a resonance. In cold atom systems, Feshbach resonances are particularly useful as they enable the tuning of the scattering length by changing the magnetic field [504], which is possible due to the difference between the magnetic moments of the closed and open channels. In practice, the realization of a many-body system with tunable interactions requires the additional condition that the relaxation rate into deep bound states (due to three-body collisions) be negligible. This condition is well satisfied by fermions [405], but in dilute bosonic gases the three-body recombinations are important and make highly correlated states unstable [492].

9.3 LIGHT-INDUCED ARTIFICIAL GAUGE FIELDS

Gauge theories are central components of our understanding of elementary particle interactions and condensed matter physics. In particular, the emergent gauge fields associated with the effective low-energy theory of certain condensed matter systems play a key role in understanding the nontrivial topological properties of many-body quantum states. For example, the adiabatic motion of quantum particles with internal structure can be described in terms of *geometric gauge potentials*, as first pointed out by Mead and Truhlar [359] and Berry [57]. This type of gauge potential can be engineered in a cold atom system by making the Hamiltonian governing the internal dynamics of the atom parametrically dependent on its position via, for example, the interaction with a position-dependent laser field. In this section, we provide a brief overview of the main ideas behind the realization of light-induced gauge fields in cold atom systems. For comprehensive reviews, see Refs. [125, 210].

9.3.1 Geometric gauge potentials

Geometric gauge potentials emerge naturally in cold atom systems when the center of mass motion of an atom is coupled to its internal degrees of freedom.

Consider an atom in the presence of a position- and time-dependent laser field. The position of the atom is given by the atomic center of mass position vector r and the corresponding momentum operator is $p = -i\hbar\nabla$. The dynamics of the internal degrees of freedom is described by the atomic Hamiltonian H_0, while the atom-light coupling is given by the position- and time-dependent term $H_1(r, t)$. Assuming that the atom moves in a (state-independent) trapping potential $V(r, t)$, the total Hamiltonian takes the form

$$H_{\text{tot}} = -\frac{\hbar^2\nabla^2}{2m} + V(r, t) + H_0 + H_1(r, t). \tag{9.16}$$

Note that the first two terms act trivially on the internal degrees of freedom of the atom. Let $|\alpha\rangle$ be the eigenstates of H_0, i.e., the atomic *bare* states. Diagonalizing the atomic Hamiltonian in the presence of the laser field, i.e., $H_0 + H_1$, generates a set of eigenstates, called atomic *dressed* states, that depend parametrically on the position of the atom and on time, $|\phi_\alpha\rangle \equiv |\phi_\alpha(r, t)\rangle$. Note that we can use the same set of labels to designate the bare and dressed states and, in the limit of vanishing atom-light coupling, $H_1 \to 0$, we have $|\phi_\alpha\rangle \to |\alpha\rangle$. The position- and time-dependent operator

$$U(r, t) = \sum_\alpha |\phi_\alpha\rangle\langle\alpha|, \tag{9.17}$$

is a unitary transformation that connects the bare and dressed basis states, $U|\alpha\rangle = |\phi_\alpha\rangle$. Note that U diagonalizes the operator $H_0 + H_1$,

$$U^\dagger(H_0 + H_1)U = \sum_\alpha \epsilon_\alpha(r, t)|\alpha\rangle\langle\alpha| \equiv \Lambda(r, t), \tag{9.18}$$

where $\epsilon_\alpha(r, t)$ are the position- and time-dependent "energies" of the atomic dressed states, $(H_0 + H_1)|\phi_\alpha\rangle = \epsilon_\alpha|\phi_\alpha\rangle$. If $|\psi\rangle$ is a solution of the time-dependent Schrödinger equation defined by the total Hamiltonian H_{tot}, the transformed state $|\psi'\rangle = U^\dagger|\psi\rangle$ will be a solution of the time-dependent Schrödinger equation defined by the transformed Hamiltonian $H'_{\text{tot}} = U^\dagger H_{\text{tot}}U - i\hbar U^\dagger\partial_t U$. Explicitly, we have

$$H'_{\text{tot}} = \frac{[p - \mathcal{A}(r, t)]^2}{2m} + V(r, t) + \Lambda(r, t) + \Phi(r, t), \tag{9.19}$$

where $\mathcal{A} = i\hbar U^\dagger\nabla U$ and $\Phi = -i\hbar U^\dagger\partial_t U$. Note that in the derivation of Eq. (9.19) we have exploited the fact that U is a unitary operator, which leads to the identities $(\nabla U^\dagger)U + U^\dagger(\nabla U) = 0$ and $U^\dagger\nabla^2 U = \nabla\cdot(U^\dagger\nabla U) + (U^\dagger\nabla U)^2$.

Next, we assume that a subset q of atomic *dressed* states is well separated in energy from the other dressed states and that the trapped atom evolves *adiabatically* within the subspace spanned by the q-states. The effective Hamiltonian that describes the adiabatic evolution of the system is $\widetilde{\mathcal{P}}_q H_{\text{tot}}\widetilde{\mathcal{P}}_q$, where $\widetilde{\mathcal{P}}_q = \sum_{\alpha\in q} |\phi_\alpha\rangle\langle\phi_\alpha|$ is the projector onto the subspace of dressed q-states.

Note, however, that the unitary operator U given by Eq. (9.17) realizes a one-to-one mapping between the dressed states $|\phi_\alpha\rangle$ and the bare states $|\alpha\rangle$. Consequently, in terms of transformed states the dressed q-subspace becomes a bare q-subspace spanned by the states $|\alpha\rangle$ with $\alpha \in q$. If $\mathcal{P}_q = \sum_{\alpha \in q} |\alpha\rangle\langle\alpha|$ is the projector onto the bare q-subspace, we have $\widetilde{\mathcal{P}}_q U = U\mathcal{P}_q$ and the effective (transformed) Hamiltonian that describes the dynamics of the system within the adiabatic approximation takes the form

$$H_{\text{eff}} = \mathcal{P}_q H'_{\text{tot}} \mathcal{P}_q = \frac{\left[\boldsymbol{p} - \boldsymbol{\mathcal{A}}^{(q)}\right]^2}{2m} + V + \Lambda^{(q)} + \Phi^{(q)} + W^{(q)}, \qquad (9.20)$$

where $\boldsymbol{\mathcal{A}}^{(q)}$, $\Lambda^{(q)}$ and $\Phi^{(q)}$ are the projections of the corresponding operators onto the q-subspace. The additional term has the form

$$W^{(q)} = \frac{1}{2m}\left[\mathcal{P}_q \mathcal{A}^2 \mathcal{P}_q - \left(\mathcal{A}^{(q)}\right)^2\right] = \frac{1}{2m}\mathcal{P}_q \boldsymbol{\mathcal{A}}(1 - \mathcal{P}_q)\boldsymbol{\mathcal{A}}\mathcal{P}_q. \qquad (9.21)$$

In terms of atomic bare states, the operators associated with the geometric vector potential $\boldsymbol{\mathcal{A}}^{(q)}$ and the scalar potentials $W^{(q)}$ and $\Phi^{(q)}$ are

$$\boldsymbol{\mathcal{A}}^{(q)} = \sum_{\alpha,\beta \in q} |\alpha\rangle \, \boldsymbol{\mathcal{A}}_{\alpha\beta} \, \langle\beta|, \qquad \boldsymbol{\mathcal{A}}_{\alpha\beta} = i\hbar \, \langle\phi_\alpha|\boldsymbol{\nabla}|\phi_\beta\rangle,$$

$$\Phi^{(q)} = \sum_{\alpha,\beta \in q} |\alpha\rangle \, \Phi_{\alpha\beta} \, \langle\beta|, \qquad \Phi_{\alpha\beta} = -i\hbar \, \langle\phi_\alpha|\frac{\partial}{\partial t}|\phi_\beta\rangle, \qquad (9.22)$$

$$W^{(q)} = \sum_{\alpha,\beta \in q} |\alpha\rangle \, W_{\alpha\beta} \, \langle\beta|, \qquad W_{\alpha\beta} = \frac{1}{2m}\sum_{\lambda \notin q} \boldsymbol{\mathcal{A}}_{\alpha\lambda} \cdot \boldsymbol{\mathcal{A}}_{\lambda\beta}.$$

Note that the geometric vector potential and the scalar potential $W^{(q)}$ are generated by the spatial dependence of the dressed states, while $\Phi^{(q)}$ is generated by their time dependence. If the q subspace is one-dimensional or if the matrices $\boldsymbol{\mathcal{A}}_{\alpha\beta}$, $\Phi_{\alpha\beta}$, and $W_{\alpha\beta}$ are diagonal, they represent *Abelian* geometric potentials. In general, however, these matrices do not commute and represent *non-Abelian* potentials. Below, we discuss two examples of setups that can be used to generate these types of geometric potentials.

9.3.2 Abelian gauge potentials: The Λ scheme

Consider a simplified atom structure consisting of three atomic states: $|g_1\rangle$, $|g_2\rangle$, and $|e\rangle$, where, typically, the first two are nearly degenerate ground states and the third is an excited state. Two laser beams couple the states $|g_1\rangle$ and $|g_2\rangle$ to the excited state resulting in the so-called Λ scheme of the atom-light coupling, which is schematically illustrated in Figure 9.6(a). We assume that the detunings of the lasers are $\delta_1 = -\delta$ and $\delta_2 = \delta$, respectively. The dynamics of the atom-light system is described by a Hamiltonian $H = H_0 + H_1$ similar to that from Eq. (9.5), but containing two interaction terms. In the rotating

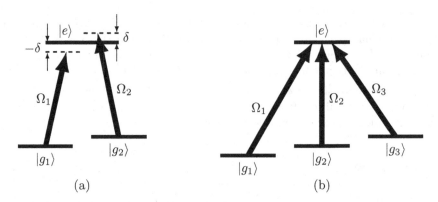

FIGURE 9.6 (a) The Λ scheme. Two laser beams induce atomic transitions $|g_1\rangle \to |e\rangle$ and $|g_2\rangle \to |e\rangle$ characterized by Rabi frequencies Ω_1 and Ω_2, respectively, and generate a dressed dark state. (b) The tripod scheme. The atomic state $|e\rangle$ is coupled resonantly to the atomic states g_i via three laser beams resulting in two degenerate dark states.

wave approximation, the effective atom-light Hamiltonian is time-independent and can be obtained following the steps leading to Eq. (9.6). Explicitly, we have

$$H' = \frac{\hbar\delta}{2}\left[|g_2\rangle\langle g_2| - |g_1\rangle\langle g_1|\right] + \frac{\hbar}{2}[\Omega_1(\mathbf{r})|e\rangle\langle g_1| + \Omega_2(\mathbf{r})|e\rangle\langle g_2| + h.c.], \quad (9.23)$$

where Ω_1 and Ω_2 are Rabi frequencies characterizing the two transitions.

The three eigenstates of H' represent the dressed atomic states $|\phi_\alpha\rangle$ of an atom-light system with Λ-type level structure. Assuming exact resonance, $\delta = 0$, we have

$$|D\rangle = \frac{|g_1\rangle - \zeta\,|g_2\rangle}{\sqrt{1+|\zeta|^2}}, \qquad |B_\pm\rangle = \frac{|g_1\rangle + \zeta^*\,|g_2\rangle}{\sqrt{2(1+|\zeta|^2)}} \pm \frac{|e\rangle}{\sqrt{2}}, \qquad (9.24)$$

where $\zeta \equiv \Omega_1/\Omega_2 = |\zeta|e^{iS}$ gives the relative magnitude and phase of the two Rabi frequencies, i.e., $|\zeta|$ and S, respectively. Note that the state $|D\rangle$, called the *dark state*, contains no contribution from the excited state $|e\rangle$. In practice, this property is important because for a system in state $|D\rangle$ the excited level does not become populated and the spontaneous decay is suppressed. Furthermore, the dark state, which is characterized by a zero eigenvalue $\epsilon_D = 0$, is well separated in energy from the other two states, which have eigenvalues $\epsilon_{B_\pm} = \pm\frac{\hbar}{2}\sqrt{|\Omega_1|^2 + |\Omega_2|^2}$. If $|\epsilon_{B_\pm}|$ exceeds the characteristic kinetic energy of the atomic motion, the atoms will evolve adiabatically within the dark state subspace $q = D$. The corresponding geometric vector potential defined by (9.22), i.e., $\mathbf{A} \equiv \mathbf{A}_{DD} = i\hbar\langle D|\mathbf{\nabla}|D\rangle$, takes the form

$$\mathbf{A} = -\hbar\frac{|\zeta|^2}{1+|\zeta|^2}\mathbf{\nabla}S. \qquad (9.25)$$

In the general language of geometric phases summarized in Chapter 1, $\mathcal{A}$ is the Mead–Berry vector potential (or Mead–Berry connection) given by Eq. (1.28). The corresponding Berry curvature vector $\mathcal{B} \equiv \Omega^D(r) = \nabla \times \mathcal{A}$, which can be interpreted as an artificial magnetic field, has the form

$$\mathcal{B} = \hbar \frac{\nabla S \times \nabla |\zeta|^2}{(1 + |\zeta|^2)^2}. \tag{9.26}$$

Note that $\mathcal{B} \neq 0$ only if the gradients of the relative intensity $|\zeta|^2$ and the relative phase S are both nonzero and not parallel to each other. This requires a nonvanishing relative orbital angular momentum of the two light beams, which can be realized, for example, using two counter-propagating Gaussian beams with an offset between their propagation axes [125, 272]. Finally, we note that in the rotating wave approximation the dressed states are time-independent and, consequently, we have $\Phi = 0$. On the other hand, the scalar potential W is, in general, nonzero and can be determined explicitly using Eq. (9.22), i.e., $W \equiv W_{DD} = \frac{1}{2m}(\mathcal{A}_{DB_+} \cdot \mathcal{A}_{B_+D} + \mathcal{A}_{DB_-} \cdot \mathcal{A}_{B_-D})$, and the explicit expressions for the dressed states given by Eq. (9.24).

9.3.3 Non-Abelian gauge potentials: The tripod scheme and spin-orbit coupling

When the atom-light system evolves adiabatically within a manifold of degenerate (or quasi-degenerate) q-states, the emerging geometric potentials given by Eqs. (9.22) are non-Abelian. A generic way to obtain a degenerate subspace is to couple $N \geq 3$ states $|g_i\rangle$, where $i = 1, 2, \ldots, N$, to a single level $|e\rangle$ via laser fields with Rabi frequencies Ω_i [125, 438]. The case $N = 3$ – the so-called *tripod scheme* – is illustrated in Figure 9.6(b). The non-Abelian gauge potentials created using this type of scheme couple the center of mass motion of the atom to its internal (spin-like) degrees of freedom, thus generating an effective "spin-orbit" coupling [193].

Consider N states $|g_i\rangle$ coupled resonantly to an excited state $|e\rangle$ via laser fields. In the rotating field approximation the effective atom-light Hamiltonian is

$$H' = \sum_{i=1}^{N} \frac{\hbar}{2}\Omega_i |e\rangle\langle g_i| + h.c. = \frac{\hbar\Omega(r)}{2}[|e\rangle\langle B(r)| + |B(r)\rangle\langle e|], \tag{9.27}$$

where $\Omega = \sqrt{|\Omega_1|^2 + \cdots + |\Omega_N|^2}$ and $|B\rangle = \frac{1}{\Omega}\sum_i \Omega_i |g_i\rangle$ is the *bright coupled* state. The dressed states are obtained by diagonalizing H'. This gives two states $|B_\pm\rangle = (|B\rangle \pm |e\rangle)/\sqrt{2}$ with energies $\epsilon_{B_\pm} = \pm\frac{\hbar}{2}\Omega$, which generalize the $|B_\pm\rangle$ states from Eq. (9.24), and $N-1$ *dark states* with energy $\epsilon_{D_n} = 0$ that are orthogonal to the bright state $|B\rangle$. These dark states provide the degenerate subspace $q = \{D_1, \ldots, D_{N-1}\}$ in which the system can evolve adiabatically.

Experimentally, the tripod scheme can be realized using, for example, the atomic transition between a ground level with angular momentum $j_g = 1$ and an excited state with $j_e = 0$. This level structure occurs in alkali-metal species such as ^{23}Na and ^{87}Rb. A proper choice of laser fields leads to a geometric vector potential $\mathcal{A}_{nm}$ with noncommuting Cartesian components. The artificial non-Abelian potential can be equivalent to the presence of an effective magnetic monopole [438] or, more interestingly, to the presence of an effective "spin-orbit" coupling [193, 486, 572].

To illustrate the generation of spin-orbit coupling (SOC) in cold-atom systems, let us consider a simple example involving a spatially uniform vector potential. We assume $N = 3$ ground states coupled resonantly to an excited state via three plane waves of equal amplitude propagating in the xy plane. The corresponding Rabi frequencies and wave vectors are

$$\Omega_j(\boldsymbol{r}) = \frac{1}{\sqrt{N}}\Omega e^{i\boldsymbol{k}_j \cdot \boldsymbol{r}}, \qquad\qquad \boldsymbol{k}_j = k(-\hat{\boldsymbol{e}}_x \cos\alpha_j + \hat{\boldsymbol{e}}_y \sin\alpha_j), \qquad (9.28)$$

where $\hat{\boldsymbol{e}}_x$ and $\hat{\boldsymbol{e}}_y$ are unit vectors along the x and y directions, respectively, and $\alpha_j = 2\pi j/N$. A convenient basis for the q subspace, which has $N - 1 = 2$ dimensions, is given by the following set of orthonormal dark states

$$|D_n\rangle = \frac{1}{\sqrt{N}} \sum_j |g_j\rangle e^{i\alpha_j n - i\boldsymbol{k}_j \cdot \boldsymbol{r}}. \qquad (9.29)$$

Substituting this expression in Eq. (9.22), we obtain the geometric vector and scalar potentials

$$\mathcal{A}_{12} = -\frac{\hbar k}{2}(\hat{\boldsymbol{e}}_x - i\hat{\boldsymbol{e}}_y) = \mathcal{A}_{21}^*, \qquad \mathcal{A}_{11} = \mathcal{A}_{22} = 0, \qquad (9.30)$$

$$W_{11} = W_{22} = \frac{\hbar^2 k^2}{4m}, \qquad\qquad W_{12} = W_{21} = 0. \qquad (9.31)$$

Note that, although these are constant potentials, the effect of the vector potential on the dynamics of the system is nontrivial because the Cartesian components of $\mathcal{A}$ do not commute. Assuming that the atoms move in a constant potential, the solutions of the Schrödinger equation defined by the effective Hamiltonian H_{eff} from (9.20) are plane waves $|\psi_{\boldsymbol{q}}(\boldsymbol{r},t)\rangle = |\varphi_{\boldsymbol{q}}\rangle \exp[i(\boldsymbol{q} \cdot \boldsymbol{r} - E_{\boldsymbol{q}}t)]$, where the amplitudes $|\varphi_{\boldsymbol{q}}\rangle$ are solutions of the eigenvalue problem $H_{\text{eff}}(\boldsymbol{q})|\varphi_{\boldsymbol{q}}\rangle = E_{\boldsymbol{q}}|\varphi_{\boldsymbol{q}}\rangle$. Note that the effective Hamiltonian is a 2×2 matrix and that the terms $\Lambda^{(q)}$ and $\Phi^{(q)}$ vanish, while V and $W^{(q)}$ are proportional to the unit matrix. Choosing for simplicity $V = -\hbar^2 k^2/4m = -W$, we have

$$H_{\text{eff}}(\boldsymbol{q}) = \frac{(\hbar\boldsymbol{q} - \mathcal{A})^2}{2m} = \frac{\hbar^2 q^2}{2m} - \frac{\hbar k}{4m}(q_x\sigma_x + q_y\sigma_y) + \frac{\hbar^2 k^2}{8m}, \qquad (9.32)$$

where $\boldsymbol{\sigma} = (\sigma_x, \sigma_y, \sigma_z)$ are Pauli matrices. Note that the vector potential is proportional to the in-plane spin-1/2 operator, $\mathcal{A} = -k(S_x\hat{\boldsymbol{e}}_x + S_y\hat{\boldsymbol{e}}_y)$, where $\boldsymbol{S} = \frac{\hbar}{2}\boldsymbol{\sigma}$. The second term in Eq. (9.32) couples the (pseudo) spin to the center of mass motion of the atom, i.e., it represents an effective spin-orbit coupling. Several other schemes were proposed to realize similar linear Dresselhaus or

Rashba type spin-orbit coupling [271, 438, 572]. Engineering such couplings in cold atom systems represents a key step in the development of quantum simulators for a variety of interesting condensed matter systems, including systems that support topologically nontrivial phases.

9.4 TOPOLOGICAL STATES IN COLD ATOM SYSTEMS

There is a large variety of theoretical proposals for realizing topological phases in cold atom systems (see, for example, [210] and references therein). Below, we briefly discuss one example and touch upon another to illustrate some of the specific ingredients, tuning knobs, and probes associated with the preparation and detection of topological phases in a cold atom environment.

9.4.1 Realization of the Haldane model with ultracold atoms

The Haldane model [231] discussed in Chapter 6 (see Section 6.2 starting on page 189) cannot be naturally realized in a condensed matter system. Cold atoms, on the other hand, appear to be better suited for its implementation [12, 209]. In essence, a state-dependent honeycomb optical lattice generated by a three-beam laser configuration traps atoms with two different internal *pseudospin* states at the nodes of two intertwined triangular lattices, as shown in Figure 9.7(a). The state-dependent potential corresponds to an effective "Zeeman shift" that has minima on the A lattice and maxima on the B lattice for pseudo-spin up atoms, while for pseudo-spin down atoms the "Zeeman shift" has opposite sub-lattice dependence. Additional lasers are used to coherently transfer atoms from one internal state to the other, thus causing laser-induced hopping [263] between nearest-neighbor (n.n.) sites.

For a sufficiently deep optical potential (and no additional lasers) only hopping between neighboring sites on each triangular lattice – i.e., next-nearest-neighbors (n.n.n.) on the honeycomb lattice – are allowed. The corresponding (second quantized) Hamiltonian takes the form

$$H_0 = -t_A \sum_{\langle\langle i_A, j_A \rangle\rangle} a_{i_A}^\dagger a_{j_A} - t_B \sum_{\langle\langle i_B, j_B \rangle\rangle} b_{i_B}^\dagger b_{j_B}, \qquad (9.33)$$

where a_{i_A} and b_{i_B} are annihilation operators for atoms at lattice sites r_{i_A} and r_{i_B} on sub-lattices A and B, respectively, and $\langle\langle i_x, j_x \rangle\rangle$ designates n.n.n. sites on the sub-lattice $x = A, B$. In general, the depths of the two optical lattices can be different, which amounts to introducing a staggered potential described by

$$H_M = -M \sum_{i_A} a_{i_A}^\dagger a_{i_A} + M \sum_{i_B} b_{i_B}^\dagger b_{i_B}. \qquad (9.34)$$

Note that H_M breaks the inversion symmetry of the honeycomb lattice with respect to the center of a hexagon.

Next, the two triangular lattices are coupled through laser-assisted tunneling, as illustrated schematically in Figure 9.7(b). The effective hopping t_{eff}

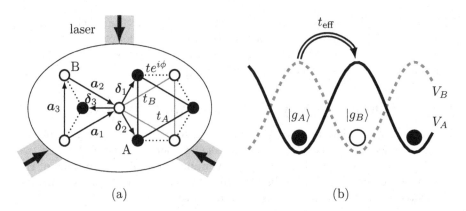

FIGURE 9.7 (a) Honeycomb lattice composed of two triangular sub-lattices A (black circles) and B (white circles) generated by a three-beam laser configuration. (b) Laser-assisted tunneling. Two internal states $|g_A\rangle$ and $|g_B\rangle$ are trapped in state-dependent optical lattices with potentials V_A and V_B, respectively. A resonant laser (not shown) couples the states and induces complex effective tunneling matrix elements $t_{\text{eff}} = -te^{i\phi}$.

contains a *Peierls phase* determined by the wave vector $\boldsymbol{q}$ of the laser [263],

$$t_{\text{eff}} = -te^{i\phi(i_A, j_B)}, \qquad \phi(i_A, j_B) = \frac{1}{2}\boldsymbol{q} \cdot (\boldsymbol{r}_{i_A} + \boldsymbol{r}_{j_B}) = -\phi(j_b, i_A). \quad (9.35)$$

The effective Hamiltonian describing the laser-assisted coupling of the two lattices has the form

$$H_{AB} = -t \sum_{\langle i_A, j_B \rangle} \left(e^{i\phi(i_A, j_B)} a_{i_A}^\dagger b_{j_B} + \text{h.c.} \right), \quad (9.36)$$

where $\langle i_A, j_B \rangle$ labels n.n. sites on the honeycomb lattice. The total Hamiltonian for the system is $H_{\text{tot}} = H_0 + H_M + H_{AB}$ and the relevant parameters are the hopping amplitudes t_A, t_B and t, the mismatch energy M, and the wave vector $\boldsymbol{q}$.

The effective Hamiltonian H_{tot} is explicitly position-dependent through the Peierls phases $\phi(i_A, j_B)$. To eliminate this dependence, it is convenient to perform the unitary transformation corresponding to $a_{i_A} \rightarrow a_{i_A} \exp(i\boldsymbol{q} \cdot \boldsymbol{r}_{i_A}/2)$ and $b_{j_B} \rightarrow b_{j_B} \exp(-i\boldsymbol{q} \cdot \boldsymbol{r}_{j_B}/2)$. The transformed Hamiltonian is characterized by real n.n. hopping and complex n.n.n. hopping, similar to the original Haldane model. After performing a Fourier transform, the Hamiltonian can be written in the standard form $H_{\text{tot}} = \epsilon(\boldsymbol{k})I_{2\times 2} + \boldsymbol{d}(\boldsymbol{k}) \cdot \boldsymbol{\sigma}$, where $\boldsymbol{\sigma} = (\sigma_x, \sigma_y, \sigma_z)$ are Pauli matrices and

$$\epsilon(\boldsymbol{k}) = -t_A f(\boldsymbol{k} - \boldsymbol{q}/2) - t_B f(\boldsymbol{k} + \boldsymbol{q}/2),$$
$$d_x(\boldsymbol{k}) = -t \,\text{Re}[g(\boldsymbol{k})], \qquad d_y(\boldsymbol{k}) = t \,\text{Im}[g(\boldsymbol{k})], \quad (9.37)$$
$$d_z(\boldsymbol{k}) = -M - t_A f(\boldsymbol{k} - \boldsymbol{q}/2) + t_B f(\boldsymbol{k} + \boldsymbol{q}/2).$$

The functions $f(\boldsymbol{k}) = \sum_\alpha \cos(\boldsymbol{k} \cdot \boldsymbol{a}_\alpha)$ and $g(\boldsymbol{k}) = \sum_\alpha \exp(-i\boldsymbol{k} \cdot \boldsymbol{\delta}_\alpha)$ are defined in terms of the primitive lattice vectors $\boldsymbol{a}_\alpha = \pm(\boldsymbol{\delta}_\mu - \boldsymbol{\delta}_\nu)$ of the honeycomb lattice shown in Figure 9.7(a). We note that the effective Hamiltonian of the cold atom system has the same structure as the original Haldane model given by Eqs. (6.22) and (6.23), with the minor difference that the "magnetic flux" configurations associated with the Peierls phases have different structures [209]. The experimental realization of a Haldane-type model in a *periodically modulated* optical honeycomb lattice was recently reported [268].

The Hamiltonian H_{tot} of the two-dimensional cold atom system is in symmetry class A and its gapped ground states are classified by a $\mathbb{Z}$ topological invariant given by the first Chern number of the occupied band. In an analog solid-state system (e.g., an integer quantum Hall fluid), the nontrivial topology is detected by measuring the Hall conductance. However, it is not clear how a similar transport measurement can be done with cold atoms. Hence, the question: How can one probe the presence of a topologically nontrivial phase in a cold atom system? A type of measurement that can be easily done with cold atoms is the so-called *time of flight* (ToF) measurement; the trapping potential is switched off and the momentum distribution of the particles is determined by imaging the expansion of the atomic cloud. For the Haldane-type model described above, the normalized vector $\boldsymbol{n}(\boldsymbol{k}) = \boldsymbol{d}(\boldsymbol{k})/|\boldsymbol{d}(\boldsymbol{k})|$ can be reconstructed from the (pseudo) spin-resolved momentum densities $\rho_{A,B}(\boldsymbol{k})$ associated with the two internal states of the atoms [12, 209]. Knowing $\boldsymbol{n}(\boldsymbol{k})$, provides us with direct access to the Berry curvature of the valence band, $\mathcal{F} = \frac{1}{2}\boldsymbol{n} \cdot (\partial_{k_x}\boldsymbol{n} \times \partial_{k_y}\boldsymbol{n})$, and to the corresponding Chern number (see Section 3.4.2 on page 91),

$$\nu = \frac{1}{4\pi} \int_{\mathbb{T}^2} d^2k\ \boldsymbol{n} \cdot (\partial_{k_x}\boldsymbol{n} \times \partial_{k_y}\boldsymbol{n}), \tag{9.38}$$

where $\mathbb{T}^2$ is the BZ torus.

This example illustrates three important potential contributions that cold atom systems could bring to the table. i) *Simplicity.* The effective low-energy physics is free from effects associated with phonons, disorder, sub-dominant interactions, and other "unwanted" sources that make life complicated in solid-state systems. Thus, cold atom systems are ideally suited for studying simple models, which is key to advancing our understanding of the physical world (including topological quantum matter). ii) *Control* and *tunability.* In solid-state systems one can easily control a few external parameters (e.g., temperature, magnetic field, etc.). With cold atoms, one can basically control all the parameters of the effective Hamiltonian. Potentially, this is a significant advantage, as it offers access to certain parameter regimes that may not be accessible in a nature-given system, facilitates the study of phase transitions (along various paths in the parameter space), and provides convenient knobs for changing the symmetry class of the Hamiltonian. iii) *Accessibility.* The cold atom environment provides certain tools for probing the topological properties of a system that are not available in solid-state physics. For example, one has direct access to the Berry curvature, as discussed above. Also, a direct measurement of the Zak phase in topological Bloch bands was recently reported [31].

9.4.2 Majorana fermions in optical lattices

We close this chapter with a few brief remarks on proposed schemes for the realization of zero-energy Majorana modes with cold atoms. To engineer a topological superconducting state one can start with fermionic atoms trapped in a 1D (or 2D) optical lattice. Synthetic spin-orbit coupling and Zeeman fields can be generated using spatially modulated laser fields [445, 568], as described in Section 9.3.3. Alternatively, one can use laser-assisted tunneling methods or Raman coupling with nonzero photon recoil. For example, one can show [266, 309] that using the Λ scheme [see Figure 9.6(a)] with two lasers propagating along $\boldsymbol{k}_1$ and $\boldsymbol{k}_2$, such that $\boldsymbol{k}_1 - \boldsymbol{k}_2 \neq 0$ is oriented parallel to the quantum wire, one can induce both (effective) spin-orbit interaction and magnetic field. To generate pairing, one can couple the system to a BEC reservoir of Feshbach molecules via radio-frequency pulses [266]. This could be regarded as the analog of the superconducting proximity effect in solid-state heterostructures.

The detection of Majorana zero modes (MZMs) can be a nontrivial task. As discussed in the previous chapter, there are several different "detection levels" determined by the type of property that is probed, starting with establishing the presence of a zero-energy mode and culminating with the demonstration of non-Abelian statistics. In cold atom systems, one can use, for example, spatially resolved radio-frequency spectroscopy [219] to obtain information about the local density of states. This would be the AMO analog of scanning tunneling spectroscopy. ToF imaging can also provide information about the existence and number of MZMs in an optical lattice setup. More specifically, the presence of MZMs is signaled by rapid oscillations of the atomic density distribution at the detector, which are associated with the nonlocal nature of the fermionic state corresponding to a pair of Majoranas localized at the opposite ends of a 1D system [309]. In addition, for a 1D Majorana system in the BDI symmetry class, the $\mathbb{Z}$ topological invariant could be measured using a ToF scheme similar to the one proposed for Chern insulators [12].

II

Quantum Information and Quantum Computation: Introductory Concepts

CHAPTER 10

Elements of Quantum Information Theory

W E MAY BE ON THE BRINK OF A QUANTUM TRANSFOR-
mation driven by our increasing ability to realize, detect, control, and
manipulate high-purity quantum states. For decades the quantum world has
crept into our lives through modern information technology, the most visible
side of science that permeates almost every aspect of our daily conduct. Yet
quantumness in its full glory remains a promise for the future, as existing
mainstream technologies are essentially based on properties of mixed quantum states realized in, e.g., metal–oxide–semiconductor heterostructures. The
scientific backbone of current technologies combines quantum theory with information theory and the theory of computation, everything being sprinkled
with a healthy dose of mathematics and brought into reality by remarkable
engineering achievements. This multidisciplinary approach has enabled us to
understand key physical phenomena and material properties, to communicate
and process information efficiently, and, ultimately, to build functional devices
that provide us with unprecedented convenience and mastery over nature, possibly with the price of being in danger to become the slaves of our computers
and gadgets. However, our technological success is built on a combination of
essentially distinct and largely independent scientific fields. Harnessing high-
purity quantum states, on the other hand, sets these previously separated
"continents of knowledge" on a collision path. The main goal of the last three
chapters of this book is to provide a bird's eye view of the "collision zone"
and prepare the reader to explore, enjoy, and perhaps exploit the newly raising
mountains. We start with a brief introduction to basic concepts in quantum
information, which also includes a summary of the operational formulation of
quantum mechanics (the "natural" language of quantum information theory)
and a synopsis of classical information theory (Chapter 10). For a thorough
treatment of concepts, problems, and techniques in quantum information theory the reader is referred to the books by Nielsen and Chuang [389] and

DOI: 10.1201/9781003226048-10

Wilde [548]. The book by Watrous [530] focuses on the mathematical theory of quantum information. An undergraduate level introduction to this subject can be found, for example, in Schumacher and Westmoreland [458]. Some of the philosophical implications of the relationship between quantum mechanics and information theory are discussed in Timpson [505]. Our survey continues with Chapter 11, where we discuss a few key aspects pertaining to quantum computation, which can be viewed as the quintessential quantum technology involving the controlled realization and manipulation of high-purity quantum states within a large Hilbert space. Finally, Chapter 12 focuses on topological quantum computation, a promising approach to overcoming the fundamental challenge associated with quantum decoherence based on harnessing the underlying robustness of topological quantum states. This also marks our final return to the realm of topological quantum matter.

10.1 INTRODUCTION

Quantum information theory is an extremely lively research area situated at the crossroads of quantum mechanics, information theory, and cryptography. The central concern of this relatively new field – what information-related tasks that *cannot* be done classically *could* be accomplished using quantum systems – is nothing but the archetypal theme of quantum mechanics (What is, essentially, the difference between the quantum and the classical worlds?) grafted onto the task-oriented spirit of information theory.

The emergence of quantum information theory and the parallel development of quantum computation have revealed that any physical theory (quantum mechanics being one example) may constitute the basis for a theory of communication and information processing. One should think *physically* about computation and communication. This idea has provided a new paradigm to be explored, beyond classical information theory and classical computation, yet the full range of new capabilities that quantum devices may posses is, perhaps, still waiting to be discovered. In turn, information theory may have something to offer to physics. Adopting an *information theoretic* perspective could reveal something about the fundamental structure of quantum theory and could provide new tools for investigating that structure. Moreover, quantum computation and quantum information could provide a new conceptual framework for dealing with complexity in a fundamentally systematic way. This could have profound implications, since fields that traditionally deal with complex systems, such as chemistry, biology, and engineering, have not yet generated anything comparable with the powerful mathematical description of nature that constitutes the core of what we call physics. Physics, on the other hand, has focused on understanding simple systems and elementary processes and is just starting to address complexity. In this chapter, we provide an elementary introduction to the key ideas underlying quantum information theory, to help the reader understand the main motivations leading to these new developments and get a clearer picture of the possible implications.

The idea of encoding and processing information using the laws of quantum mechanics may nowadays look as "natural," but in reality it captured the attention of physicists and computer scientists only after a series of conceptual breakthroughs that demonstrated i) the existence (or the strong possibility) of a fundamental difference between the classical and quantum regimes and ii) the feasibility of quantum information processing. One turning point is represented by Bell's work [41, 42], which showed that the degree of correlation between spatially separated entangled quantum systems exceeds the values predicted by any theory that describes the system in terms of classical variables. This discovery conduced to intense research into problems concerning the foundation of quantum mechanics, which, on the one hand, led to major experimental advances in controlling single quantum systems and, on the other, revealed the affinity of quantum mechanics for ideas coming from classical information theory. Another important step was the realization that quantum properties, such as the existence of entangled states, can be exploited and used in practical applications. Quantum cryptography (particularly quantum key distribution) is the first remarkable example of how an information-related task can be better performed when taking full advantage of the structure of the natural world.

The fundamental ideas that have contributed to the development of quantum information theory originate from quantum mechanics, on the one hand, and from classical information theory and cryptography, on the other. In addition, since (most likely) quantum computation will only work in combination with quantum error-correction, there is an intrinsic relationship between quantum computation and quantum information theory, a relationship much more intimate than its classical counterpart involving computer science and information theory. To provide some perspective, we begin with a sketch of the historical strands that have converged into the development of quantum information theory.

The modern theory of *quantum mechanics* was essentially created in the 1920s after about a quarter century of profound crisis marking the limits of classical physics. Most generally, quantum theory can be viewed as a mathematical and conceptual *framework* for constructing physical theories, such as, for example, the nonrelativistic quantum mechanics and quantum electrodynamics. The first steps toward a quantum theory of information came almost half a century later, in 1964, with Bell's analysis [41] of the Einstein–Podolsky–Rosen (EPR) thought experiment [152]. Bell's work has shown that the correlations between two quantum systems that have interacted in the past are inconsistent with the laws of classical physics. This profound insight was subsequently confirmed experimentally by a sequence of improved tests, starting with the 1972 work of Freedman and Clauser [182] and the 1982 experiment by Aspect, Dalibard, and Roger [30]. These developments generated a serious effort to understand the foundations of quantum mechanics, including its possible relation with information theory, and led to remarkable experimental advances, including the possibility of controlling single quantum systems, such as a single atom in an "atom trap," which host high purity quantum states. This ability to realize and control high purity quantum states is also a critical requirement for quantum computation. The basic concepts associated with the operational formulation of quantum mechanics (the natural language for quantum information theory) are summarized below (Sec. 10.3).

Information theory has its roots in a remarkable pair of papers published by Claude Shannon in 1948 [466]. Shannon revolutionized our understanding of communication by introducing the technical, mathematically defined concept of *information* (which differs substantially from the corresponding everyday term) and by providing answers to two basic questions: i) how much resources are required to send a certain amount of information over a communication channel, in particular what is the irreducible complexity below which a signal cannot be compressed, and ii) how can one transmit information so that it is protected against noise in the communication channel? The answers to these questions represent the content of two fundamental theorems in information theory, the *noiseless coding theorem*, which provides the mathematical tool for assessing data compaction (i.e., the limits for lossless data compression), and the *noisy coding theorem*, which establishes the ultimate transmission rate for reliable communication over a noisy channel. Reliable transmission in the presence of noise requires *error-correcting* codes for protecting the information being sent and Shannon's theorem gives the upper limit for the protection that can be supplied by such codes. A synopsis of classical information theory is provided in Sec. 10.2.

Cryptography is the venerable science of safe information processing in a world of mistrust. A well-known cryptographic problem concerns the safe communication between two parties in the possible presence of a malevolent third party. "Traditional" cryptographic protocols are based on a *private key* that is shared by the two parties and used for encrypting and decrypting the messages transmitted over an unsafe channel. Such cryptosystems face the basic problem of key distribution – the key may be intercepted by the malevolent third party and used to decrypt some messages. The first ideas for solving this problem using *quantum key distribution* were proposed by Stephen Wiesner in the late 1960s (but not accepted for publication!) and expanded almost 15 years later by Bennet and Brassard [48]. A second major type of cryptosystem – the *public key* or *asymmetric key* cryptosystem – is based on one party having a *public key* that allows encryption and is made available to the general public and a *private key* that allows decryption and is kept secret. There is an algorithm relating the public and private keys, but running it is a computationally hard problem, e.g. factoring a large integer. The ideas behind public key cryptography were (publicly) proposed during mid 1970s by Diffie and Hellman [143] and by Ralph Merkle [360] and later refined by Rivest, Shamir, and Adleman [433] who developed the RSA cryptographic system (currently the most widely used public key cryptosystem). Similar ideas were developed a few years earlier by scientists working for a British intelligence agency. Since the security of the RSA system relies on factoring being a hard computational problem, the possibility of breaking the cryptographic codes by running Shor's algorithm on a quantum computer accounts for much of the current interest in quantum information and quantum computation.

The development of *quantum information theory* can be traced back to the 1960s and 1970s, with research inspired by the problem of transmitting classical data using quantum systems, e.g., coherent states generated by lasers [205, 206]. The existence of a fundamental bound for our ability to access classical information from a quantum system was conjectured by Gordon in 1964 [213] and later stated without proof by Levitin [330]. The formal proof that this bound (now known as the *Holevo bound*) holds was given in 1973 by Holevo [248]. Three years earlier, Wiesner introduced the visionary notion of "quantum money," but, unfortunately, his work was not accepted for publication until 1983 [541]. Wiesner's idea was instrumental to the development of the so-called BB84 protocol for quantum key distribution, which was proposed by Bennett and Brassard in 1984 [48]. Two years earlier, a profound, although apparently simple result – the *no-cloning theorem* showing the impossibility of universally cloning quantum states – was demonstrated by Wootters and Zurek [555] and by Dieks [141]. A different protocol for quantum key distribution, based on entangled states, was introduced by Ekert in 1991 [154]. The following years witnessed the development by Bennett and several collaborators of the superdense coding protocol [51] and the teleportation protocol [49]. Both these protocols rely heavily on quantum entanglement. In 1995 Ben Schumacher [457] introduced the concept of *quantum information* and provided the quantum analog of Shannon's noiseless coding theorem. No analog of the noisy coding theorem has yet been proven, but the theory of quantum error-correction [140, 336] has rapidly developed since 1996, starting with the work of Calderbank and Shor [93] and Steane [491]. This development has played a crucial role in establishing the feasibility of quantum computation. Note that, given the extreme fragility of coherent quantum states in the presence of noise, the feasibility of quantum computation was a serious initial concern. Quantum error correction was further advanced by the discovery of the so-called *stabilizer codes* by Calderbank et al. [92] and by Gottesman [215]. Another important step forward was the demonstration by Shor [477] and Kitaev [291] that error-correction can be achieved using imperfect corrective operations, which led to the general concept of *fault tolerant* computation [416]. A particularly interesting direction is the development of *topological stabilizer codes* [82, 297], which are able to attain superior protection from decoherence by encoding the quantum information using topologically nontrivial states.

This brief historical perspective does not address the latest developments in quantum information theory. The field has visibly matured and the research activity has increased significantly, focusing more and more on practical aspects relevant to quantum technologies. More recent examples of remarkable results include the discovery of the state-merging protocol [251], the realization that certain noisy quantum channels with zero individual quantum capacity have a nonzero joint quantum capacity [481], or finding the limits imposed on superresolution techniques [510] – schemes that allow one to increase imaging precision beyond the typical diffraction limit – by noise stemming from the cross-talk among different modes used experimentally [202]. In addition, we have not mentioned the main open problems in this field, some of them having major breakthrough potential [252]. Nonetheless, our incomplete survey clearly reveals the main themes of quantum information theory and points out some of the critical tasks that need to be addressed. We note here that the term *quantum information* is used either broadly, encompassing all subjects related to information processing using quantum states, or in a more specialized manner, as referring to the study of *elementary* quantum information processing tasks. We will use the terms *quantum information theory* or *quantum Shannon theory* to refer to the specialized meaning. The basic goals in this field are:

1. Identify and characterize the static information resources made available by quantum mechanics. How can one encode information using quantum states and what are the measures of this information? Basic examples include quantum states in a two-dimensional Hilbert space (the so-called *qubit*), classical *bits* (classical physics is a special limit of quantum behavior), and entangled Bell-type states shared by two spatially separated parties. The von Neumann entropy (which generalizes the Shannon entropy) provides a method of quantifying the information content.

2. Identify the dynamical processes of quantum mechanics that can be exploited to perform quantum information tasks. What types of "information processing" are available in quantum mechanics? Basic examples include unitary transformations representing *quantum gates* (the quantum correspondent of classical logic gates), measurements (which represent the interface between the quantum and the classical worlds, "translating" quantum information into classical information), transmission of information along quantum channels (formally represented by completely positive trace-preserving maps), etc.

3. Quantify the resource requirements necessary for performing elementary dynamical processes using static resources. For example, what are the resources necessary to achieve optimal compression of a quantum information source, i.e., what is the fundamental limit on quantum data compression? Or the quantum version of the noisy channel problem: What are the minimal resources required to reliably transfer quantum information over a noisy communications channel?

We emphasize that in quantum information theory there are certain types of resources (e.g. entanglement) that have no classical correspondent. This translates in a larger variety of problems that need to be address. An incomplete list of problems include: constructing tools for the characterization and quantification of resources (e.g., *entropy* and *distance measures* – how much information is encoded in a set of quantum states and how "close" two quantum states are?); source coding (e.g., *Schumacher compression* – what is the limit on quantum data compression?); determining the accessible information (e.g., the *Holevo bound* – the upper bound to the amount of information which can be known about a quantum state); determining the classical capacity of a quantum channel (how much classical information can be transmitted over a quantum channel?); determining the quantum channel capacity (how much quantum information can be reliably transmitted over a noisy quantum channel?); optimizing the use of "unique" quantum resources (e.g., *entanglement* – what can be achieved using entanglement as a resource?); identifying possible applications (e.g., the *quantum key distribution*, a protocol that enables secure distribution of private information). Some of the basic aspects of these problems are addressed in this chapter. An elementary discussion of quantum error correction is provided in Chapter 11.

As a closing remark, we note that the development of quantum computation and quantum information have clearly revealed the need to think physically about computation and, more generally, about information processing. A really exciting possibility would be to discover a fundamental information-theoretic principle that constrains all possible physical theories. Something like "No physical theory can allow an efficient solution of class X computational problems." This would be a huge step in the series of grand syntheses that includes action principles, symmetry principles, and thermodynamics principles, which are at the core of our understanding of the physical world.

10.2   CLASSICAL INFORMATION THEORY

What is information? The answer is not simple, as this term has a large variety of technical and nontechnical uses, including different everyday life connotations. It is, therefore, a crucial task to develop a concept of "information" that i) is general enough to cover all technical aspects related to transmitting messages and processing data and ii) can be made quantitative. This task was brilliantly accomplished by Shannon in the context of addressing the technical problem of communication. In Shannon's own words [466], "*the fundamental problem of communication is that of reproducing at one point either exactly or approximately a message selected at another.*" From this perspective, information is what a communication protocol aims to transmit, i.e., something produced by a source that has to be reproduced at the destination. For concreteness, let us imagine that our friend Bob is preparing to leave his windowless lab carrying some sensitive piece of equipment, when he receives the following message from Alice: "it rains." *What* is the information here, *how much* of it was transmitted, and what are the *fundamental problems* concerning the communication process? These key questions were first asked and answered in Shannon's 1948 seminal work.

Information is contextual. How much information is acquired by Bob through successfully receiving Alice's message depends on his prior knowledge and expectations. Learning that "it rains" provides more information if, for example, there were three other equally likely possibilities (e.g., "it's sunny," "it's windy," "it snows") than in the case when there is only one (equally likely) alternative (e.g., "it doesn't rain"). The key aspect here is that the actual message is one *selected from a set* of possible messages. We express this idea mathematically by saying that the message $a =$ "it rains" belongs to a set of possible (weather-related) messages $\mathcal{A} = \{a_1, a_2, \dots\}$. Furthermore, Alice's message carries almost no information if the lab is somewhere in a jungle and the communication occurs during the rainy season ("yeah, of course it rains"). By contrast, Bob would be extremely surprised to hear this message if the lab were located in the Atacama desert. To quantify this degree of "surprise," we ascribe an *a priori* probability $p_k \in [0, 1]$ to every possible message $a_k \in \mathcal{A}$ and define a measure of the information carried by a_k using a (decreasing) function of this probability. In other words, the less likely the message, the more information it carries. Following Shannon, we define the *information content* of a_k as

$$I(a_k) = -\log(p_k). \qquad (10.1)$$

The quantity $I(a_k)$, which is also called *self-information* or *surprisal*, is measured in units determined by the base of the logarithm in Eq. (10.1). We adopt the convention $\log(x) = \log_2(x)$, which corresponds to measuring the information content in *bits*. One bit represents the information content of a message that occurs with 50% probability, e.g., the outcome of a (fair) coin flip. The information source is completely described by the ensemble $A = \{a_1, a_2, \dots; p_1, p_2, \dots\}$ of possible messages to be transmitted and corresponding *a priori* probabilities. To characterize the information source, we define the expected information content of the information source as

$$H(A) = -\sum_k p_k \log(p_k), \qquad (10.2)$$

where the logarithm (here and throughout this chapter) is understood to have base two. The quantity defined in Eq. (10.2) is called *entropy* (or *Shannon entropy*) and plays a crucial role in information theory. Note that the entropy of an information source that can send M possible messages, $\mathcal{A} = \{a_1, \dots, a_M\}$, is maximum when all messages have the same probability $p_k = 1/M$, explicitly $H(A) = \log(M)$. Skewed probability distributions have lower entropies, which means that the corresponding sources contain less information.

Information is coded. Information is not a type of "physical substance," a material "thing," but an abstract concept that captures something related to the *structure* of physical matter. The message "it rains" will provide Bob with the same information regardless of how he receives it: as a modulated sound wave (e.g., Alice shouts from outside the lab), as ink spots on paper or dark pixels on a screen (e.g., Alice sends Bob an email), or as electromagnetic pulses (e.g., Alice calls Bob). In general, the same message is encoded in several different ways as it travels from the *source* (Alice) to the *receiver* (Bob), e.g., in the

FIGURE 10.1 The "standard model" of a communication channel. An information source randomly delivers symbols from an alphabet. The source encoder compresses the data from the source by eliminating redundancy, while the channel encoder adds redundancy to protect the transmitted signal against noise. The received values are decoded to extract the information.

case of a phone call, as sound waves, electromagnetic waves, and sound waves again. Moreover, Alice could have transmitted the same information by saying "il pleut," "es regnet," or "plouă," provided Bob possesses the right decoder. Hence, the differences among various distinct ways to encode information stem from i) information being stored and transmitted using different physical media (e.g., ink spots, sound waves, and electromagnetic pulses) and ii) the use of different codes (e.g., different languages). These aspects are intimately interconnected in the context of the technical problem of communication (see Figure 10.1). Engineered communication channels, such as, for example, telephone lines, represent finite resources characterized by a limited capacity for transmitting information (e.g., a maximum number of bits per second). To maximize the flux of information transmitted through the channel, it is essential to optimize the coding of this information. In other words, one has to address the following question: how much can one compress the transmitted data without loosing any information? The answer to this fundamental problem in information theory represents the content of Shannon's source coding theorem (also called the noiseless coding theorem). In general, a message can be viewed as a string $x_{k_1} x_{k_2} \ldots x_{k_L}$ of symbols (called *letters*) chosen from an *alphabet* $\mathcal{X} = \{x_1, \ldots, x_M\}$. The letters themselves are messages of length $L = 1$. We assume that the letters in the message are *statistically independent* and occur with *a priory* probabilities p_k, $1 \leq k \leq M$, so that the information source is characterized by the ensemble $X = \{x_1, \ldots, x_M; p_i, \ldots, p_M\}$. As a concrete example, suppose that Alice wants to communicate to Bob the results of a quantum measurement that has four possible outputs, $x_k \in \mathcal{X} = \{a, b, c, d\}$, occurring with probabilities $p_a = 1/4$, $p_b = 1/8$, $p_c = 1/8$, and $p_d = 1/2$. The experiment is repeated many times and the results are sent to Bob using a noiseless channel that only accepts bits, i.e., zeros and ones. Consequently, Alice has to encode the information using the binary alphabet $\mathcal{Y} = \{0, 1\}$. A straightforward coding scheme is the following

$$a \to 00, \quad b \to 01, \quad c \to 10, \quad d \to 11, \tag{10.3}$$

i.e., each letter of the alphabet $\mathcal{X}$ is represented using the binary alphabet $\mathcal{Y}$ as a two-letter codeword. Using this scheme, the message "*daadcbdd*" will be received by Bob as "1100001110011111." The performance of the coding scheme can be measured by the *expected* length of a codeword, $\langle L \rangle = \sum_k p_k L_k$, where L_k is the length of the codeword for x_k. More generally, if we consider a large number of messages (or a long message) consisting of $n \gg M$ letters, we can define the *compression rate* as the average number of code bits per letter. For the above example, the expected length is $\langle L \rangle = 2$, the same as the compression rate. Can one do better than that? The answer is yes, and the basic idea is to use a *variable-length* code: shorter codewords for letters (i.e., measurements outputs) that occur with high probability and longer codewords for letters that occur with low probability. For example, the coding scheme

$$ a \rightarrow 00, \quad b \rightarrow 010, \quad c \rightarrow 011, \quad d \rightarrow 1, \tag{10.4} $$

is characterized by an expected codeword length $\langle L \rangle = \frac{7}{4} < 2$. Using this code, the message "*daadcbdd*" becomes the sequence "10000101101011," which can be unambiguously parsed and decoded by Bob. Note that the length of each codeword is equal to the information content $I(x_k)$ of the corresponding letter and the expected length (which is the same as the compression rate) is equal to the entropy of the source, $H(X) = -\sum_k p_k \log(p_k) = \frac{7}{4}$. Also, note that the letters of the binary code occur with equal probabilities, $p_0 = p_1 = \frac{1}{2}$.

The optimized coding scheme in (10.4) is possible because the probability of each symbol x_k is the reciprocal of a power of two. What about the general case? The basic idea is that Alice should group the data to be sent into long sequences of $n \gg M$ letters and, instead of encoding each letter separately, encode the sequence. There is a total of M^n possible sequences, where M is the number of letters in the alphabet $\mathcal{X}$. However, in the limit of large n, a randomly chosen sequence will likely be a so-called *typical sequence*, which is characterized by a number of occurrences of the letter x_k that is proportional to the corresponding *a priori* probability, i.e., $n_1 = np_1$ occurrences for x_1, $n_2 = np_2$ for x_2, etc. The number of typical sequences is $N_{TS} = n!/n_1! \, n_2! \ldots n_M!$. Using the Stirling approximation, $\log n! \approx n \log n - n$, we have

$$ \log N_{TS} \approx n \log n - n - \sum_{k=1}^{M} [np_k \log np_k - np_k] = -n \sum_{k=1}^{M} p_k \log p_k = nH(X), \tag{10.5} $$

where $H(X)$ is the entropy of the information source. In other words, for any skewed probability distribution (which implies $H(X) < \log M$), the number of typical sequences, $N_{TS} = 2^{nH(X)}$, is exponentially smaller than the total number of sequences, $M^n = 2^{n \log M}$. Furthermore, the probability that an actual message is a specific typical sequence is $(p_1)^{n_1}(p_2)^{n_2} \ldots (p_M)^{n_M} \approx 2^{-nH(X)} = 1/N_{TS}$. Note that this probability is the same for all typical sets. Hence, the probability that an emitted sequence is atypical vanishes in the limit $n \rightarrow \infty$, although the number of possible atypical sequences is, in general, exponentially larger than N_{TS}. Consequently, Alice could just encode the typical sequences and ignore the rest. The encoding of the typical sequences can be done using binary strings of length $nH(X)$. There are $2^{nH(X)}$ such binary codewords, the same as the number of typical sequences. The corresponding compression rate (average number of code bits per letter) is $H(X)$. This result, which establishes the limits to possible data compression, represents Shannon's source coding theorem. There exists a coding scheme that can achieve lossless data compression with a rate greater than, but arbitrarily close to the Shannon entropy. If the compression rate is less than $H(X)$, one can virtually be certain that information is lost.

Information is sensitive to noise. Let us return to our friends, Alice and Bob. Suppose that the communication takes place in a noisy environment. To ensure that Bob gets the correct message, Alice can i) speak louder or/and ii) repeat the message several times. If the message is sent over a binary channel, there is a finite probability for a 0 to become 1, or vice versa, due to the presence of random ambient noise and imperfections in the signaling process. In this case, "speaking louder" would imply engineering a new, more reliable channel, which may not be an option for Alice and Bob. On the other hand, "repeating the message," i.e., redundantly encoding the information, increases the probability that Bob correctly determines the message sent by Alice without altering the physical channel (Figure 10.1). However, this solution comes at a price: one needs to use more physical bits in the code, which reduces the rate of the encoding scheme, i.e., the rate at which information is transmitted through the channel. Is there an optimal way to encode the information so that one does not exceed a certain probability of error while maintaining a good communication rate?

The answer, which represented a breakthrough in communication science, is provided by Shannon's second important theorem, the *noisy channel coding theorem*. In essence, Shannon demonstrated that, as long as there is a nonzero correlation between input and output, a channel can be used for arbitrarily reliable communication at nonzero rate. He also found an expression for the optimal rate that can be achieved.

To appreciate the significance of Shannon's result without delving into the mathematical aspect of the problem, let us consider a simple example. Assume that Alice and Bob use a so-called *binary symmetric channel*, i.e., they use a binary alphabet with 1 and 0 occurring with *a priory* probabilities $p_0 = p_1 = \frac{1}{2}$ and a channel that acts on each bit independently flipping it with probability p. To protect the data, Alice uses the following encoding scheme

$$0 \longrightarrow 000, \qquad 1 \longrightarrow 111, \tag{10.6}$$

i.e., she repeats each bit three times. Some of the transmitted bits will be flipped during transmission, so Bob applies a *majority rule* to decode the message, e.g., codewords containing more zeros than ones are decoded as "0." Consequently, when Alice transmits a "0" (using the codeword "000"), Bob interprets correctly the message if he receives "000," which happens with probability $(1-p)^3$, or if he receives one of the combinations "001," "010," and "100," each occurring with probability $p(1-p)^2$. On the other hand, if two or three bits are flipped during transmission (e.g., if Bob receives "101"), the message is decoded incorrectly. The probability of error is $P_e(p) = 3p^2 - 2p^3$. This is less that p, the probability of error without coding, when $0 < p < 1/2$. However, this comes at a price: for each bit of useful information one has to transmit three bits. As a measure of the efficiency of the coding scheme, we define the *rate* R as the number of information bits per bit transmitted

$$R = \frac{k}{n}. \tag{10.7}$$

For the above example the rate is $R = 1/3$. To further reduce the probability of error, Alice could use a *concatenation scheme*: each transmitted bit from (10.6) is represented by a three-bit codeword, e.g., $0 \longrightarrow 000\ 000\ 000$. The probability of error becomes $P_e(P_e(p)) = \mathcal{O}(p^4)$, but the rate drops to $R = 1/9$. Continuing the concatenation procedure makes the probability of error arbitrarily small, but this reliable communication will be at a rate that approaches zero.

In the light of this discussion, Shannon's theorem may come as a surprise. In spite of what the above example seems to suggest, it is in fact possible to attain reliable communication at nonzero rate. The key idea is to consider long strings of n bits carrying the information contained in k data bits at a rate $R = k/n$ and take the limit $n \to \infty$ (with R constant). To encode the data, we choose 2^k codewords among the available 2^n long strings. During transmission, about np bits will flip due to noise, hence each given codeword will "diffuse" into one string from a certain set of about $n!/(np)!(n - np)! \approx 2^{nH(p)}$ typical output strings, where $H(p) = -p \log p - (1 - p) \log(1 - p)$ is the entropy function. To avoid errors, the codewords should be "far apart," so that the "diffusion sets" of two different codewords do not overlap. To fit $2^k = 2^{nR}$ "diffusion sets" (each of size $2^{nH(p)}$) without overlap, one has to satisfy the condition $2^{nR}2^{nH(p)} \leq 2^n$, which implies

$$R \leq 1 - H(p) \equiv C(p). \tag{10.8}$$

The quantity $C(p)$, called the *capacity* of the channel, represents the upper bound of the rate at which information can be reliably transmitted over the noisy channel. Shannon's theorem, the central result of classical information theory, demonstrates the existence of error-correcting codes[1] that allow reliable communication at a rate less than (but arbitrarily close to) the channel capacity, $k/n < C(p)$ with n sufficiently large.

10.3 OPERATIONAL QUANTUM MECHANICS

Operational formulations of quantum mechanics [91, 404] associate the basic concepts of the theory to lists of instructions to be executed in the laboratory.

---

[1] An error correcting code is an algorithm for adding redundant data to a message, so that it can be correctly recovered even when errors occur during transmission. In essence, the code provides a recipe for ascribing 2^k *robust* codewords using the available 2^n strings.

We use this operational approach to overview the basic formalism because it provides the natural language for quantum information theory and preserves a transparent relation with the postulates of quantum mechanics. We note that, more generally, the formal refinements developed within this approach provide key insight for constructing an axiomatic basis for the theory and a solid benchmark for the realist formulations of quantum mechanics.

In the operational approach, notions such as quantum system and quantum state represent useful abstractions that are not in direct correspondence with something that really exists in nature. Instead, a quantum system is defined by an *equivalence class of preparations* – understood as equivalent, well specified macroscopic procedures for producing what one typically calls an electron or a photon with a certain polarization. While every experiment starts with a *preparation*, the final step is represented by the *measurement*, a completely specified experimental procedure that supplies previously unknown information to the observer. *Preparations* and *measurements* are viewed as the primitive notions of quantum theory. The theory itself provides a mathematical representation of these experimental procedures and the rules for calculating the probabilities of the outcomes of every conceivable measurement following every conceivable preparation. In general, it is useful to regard quantum experiments as three-step processes consisting of *preparation* (P), *transformation* (T), and *measurement* (M). The intermediate step, corresponding to the evolution of the quantum state, can be effectively incorporated into either the preparation or the measurement, as we will explicitly show below. In quantum information theory, the state preparation (or initialization), the quantum operations, and the measurement, corresponding to P, T, and M, respectively, are the three basic steps of a quantum information processing protocol.

To clarify these general ideas, let us consider the idealized Stern-Gerlach experiment [201] illustrated in the top panel of Figure 10.2. For this experiment, the *preparation* consists of evaporating silver atoms in a furnace and passing them through a velocity selector, both these devices being symbolized by the left box. The *transformation* corresponds to the evolution of the atoms that emerge from the selector, pass through an inhomogeneous magnetic field, and approach the detection screen. Finally, the *measurement* corresponds to the atoms striking the detector and producing impact marks inside two narrow strips. From an operational point of view, each of these steps represents a well defined list of procedures to be followed in the laboratory. Schematically, they can be represented as "black boxes" with classical or quantum mechanical inputs and outputs, as shown in the bottom panel of Figure 10.2.

10.3.1 Noiseless quantum theory

The first task of quantum theory is to provide a mathematical representation of the experimental procedures. Thus, the output of the preparation P is represented by a ray in a certain complex Hilbert space $\mathcal{H}$, i.e., the state vector $|\psi\rangle$. In addition, as shown in Figure 10.2, P has a classical control (e.g., the

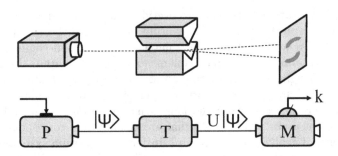

FIGURE 10.2 (*Top*) Idealized Stern-Gerlach experiment: silver atoms evaporated in a furnace emerge from a velocity selector (left), pass through an inhomogeneous magnetic field (center), and hit a detector (right) generating two narrow strips. (*Bottom*) Schematic representation of the basic steps in a typical (noiseless) quantum experiment: preparation (P), transformation (T), and measurement (M). The thin lines and thick arrows denote quantum information and classical information, respectively.

on/off switch). In the language of quantum information theory, P uses classical information as an input and outputs quantum information. By contrast, the transformation unit T has quantum information as both input and output. Mathematically T is represented by the action of a unitary operator U on the state vector and the transformation corresponds to $|\psi\rangle \rightarrow |\psi'\rangle = U|\psi\rangle$. In quantum information theory [548] this reversible unitary evolution represents a *quantum operation*. The final step, the measurement M, takes the quantum information from T as an input and generates a classical output – the classical variable k representing the result of the measurement. We note that certain measurement schemes allow for repeated measurements, i.e., M has an additional quantum information output that can be used as input for another measurement, as suggested by the symbol for measurement in Figure 10.2.

To identify the mathematical representation of M, we recall that the result of every measurement is an eigenvalue a_k of a certain Hermitian operator A associated with M and acting on the Hilbert space $\mathcal{H}$. For concreteness, let us assume that the eigenvalue problem for A reads

$$A|k, j\rangle = a_k|k, j\rangle, \tag{10.9}$$

where j labels the degeneracy of a_k. Also, recall that the second task of quantum theory is to provide rules for calculating the probability $P_k^{(PTM)}$ of the outcome a_k when performing the measurement M on the preparation P followed by the transformation T. This probability is given by the Born rule,

$$P_k^{(PTM)} = \sum_j |\langle k, j|\psi'\rangle|^2, \tag{10.10}$$

where $|\psi'\rangle = U|\psi\rangle$. It is convenient to express Eq. (10.10) in terms of the projection operator $\Pi_k = \sum_j |k, j\rangle\langle k, j|$ onto the subspace corresponding to the eigenvalue labeled by k. We have

$$P_k^{(PTM)} = \langle \psi'|\Pi_k|\psi'\rangle. \tag{10.11}$$

Here, our interest is not in the eigenvalues a_k themselves, but rather in the probability $P_k \equiv P_k^{(PTM)}$ of having an eigenvalue labeled by k as an outcome of M. According to Eq. (10.11), this probability is completely determined by the set of Π_k projectors, called von Neumann's *projection valued measure* (PVM). Consequently, we represent the measurement M by the PVM corresponding to the set Π_k. Note that, in fact, M corresponds to an equivalence class of measurements associated with different Hermitian operators A that have the *same eigenvectors* and eigenvalues labeled by the *same set* of k labels. For example, measurements associated with the Hermitian operators A (with eigenvalues a_k) and A^3 (with eigenvalues a_k^3) belong to the same equivalence class M and are represented by the same PVM.

The conclusion, so far, is that the mathematical representation of the output of the preparation P is a state vector $|\psi\rangle$, the transformation T is associated with a unitary operator U, and the measurement corresponds to a PVM Π_k, while the probability for having the output k is given by Eq. (10.11). We note that the transformation T can be combined with the preparation P, which results in an effective preparation $P' = PT$. Similarly, it can be merged into the measurement M and the result is an effective measurement $M' = TM$. These operations do not affect the probabilities for the outcomes of measurements and we have $P_k^{(PTM)} = P_k^{(P'M)} = P_k^{(PM')}$ or, explicitly,

$$\langle \psi|U^\dagger \Pi_k U|\psi\rangle = \langle \psi'|\Pi_k|\psi'\rangle = \langle \psi|\Pi_k'|\psi\rangle. \tag{10.12}$$

From Eq. (10.12) one can easily identify the mathematical representations of the outcome of the effective preparation P' and of the effective measurement M' as being $|\psi'\rangle = U|\psi\rangle$ and $\Pi_k' = U^\dagger \Pi_k U$, respectively. Note that incorporating the evolution into the state vector corresponds to the Schrödinger picture, while associating it with the PVM is equivalent to using the Heisenberg picture.[2]

10.3.2   Noisy quantum theory

We have implicitly assumed, so far, that there is no loss of information associated with any of the steps involved in a quantum experiment. However, this is not the case in real-life experiments, which typically involve classical stochastic processes and an uncontrolled coupling between the system of interest and

---

[2]Basically, in the Schrödinger picture (or Schrödinger representation) the state vectors are time-dependent, while the operators are constant. By contrast, in the Heisenberg picture the time dependence is incorporated into the operators [361, 443].

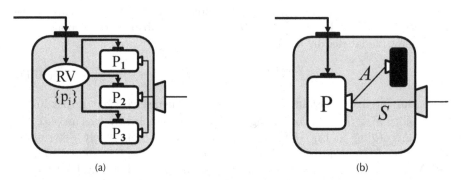

FIGURE 10.3 Schematic representation of a generalized preparation as an ensemble of preparations (left) and as the preparation of a composite system (right). (a) A preparation P_i from the ensemble is selected with probability p_i by a random variable (RV). (b) A composite system is prepared in a pure state $|\Psi^{(\mathcal{SA})}\rangle$ and the ancillary component $\mathcal{A}$ is discarded by performing a "measurement of the identity."

other quantum systems that we cannot access. These can be viewed as sources of noise and they may have an impact on any of the basic steps in a real-life experiment. The natural question is how can one generalize the formalism to incorporate the effect of noise generated by classical and quantum sources.

Generalized preparations

We start with the preparation, more specifically with the question: what is the mathematical representation of the most general preparation? To model the presence of stochastic processes, we assume that the preparation is randomly selected from an ensemble of N possible preparations according to the value of a classical random variable. As shown in Figure 10.3(a), the preparation P_i represented by $|\psi_i\rangle$ is realized when the random variable takes the value i, which happens with a probability p_i. Let us assume that the experiment is finalized by a measurement M represented by the set of projectors Π_k. Since we do not have access to the random variable, the probability of getting the outcome k will be given by the ensemble average $P_k^{(PM)} = \sum_i p_i P_k^{(P_i M)}$. Explicitly, we have

$$P_k^{(PM)} = \sum_i p_i \langle \psi_i | \Pi_k | \psi_i \rangle = \text{Tr}\left(\rho \Pi_k\right), \qquad (10.13)$$

where $\text{Tr}(\dots)$ represents the trace and ρ is the *density operator*,

$$\rho = \sum_i p_i |\psi_i\rangle\langle\psi_i|. \qquad (10.14)$$

It is straightforward to show that the density operator is Hermitian, $\rho = \rho^\dagger$, and satisfies the following properties

$$\langle \psi | \rho | \psi \rangle \geq 0 \qquad \forall | \psi \rangle \in \mathcal{H}, \tag{10.15a}$$

$$\text{Tr}(\rho) = 1, \tag{10.15b}$$

$$\text{Tr}(\rho^2) \leq 1. \tag{10.15c}$$

The preparation described by a density operator with $\text{Tr}(\rho^2) < 1$ is called a *mixed state*, while a preparation represented by a single vector $|\psi_{i_0}\rangle$, which implies $p_i = \delta_{ii_0}$ and $\rho = |\psi_{i_0}\rangle\langle\psi_{i_0}|$, is called a *pure state* and is characterized by $\text{Tr}(\rho^2) = 1$. It can be shown that any density operator admits a convex decomposition in pure states, which means that there always exists an ensemble $\{p_i, |\psi_i\rangle\}$, with $|\psi_i\rangle$ being pure states and p_i probabilities, i.e., $p_i \geq 0$ and $\sum_i p_i = 1$, that satisfies Eq. (10.14). Note that the decomposition is not unique.

A density operator defined by the properties in Eq. (10.15) is the mathematical representation of the most general preparation. However, there is an alternative viewpoint concerning the source of noise in quantum systems which involves the coupling of the system of interest with another quantum system that one cannot access. Let us consider the preparation of a composite system consisting of subsystems $\mathcal{S}$, the system of interest, and $\mathcal{A}$, an *auxiliary* (or *ancilla*) system that we completely disregard. The Hilbert space associated with the composite system is $\mathcal{H}^{(\mathcal{S})} \otimes \mathcal{H}^{(\mathcal{A})}$, where $\mathcal{H}^{(\mathcal{S})}$ and $\mathcal{H}^{(\mathcal{A})}$ are the Hilbert spaces for the main system and the ancilla, respectively. Having no information about $\mathcal{A}$ is equivalent to a measurement of the identity operator $I^{(\mathcal{A})}$. Indeed, the identity operator can be viewed as a trivial PVM with a single outcome that is always realized, as evident from Eq. (10.13) upon the substitution $\Pi_k \to I$ and using the property $\text{Tr}(\rho) = 1$. Consequently, "measuring the identity" provides no information.

Let us assume that the preparation of the composite system is represented by the state vector $|\Psi^{(\mathcal{SA})}\rangle$ and that a measurement $\Pi_k^{(\mathcal{S})}$ is performed on the main system, as shown in Figure 10.3(b). The actual measurement of the composite system is represented by the direct product $\Pi_k^{(\mathcal{S})} \otimes I^{(\mathcal{A})}$, where the two operators act on $\mathcal{H}^{(\mathcal{S})}$ and $\mathcal{H}^{(\mathcal{A})}$, respectively. The probability of an outcome k is

$$P_k^{(PM)} = \langle \Psi^{(\mathcal{SA})} | \Pi_k^{(\mathcal{S})} \otimes I^{(\mathcal{A})} | \Psi^{(\mathcal{SA})} \rangle. \tag{10.16}$$

Next, let $|\psi_i\rangle$ and $|\varphi_m\rangle$ be basis vectors in $\mathcal{H}^{(\mathcal{S})}$ and $\mathcal{H}^{(\mathcal{A})}$, respectively. The vector representing the preparation of the composite system can be written in terms of these basis vectors as $|\Psi^{(\mathcal{SA})}\rangle = \sum_{i,m} \alpha_{im} |\psi_i\rangle \otimes |\varphi_m\rangle$, where α_{im} are complex coefficients. The key observation is that $\Lambda_{ij} = \sum_m \alpha_{im}\alpha_{jm}^*$ is a Hermitian matrix that can be diagonalized by a unitary transformation $U^{(\Lambda)}$. Consequently, one can always change the basis in $\mathcal{H}^{(\mathcal{S})}$, $|\psi_i\rangle \to |\widetilde{\psi}_i\rangle$, so that $|\psi_i\rangle = \sum_j U_{ij}^{(\Lambda)} |\widetilde{\psi}_j\rangle$ and $\widetilde{\alpha}_{im} = \sum_j U_{ij}^{(\Lambda)} \alpha_{jm}$, which makes the corresponding Hermitian matrix diagonal, $\widetilde{\Lambda}_{ij} = \sum_m \widetilde{\alpha}_{im}\widetilde{\alpha}_{jm}^* = \delta_{ij} \sum_m |\widetilde{\alpha}_{im}|^2$. Using this basis and the fact $\langle\varphi_m|I^{(\mathcal{A})}|\varphi_n\rangle = \delta_{nm}$, Eq. (10.16) becomes

$$P_k^{(PM)} = \sum_i \left(\sum_m |\widetilde{\alpha}_{im}|^2\right) \langle\widetilde{\psi}_i|\Pi_k^{(\mathcal{S})}|\widetilde{\psi}_i\rangle. \tag{10.17}$$

Finally, we observe that Eq. (10.17) has the same form as the first equality in Eq. (10.13) with $p_i = \sum_m |\widetilde{\alpha}_{im}|^2$ being positive quantities that satisfy the condition $\sum_i p_i = 1$ dictated by the normalization of $|\Psi^{(\mathcal{SA})}\rangle$. We conclude that the effective preparation of $\mathcal{S}$ involving an ancillary system about which we have no information is represented mathematically by a density operator ρ that acts on $\mathcal{H}^{(\mathcal{S})}$ and satisfies the properties in Eq. (10.15).

A few comments are in order. First, we point out that the density operator can also be written as a partial trace over the degrees of freedom of the auxiliary system, $\rho = \mathrm{Tr}_{\mathcal{A}}\left(|\Psi^{(\mathcal{SA})}\rangle\langle\Psi^{(\mathcal{SA})}|\right)$, as one can establish starting from Eq. (10.16) and using the relation $\mathrm{Tr}_{\mathcal{SA}}(\dots) = \mathrm{Tr}_{\mathcal{S}}\mathrm{Tr}_{\mathcal{A}}(\dots)$. The pure state $|\Psi^{(\mathcal{SA})}\rangle$ is called a *purification* of ρ. Second, we note that ρ represents a mixed state if $\mathcal{S}$ and $\mathcal{A}$ are entangled and a pure state otherwise. Indeed, if the state vector of the composite system can be written as a direct product, $|\Psi^{(\mathcal{SA})}\rangle = |\psi^{(\mathcal{S})}\rangle\otimes|\varphi^{(\mathcal{A})}\rangle$ (i.e., the two systems are not entangled), it follows from our first observation that $\rho = |\psi^{(\mathcal{S})}\rangle\langle\psi^{(\mathcal{S})}|$. Physically, these properties imply that the lack of information involved in the preparation stems from the entanglement of the system with another quantum system (e.g., the environment) to which we have no access. Third, one can prove [389] the following *purification theorem*: every density operator ρ acting on $\mathcal{H}^{(\mathcal{S})}$ admits a purification $|\Psi^{(\mathcal{SA})}\rangle \in \mathcal{H}^{(\mathcal{S})}\otimes \mathcal{H}^{(\mathcal{A})}$ on a larger system. The purification is not unique.

Generalized measurements

Standard von Neumann projective measurements describe ideal situations in which noise, such as, for example, the loss of information due to faulty detectors, is completely absent. To incorporate noise, one can generalize the concept of measurement following the main ideas developed in the context of the discussion about generalized preparations. As suggested by the schematic representations in Figure 10.4, one can adopt two points of view: i) the noise is due to the lack of information about the measurement that is actually performed and ii) the noise stems from the fact that one also measures an auxiliary (or ancilla) system $\mathcal{A}$, in addition to the system of interest $\mathcal{S}$.

Consider the first point of view: there are m sets of PVMs, $\Pi_k^{(i)}$, with the same number of possible outcomes and, depending on the value of a random variable, a measurement with an outcome k is performed by the set $\Pi_k^{(i)}$ with probability p_i [see Figure 10.4(c)]. The probability of having the outcome k is $P_k = \sum_i p_i \mathrm{Tr}\left(\rho\Pi_k^{(i)}\right) = \mathrm{Tr}\left(\rho E_k\right)$, where the operator $E_k = \sum_i p_i \Pi_k^{(i)}$ is the mathematical representation of the noisy effective measurement. The set E_k is called a *positive operator valued measure* (POVM) [404, 417]. Unlike projective measurements, which satisfy the orthogonality condition $\Pi_k\Pi_{k'} = \delta_{kk'}\Pi_k$, the generalized operators E_k are nonorthogonal. One consequence of this property

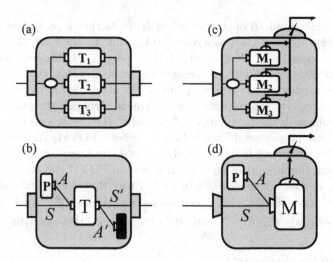

FIGURE 10.4 Schematic representation of generalized transformations (a-b) and generalized measurements (c-d). The source of noise is either our incomplete knowledge of the transformation (or measurement) that is actually performed (a-c), or the fact that the system of interest ($\mathcal{S}$) is transformed (or measured) together with an auxilliary system ($\mathcal{A}$) about which we have no information (b-d).

is that there is no preparation that would make the outcome of the measurement completely predictable. By contrast, a measurement represented by a (standard) PVM gives the outcome k with certainty when measuring the pure state represented by the state vector $|\psi_k\rangle$ or, equivalently, by $\rho = |\psi_k\rangle\langle\psi_k|$.

Before summarizing the properties of E_k, we note that the interpretation of a POVM as a mixture of PVMs is not always possible. A generic POVM can be realized using the ancilla construction illustrated in Figure 10.4(d). The system $\mathcal{S}$ is measured together with an auxiliary system $\mathcal{A}$ prepared in a state represented by $\rho^{\mathcal{A}}$ using the projective measurement $\Pi_k^{(\mathcal{SA})}$. Assume that $\mathcal{S}$ is obtained using a preparation $\rho^{\mathcal{S}}$. The probability of an outcome k is $P_k = \text{Tr}_{\mathcal{SA}}\left[\Pi_k^{(\mathcal{SA})}\left(\rho^{\mathcal{S}} \otimes \rho^{\mathcal{A}}\right)\right] = \text{Tr}_{\mathcal{S}}\left[\rho^{\mathcal{S}}\text{Tr}_{\mathcal{A}}\left(\Pi_k^{(\mathcal{SA})}\rho^{\mathcal{A}}\right)\right]$. Consequently, the generalized measurement is represented by

$$E_k = \text{Tr}_{\mathcal{A}}\left(\Pi_k^{(\mathcal{SA})}\rho^{\mathcal{A}}\right). \tag{10.18}$$

Note that E_k is an operator acting on the Hilbert space $\mathcal{H}^{(\mathcal{S})}$ associated with $\mathcal{S}$. Using Eq. (10.18) one can demonstrate the following properties:

$$P_k = \text{Tr}(\rho E_k), \tag{10.19a}$$

$$\langle\psi|E_k|\psi\rangle \geq 0 \qquad \forall|\psi\rangle \in \mathcal{H}, \tag{10.19b}$$

$$\sum_k E_k = I, \tag{10.19c}$$

where $\mathcal{H} \equiv \mathcal{H}^{(S)}$ and all operators act of $\mathcal{H}$. Note that a projective measurement Π_k also satisfies the properties (10.19) but, in addition, the operators Π_k are orthogonal and idempotent, $\Pi_k \Pi_{k'} = \delta_{kk'} \Pi_k$. Orthogonality ensures that for certain preparations (corresponding to pure states $|\psi_k\rangle$) the outcome of the measurement is certain, as pointed out above, while idempotence expresses the property that repeated measurements generate the same outcome, in accordance with the postulates of "standard" (i.e., noiseless) quantum mechanics. These properties do not hold for a generic POVM. We conclude that a general measurement is represented by a positive operator valued measure E_k satisfying the properties (10.19). Any desired POVM can be generated by extending the Hilbert space $\mathcal{H}^{(S)}$ in such a way that, in the extended space $\mathcal{H}^{(S)} \otimes \mathcal{H}^{(A)}$ there exists a set of orthogonal projectors $\Pi_k^{(SA)}$ such that E_k is the result of projecting $\Pi_k^{(SA)}$ onto $\mathcal{H}^{(S)}$ (Neumark's theorem [198, 404]).

Generalized transformations

The evolution of a quantum system in the standard formulation of quantum theory is represented by unitary maps, as discussed in Section 10.3.1. Generalizing these transformations to account for stochastic processes and decoherence, or, in other words, for the loss of quantum information, can be achieved using the same ideas already discussed in the context of generalized preparations and measurements. As shown schematically in Figure 10.4(a), the first point of view involves an ensemble of unitary transformations, $\{p_i, U_i\}$, each transformation U_i being realized with a certain probability p_i. Assuming that the transformation applies to a preparation ρ and is followed by a measurement E_k, the probability of an outcome k is $P_k = \sum_i p_i \mathrm{Tr}\left(U_i \rho U_i^\dagger E_k\right)$. This probability can be expressed in terms of a generalized transformation $\mathcal{T}$ as $P_k = \mathrm{Tr}\left[\mathcal{T}(\rho) E_k\right]$, where $\mathcal{T}(\cdot)$ is a so-called *superoperator* defined as a linear map from the space of operators on $\mathcal{H}$ to the space of operators on $\mathcal{H}'$, where, in general, $\mathcal{H}$ and $\mathcal{H}'$ can be different Hilbert spaces,

$$\mathcal{T}(\cdot) = \sum_i p_i U_i(\cdot) U_i^\dagger. \tag{10.20}$$

Note that Eq. (10.20) implies $\mathcal{H} = \mathcal{H}'$. However, this is not the most general representation of a superoperator. Figure 10.4(b) illustrates schematically the evolution of an open quantum system that interacts with an auxiliary system, typically the environment. The evolution of the composite system is represented by the unitary operator $U^{(SA)}$, while the effective transformation $\mathcal{S} \to \mathcal{S}'$ is given by a partial trace over the degrees of freedom of the evolved ancilla $\mathcal{A}'$,

$$\mathcal{T}(\cdot) = \mathrm{Tr}_{\mathcal{A}'}\left[U^{(SA)}\left(\cdot \otimes \rho^{(A)}\right)\left(U^{(SA)}\right)^\dagger\right], \tag{10.21}$$

where $\rho^{(A)}$ is the density operator associated with the auxiliary system. Note that, in general, $\mathcal{A}' \neq \mathcal{A}$ and $\mathcal{S}' \neq \mathcal{S}$.

The properties of the superoperator $\mathcal{T}$ defined by Eq. (10.21) guarantee that, when applied to a density operator, the result is a density operator. More specifically, the superoperator $\mathcal{T}$ is i) convex-linear, ii) trace-preserving, and iii) completely positive. Formally, these properties can be expressed as follows

$$\mathcal{T}\left(\sum_i w_i \rho_i\right) = \sum_i w_i \mathcal{T}(\rho_i) \qquad \forall w_i \geq 0, \tag{10.22a}$$

$$\text{Tr}\left[\mathcal{T}(\rho)\right] = 1 \qquad \forall \rho \text{ with } \text{Tr}(\rho) = 1, \tag{10.22b}$$

$$\left(\mathcal{T} \otimes I^{(\mathcal{A})}\right)\left(\rho^{(\mathcal{SA})}\right) > 0 \qquad \forall \rho^{(\mathcal{SA})} \in \mathcal{H}^{(\mathcal{S})} \otimes \mathcal{H}^{(\mathcal{A})}, \tag{10.22c}$$

where $I^{(\mathcal{A})}$ is the identity operator on $\mathcal{H}^{(\mathcal{A})}$. Note that an operator ρ acting on $\mathcal{H}$ is positive if $\langle \psi | \rho | \psi \rangle \geq 0$ for any vector $|\psi\rangle \in \mathcal{H}$. The condition that $\mathcal{T}$ be completely positive is stronger than the positivity condition, which only requires that any positive operator ρ be mapped to a positive operator $\mathcal{T}(\rho)$. A map satisfying the conditions (10.22), often called a *completely positive trace preserving map*, is the mathematical representation of a general transformation or quantum operation. We note that a completely positive trace preserving map can be represented in terms of so-called Kraus operators [310], $K_i : \mathcal{H}^{(\mathcal{S})} \to \mathcal{H}^{(\mathcal{S}')}$, satisfying $\sum_i K_i^\dagger K_i = I^{(\mathcal{S})}$. Using the Kraus representation, the quantum operation on the preparation ρ represented by the superoperator $\mathcal{T}$ takes the form

$$\mathcal{T}(\rho) = \sum_i K_i \rho K_i^\dagger. \tag{10.23}$$

We conclude this section with a summary of the main results. Operational quantum mechanics is formulated in terms of basic concepts that correspond to physical procedures to be followed in the laboratory : *preparations, transformations* (also called *evolutions* or *quantum operations*), and *measurements* (also called *tests*). The role of quantum theory is to i) provide a mathematical representation of these procedures, and ii) to predict the probability of a certain outcome for the measurement M given the preparation P and the transformation T. In the standard framework, which corresponds to noiseless processes, preparations are represented by vectors in a complex Hilbert space, $\psi \in \mathcal{H}$, transformations are represented by unitary operators, $U : \mathcal{H} \to \mathcal{H}$, and measurements are represented by a set of projectors Π_k. To incorporate stochastic processes and interactions with other quantum systems that we cannot access, which leads to quantum decoherence, the mathematical representations are generalized as follows: preparation $\to \rho$, density matrix satisfying the properties in Eq. (10.15), transformation $\to \mathcal{T}$, completely positive trace preserving map satisfying Eq. (10.22), and measurement $\to E_k$, positive operator valued measure with properties described by Eq. (10.19). Finally, the general rule for calculating the probability of an outcome k, which corresponds to $P_k = \langle \psi | U^\dagger \Pi_k U | \psi \rangle$ in the standard formulation, is

$$P_k = \text{Tr}\left[\mathcal{T}(\rho) E_k\right]. \tag{10.24}$$

10.4   QUANTUM INFORMATION THEORY: BASIC CONCEPTS

Quantum information theory may be seen as an extension of its classical counterpart that focuses on the challenges and opportunities that quantum behavior provide for communication and computation. What are the new primitive resources made available by quantum systems and what can one do with them that one could not do using classical resources?

10.4.1   Quantum bits

In Sec. 10.2, we have defined the *bit* as the basic unit of classical information. We have also seen that information is always "encoded" in the states of various physical systems. Consequently, the term *bit* is often used to refer to the physical system that "carry" the information – in this case a system that can have either one of two distinct classical states. Formally, the term may designate an abstract pair of states (typically labeled as "0" and "1," "+" and "−," "yes" and "no," etc.), without reference to a specific physical system (e.g., a coin, two distinct voltage levels in a circuit, two directions of the magnetization in a ferromagnet, etc.).

The *quantum bit* or the *qubit* – the quantum analog of the classical bit – is a mathematical object representing a two-state quantum system. If $|0\rangle$ and $|1\rangle$ are two possible (orthogonal) states for the qubit (physically realized, for example, as the "spin-up" and "spin-down" states of a spin-$\frac{1}{2}$ quantum system), a generic state is given by the superposition

$$|\psi\rangle = \alpha|0\rangle + \beta|1\rangle, \tag{10.25}$$

where $|\alpha|^2 + |\beta|^2 = 1$ and the orthogonal states $|0\rangle$ and $|1\rangle$, which represent the so-called *computational basis*, are normalized. The complex coefficients α and β can be expressed (up to an irrelevant overall phase) in terms of two angles as $\alpha = \cos\frac{\theta}{2}$ and $\beta = e^{i\varphi}\sin\frac{\theta}{2}$. The mapping $|\psi\rangle \rightarrow (\theta, \varphi)$ gives a representation of the two dimensional Hilbert space of a qubit by a unit sphere (called the *Bloch sphere*) with the North and South poles corresponding to $|0\rangle$ and $|1\rangle$, respectively. For convenience, we also introduce the vector representation:

$$|0\rangle \equiv \begin{bmatrix} 1 \\ 0 \end{bmatrix}, \qquad |1\rangle \equiv \begin{bmatrix} 0 \\ 1 \end{bmatrix}, \qquad |\psi\rangle = \alpha|0\rangle + \beta|1\rangle \equiv \begin{bmatrix} \alpha \\ \beta \end{bmatrix}. \tag{10.26}$$

How much information is represented by a qubit? Naively, the existence of an infinite number of superpositions (i.e., points on the Bloch sphere) suggests that qubits contain significantly more information than classical bits. However, we should be careful here and make a distinction between *specification* information – the amount of information (in bits) required to specify a quantum state (or a sequence of states) – and *accessible* information – how much classical information can be obtained by measuring the qubit (or qubits). The first can be arbitrarily large (one has to specify a point on the Bloch sphere out of infinitely many possible points); the second, however, is limited because in general one cannot reliably identify the state $|\psi\rangle$ by measuring it. In fact, one can show (see Section 10.7) that the maximum amount of information that can be obtained by measuring a qubit is *one bit*, the same as in the classical case. Finally, we note that the information-theoretic sense of the term "qubit" as a measure of quantum information will be set on firm ground in the context of the Schumacher compression (see Section 10.6), the quantum correspondent of Shannon's noiseless coding theorem.

Composite systems and entangled states

Consider now a quantum system containing $n > 1$ qubits. The corresponding 2^n dimensional Hilbert space is the direct product of the Hilbert spaces associated with the qubits, $\mathcal{H} = \mathcal{H}_1 \otimes \mathcal{H}_2 \otimes \cdots \otimes \mathcal{H}_n$. A convenient basis, called the *computational basis*, consists of all 2^n strings of the form

$$|x_1 x_2 \ldots x_n\rangle = |x_1\rangle \otimes |x_2\rangle \otimes \cdots \otimes |x_n\rangle, \qquad (10.27)$$

where $x_i \in \{0, 1\}$ and $|x_i\rangle \in \mathcal{H}_i$. A generic state of the n-qubit system can be written as linear combination of basis states with 2^n complex coefficients (also called *amplitudes*). Note that the corresponding specification information is enormous even for moderate values of n; for comparison, the estimated total number of atoms in the universe is of the order 2^{270}.

Let us now consider a generic two-qubit state of the form

$$|\psi\rangle_{12} = \alpha_{00}|00\rangle + \alpha_{01}|01\rangle + \alpha_{10}|10\rangle + \alpha_{11}|11\rangle, \qquad (10.28)$$

with $\sum_{i,j \in \{0,1\}} |\alpha_{ij}|^2 = 1$. The bipartite state $|\psi\rangle_{12}$ is said to be a *product state* if it can be written as $|\psi\rangle_{12} = |\psi_1\rangle \otimes |\psi_2\rangle$, where $|\psi_1\rangle = a_0|0\rangle + a_1|1\rangle$ and $|\psi_2\rangle = b_0|0\rangle + b_1|1\rangle$. For a product state, the amplitudes in Eq. (10.28) satisfy the condition $\alpha_{ij} = a_i b_j$. A bipartite state $|\psi\rangle_{12}$ that is not a product state is called *entangled*. Important examples of two-qubit entangled states are the so-called *Bell states* (or *EPR pairs*),

$$|\Psi^\pm\rangle_{12} = \frac{|01\rangle \pm |10\rangle}{\sqrt{2}}, \qquad |\Phi^\pm\rangle_{12} = \frac{|00\rangle \pm |11\rangle}{\sqrt{2}}. \qquad (10.29)$$

Bell has shown [41, 42] that the correlations between the outcomes of (independent, spatially separated) measurements performed on two entangled qubits are stronger than any correlation that is classically permitted. This is a consequence of the fact that entangled states possess global properties that are irreducible to local features. Consequently, *shared entanglement* (e.g., Alice and Bob sharing a pair of qubits in an entangled state) represents a new type of (static) information resource that allows information processing beyond what is classically possible. Paradigmatic examples are *superdense coding* [51] and *teleportation* [49] (see Section 10.8).

For multiple qubit systems with $n > 2$ the classification with respect to entanglement is richer. Again, we define $|\psi\rangle$ as being a *product state* (or *fully*

classical control

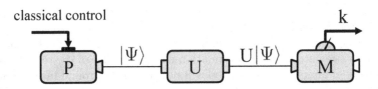

FIGURE 10.5 Schematic representation of typical (noiseless) quantum information processing protocol: state preparation (P), unitary operation (U), and measurement (M). The thin lines and thick arrows denote quantum information and classical information, respectively.

n-partite separable) if and only if $|\psi\rangle = |\psi_1\rangle \otimes \cdots \otimes |\psi_n\rangle$. In addition, we can define separability with respect to various partitions, for example bipartite partitions. A state $|\psi\rangle$ is said to be *bipartite separable* (with respect to the partition $\{i_1, \ldots, i_m\}, \{j_1, \ldots, j_{n-m}\}$) if it can be written as $|\psi\rangle = |\psi_i\rangle \otimes |\psi_j\rangle$, where $|\psi_i\rangle \in \mathcal{H}_{i_1} \otimes \cdots \otimes \mathcal{H}_{i_m}$ and $|\psi_j\rangle \in \mathcal{H}_{j_1} \otimes \cdots \otimes \mathcal{H}_{j_{n-m}}$. Finally, an *n*-qubit state $|\psi\rangle$ is said to be *genuinely n-partite entangled* if and only if $|\psi\rangle$ is *not separable* with respect to any bipartite partition.

10.4.2 Quantum operations

Qubits and shared entanglement are basic (static) resources in quantum information theory. The question now is how does one "encode" classical information by initializing the state of a quantum system? Furthermore, how is quantum information processed and how does one extract the result (as classical information)?

A typical quantum information processing protocol is illustrated schematically in Figure 10.5. A state $|\psi\rangle$ of the quantum system is *initialized* (i.e., *prepared*) based on classical information provided by a control device. The state is then evolved by performing some *quantum operations*. These operations (also called *quantum gates*) are represented by unitary operators. Finally, a (projective) measurement is performed, which provides the classical output. This measurement is represented by a set of projectors, i.e., a *projection valued measure* (PVM). The reader has probably recognized the three basic steps of a quantum experiment – preparation (represented by the state vector $|\psi\rangle$, transformation (unitary operator U), and measurement (PVM Π_k) – as described in the language of (noiseless) operational quantum mechanics. One can also anticipate that the effects of noise can be incorporated by generalizing these steps, so that preparations are represented by *density operators*, measurements by *positive operator valued measures* (POVMs), and noisy evolutions (i.e., transformations) by *completely positive trace preserving maps* (see Section 10.3.2).

Let us focus for a moment on the noiseless evolution of quantum states by considering simple unitary operations performed on qubits (also called quantum *logic gates*), the elementary dynamical processes used for manipulating quantum information. Classically, there is only one (nontrivial) single bit logic gate: the NOT gate. It is natural to define its quantum analog as the unitary operator that evolves the state $|0\rangle$ into the state $|1\rangle$, and vice versa. Using the vector representation (10.26) for the states of the qubit, the unitary operator X corresponding to the NOT gate is represented by the Pauli matrix σ_x. We have

$$X \equiv \begin{bmatrix} 0 & 1 \\ 1 & 0 \end{bmatrix}, \qquad X \begin{bmatrix} \alpha \\ \beta \end{bmatrix} = \begin{bmatrix} \beta \\ \alpha \end{bmatrix}. \qquad (10.30)$$

Notice that the quantum NOT gate corresponds to a rotation in the Hilbert space of the qubit, more specifically a π rotation about the x-axis on the Bloch sphere. In fact, an arbitrary single qubit unitary operator can be written as $U = e^{i\alpha} R_{\hat{n}}(\theta)$, where $\alpha \in [0, 2\pi]$ and $R_{\hat{n}}(\theta)$ represents a rotation by θ about the $\hat{n}$ axis. The rotation can be represented as

$$R_{\hat{n}}(\theta) = \exp\left(-i\theta\hat{n} \cdot \frac{\vec{\sigma}}{2}\right) = \cos\left(\frac{\theta}{2}\right) I - i \sin\left(\frac{\theta}{2}\right)(n_x X + n_y Y + n_z Z),$$

$$(10.31)$$

where I is the 2×2 identity matrix and $\vec{\sigma} = (X, Y, Z)$ are the Pauli matrices. We have $X = e^{i\pi/2} R_x(\pi)$ and similar relations hold for Y and Z.

In addition to the Pauli matrices, there are other important single qubit unitary operations, such as, for example, the *Hadamard* gate (see Figure 10.6)

$$H \equiv \frac{1}{\sqrt{2}} \begin{bmatrix} 1 & 1 \\ 1 & -1 \end{bmatrix}. \qquad (10.32)$$

It is easy to check that H transforms the computational basis into the so-called "$+/-$" or Hadamard basis: $|0\rangle \rightarrow |+\rangle \equiv (|0\rangle + |1\rangle)/\sqrt{2}$ and $|1\rangle \rightarrow |-\rangle \equiv (|0\rangle - |1\rangle)/\sqrt{2}$. Two other frequently used gates are the so-called *phase* (S) and $\pi/8$ (T) gates,

$$S \equiv \begin{bmatrix} 1 & 0 \\ 0 & i \end{bmatrix} \qquad \text{and} \qquad T \equiv \begin{bmatrix} 1 & 0 \\ 0 & e^{i\pi/4} \end{bmatrix}. \qquad (10.33)$$

Note that the Pauli-Z gate can be implemented by applying the phase gate twice, $Z = S^2$. In turn, we have $S = T^2$.

How can one generalize these unitary operations for multiple qubits? First, let us note that a direct correspondence with standard classical gates, such as AND, OR, exclusive OR (XOR), etc., is not possible because these gates are *irreversible* (i.e., *noninvertible*), while quantum unitary evolution is always reversible. Second, we point out an important result in classical computation theory concerning the existence of a *universal gate*: any Boolean function (i.e., function on bits) can be implemented using a combination of NAND gates[3] alone. Finally, we note that one of the most important types of *controlled operation* in classical computation corresponds to the sequence: *if x is true*,

---

[3] The NAND gate (negative AND) generates the output 0 (false) if both inputs are true and the output 1 otherwise.

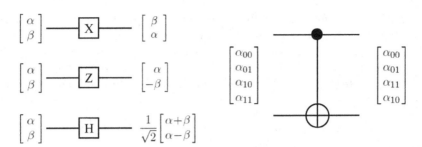

FIGURE 10.6 Schematic representation of some frequently used quantum logic gates and their action on generic single- and two-qubit states. *Left*: Pauli-X (quantum NOT), Pauli-Z, and Hadamard gates. *Right*: CNOT gate.

do the transformation $y \rightarrow y'$. It turns out that this type of operation can be implemented as a quantum gate with two input qubits, known as the *control* qubit $|\psi_C\rangle$ and *target* qubit $|\psi_T\rangle$, and plays a key role in quantum computation.

Assume that U is a single-qubit unitary operation. A *controlled*-U operation corresponds to applying U to the target qubit only if the control qubit is set (otherwise $|\psi_T\rangle$ is left unchanged). The prototypical quantum controlled gate is the *controlled*-NOT (or CNOT) gate (see Figure 10.6). Using the computational basis $|xy\rangle$ corresponding to Eq. (10.28), where $|x\rangle$ and $|y\rangle$ represent the states of the control and target qubits, respectively, the matrix representation of CNOT and its action on a generic two-qubit state are

$$U_{CN} \equiv \begin{bmatrix} 1 & 0 & 0 & 0 \\ 0 & 1 & 0 & 0 \\ 0 & 0 & 0 & 1 \\ 0 & 0 & 1 & 0 \end{bmatrix}, \qquad U_{CN} \begin{bmatrix} \alpha_{00} \\ \alpha_{01} \\ \alpha_{10} \\ \alpha_{11} \end{bmatrix} = \begin{bmatrix} \alpha_{00} \\ \alpha_{01} \\ \alpha_{11} \\ \alpha_{10} \end{bmatrix}. \qquad (10.34)$$

We note that another important example of two-qubit operation is the controlled-Z gate, which corresponds to $U = Z$ and, consequently, changes the sign of the state $|11\rangle$. Remarkably, unlike the CNOT gate, the action of the controlled-Z gate does not change if we switch the control and target qubits. This symmetry property makes the controlled-Z gate more natural for physical implementation, since interactions between identical particles are symmetric.

Similar to classical gates, quantum logic gates can be combined into *quantum circuits*[4]. In general, a quantum circuit can be viewed as a model for quantum computation in which a certain information processing task is realized as a sequence of quantum gates. To design and study quantum circuits, it is convenient to use symbol notations for the gates, as illustrated in Figure 10.6. Finally, we note that the importance of the CNOT gate stems from a remarkable universality result: any multiple qubit logic gate can be implemented

---

[4]See Chapter 11 for more details.

as a combination of CNOT and single qubit gates, i.e., CNOTs and unitary single qubit operations form a *universal set* for quantum computation. In addition, one can show that single qubit operations can be *approximated* with arbitrary precision using a *finite set* of gates. However, this approximation cannot always be done *efficiently*.

10.4.3 No cloning

Classical information processing protocols take full advantage of the fact that the value of a bit can be reproduced by generating an arbitrary number of copies. By contrast, it is impossible to clone an *unknown* quantum state. This statement, known as the *no-cloning theorem* [141, 555], implies that, given a system in a quantum state $|\psi\rangle$, it is impossible to end up with two or more systems in state $|\psi\rangle$, unless the state is already known. Note that the statement does not exclude the possibility of swapping the state $|\psi\rangle$ from one system to another.

Suppose that we have a "quantum copy machine" that takes the "data" (an unknown quantum state $|\psi\rangle$) and a blank "copy" (some standard pure state $|\emptyset\rangle$) and generates a copy of $|\psi\rangle$ through the unitary evolution U,

$$|\psi\rangle \otimes |\emptyset\rangle \longrightarrow U[|\psi\rangle \otimes |\emptyset\rangle] = |\psi\rangle \otimes |\psi\rangle. \tag{10.35}$$

Consider now two arbitrary pure states $|\psi\rangle$ and $|\phi\rangle$ that have been copied using this procedure, i.e., $U[|\psi\rangle \otimes |\emptyset\rangle] = |\psi\rangle \otimes |\psi\rangle$ and $U[|\phi\rangle \otimes |\emptyset\rangle] = |\phi\rangle \otimes |\phi\rangle$. Taking the inner product of these two equations we have $\langle\psi|\phi\rangle = (\langle\psi|\phi\rangle)^2$. However, this equality holds only for $|\psi\rangle = |\phi\rangle$ or when the two states are orthogonal. Hence, a general purpose quantum copy machine is impossible.

It is interesting to note the connection between no-cloning and two defining feature of quantum mechanics. First, there is a link with quantum measurement: no-cloning is logically equivalent to the impossibility of determining the (unknown) state of a single quantum system by performing a measurement on that state. Indeed, if one could determine an unknown state, one could also prepare an arbitrary number of copies of the (now known) state. On the other hand, if cloning were possible, one could produce a large number of copies of the unknown state, then measure them using various measurements, which would allow determining the state with arbitrary precision.

The second connection involves the EPR thought experiment and the impossibility of superluminal signaling. Suppose that Alice and Bob, who are separated by a large distance, share EPR pairs of the form $\frac{1}{\sqrt{2}}(|00\rangle + |11\rangle)$. On each pair, Alice is performing a measurement in either the computational basis $\{|0\rangle, |1\rangle\}$ or the $+/-$ basis $\{|+\rangle, |-\rangle\}$. If cloning were possible, Bob could produce a certain number of copies of his state (after Alice has done the measurement), then measure half of them in the computational basis and half in the $+/-$ basis. This would unambiguously tell Bob what was the basis for the measurement done by Alice, hence Alice and Bob could communicate faster than light. However, quite remarkably, the no-cloning theorem ensures the consistency of nonrelativistic quantum mechanics with relativity.

10.5 ENTROPY AND INFORMATION

In Section 10.2, we have introduced the *Shannon entropy* as a measure of classical information. In essence, if $X = \{x_1, x_2, \ldots; p_1, p_2, \ldots\}$ represents a random variable that takes the value x_i with probability p_i (e.g., letters generated with certain probabilities by an information source), the Shannon entropy $H(X)$ quantifies how much information is gained, on average, by learning the value of X. This corresponds to the amount of uncertainty one has about X before one learns its value. Other important classical information measures, e.g., the *mutual information*, quantify the correlations between random variables.

Here, we discuss several information measures that are used to quantify the amount of information and correlation associated with quantum systems. While most of the definitions are similar to their classical counterparts, the properties of these measures may exhibit radical departures from the expected classical behavior, such as, for example, the *conditional quantum entropy* being negative for certain quantum states and the *mutual information* of entangled qubits exceeding the classical limit. We emphasize that, despite the similarity between the mathematical machinery used in quantum information theory and the mathematics of Shannon's theory, the conceptual contents are sometimes radically different. These differences stem from fundamental quantum properties that have no classical correspondent; in particular, we point out the following features: i) nonorthogonal quantum states cannot be perfectly distinguished and ii) quantum correlations can exceed the classical limit.

Von Neumann entropy. Assume that a source (Alice) prepares quantum states $|x\rangle$ chosen from an ensemble, $x \in \{x_1, x_2, \ldots, x_n\}$, each state occurring with *a priori* probability p_x. From the point of view of an observer (Bob) with no knowledge about which state was prepared, the source is completely characterized by the density operator $\rho = \sum_x p_x |x\rangle\langle x|$. We define the *von Neumann* (or *quantum*) *entropy* – the quantum analog of the Shannon entropy – as

$$S(\rho) = -\text{Tr}(\rho \log \rho), \tag{10.36}$$

where Tr represents the trace and the base of the logarithm is 2. If $\{|\alpha\rangle\}$ are the (normalized) eigenstates of ρ and $\lambda_\alpha = \sum_x p_x |\langle x|\alpha\rangle|^2$ the corresponding eigenvalues, so that $\rho = \sum_\alpha \lambda_\alpha |\alpha\rangle\langle\alpha|$, we can write the entropy as

$$S(\rho) = -\sum_\alpha \lambda_\alpha \log \lambda_\alpha \equiv H(A), \tag{10.37}$$

where $H(A)$ is the Shannon entropy of the ensemble $A = \{\alpha; \lambda_\alpha\}$. Consequently, if the source generates mutually orthogonal pure states, we have $X \equiv \{x; p_x\} = A$ and $S(\rho) = H(X)$, i.e., the quantum source reduces to a classical one. In general, as we will show in Section 10.6, the von Neumann entropy quantifies the incompressible information content of the source, i.e., the minimum number of qubits per letter needed to reliably encode the information generated by a source. Thus, the von Neumann entropy provides meaning to the information-theoretic notion of *qubit* as the fundamental unit of quantum information (i.e., the quantum analog of the classical *bit*). Note that this concept is different from that of *physical qubit*, which corresponds to a two-state quantum system, and its mathematical representation (vectors from a two-dimensional Hilbert space).

Several frequently used basic properties of the quantum entropy are listed below without proof. Most of the proofs are straightforward; more details can be found in, for example, Chapter 9 of Peres [404], Chapter 11 of Wilde [548], or the review article by Wehrl [531].

1. **Positivity.** The entropy is nonnegative, $S(\rho) \geq 0$, for any ρ.

2. **Minimum.** The entropy of a pure state $\rho_\psi = |\psi\rangle\langle\psi|$ is zero, $S(\rho_\psi) = 0$.

3. **Maximum.** The entropy is maximum, $S(\rho) = \log d$, for randomly chosen quantum states, i.e., when $\rho = \frac{1}{d}I$, where I is the identity operator on the d-dimensional Hilbert space.

4. **Concavity.** Mixing states enhances the ignorance about preparation, i.e., it increases the von Neumann entropy, $S(\rho) \geq \sum_k p_k S(\rho_k)$, where $\rho = \sum_k p_k \rho_k$.

5. **Entropy of preparation.** If a pure state is drawn randomly from the ensemble $\{|x\rangle; p_x\}$, so that $\rho = \sum_x p_x |x\rangle\langle x|$, then $S(\rho) \leq H(X)$, with equality if the states $|x\rangle$ are mutually orthogonal.

6. **Entropy of measurement.** Consider the measurement of observable $A = \sum_y \alpha_y |\alpha_y\rangle\langle\alpha_y|$ in state ρ, so that the outcome α_y occurs with probability $p_y = \langle\alpha_y|\rho|\alpha_y\rangle$. Then, $S(\rho) \leq H(Y)$, where $Y = \{\alpha_y; p_y\}$ is the ensemble of measurements outcomes; equality obtains when A and ρ commute.

7. **Subadditivity.** Consider a bipartite system AB in the state ρ_{AB}. Then, $S(\rho_{AB}) \leq S(\rho_A) + S(\rho_B)$, where $\rho_A = \text{Tr}_B[\rho_{AB}]$ and $\rho_B = \text{Tr}_A[\rho_{AB}]$, with equality for uncorrelated systems, $\rho_{AB} = \rho_A \otimes \rho_B$.

8. **Triangle inequality.** For a bipartite system, we have $S(\rho_{AB}) \geq |S(\rho_A) - S(\rho_B)|$. If ρ_{AB} is a pure state, $S(\rho_{AB}) = 0$ and $S(\rho_A) = S(\rho_B)$.

Some of these properties have very similar classical analogs. The distinction between quantum and classical information is nicely illustrated, for example, by the contrast between the triangle inequality and its classical counterpart, $H(X, Y) \geq H(X), H(Y)$. Classically, there is more information to be found in the whole system than in any of its parts (i.e., the whole is less predictable than its parts). Quantum mechanically, the whole may be completely known, yet the measurement outcomes for an entangled subsystem remain unpredictable.

Conditional quantum entropy. Consider the bipartite quantum system AB. The *conditional quantum entropy* $S(A|B)$ is defined as the difference of the *joint quantum entropy* $S(A, B) \equiv S(\rho_{AB})$ and the *marginal entropy* $S(B) \equiv S(\rho_B)$,

$$S(A|B) = S(A, B) - S(B). \qquad (10.38)$$

This quantity is the quantum analog of the (classical) conditional entropy, $H(X|Y) = H(X, Y) - H(Y)$, which is a measure of how uncertain we are, on average, about the value of X, given that we know the value of Y. Of course, $H(X|Y) \geq 0$. By contrast, we can sometimes be more certain about the joint state of a quantum system than we can be about any one of its parts. For example, if AB is a system of two qubits in the entangled state $(|00\rangle + |11\rangle)/\sqrt{2}$, we have $S(A, B) = 0$ (since ρ_{AB} is a pure state) and $S(B) = 1$ (since $\rho_B = \frac{1}{2}I$). This results in a negative value of $S(A|B)$. The negative of the conditional entropy, called *coherent information*, is a measure of quantum correlations, i.e., a measure of the extent to which we know less about part of a system than we do about its whole. Note that the quantum conditional entropy can also be defined operationally as a measure of the quantum communication *cost* (if $S(A|B) > 0$) or *surplus* (if $S(A|B) < 0$) when performing *quantum state merging* [251].

Quantum mutual information. The (classical) *mutual information*, defined as $\overline{H(X:Y) = H(X) + H(Y) - H}(X, Y)$, represents the standard informational measure of correlations in the classical world. In essence, $H(X:Y)$ quantifies how much information X and Y have in common. The quantum analog of this quantity – the *quantum mutual information* of a bipartite state ρ_{AB} – is

$$S(A:B) = S(A) + S(B) - S(A, B) = S(A) - S(A|B). \qquad (10.39)$$

One can show [548] that the quantum mutual information of any bipartite quantum state is positive, $S(A:B) \geq 0$. We also note that discarding quantum systems never increases mutual information, i.e., $S(A:B) \leq S(A:B, C)$ for any composite system ABC, and that quantum operations never increase quantum information, i.e., $S(A':B') \leq S(A:B)$, where $S(A':B')$ is the mutual information after a trace-preserving operation $\mathcal{T}$ (see Section 10.3.2) was applied to subsystem B.

10.6 DATA COMPRESSION

Information is measured by the amount of communication needed to convey it. Classically, the *bit* is shown to be a useful measure of information in the context of Shannon's noiseless coding theorem [466], which establishes how much one can compress a long message without loosing any information, i.e., what is the minimum number of binary symbols necessary for encoding it. *Schumacher compression* [457], the quantum analog of Shannon's noiseless coding theorem, provides an operational interpretation of the *von Neumann entropy* as the fundamental limit on the rate of quantum data compression and establishes the *qubit* as a measure of the amount of quantum information "contained" in a quantum information source.

Suppose a sender (Alice) and a receiver (Bob) have access to a quantum information source and a noiseless quantum channel. Can they optimize the use of these resources by transmitting compressed quantum information? The answer is yes, but, before we go into more details, let us define the four basic steps of a generic quantum compression protocol: state preparation, encoding, transmission, and decoding.

State preparation. The basic type of quantum information source is one that outputs a sequence of systems in particular (pure) quantum states, $|\psi_{x^n}\rangle = |\psi_{x_1}\rangle \otimes \cdots \otimes |\psi_{x_n}\rangle$, according to the ensemble $\{|\psi_x\rangle; p_x\}$. From the perspective of someone ignorant of the actual states, they are characterized by the density operator $\rho = \sum_x p_x |\psi_x\rangle\langle\psi_x|$. Hence, the quantum source can be formally described by a Hilbert space $\mathcal{H}$ and a density matrix ρ acting on that Hilbert space. The compression operation will take the states $|\psi_{x^n}\rangle \in \mathcal{H}^{\otimes n}$, where $x^n = x_1 x_2 \ldots x_n$, to states in a lower dimensional *compressed space* in such a way as to ensure the possibility of faithful recovery.

A different type of quantum source always outputs systems in a particular *mixed* state ρ as a result of these systems being only a part of a larger system which is in a pure state. The mixed nature of the state is due to entanglement between the "visible" subsystem A and the inaccessible reference system R. The (pure) states of the composite system, which represent a *purification* of the density operator ρ, are

$$|\Phi_\alpha\rangle_{RA} = \sum_x \sqrt{p_x} |\phi_{x\alpha}\rangle_R |\psi_x\rangle_A. \qquad (10.40)$$

In this case the source produces a sequence $|\Phi_{\alpha^n}\rangle_{RA} = |\Phi_{\alpha_1}\rangle_{RA} \otimes \cdots \otimes |\Phi_{\alpha_n}\rangle_{RA}$ and the task is to optimally encode the quantum information of the source in such a way as to faithfully transfer the *entanglement* of the output system with the reference system. A successful compression-transmission-decompression protocol will provide Bob with a sequence $|\Phi_{\alpha^n}\rangle_{RB} = |\Phi_{\alpha_1}\rangle_{RB} \otimes \cdots \otimes |\Phi_{\alpha_n}\rangle_{RB}$, where $|\Phi_\alpha\rangle_{RB} = \sum_x \sqrt{p_x} |\phi_{x\alpha}\rangle_R |\psi_x\rangle_B$, similar to that prepared by Alice.

Encoding. The output sequence generated by the source is encoded onto a quantum system Σ_C^n according to some *completely positive trace preserving* compression map[5] $\mathcal{C} : \mathcal{H}^{\otimes n} \to \mathcal{H}_C^n$, where the compressed Hilbert space $\mathcal{H}_C^n$ has dimension 2^{nR}, R being the rate of compression.

Transmission. The quantum system Σ_C^n is transmitted to the receiver (Bob) using $n(R + \delta)$ noiseless qubit channels, where δ is an arbitrarily small positive number.

Decoding. Bob decodes the message by applying a decompression map $\mathcal{D} : \mathcal{H}_C^n \to \mathcal{H}^{\otimes n}$. The protocol is successful if the decompressed state, e.g., $|\Phi_{\alpha^n}\rangle_{RB}$, is close (in trace distance[6]) to the original state.

---

[5] See Section 10.3.2 for the definition and basic properties.

[6] Given any two operators M and N, the trace distance between them is $\|M - N\| = \mathrm{Tr}\left\{\sqrt{(M-N)^\dagger(M-N)}\right\}$. If ρ and σ are density operators, $0 \le \|\rho - \sigma\| \le 2$. The lower bound applies when the two quantum states are equivalent (then, no measurement can

10.6.1   Schumacher's noiseless quantum coding theorem

Consider a quantum information source described by the density operator ρ acting on the Hilbert space $\mathcal{H}$. A reliable compression scheme of rate R exists if and only if $R < S(\rho)$, where $S(\rho)$ is the von Neumann entropy that characterizes the source.

This theorem [269, 457], which represents the quantum analog of Shannon's noiseless coding theorem [466], establishes the von Neumann entropy as the fundamental limit on the rate of quantum data compression. The technical details of the proof can be found in, for example, Refs. [389] and [548]. The key idea is to promote the classical notion of a *typical sequence*[7] to that of a *typical subspace*. More specifically, a quantum state $|\psi_{x^n}\rangle = |\psi_{x_1}\rangle|\psi_{x_2}\rangle \ldots |\psi_{x_n}\rangle$ is said to be a *typical state* if $x^n = x_1 x_2 \ldots x_n$ is a classical *typical sequence*. The *typical subspace* $\mathcal{H}_C$ is defined as the subspace spanned by all typical states $|\psi_{x^n}\rangle$. The crucial distinction between the classical case and the Schumacher compression stems from the fact that the number of typical sequences (hence, the number of typical states) is $2^{nH(p)}$, where $H(p)$ is the Shannon entropy associated with the probability distribution $\{p_x\}$, while the dimension of the typical subspace is, in general, lower, because the states $|\psi_{x^n}\rangle$ are not necessarily orthogonal. Specifically, we have $\dim(\mathcal{H}_C^n) = \text{Tr}(\Pi_C^n) = 2^{nS(\rho)}$, where Π_C^n is the projector onto the typical subspace and $S(\rho)$ is the von Neumann entropy that characterizes the source.

Since nearly all long messages have nearly unit overlap with the typical subspace, it is enough if Alice, by applying the compression map $\mathcal{C}$, encodes the typical subspace components of these states (using $nS(\rho)$ qubits) and ignores the rest. Next, she sends the coded message to Bob, who reconstructs the original message by applying the decompression map $\mathcal{D}$. Note that Alice could have done an effectively classical compression by encoding the (typical) sequences $x_1 x_2 \ldots x_n$ using $2^{nH(p)}$ mutually orthogonal quantum states. Bob would have no problem in reconstructing the original states with arbitrarily high accuracy in the asymptotic limit $n \to \infty$. However, if the "letters" $|\psi_x\rangle$ are drawn from an ensemble of nonorthogonal states, $H(p) > S(\rho)$, hence this classical-type compression is not optimal as there is an additional redundancy associated with the indistinguishability of nonorthogonal states.

Finally, concerning the implementation of this protocol, we note that any encoding circuit must be completely reversible. Also, the original state has to be destroyed in the compression process, otherwise Bob would end up possessing a copy of Alice's state, which is not allowed by the no-cloning theorem. Performing noiseless coherent operations over a large enough number of qubits (to implement the compression protocol) is something extremely far from what is currently accessible in experiment. Nonetheless, Schumacher's theorem

---

distinguish ρ from σ), while the upper bound corresponds to ρ from σ having support on orthogonal subspaces (there exists a measurement that can perfectly distinguish between the two states). For pure states, $\rho = |\psi\rangle\langle\psi|$, $\sigma = |\phi\rangle\langle\phi|$, we have $||\rho - \sigma|| = 2 - 2|\langle\psi|\phi\rangle|^2$.

[7]See Section 10.2 for details.

provides the first important quantum information-theoretic result and the useful insight that, when dealing with quantum protocols, it may be enough to only consider a high-probability subspace, instead of the whole Hilbert space.

10.7 ACCESSIBLE INFORMATION

The ideas discussed in the previous section have close correspondents in the classical world. Next, we focus on a feature of quantum information that illustrates a striking contrast with classical information. The question we address is the following: given a message constructed from an alphabet of quantum states, how much *classical* information can one extract from it?

Suppose that Alice produces (long) messages by preparing either pure or mixed quantum states drawn from the ensembles $\mathcal{E}_\psi = \{|\psi_x\rangle; p_x\}$ and $\mathcal{E}_\rho = \{\rho_x; p_x\}$, respectively. The amount of classical information (per letter) encoded in these messages is given by the Shannon entropy $H(X)$ corresponding to the probability distribution $p_{x_1}, p_{x_2}, \ldots, p_{x_n}$ of the random variable $X = \{x_1, x_2, \ldots, x_n\}$. To extract (some of) this information, Bob is free to perform the measurement of his choice, generally a POVM[8] $\{E_y\} = \{E_{y_1}, \ldots, E_{y_m}\}$. Note that Bob can choose to perform collective measurements on all n letters. The measurement outcomes $Y = \{y_1, \ldots, y_m\}$ will be characterized by a certain probability distribution $\{p_{y_1}, \ldots, p_{y_m}\}$. For example, given the preparation x, Bob will obtain the measurement outcome y with conditional probability

$$p(y|x) = \langle \psi_x | E_y | \psi_x \rangle \quad \text{or} \quad p(y|x) = \text{Tr}(\rho_x E_y). \tag{10.41}$$

The amount of information (per letter) that Bob has gained by doing the measurement is given by how much information X and Y have in common, i.e., by the *mutual information*,

$$H(X:Y) = \sum_{x,y} p_{xy} \log \frac{p(y|x)}{p_y}, \tag{10.42}$$

where $p_{xy} = p(y|x)p_x$ and $p_y = \sum_x p_{xy}$. The POVM that maximizes the information gain is called the *optimal measurement*. Again, for n-letter messages this may be a collective measurement. The maximal information gain represents the *accessible information* of the ensemble $\mathcal{E}_\psi$ (or $\mathcal{E}_\rho$).

Suppose now that Alice sends Bob effectively classical messages using mutually orthogonal states drawn from an ensemble $\{|\psi_x\rangle; p_x\}$. Bob can perform the orthogonal measurement $\Pi_y = |\psi_y\rangle\langle\psi_y|$, which has conditional probability $p(y|x) = \delta_{x,y}$. This implies $p_{xy} = p_x \delta_{x,y}$ and $p_y = p_x$, so that $H(X:Y) = H(X)$. In other words, Bob has complete access to the (classical) information sent by Alice. In fact classically the accessible information is always the same as the entropy of the source $H(X)$.

---

[8]See Section 10.3.2.

Next, consider the case when Alice prepares pure quantum states that are nonorthogonal to each other. No measurement done by Bob can perfectly distinguish between two such states and, as a result, the maximum information gain will be less than $H(X)$. In fact, we would expect the accessible information not to exceed the von Neumann entropy of the source, $S(\rho) < H(X)$. Interestingly, this reduced accessibility (as compared with the classical case) can be viewed as a statement equivalent to the no-cloning theorem. Indeed, assume that Alice prepares the nonorthogonal states $|\psi\rangle$ and $|\phi\rangle$ with probabilities p and $1 - p$, respectively. Alice sends Bob one state. If the accessible information were $H(p)$, Bob could unambiguously identify the state by doing some measurement, then he could prepare copies of that state (in violation of the no-cloning theorem). On the other hand, if cloning nonorthogonal states were possible, Bob could produce multiple copies of whatever state Alice sends him and generate the state $|\psi\rangle^{\otimes n}$ or the state $|\phi\rangle^{\otimes n}$. These two states are nearly orthogonal and can be distinguished with arbitrarily high accuracy by a projective measurement, which would imply that the accessible information is $H(p)$ (in violation of the reduced accessibility statement).

The most general situation involves the preparation of mixed states drawn from an ensemble $\mathcal{E}_\rho = \{\rho_x; p_x\}$. Equivalently, Alice may try to send pure states through a noisy quantum channel, but, as a result of decoherence, Bob receives some mixed states that he has to decode. In this case, the accessible information is limited by the so-called *Holevo information*, $\chi(\mathcal{E}_\rho) = S(\rho) - \sum_x p_x S(\rho_x)$, where $\rho = \sum_x p_x \rho_x$.

10.7.1 The Holevo bound

Suppose Alice prepares a quantum state drawn from the ensemble $\mathcal{E} = \{\rho_x; p_x\}$, then Bob performs a measurement described by the POVM $\{E_y\}$ with possible outcomes $Y = \{y_1, \ldots, y_m\}$. For any such measurement we have

$$H(X:Y) \leq S(\rho) - \sum_x p_x S(\rho_x) \equiv \chi(\mathcal{E}), \qquad (10.43)$$

where $\rho = \sum_x p_x \rho_x$. Equation (10.43) establishes an upper bound on the accessible information, called the *Holevo bound* [248]. The proof can be found in, for example, Nielsen and Chuang [389]. Note that for pure quantum states $\rho_x = |\psi_x\rangle\langle\psi_x|$, we have $S(\rho_x) = 0$ and the Holevo bound becomes $\chi(\mathcal{E}) = S(\rho)$, as discussed above. Also, suppose that $\mathcal{E}$ is an ensemble of mutually orthogonal *mixed* states, i.e., if $x \neq x'$ ρ_x and ρ_y have support on mutually orthogonal subspaces of the Hilbert space and $\text{Tr}(\rho_x \rho_y) = 0$. Then, one can easily show that $S(\rho) = H(X) + \sum_x p_x S(\rho_x)$, which implies $\chi(\mathcal{E}) = H(X)$. In other words, classical information encoded using mutually orthogonal mixed states (or mutually orthogonal pure states), which can be distinguished from one another with arbitrarily high accuracy, can be fully recovered by performing an optimal (generalized) measurement. By contrast, if classical information is encoded using nonorthogonal quantum states, it cannot be fully recovered.

10.8 ENTANGLEMENT-ASSISTED COMMUNICATION

Shared entanglement is an information resource that has no classical correspondent. Hence, the natural question: what can one do with it that is not possible classically? We already know that the irreducible nonlocal nature of a pair of quantum systems in an entangled state leads to the violation of Bell's inequalities. But what communication and computational tasks can one accomplish using this global property? This task-oriented approach has been extremely successful and has led to the development of quantitative measures of entanglement and detailed criteria for detecting and characterizing it [86, 153]. The prototypical examples of entanglement-assisted communication are *superdense coding* [51] and *teleportation* [49].

In Section 10.4, we have defined an entangled state as a state that is *not separable*. A pure state $|\Psi\rangle$ (or a mixed state ρ) is said to be separable if the system can be partitioned into two subsystems A and B so that

$$|\Psi\rangle = |\phi\rangle_A \otimes |\psi\rangle_B \qquad \text{or} \qquad \rho = \sum_i \alpha_i \, \rho_A^i \otimes \rho_B^i, \qquad (10.44)$$

where $\alpha_i > 0$ and $\sum_i \alpha_i = 1$. In the particular case of bipartite systems AB, pure states can always be written using the Schmidt decomposition as

$$|\Psi\rangle_{AB} = \sum_i \sqrt{p_i} \, |\phi_i\rangle_A \otimes |\psi_i\rangle_B, \qquad (10.45)$$

where $\{|\phi_i\rangle_A\}$ and $\{|\psi_i\rangle_B\}$ are orthonormal bases for subsystems A and B, respectively, and p_i are the nonzero values of the reduced density matrix. The state $|\Psi\rangle_{AB}$ is separable if there is only one coefficient in the Schmidt decomposition (10.45); otherwise, $|\Psi\rangle_{AB}$ is entangled. Operational criteria for entanglement in a few other particular cases have been determined [253]. However, at the time of this writing there is no general theory of entanglement for multi-partite systems.

10.8.1 Superdense coding

Classical information can be stored and transmitted using qubits. For example, the classical bit string $x_1, x_2, \ldots, x_n$ (where $x_i \in \{0, 1\}$) can be transmitted by preparing n qubits in the (direct product) state $|x_1, x_2, \ldots, x_n\rangle$, sending them through a (noiseless) quantum channel, and measuring them in the computational basis. A rather surprising communication protocol is made possible by the laws of quantum mechanics through the use of shared entanglement.

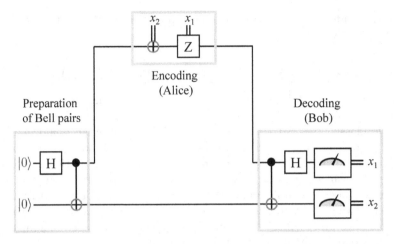

FIGURE 10.7 Schematic representation of the superdense coding protocol.

Suppose that Alice and Bob share a *maximally entangled*[9] state of two qubits, e.g. the state $|\Phi^+\rangle_{AB} = (|00\rangle + |11\rangle)/\sqrt{2}$, with Alice initially possessing qubit A and Bob qubit B. The preparation can be done starting with the state $|00\rangle$ by applying a Hadamard gate, followed by a CNOT, as illustrated in Figure 10.7. Superdense coding [51] is a protocol that enables Alice to transmit to Bob *two* classical bits by sending him one qubit (qubit A). The protocol has the following three steps:

(1) Alice applies one of the unitary operators I, X, Z, or $iY = ZX$ to qubit A, conditional on the classical bits x_1 and x_2 that she wants to transmit (see Fig. 10.7). This local operation changes the (global) properties of the joined system by evolving the entangled state $|\Phi^+\rangle_{AB}$ into $|\Phi^+\rangle_{AB}$, $|\Psi^+\rangle_{AB}$, $|\Phi^-\rangle_{AB}$, or $|\Psi^-\rangle_{AB}$, respectively. These Bell-EPR states are defined in Eq. (10.29).

(2) Alice sends qubit A to Bob using a noiseless qubit channel.

(3) Once in possession of both qubits, Bob performs a measurement in the Bell basis and determines the Bell-EPR state produced by Alice, which encodes two bits of classical information.

The details of the protocol are illustrated schematically in Figure 10.7. The maximally entangled state is prepared starting with each of the two qubits in the $|0\rangle$ state by applying a Hadamard gate to one of the qubits, $|0\rangle \rightarrow (|0\rangle + |1\rangle)/\sqrt{2}$, followed by a CNOT, $(|00\rangle + |10\rangle)/\sqrt{2} \rightarrow (|00\rangle + |11\rangle)/\sqrt{2}$. The two entangled qubits are then distributed to Alice and Bob. Suppose now that Alice wants to transmit to Bob two classical bits, x_1 and x_2. First, she applies a unitary transformation U to her qubit, conditional on the bit string "$x_1 x_2$," and thus changes the entangled state. There are four possibilities: i)

---

[9] A pure state $|\psi\rangle_{AB} \in \mathcal{H}_A \otimes \mathcal{H}_B$, with $\dim(\mathcal{H}_A) \leq \dim(\mathcal{H}_B)$, is maximally entangled if the reduced density matrix is proportional to the identity operator, $\rho_A = \mathrm{Tr}_B |\psi\rangle_{AB\,AB}\langle\psi| = \frac{1}{\dim(\mathcal{H}_A)} \mathbf{1}_A$.

The bit string is "00"; Alice does nothing and the entangled state remains $|\Phi^+\rangle_{AB}$. ii) The bit string is "10"; Alice applies the phase flip Z and the entangled state becomes $|\Phi^-\rangle_{AB}$. iii) The bit string is "01"; Alice applies the NOT gate and produces the state $|\Psi^+\rangle_{AB}$. iv) The bit string is "11"; Alice applies the ZX transformation to her qubit and the entangled state becomes $|\Psi^-\rangle_{AB}$. Next, Alice sends her qubit to Bob.

To decode Alice's message, Bob applies a CNOT operation to the entangled pair, followed by a Hadamard gate on the control qubit, then measures the two qubits (in the computational basis). The unitary operations performed by Bob correspond to the following evolutions: i) $|\Phi^+\rangle_{AB} \xrightarrow{\text{CNOT}} |+0\rangle \xrightarrow{H \otimes I} |00\rangle$; ii) $|\Phi^-\rangle_{AB} \xrightarrow{\text{CNOT}} |-0\rangle \xrightarrow{H \otimes I} |10\rangle$; iii) $|\Psi^+\rangle_{AB} \xrightarrow{\text{CNOT}} |+1\rangle \xrightarrow{H \otimes I} |01\rangle$; iv) $|\Psi^-\rangle_{AB} \xrightarrow{\text{CNOT}} |-1\rangle \xrightarrow{H \otimes I} |11\rangle$. Hence, the outputs of the final measurements are exactly the classical bits x_1 and x_2 sent by Alice.

Using superdense coding Alice is able to transmit two classical bits, while sending Bob only one qubit. The key to understanding this apparent violation of the Holevo bound is to recognize that the information is not "carried" by Alice's qubit, but rather by the entangled pair as a whole. The remarkable quantum effect revealed by this protocol is based on the nontrivial impact of local operations (i.e., the unitary operations done by Alice) on the irreducibly global properties of the entangled system – the original joint state is transformed into an orthogonal state. Finally, we note that superdense coding enables secure communication: no measurement of Alice's qubit (intercepted by a third party) could reveal the encoded information without possession of the entangled partner (i.e., Bob's qubit).

10.8.2 Quantum teleportation

Quantum teleportation [49] is a protocol in which a certain unknown quantum state is destroyed at one location (Alice's lab) and recreated at a distant location (Bob's lab). First, two important remarks: i) what is transported from Alice to Bob is a *quantum state* $|\phi\rangle$, not *matter* (i.e., not a *physical qubit*); ii) the unknown state is not *cloned*, i.e., the quantum information that is generated in Bob's lab has disappeared from Alice's lab.

Teleportation addresses the following problem. Suppose that Alice wants to communicate to Bob the quantum state $|\phi\rangle$ of a single qubit without actually sending him the physical qubit. If Alice knows the state (e.g., $|\phi\rangle = \alpha|0\rangle + \beta|1\rangle$), she can simply send Bob the classical information corresponding to the coefficients (α, β) and he would be able to prepare the state using his own physical qubit. Note, however, that reconstructing the state with high accuracy requires a huge amount classical information, since the two coefficients are continuous variables. But what if the state is unknown? Measuring it is not a solution because, in general, $|\phi\rangle$ is a linear combination of the computational basis and cannot be determined with certainty by performing one measurement. Cloning it and doing multiple measurements is also impossible. Remarkably, there is a solution to this problem that, similarly to

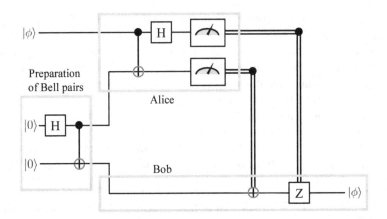

FIGURE 10.8 Schematic representation of the quantum teleportation protocol.

superdense coding, involves the use of quantum entanglement as an information resource. Specifically, assuming that Alice and Bob share an EPR pair, they can use the following three-step protocol:

(1) Alice interacts the qubit carrying the (unknown) state $|\phi\rangle = \alpha|0\rangle + \beta|1\rangle$ with her half of the entangled pair, then applies a CNOT followed by a Hadamard gate on the control qubit and, finally, measures the two qubits in her possession. The possible results are "00," "01," "10," and "11."

(2) Alice sends the results (i.e., two bits of classical information) to Bob.

(3) Bob performs one of four operations (I, X, Z, or $iY = ZX$) on his half of the EPR pair, conditional on the information received from Alice, and recovers the original state $|\phi\rangle$.

The details of the protocol are illustrated schematically in Figure 10.8. At the beginning of the protocol, the three-qubit quantum state is $|\psi_1\rangle = |\phi\rangle \otimes |\Phi^+\rangle_{AB} = (\alpha|000\rangle + \beta|100\rangle + \alpha|011\rangle + \beta|111\rangle)/\sqrt{2}$, where the first two qubits belong to Alice and the rightmost qubit to Bob. Applying the CNOT gate on the first two qubits evolves the state into $|\psi_2\rangle = (\alpha|000\rangle + \beta|110\rangle + \alpha|011\rangle + \beta|101\rangle)/\sqrt{2}$. Finally, applying the Hadamard gate generates $|\psi_3\rangle = [|00\rangle(\alpha|0\rangle + \beta|1\rangle) + |10\rangle(\alpha|0\rangle - \beta|1\rangle) + |01\rangle(\alpha|1\rangle + \beta|0\rangle) + |11\rangle(\alpha|1\rangle - \beta|0\rangle)]/2$. Once Alice has measured her qubits, Bob's qubit will be in one of the states $\alpha|0\rangle + \beta|1\rangle$ (if the results of the measurements are "00"), $\alpha|0\rangle - \beta|1\rangle$ (for "10"), $\alpha|1\rangle + \beta|0\rangle$ (for "01"), or $\alpha|1\rangle - \beta|0\rangle$ (for "11"). After receiving from Alice the (classical) information concerning the measurement outcome, Bob applies the appropriate unitary operation – i.e., I for "00," Z for "10," X for "01," and ZX for "11" – and recovers the state $|\phi\rangle$.

It is worth noting that quantum teleportation does not enable faster than light communication, because the protocol cannot be completed before Bob receives the measurement results from Alice, which implies transmission over a classical communication channel. Also, as in the case of superdense coding, the "driving force" behind teleportation is the ability to induce global changes in the properties of the entangled system by performing local operations. Finally, we note that teleportation reveals the possibility of using quantum information resources interchangeably: one shared EPR pair plus two classical bits represent a communication resource at least the equal of one qubit.

10.9 QUANTUM CRYPTOGRAPHY

Quantum cryptography [204], more precisely *quantum key distribution* [48, 541], represents the most remarkable practical application of quantum information that can be implemented using current technology. While quantum cryptography is primarily concerned with quantum key distribution, it also includes other ideas, such as those related to *bit commitment* protocols[10] [77]. Below, we focus on the basic ideas behind quantum key distribution. We note that this field is not concerned with the actual transmission of secret messages, but rather with the secure distribution of a *private key*, which is then used for encoding and transmitting secret messages classically. In essence, quantum cryptography builds upon the impossibility of perfectly distinguishing nonorthogonal quantum states. In other words, quantum cryptography exploits the fundamental laws of physics – in particular the existence of incompatible physical quantities and the noncommutativity of the corresponding observables – to create private key bits between two parties (Alice and Bob) using protocols that are *provably* secure. Any attempt by Eve to gain access to this key necessarily results in a detectable disturbance of the quantum state that reveals the eyedropper's presence.

10.9.1 Quantum key distribution

As briefly mentioned in Section 10.1, there are two main cryptographic techniques: *private-key* (or *symmetric*) cryptography and *public-key* (or *asymmetric*) cryptography. The first method is based on the two parties (e.g., Alice and Bob) sharing a secret key that is used for both encryption and decryption. For example, the so-called *one-time pad* technique involves Alice and Bob sharing a secret random bit string (the *key*) of the same length as the message to be transmitted. Alice encrypts the message by adding (modulo 2) the key bits to the message bits, while Bob decrypts it by subtracting the key (modulo 2). The encrypted message is a random string that provides no information to Eve, if she intercepts it. However, the key has to be discarded after one use, because Eve could start identifying the secret bit string by comparing messages encoded using the same key. The main drawback of this method is that the two parties have to be able to secretly share the key, a problem as

---

[10]In essence, bit commitment addresses the following problem: Alice casts a vote and Bob, while being sure that the vote was casted, cannot learn Alice's decision before a certain time chosen by her. How can this be done in such a way that Bob cannot (yet) know Alice's vote, yet she cannot change her mind between the moment she casts the vote and the moment she reveals it?

difficult as the original problem one wants to address, i.e., realizing secure communication.

Public-key cryptography is based on the idea of generating a pair of keys using *one-way* functions – functions that are easy to calculate (given the arguments) but hard to invert. For example, multiplying two large prime numbers is easy, but factoring the product appears to be hard (i.e., it is a computationally intractable problem on a classical computer, as far as we know). Alice generates the pair of keys and broadcasts the public key (e.g. the large product), which is then used by Bob to encrypt the message. To decrypt the message, Alice uses her private secret key (i.e., the prime factors). Eve, on the other hand, has no access to the information transmitted by Bob unless she is able to perform the factorization. This method eliminates the problem of secretly sharing the key, but relies on the computational intractability of inverting a one-way function. The existence of adequate one-way functions is an unsolved problem. For example, Eve could break the most widely used public key protocol (the RSA[433]), which is based on the apparent difficulty of factoring large numbers, if she manages to find an efficient classical factorization algorithm (assuming that it exists) or if she gets a quantum computer.

Quantum key distribution is a *provably secure* protocol that allows two parties to generate a private key without the need to meet in secret or share secret information beforehand. The basic idea behind this technique exploits two fundamental properties of quantum systems: the nocloning theorem – Eve cannot clone any qubit that she may have intercepted (see Section 10.4.3) – and the impossibility of distinguishing between two nonorthogonal quantum states without disturbing them. Suppose that Eve is trying to obtain information about the nonorthogonal states $|\psi\rangle$ and $|\phi\rangle$. She does not know the bases in which these states were prepared, so measurement is not an option. She may try to unitarily interact the states with ancilla qubits prepared in a standard state $|u\rangle$. Assuming that this process does not disturb the states, one has $|\psi\rangle|u\rangle \to |\psi\rangle|v_\psi\rangle$ and $|\phi\rangle|u\rangle \to |\phi\rangle|v_\phi\rangle$. However, unitary transformations preserve the inner product, which implies $\langle v_\psi|v_\phi\rangle = 1$, i.e., the final states of the ancilla are the same. Eve cannot acquire any information about the identity of the original states without perturbing at least one of them.

These ideas, first presented in 1984 by Bennett and Brassard [48], are incorporated into the following protocol (dubbed "BB84"):

(1) Alice prepares a large number $4n$ of qubits in one of the states corresponding to the σ_z basis (i.e., $|0\rangle$ or $|1\rangle$) or the σ_x basis (i.e., $|+\rangle$ or $|-\rangle$), randomly chosen, records each preparation, and sends the qubits to Bob via a quantum channel. Associating the states $|0\rangle$ and $|+\rangle$ with 0; $|1\rangle$ and $|-\rangle$ with 1, Alice is in possession of a random string of $4n$ classical bits.

(2) Bob measures the qubits received from Alice choosing the measurement basis randomly between σ_z and σ_x and records the measurement bases and the outcomes. About half of the qubits will be measured in the same basis Alice used to prepared them; Bob will ascribe to these outcomes the same bit values as Alice, but the other half will generate uncorrelated bit values.

(3) Alice and Bob announce the basis they chose for each qubit using a public classical channel. The bit values corresponding to different preparation and measurement bases are disregarded. The coincidences in their independent random choices of basis provide Alice and Bob with classical bit strings of length approximately $2n$ that, in the absence of noise or eavesdropping on the transmitted qubits, are identical.

(4) Alice and Bob check for Eve's presence. Assume that an eavesdropper intercepts the qubits sent by Alice and tries to find information about the sequence being sent. This will create disturbances that can be easily identified. For example, if Eve measures the qubits *en route* between Alice and Bob using either the σ_z or the σ_x basis, half of the time she will project the state into the other basis than the one chosen by Alice, which implies a 25% chance for Alice and Bob to ascribe different bit values when choosing the same basis. This anomaly, which reveals Eve's presence, can be detected by Alice announcing the values of n randomly chosen bits from her key. If these values agree with Bob's corresponding bit values, there was no eavesdropping; the checked bits are discarded and the remaining n bits constitute a secret random key shared by Alice and Bob. If the bit values made public by Alice do not coincide with Bob's values, the protocol is aborted.

Once the protocol is successfully completed, Alice and Bob share a private key that can be used for one-time pad encryption. Alice can send the encrypted message over a public channel with no risk of being decrypted by anybody, except Bob. A few remarks on realistic implementations and possible variations of this protocol are warranted. First, we note that in practice the protocol has to allow for the presence of noise and for Eve adopting a more subtle strategy (e.g., not intercepting all the qubits). The key may be retained if the error rate in step (4) is well below 25%, but Alice and Bob will have to perform *information reconciliation* and *privacy amplification* protocols on the remaining n bits [389]. The information reconciliation protocols are designed to increase the correlation between Alice's and Bob's strings, while (possibly) providing Eve with minimal information. For example Alice and Bob can check the parity of randomly chosen pairs of bits; if the corresponding pairs have the same parity, they keep the first bit and disregard the second, otherwise both bits are discarded [204]. After reconciliation, Alice and Bob will share the same key, within some acceptable error rate. Finally, privacy amplification [389] can systematically reduce the correlation between the string shared by Alice and Bob and the string that Eve may have acquired.

Second, we emphasize that the protocol described above involves the use of both a quantum channel (for transmitting the qubits) and a public classical channel (for broadcasting basis choices, bit values, parities, etc.). It is assumed that Eve may have access to the quantum channel and possibly perturb the qubits, but that she cannot influence the data sent over the public channel. The key feature of this protocol is the fact that Eve cannot acquire any information about the transmitted qubits without perturbing them, i.e., without revealing her presence.

Finally, we note that another quantum key distribution approach [154, 50] involves the use of shared EPR states. If Alice and Bob share a large number of entangled pairs, they can measure each system using the σ_z or σ_x basis, chosen randomly; this will provide each of them with a random sequence of 0 and 1 outcomes. The protocol proceeds as before, with a public announcement of the random basis choices, followed by the selection of bit values corresponding to measurements in the same basis, the check for Eve's presence, and the completion of information reconciliation and privacy amplification protocols.

CHAPTER 11

Introduction to Quantum Computation

UANTUM COMPUTATION IS NOT AN IMMINENT ACHIEVEment. We will not wake up tomorrow and find quantum computers on our desks. At the time of this writing, IBM has just unveiled the Condor – a chip containing 1,121 superconducting qubits. This number will likely increase (perhaps spectacularly) in the near future. However, the qubits have a relatively large error rate and, in the absence of a breakthrough advance that significantly lowers this rate, one would need hundreds of physical qubits to realize an error-corrected logical qubit. Predicting how long it will take to achieve the controlled manipulation of thousands of (logical) qubits (i.e., a number of qubits consistent with the "useful" implementation of known quantum algorithms) is probably not a rational undertaking, at least for a scientist. The most pessimistic experts regard the lively research activity in quantum computation as a waste of time and resources. But the pervading mood in this field is one of cautious optimism, particularly considering the possibility of achieving useful tasks (e.g., simulating quantum systems) with a sufficiently large number of "imperfect" qubits. Expediency may be responsible for some of this optimism; a good feeling about the prospects for quantum computation certainly helps when it comes to publishing papers and getting funded. Ultimately, however, the responsibility for the wise distribution of resources rests with the funding agencies; scientists, on the other hand, are called to explore every aspect of real and possible worlds that looks fascinating and worthy of their curiosity. From this perspective, quantum computation (defined broadly) offers plenty of opportunities for producing "good science." The problems brought to the fore by the development of this field range from basic questions regarding the foundations of quantum mechanics and unsolved problems in computational complexity, to the investigation of new phases of matter (e.g., topological quantum phases), the quest for "complete" control over quantum systems, and the critical need for significant advances in

DOI: 10.1201/9781003226048-11

materials science and device fabrication (e.g., engineering nanoscale heterostructures with atomic precision). Progress in these areas will naturally open new possibilities for interesting applications. Precisely identifying these applications is premature, but building a (general purpose) quantum computer will not necessarily be among the first of them.

11.1 INTRODUCTION

Modern *computer science* was founded on the work of Alan Turing, being heralded by a remarkable paper in 1936 [513]. The origins of algorithmic ideas are lost in the depths of history and many of the key elements of a modern computer were conceived a century earlier by Charles Babbage (1791-1871), who designed the *analytical engine* (a proposed mechanical general-purpose computer); but it was Turing who clarified and developed in detail the abstract notion of a programmable computer. He elaborated a simple computational model, now known as the *Turing machine*, and showed that there exists a *universal Turing machine* (UTM) that can be used to simulate any other Turing machine, completely capturing what it means to perform an algorithm. In other words, if an algorithm can be performed on some piece of hardware (e.g., a modern PC), there is an equivalent algorithm for the UTM that fulfills the same task, a statement known as the *Church–Turing thesis*.

It is worth noting that Turing's work was motivated by an abstract, fundamental mathematical question, which is perhaps surprising from today's perspective, considering the almost self-evident association between computers and practical applications. During the first decades of the 20th century David Hilbert had emphasized the importance of investigating the nature of mathematics [247]. Rather than only being concerned with proving or disproving mathematical propositions, one should consider what general type of propositions are even amenable to mathematical proof. Hilbert, as most other mathematicians, hoped that mathematics is complete, meaning that, within a consistent systems of axioms, every conjecture could, in principle, be proven true or false. This hope was brought to an end by Kurt Gödel [195], who established the existence of undecidable mathematical propositions – propositions that cannot be either proved or disproved. Gödel's work on Hilbert's decision problem was continued by Alonzo Church [111] and Alan Turing, who were looking for alternate techniques for addressing this type of problem. The *Turing machine* is a theoretical construct that establishes the idea of automated mathematical proof, something way beyond the scope of an "arithmetic calculator." Using this construct, Turing was able to show that decidability can be formulated as a "halting problem," i.e., a question of the type "given an input, would a Turing machine ever halt?" Assuming that there is a general algorithm that can solve the halting problem for any input leads to a contradiction, which means that the halting problem is *uncomputable*.[1] This result, which is similar in spirit with Gödel's theorem, reveals mathematics as being a rich and, in some sense, "free" domain containing many different ideas that cannot be encapsulated within a single "grand algorithm."

While uncomputable problems represent impossible tasks for a computer, more generally one faces the critical problem concerning the efficiency of various algorithms. In essence, an *efficient* algorithm is solvable in polynomial time, i.e., it requires a number of steps that grows polynomially with the size of the input. By contrast, an *inefficient* algorithm requires a number of steps that grows exponentially with the size of the problem to be solved. Any computational task that admits an efficient algorithm is deemed tractable and belongs to a certain complexity class (e.g., class P in the case of decision problems that can be solved on a deterministic Turing machine in polynomial time), while those that do not satisfy this condition are said to be "hard" and belong to a different complexity class. Classifying computational problems according to their inherent difficulty is the focus of *computational complexity theory* [402].

From this perspective, quantum mechanics seems to be computationally hard, because one has to deal with (exponentially large) Hilbert spaces. In 1982 Feynman suggested [166] that building computers based on the principles of quantum mechanics could allow us to overcome these difficulties. What Feynman had in mind was a *quantum simulator*, a purpose-built quantum system that could efficiently simulate the physical behavior of other quantum systems. Such a task is believed to be impossible for a classical computer. After all, one can argue that Turing's model of computation contains implicitly classic assumptions, hence it is not the most general possible model. This idea was explored by David Deutsch [138], who attempted to define a computational device capable of efficiently simulating arbitrary physical systems – the *universal quantum computer*. The fundamental question is whether a quantum computer can efficiently solve computational problems that are hard for classical computers, e.g. for a (probabilistic) Turing machine. At the time of this writing there is no rigorously proven answer to this question. However, there are strong reasons to believe that the answer is affirmative. A spectacular result indicating that quantum computers could be more powerful than Turing machines is Peter Shor's 1994 demonstration [476, 478] that the problem of finding the prime factors of an integer can be solved efficiently on a quantum computer. There is no known classical algorithm for this problem that runs in polynomial time. Shor's discovery plays a conceptual role similar to that

---

[1]An *undecidable problem* is a decision problem for which there is no algorithm (i.e., Turing machine) that always gives a correct yes-or-no answer. *Uncomputable functions* are mathematically well defined functions that no Turing machine (TM) can compute. The existence of uncomputable functions can be established using a cardinality argument: every TM can be thought of as a finite string over some finite alphabet, hence the set of all possible TMs is *countably* infinite; on the other hand, the set of all *characteristic functions* $f : \mathbb{N} \to \{0, 1\}$ is *uncountably* infinite; hence, there are not "enough" TMs to compute all characteristic functions.

played by Bell's inequalities: computation based on manipulating quantum states might represent a new type of computation capable of solving efficiently certain (relevant) computational tasks that are classically intractable. It is tempting to see a parallel between the quantum complexity class being larger than the set of problems that can be efficiently solved using classical computers (see below, Figure 11.2) and the set of quantum behaviors being larger than the set of local behaviors.[2]

So far, there are only a few other examples of computational problems that admit efficient quantum algorithms but have no known efficient classical solutions. Should the impossibility of an efficient classical solution be proven for any of these problems, it would demonstrate that the quantum classification of computational complexity is different from the classical classification – a result that would place computer science on a different foundation. However, quantum computation involves large scale interference effects, which are notoriously sensitive to noise and imperfections. So, is it feasible, even in principle? The discovery of quantum error correction has represented a decisive step toward an affirmative answer to this question, although some skepticism persists [135, 148, 432, 516]. The concept of fault-tolerance demonstrates that error correction can be achieved even using imperfect corrective operations, provided that the degree of imperfection is not "too high." The next piece of the puzzle is the idea to implement fault-tolerance using topological quantum states. Encoding information and performing operations using the topology of the system offers, in principle, significant protection against errors associated with processing quantum information (e.g., performing unitary transformations) or errors caused by interactions with the environment. A better understanding of topological quantum matter and, most importantly, an enhanced ability to control topological states are key requirements for progress along this direction. Whether or not this will lead to the realization of an actual quantum computer may still depend on the discovery of some as-yet-unknown quantum phase or of a new fundamental limitation. Either way the fight is worth fighting and the reward great (although not necessarily the expected one).

But what do we mean by *quantum computation*? Basically, it is the process of accomplishing a (classical) computational task using the laws of quantum mechanics. In essence, this involves three distinct steps: 1) Initialization – classical information (the input) is encoded as quantum information into a quantum state $|\psi_i\rangle \in \mathcal{H}$, where $\mathcal{H}$ is a large Hilbert space. 2) Controlled evolution – the quantum information is processed by evolving the initial state $|\psi_i\rangle$ into a final state $|\psi_f\rangle$. The evolution can be realized by applying a unitary operator, $|\psi_f\rangle = U|\psi_i\rangle$, or by performing certain measurements, or by combining unitary evolutions and measurements. 3) Readout (measurement) – classical information (the output of the computation) is extracted from the final

---

[2]A *behavior* represents the set of joint probabilities that characterize the outcomes of measurements in a Bell-type experiment. Quantum and local *behaviors* are predicted by the quantum theory and by local hidden variable theories, respectively.

quantum state by measuring $|\psi_f\rangle$ in a certain basis. Note that sometimes quantum computation is narrowly understood as the quantum information processing phase (i.e., step 2). The basic intuition leading to the concept of quantum computation is based on the fact that every computation is performed by a physical machine (i.e., a computer) and, consequently, is essentially a physical process. From this perspective, an isolated quantum object may be regarded as a *dedicated quantum computer* that computes its own behavior in real time, i.e., evolves its quantum state according to a unitary operator generated by its Hamiltonian.

There are different quantum computational models that have been shown to be equivalent, in the sense that they can simulate each other with only polynomial overhead. The models are distinguished by the way classical information is encoded and, most importantly, by the elementary processes in which the evolution from $|\psi_i\rangle$ to $|\psi_f\rangle$ is decomposed. Historically, the first quantum computational model – the *quantum Turing machine* – was proposed by David Deutsch in 1985 [138], based on earlier work by Benioff [45], and represents the quantum generalization of the classical Turing machine [513]. In practice, the model that provides a straightforward architecture for building a quantum computer and an intuitive framework for designing quantum algorithms is the *quantum circuit model* (see Sec. 11.3). The equivalence between the circuit model and the Turing machine model was established by Yao in 1993 [562]. Within the circuit model, the input is encoded as a specific quantum state of n qubits and the evolution, which corresponds to a certain unitary operator U acting on a 2^n-dimensional Hilbert space, is decomposed into a sequence of few-qubit gates. The step-by-step set of instructions specifying the sequence of quantum gates that realizes the desired unitary evolution U (and, implicitly, the desired output, when measuring the final state $|\psi_f\rangle$) is called the *quantum algorithm*. Hence, within this model, a quantum computation on n qubits can be viewed as a rotation (i.e., unitary transformation) in a 2^n-dimensional space that is specified by a certain quantum algorithm (i.e., sequence of quantum gates). A *universal* quantum computer is a machine running algorithms that are written in terms of an elementary set of quantum gates capable of generating *all* possible unitary operations.

Oftentimes in the literature when saying "quantum computer" people actually mean "quantum circuit." Nonetheless, several other models of quantum computation have been proposed. Although these models are mathematically equivalent to the circuit model, it is useful to investigate them as they may offer better alternatives for practical implementation or intuition for constructing new algorithms. One alternative model is the so-called *one-way quantum computation*, which, basically, involves two steps. First, a particular highly entangled initial state, known as a "cluster state," is prepared. Then, the computation involves just a long sequence of single-qubit measurements applied to the cluster state. The exact sequence of bases in which the measurements are performed depends upon the results of intermediate measurements.

A key lessons that this model teaches us is that quantum computation is not something that necessarily involves only unitary operations.

Another model, in which the algorithm is generated by an adiabatic process, is the so-called *adiabatic quantum computation* [127]. Basically, the input state is the (nondegenerate) ground state of an initial Hamiltonian H_i, which is adiabatically evolved into a final Hamiltonian H_f by slowly varying a certain parameter (or a set of parameters) λ. The desired solution of the computational problem is given by the (nondegenerate) ground state of H_f. Note that encoding information into the ground state of the Hamiltonian is meaningful provided the ground state is protected at all times (i.e., for all values of λ) by a finite energy gap.

Of particular significance for our specific purpose is *topological quantum computation* [384] (see Chapter 12 for details). In this case, quantum information is encoded in topologically nontrivial quantum states containing multi-quasiparticle excitations (called *anyons*), which can be created in pairs, moved apart, and annihilated when brought back together. Performing unitary operations involves braiding the anyons. Remarkably, the unitary operation generated by a process involving the creation, movement, and annihilation of anyons depends solely on the topology of the path traversed by the quasiparticles and not on the details of the path. As a result, the system is naturally insensitive to any local noise experienced by the quasiparticles. The basic ideas behind topological quantum computation will be discussed in more detail in the next chapter.

Finally, we note that quantum computation is an active research area and new computational models are still emerging. At the same time, there are sustained efforts to identify quantum algorithms that offer significant speedups over their classical counterparts. For guiding these efforts and shedding more light on the essential difference between classical and quantum computing, there is a need for systematic studies of computing models [391, 525], as well as investigations of a possible "grand unification" of quantum algorithms [356].

Quantum computation can be viewed as an attempt at redefining what it means to compute. Classically, a computation is a process that we can perform on a Turing machine. However, quantum systems are very hard to simulate on a classical computer. Could it be that exploiting the laws of quantum mechanics one could perform efficiently certain computational tasks that classically are hard to perform? This thought inspired David Deutsch to formally define the notion of *quantum computer* – a machine that runs computational processes based on the laws of physics, which are believed to be the laws of quantum mechanics. In the words of David Deutsch, "what computers can or cannot compute is determined by the laws of physics alone, and not by pure mathematics." The basic idea can be summarized as the following generalization of the strong Church–Turing thesis:

> **Church–Turing–Deutsch principle**: *Any physical process can be efficiently simulated on a quantum computer.*

A few remarks are warranted at this point. First, without the specification "efficiently," this principle would be the analog of the standard Church–Turing thesis[3], with the notion of "physical process" replacing the rather vague "algorithmic process" and the quantum computer taking the place of the universal Turing machine. These modifications express Deutsch's aspiration to be able to formally derive this principle using the laws of physics. At the time of this writing, this is one of the most important open problems in quantum information science. Second, if classical computers cannot efficiently simulate quantum systems, it is natural to assume that they cannot efficiently simulate machines operating according to the laws of quantum mechanics, i.e., quantum computers. Should this be the case, quantum computers would not be polynomially equivalent to a (probabilistic) Turing machine. This nonequivalence would represent a violation of the strong Church–Turing thesis. Third, we should emphasize the importance of the Church–Turing–Deutsch principle for both computer science and physics. On the one hand, this principle could place computational complexity on a stronger foundation by connecting some of its basic concepts with the fundamental laws of physics. On the other hand, it would lead to the remarkable conclusion that the whole diversity of physical processes found in nature can be efficiently simulated by a universal computing device, which a priori is not an obvious possibility.

The fundamental tasks of quantum computation can be summarized as follows:

I. Develop quantum *computational models*. The fundamental model of quantum computation is the *quantum circuit model*, which defines a computation as a sequence of *quantum gates* – the quantum analog of classical logic gates. It can be demonstrated that there is a small set of so-called *universal gates* with the property that any quantum computation can be expressed in terms of these gates.

II. Design *quantum algorithms*. The fundamental underlying question concerns the ultimate limits of quantum computation. What class of computations can be performed using a quantum computer? Are there computational tasks that a quantum computer can perform better than a classical computer? So far, there are only a few examples of quantum algorithms that provide an advantage over known classical algorithms. Examples include the Deutsch–Jozsa and Shor algorithms, quantum search algorithms (e.g., the Grover algorithm), and quantum simulations (simulating a physical quantum mechanical system).

III. Develop *quantum error-correction* codes. The underlying fundamental question is: how can one reliably perform quantum information processing in the presence of noise? The concept of *fault-tolerant quantum computation* reveals that the encoding and decoding of quantum states can tolerate errors and, in addition, one can perform logical operations on encoded quantum states using imperfect underlying gate operations. An important result is the so-called *threshold theorem*: it is possible to

---

[3]See below, Section 11.2.

efficiently perform an arbitrarily large quantum computation provided the noise in individual quantum gates is below a certain constant threshold.

IV. Identify specific *physical systems and quantum properties* to be used for the *realization* of quantum computers. The basic requirements for quantum computation concern the abilities to a) robustly represent quantum information, b) prepare the initial states, c) perform universal unitary transformations, and d) measure the output. What physical systems are potentially good candidates for satisfying these requirements? Since the noise level of elementary components should not exceed a certain level (according to the threshold theorem), exploiting the intrinsic robustness of topological quantum states appears as a promising path forward.

In the remainder of this chapter, we first provide a brief summary of the classical theory of computation, focusing on basic concepts regarding computational models (e.g., the Turing machine) and computational complexity theory (Section 11.2), then we address a few key aspects pertaining to tasks I-III. More specifically, in Section 11.3, we provide a basic description of the most widely used model of quantum computation, the quantum circuit model, followed (in Section 11.4) by a brief discussion of the most important quantum algorithms. We end with a summary of the main ideas behind quantum error correction (Section 11.5). Basic aspects regarding quantum computation using topological quantum states (which are mostly related to task IV) are presented in Chapter 12. For a captivating general introduction to quantum computation the reader is referred to the book by Scott Aaronson [2]. Detailed technical treatments of various aspects of this subject can be found in the comprehensive textbook by Nielsen and Chuang [389] and a number of more recent publications, e.g., Refs. [56, 246, 279, 317, 496, 554].

Concluding our introductory remarks, we emphasize that building a quantum computer involves considerable practical difficulties stemming, in essence, from the requirement of imposing strict limits on quantum decoherence. Not only that we do not currently have a quantum computer, but there still are significant gaps in our understanding of the basic physics underlying some of the potential quantum computing platforms and monumental technological challenges that need to be overcome. Moreover, there are limitations for quantum computation that have nothing to do with the practical difficulties. Although it was not rigorously proved, it appears that quantum computers would solve *certain specific problems* (e.g., factoring integers) dramatically faster than classical computers, but for most problems (e.g., playing chess and proving theorems) the improvement over conventional computers would be modest. In this context, we note that the "classically-hard" problems include the so-called NP-complete (NP-C) problems. Cook, Karp and Levin have shown [119, 277, 326] that, if an efficient algorithm were found for any NP-C problem, it could be adapted to efficiently solve all the other NP problems (implying that $P = NP$). However, it is strongly believed that, on the one hand, $P \neq NP$, and, on the other hand, quantum computers cannot solve NP-C problems in polynomial time. Could then the quest for quantum computation be essentially motivated by the goal of breaking public key encryption using Shor's algorithm? Probably not. It is true that the very possibility of "killer apps" (such as, e.g., running Shor's algorithm) makes quantum computation a race that no significant player can afford to loose. However, if quantum computers will become a reality, their most likely initial use will involve the (efficient) simulation of quantum systems, an achievement that could dramatically advance research in certain areas of condensed matter physics, as well as chemistry, nanotechnology, and several other fields. Most importantly (from a basic science perspective), there is the possibility that quantum computation cannot be realized for some fundamental reason. In the words of Scott Aaronson [1], "... *the most exciting possible outcome of quantum computing research would be to discover a fundamental reason why quantum computers are not possible. Such a failure would overturn our current picture of the physical world, whereas success would merely confirm it."*

11.2 CLASSICAL THEORY OF COMPUTATION

What is a *computation* and what is an *algorithm*? What are the *limitations* of computation, e.g., what is computable, and what *resources* are necessary to achieve a given computational task? Answering these fundamental questions is the main object of the theory of computation. Intuitively, an algorithm is understood as a self-contained step-by-step set of operations to be performed for calculating a function. Algorithmic computations transform a finite input (given at the start of the computation) into a finite output (available at the end of the computation) using a finite number of steps (i.e., finite amount of time). Historically, the notion of algorithm[4] goes back to antiquity, but the modern concept was formalized in the 1930s by Gödel (in terms of recursive functions [195]), Church (λ–*calculus* [111]), and Turing (*Turing machine* [513]) in response to Hilbert's conjecture [247] that any mathematical proposition could be decided (proved true or false) by mechanistic logical methods (the so-called *Entscheidungsproblem*).

Having constructed a *model of computation* – such as, for example, the Turing machine – the fundamental question concerns the resources required to perform a given computational task. First, one has to understand what computational tasks are possible, which constitute the object of *computability theory*; second, one needs to identify the limitations on our ability to accomplish a given computational problem. Intuitively, a computational problem is understood to be "hard" if the amount of required resources (e.g., number of necessary algorithmic steps) rises exponentially with the size of the problem (i.e., the amount of information required to specify the input). Classifying computational problems according to their degree of "difficulty" and proving

---

[4]The word "algorithm" comes from the Latin translation of the name al-Khwarizmi (Abu Ja'far Muhammad ibn Musa) – a 9th-century Persian mathematician, astronomer, and geographer.

that a given problem belongs to a certain complexity class are the main tasks of *complexity theory*. Computational complexity [402] provides *lower bounds* on the time and space resources required by the best possible algorithm that can solve a given task. This is complementary to algorithm design [303], since the best possible algorithm may not be known.

Below, we first introduce the Turing machine as a fundamental model for computation, then we sketch a few key concepts in computational complexity. We close this section with some remarks on the energy resources required to perform a computation.

11.2.1 Computational models: The Turing machine

To capture in a mathematical definition the intuitive idea of an algorithm and show that there is no algorithm that can solve all mathematical problems, Turing introduced a model of *automatic machines* [513], now known as *Turing machines*. As shown in Figure 11.1, a Turing machine has four basic elements: (1) a one-dimensional, infinitely long *erasable tape* divided into cells and capable of storing one symbol per cell; (2) a *read-write head* capable of reading/writing one symbol at the current location and moving (left or right) on the tape; (3) a *finite state control* that coordinates the operations of the machine and has a bounded number of available states; and (4) a *transition table* or a *program*, which, given the current state and the symbol at the current location on the tape, specifies the new state and the next action of the read-write head. We assume that the tape stretches off to infinity in one direction and we label the cells by integers $i \in \{0, 1, 2, \dots \}$. The cells contain symbols from a finite alphabet Σ that includes the "blank" symbol (which we denote by "b"). The *state control* mechanism consists of a finite set of internal states $Q = \{q_s, q_h, q_1, q_2, \dots, q_m\}$ and represents the central processing unit. It also provides temporary information storage. The special state q_s (the *starting state*) corresponds to the beginning of the computation; the operation is completed when the machine reaches the *halting state* q_h.

The Turing machine (TM) always starts in the initial state q_s with the head over the leftmost cell $i = 0$ and the tape containing a finite number of nonblank symbols (the input). We assume that the (finite) input is written on the tape starting with cell $i = 1$. The computation proceeds in a step-by-step manner according to the transition table. At each step of the computation the head reads the symbol $x \in \Sigma$ from the current cell then, depending on this reading and the current internal state $q \in Q$, the program indicates i) the new internal state q', ii) the new symbol $x' \in \Sigma$ to be written on the current cell, and iii) the movement s of the tape head, which can be $s = +1$ (move right one cell), $s = -1$ (move left), or $s = 0$ (stand still). The program (or transition table) is a finite list of instruction lines of the form $\langle q, x; q', x', s \rangle$. During each computation cycle, the TM identifies the program line containing q and x and executes it. This can lead to four possible outcomes. 1) The TM loops or runs forever and does not halt. 2) The TM tries to move left from the cell $i = 0$:

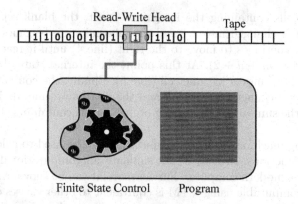

FIGURE 11.1 The basic structure of a Turing machine.

the TM halts and fails (i.e., the output is undefined). 3) The pair (q, x) is not in the list of instructions (i.e., the instruction is not applicable): the TM halts and fails. 4) The TM reaches the halting state q_h: the computation is successful and the output is given by the (nonblank) tape content.

As a simple example, let us consider the addition of two nonnegative integers. The simplest implementation uses a *unary* numeral system in which a number n is represented by a string containing one symbol (e.g., "1") repeated n times, for example, $2 \to 11$, $3 \to 111$, etc. The symbols on the tape belong to the alphabet $\Sigma = \{b, 1\}$ and the input (i.e., the numbers m and n) is represented by two strings of cells (of lengths m and n, respectively) containing the symbol "1" separated by a blank cell. We assume that the first string starts at cell $i = 1$ (i.e., the cell $i = 0$ is blank). A zero is represented by a blank cell. With these conventions, the inputs for the additions $1 + 2$, $0 + 4$, and $3 + 0$ are $[b, 1, b, 1, 1, b, b, \ldots)$, $[b, b, b, 1, 1, 1, 1, b, b, \ldots)$, and $[b, 1, 1, 1, b, b, \ldots)$, respectively, where $b, \ldots)$ is an infinite string of blank cells. The program contains the following lines:

$$
\begin{aligned}
1: & \quad \langle q_s, b; \ q_1, b, +1 \rangle \\
2: & \quad \langle q_1, 1; \ q_1, 1, +1 \rangle \\
3: & \quad \langle q_1, b; \ q_2, 1, +1 \rangle \\
4: & \quad \langle q_2, 1; \ q_2, 1, +1 \rangle \\
5: & \quad \langle q_2, b; \ q_3, b, -1 \rangle \\
6: & \quad \langle q_3, 1; \ q_h, b, 0 \rangle
\end{aligned}
$$

Note that the line numbers are used for reference and are not part of the program. At the start, the internal state is q_s and the tape head reads a blank from the leftmost cell $i = 0$; the first cycle is completed by the TM executing program line 1, i.e., the state changes to q_1 and the head moves to the right. During the next m cycles the TM executes line 2 as the head moves through the cells containing the integer m. Next, the blank separating the input integers is changed into a 1 and the internal state becomes q_2 (line 3). The tapehead continues to move to the right (line 4) until it reaches a blank cell (the cell $i = m + n + 2$). At this point, the internal state changes to q_3 and the tape head moves left one cell (line 5). Finally, the content of the cell $i = m + n + 1$ is changed to a blank and the TM halts (line 6). The output, representing the sum $m + n$, is a string of $m + n$ cells containing the nonblank symbol "1."

The Turing machine model of computation can be used to calculate more complicated functions. In fact, it can simulate all computational operations performed by a modern computer. Furthermore, it can be shown that the class of functions computable using a TM is equivalent to other classes of functions that were introduced to define computability, e.g., the *recursive functions* defined by Gödel and the λ-*calculus* invented by Church. Other computational models, such as the *circuit model*,[5] which has significant practical importance and is most useful in the context of quantum computation, were shown to be equivalent to the Turing machine model. This equivalence suggests that an algorithm is, in fact, what a Turing machine implements. In other words, a function is computable if and only if it can be solved by a Turing machine. This is known as the *Church–Turing thesis*:

> **Church–Turing thesis:** *Every function which would naturally be regarded as computable by an algorithm can be computed by a Turing machine.*

Note that this is a conjecture (hypothesis), not a theorem. The thesis is supported by many examples of algorithmic computations that can be simulated by a Turing machine; there is no known counterexample, which would disprove the thesis. Also note that we didn't say anything about the efficiency of the computation, an issue that we will address later.

There are different variations of the basic TM model described above; for example, one can define Turing machines equipped with multiple tapes. One can also include randomness in the model, e.g., choose a new internal state randomly (out of two possible state) by "tossing a coin." Nonetheless, one can show that all these models of computation are equivalent in the sense that different types of machines can simulate each other, although, possibly, with some computational overhead. Let us note that two machines from the same class (e.g., single tape deterministic Turing machines) correspond to different algorithms and have different programs and, possibly, different numbers of internal states. The information about the transition tables and internal states can be encoded, so that each Turing machine T is uniquely specified by a binary number $d[T]$ called the Turing number of machine T. Then, one can imagine a *universal Turing machine* U with the property that on input $d[T]$

---

[5]The circuit model consists of logical gates (e.g., NOT, AND, OR, etc.) connected by *wires*; the gates perform simple logical operations (i.e., simple computations) and the wires carry the information.

followed by any finite string of symbols x (separated by a blank) it gives the same output as the output $T(x)$ given by T on input x,

$$U(d[T], x) = T(x). \tag{11.1}$$

In other words, the universal Turing machine (UTM) is capable of simulating any other Turing machine. The UTM has fixed program and state control (analogous to the fixed hardware of a modern computer) and the action to be taken – the TM to be simulated, i.e., algorithm to be implemented – is determined by the string $d[T]$ stored on the tape (analogous to the program to be executed by a modern computer, which is stored in the memory).

How did Turing use these ideas to solve Hilbert's problem? Consider the following question: does the Turing machine T with Turing number $d[T]$ halt if it were fed $d[T]$ as input? Let us assume that there is an algorithm that solves this problem. In other words, there is an algorithm that, given *any* Turing machine T, could tell us whether or not T halts when fed $d[T]$ as input. Then, it is possible to make a Turing machine H with the property that $H(d[T])$ halts if and only if $T(d[T])$ does not halt. However, for $T \to H$ this implies that $H(d[H])$ halts if and only if $H(d[H])$ does not halt, which is a contradiction. It follows that the initial assumption about the existence of an algorithm that solves the halting problem is false. We say the halting problem is *uncomputable*, i.e., cannot be solved algorithmically.

11.2.2 Computational complexity

Calculating uncomputable functions represents an impossible task for a computer. However, there are other, nonabsolute limitations on the computational tasks that a computer can perform. These limitations are dictated by the time and space resources required to perform a computation. *Computational complexity* studies these time (i.e., number of computational cycles) and space (i.e., memory) requirements and determines *lower bounds* on the resources needed to run the *best* (i.e., most efficient) algorithm that solves a given problem. Note that this algorithm may not be known. Based on the proven lower bounds, each computational task is placed into a certain *complexity class*.

There are many complexity classes [27, 402]. In essence, when defining a complexity class one has to specify the resource of interest (time, space, ...), the underlying computational model (deterministic TM, probabilistic TM, quantum circuit model, ...), and the type of computational problem we want to address (decision problem, optimization problem, ...). Finally, one has to distinguish between *easy* and *hard* problems based on how the resources required to solve them scale with the size n of the input. The most basic distinction is between polynomial and exponential dependence on n, although, in practice, finer distinctions may be of interest. A computational problem is deemed *easy*, (or *tractable*) if there exists an algorithm for solving the problem that uses polynomial resources; the problem is regarded as *hard*, (or *intractable*) if the best possible algorithm requires exponential resources.

Let us focus on computational problems that can be formulated as *decision problems*, i.e., questions that can be answered yes or no. For example, the *primality* decision problem: is a given integer n a prime number? To address this type of problem, it is convenient to slightly modify the definition of the Turing machine by replacing the halting state q_h with two states: the *accept* (or *yes*) state q_A and the *reject* (or *no*) state q_R, which represent the "yes" and "no" answers, respectively. Complexity cases can be defined using various computational models, including many types of Turing machines, such as *deterministic* TMs, *probabilistic* TMs, *nondeterministic* TMs, *quantum* TMs, etc. These machines can simulate each other, but some may require more resources than others. We note that, unlike deterministic TMs (which have unique transitions at each step of the computation), a probabilistic TM can randomly choose the transition at each step (from two or more available transitions) according to some probability distribution. Similarly, a nondeterministic TM has a set of rules that prescribes more than one action for a given situation. However, by contrast with the probabilistic TM, one can view the nondeterministic TM as branching into different computational paths and simultaneously exploring them. This branching captures an essential feature of many mathematical models we want to analyze. The decision problem is solved if q_A is reached along one of the paths (and in this case the answer is "yes"), or if all the paths end with q_R (the answer being "no"). Figure 11.2(a) illustrates schematically the structure of the computational models corresponding to different types of Turing machines. We emphasize that the nondeterministic TM is not a model of physically realizable computation, but rather an abstract machine useful in the study of computational complexity.

During the 1970s, it was noticed by the computer science community that any problem solvable in n steps (i.e., using n elementary operations) within a certain "realistic" computational model, can also be solved in at most $p(n)$ steps on a probabilistic Turing machine, where p is a polynomial function. Consequently, if the problem is easy in some "realistic" computational model, it will be easy for a probabilistic TM. This led to the following "stronger" version of the Church–Turing thesis:

Modified Church–Turing thesis: *Every realistic (or "reasonable") model of computation can be simulated* efficiently *(i.e., with at most a polynomial increase in the number of required steps) using a probabilistic Turing machine.*

This thesis has been challenged from several different directions. From the perspective of classical computer science, for example, it has been pointed out that the thesis is based on implicitly assuming algorithmic computational models: a (given) finite input is transformed (inside a closed-box) into a finite output in a finite amount of time. However, one can argue [208] that this type of model does not apply to all possible computations; an example of nonalgorithmic computation could be an interactive process (instead of a

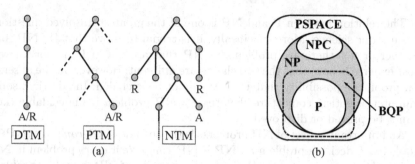

FIGURE 11.2 (a) Schematic representation of a (successful) computation within different Turing machine models: deterministic TM (DTM – one possible branch), probabilistic TM (PTM – multiple possible branches, one actual branch per run), and nondeterministic TM (NTM – computational tree). In general, a computational branch may end with an accept (or reject) state, or it may halt and fail, or it may never halt. (b) Suspected relationship of the quantum complexity class BQP with basic classical complexity classes.

function-based input-to-output transformation) in which communication (interaction) with the outside world happens during the computation. For us, the interesting challenge comes from quantum computation: if, for example, the factorization problem[6] does not have an efficient classical solution, then a quantum computer cannot be efficiently simulated using a probabilistic Turing machine.

Returning to our discussion regarding complexity classes, let us consider time (i.e., the number of elementary computational steps) as the resource of interest. A computational problem is said to be in class **P** if there is a (deterministic) Turing machine that can solve it in polynomial time $\mathcal{O}(p(n))$, where p is a polynomial function and n is the size of the input. Class **P** is considered to be the set of "efficiently" solvable problems. In addition to problems in class **P**, some decision problems may be hard to solve, but the instances where the answer is "yes" have efficiently verifiable proofs, i.e., proofs that can be *verified* by a *deterministic* TM in polynomial time. Consider, for example, the factoring decision problem: given an integer n and $k < n$, does n have a nontrivial factor smaller than k? If someone provides such a factor, we call it a *witness* to the fact that n has a factor that is smaller than k. The (hard) computational problems that admit easily checkable witnesses belong to an important complexity class known as **NP**. We note that an alternative definition (which is equivalent to the verifier-based definition) characterizes the **NP** class as the set of decision problems that can be solved by a *nondeterministic* TM in polynomial time; hence the name – nondeterministic **polynomial** time (**NP**) problems.

---

[6]Shor's algorithm provides an efficient solution for the factorization problem on a *quantum computer*. See Section 11.4.3 for details.

The relation between **P** and **NP** is one of the greatest unsolved questions in computer science. More specifically, it is trivial to prove that **P**⊆**NP**, but whether or not there are problems in **NP** that are *not* in **P**, is not known. Most experts believe that the two classes are different. However, in the absence of a proof, the possibility that **P**=**NP** cannot be excluded. In addition, some computer scientists conjecture that the **P**=**NP** problem is undecidable (i.e., cannot be proved or disproved).

An important subset of **NP** problems is the so-called **NP**-*complete* (**NPC**) problems. A decision problem $p \in$ **NP** is **NP**-*complete* if *every* problem in **NP** is *reducible* to p in polynomial time. An example of **NP**-*complete* problem is the *traveling salesman problem* (TSP): given a list of cities, the distances between each pair of cities, and a bound B, is there a route that starts and ends at a given city, visits every city exactly once, and has a total length less than B? Note that if *any* **NP**-*complete* problem can be solved *efficiently* (i.e., in polynomial time), then every problem in **NP** can (i.e., **NP**=**P**). On the other hand, if **NP**≠**P**, one can show that there are problems $p \in$ **NP** (called **NP**-intermediate) that are outside **P** but are not **NP**-complete. The factoring problem is believed to be an example of such a problem.

What is the place of quantum complexity classes within this picture? We can define **BQP** as the class of computational problems that can be solved on a quantum computer in polynomial time, if a bounded probability of error is allowed. It is known that **P**⊆**BQP**, i.e., quantum computers could efficiently solve problems in **P**. It is also known that quantum computers cannot solve efficiently problems that are outside the class **PSPACE** – decision problems solvable on a Turing machine using polynomial space (i.e., a number of working bits that grows polynomially with the size of the input), with no time limitation. Where exactly **BQP** lies between **P** and **PSPACE** it is not known. The most prevalent belief concerning the relationship of **BQP** with the main classical complexity classes is represented schematically in Figure 11.2(b).

11.2.3 Energy and computation

Everybody knows that a laptop may become very hot during operation. Hence, the natural question: what are the energy requirements for performing a given computational task? Surprisingly, the answer is *none*. In other words, computation can be done *in principle* without energy consumption. The key condition that needs to be satisfied is *reversibility*: a *reversible* computation can be performed (in principle) without energy consumption. The possibility of reversible computation may not be obvious when thinking in terms of the Turing machine model. On the other hand, using the *circuit model*, which is equivalent to the TM in terms of computational power, one can link reversible computation to having reversible logic gates. The NOT gate,[7] for example, is manifestly reversible (given the output $x_{out} = x$ one can uniquely determine the input $x_{in} = \bar{x}$). On the other hand, the AND gate is *irreversible* – given the output $x_{out} = 0$, the input could have been $(0,0)$, $(0,1)$, or $(1,0)$. Nonetheless, one can show that the AND gate (as well as other irreversible gates) can be replaced by reversible gates, such as, for example the *Fredkin* and *Toffoli* gates [176] – universal reversible logic gates with 3-bit inputs and outputs. So, in principle, reversible computation without energy consumption is possible. However, all models of reversible computation are highly sensitive to noise and, consequently, require some form of error correction. In turn, the error-correction process generates additional information (e.g., from measurements done on the system to establish whether or not an error has occurred), which, eventually, has to be discarded. This erasure of information, which is an irreversible, dissipative process, would cost energy.

In 1961 Rolf Landauer pointed out [315] that there is a connection between irreversibility in the computation process (e.g. operations involving erasure of information) and energy consumption. This connection is captured by the following principle:

> **Landauer's principle**: *Consider a computational operation in which a single bit of information is erased. The entropy of the environment increases by at least $k_B \ln 2$, i.e., the amount of energy dissipated into the environment (having temperature T) is at least $k_B T \ln 2$.*

The principle was verified experimentally in 2012 [33]. We note that Landauer's principle provides *lower bounds*. In fact, existing computers dissipate hundreds of times more than the Landauer limit $k_B T \ln 2$ for each elementary logical operation. Here, however, we are not concerned with the potential relevance of this principle for computer technologies. Instead, we emphasize that Landauer's principle provides a new perspective on the profound relationship between information and physics. We have already seen that information is always encoded in the states of physical systems, while computation involves information processing that is carried out on physically realizable devices. Landauer's principle suggests that general physics principles (e.g., thermodynamic principles) constrain our ability to manipulate information.

A suggestive illustration of this idea is provided by the solution to the Maxwell's demon "paradox." In 1871 Maxwell proposed a thought "experiment" that apparently violates the second law of thermodynamics. Consider a chamber separated into two compartments by a middle partition having a small controllable door. The chamber contains a gas, initially at equilibrium, and the door is controlled by a little "demon." The demon measures the velocities of the molecules and separates the fast and slow moving molecules by appropriately opening and closing the door. These actions can be done, in

---

[7]The NOT gate takes one bit $x \in \{0,1\}$ as input and outputs $\bar{x}$ (i.e., inverts x, $x = 1 \rightarrow \bar{x} = 0$, etc.). The AND gate takes two bits as input and outputs one: $(1,1) \rightarrow 1$ and $(x, x') \rightarrow 0$ if $x = 0$ or $x' = 0$.

principle, with arbitrarily low energy cost. The final result is a chamber with a "hot" compartment and a "cold" compartment having a total entropy lower than the initial (equilibrium) value, in apparent violation of the second law of thermodynamics. Progress toward solving this problem was made by Szilard in 1929 [497], but the final answer came almost 50 years later [46]. In essence, the demon has to collect information regarding the movement of the molecules. Eventually, this information has to be erased (to bring the "environment" to its initial state) and, according to Landauer's principle, this erasure increases the entropy of the combined system (i.e., chamber + demon + environment) so that the net entropy balance satisfies the constraints imposed by the second law of thermodynamics.

From this perspective, if the laws of physics have indeed an impact on how information can be manipulated, a natural question arises: how deep is this connection between information and physics? In particular, since the fundamental laws that govern physical systems are, as far as we know, the laws of quantum mechanics, what is the impact of these laws on the ability to transmit information and to perform computational tasks?

11.3 QUANTUM CIRCUITS

Quantum computation can be viewed as a three-step process involving 1) the preparation of an n-qubit input in the *computational basis* (i.e., the basis associated with the single-qubit states $|0\rangle$ and $|1\rangle$ – see Section 10.4.1), 2) the application of a sequence of single- and two-qubit quantum logic gates (i.e., unitary operators), and 3) the measurement of the n-qubit final state in the computational basis. Note that in this "standard model" of quantum computation the input is restricted to computational basis states, which corresponds to encoding *classical information* using, essentially, strings of "zeros and ones." Of course, the output is another strings of "zeros and ones" (corresponding to the n-qubit measurement in the computational basis) that encodes classical information. Also note that this model is developed in close analogy to the classical circuit model of computation. There are, however, a few important differences. The obvious one is that the basic unit – the classical bit – is replaced by its quantum correspondent, the qubit. In addition, the classical logic gates are spread out in *space* and connected by wires, which results in a two-dimensional network. By contrast, the quantum gates correspond to interactions that are turned on and off in *time*. Each gate acts on a few qubits (typically one or two), which evolve in time but have fixed positions, resulting a "network" that has time as one of its dimensions (see Figure 11.3). Finally, unlike classical machines, a quantum computer requires external control by a classical computer running in parallel, which has to place the quantum gates and decide when to do the measurement.

The importance of the circuit model is twofold. On the one hand, this language provides a variety of tools that help simplify algorithm design and provide conceptual insight. On the other hand, it enables us to quantify the

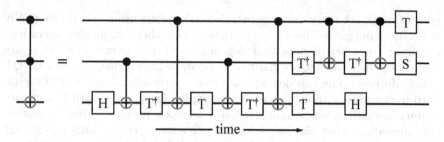

FIGURE 11.3 Quantum circuit showing the implementation of the (quantum) Toffoli gate using Hadamard, phase (S), CNOT, and $\pi/8$ (T) gates.

cost of algorithms in terms of necessary number of gates or circuit depth. The tools provided by the circuit model can be viewed as a primitive form of high-level programming language that facilitates the development of quantum algorithms. We note that almost all known quantum algorithms were developed in the quantum circuit model (a notable exception is the algorithm for evaluating Jones polynomials, which was developed using the topological quantum computation model [181]).

A specific quantum algorithm is a sequence of elementary operations (quantum logic gates) applied to a system of n qubits, which generates a particular unitary transformation $U \in U(2^n)$, where 2^n is the dimension of the Hilbert space corresponding to n qubits. The operator U acts on the state $|\psi_i\rangle$ that encodes the input information and generates the state $|\psi_f\rangle = U|\psi_i\rangle$ containing the desired solution of the computational problem. The output is extracted by performing a projective measurement on $|\psi_f\rangle$. Typically, we take $|00\ldots0\rangle$ as the initial state and consider the encoding of the input, $|\psi_i\rangle = U_{\psi_i}|00\ldots0\rangle$, as an input-dependent part of the algorithm, which now corresponds to the "total" unitary transformation UU_{ψ_i}.

The physical implementation of quantum gates is not an easy task. Therefore, it is critical to be able to realize arbitrary unitary transformations using a finite set of gates, each acting on just a few qubits at a time. A specific set of quantum gates that is sufficient to generate all unitary operators $U \in U(2^n)$ with arbitrary precision is called *universal*. Note that a quantum gate acting on m qubits corresponds to a unitary matrix that rotates the vectors from a particular 2^m-dimensional subspace and leaves unaffected the orthogonal complement of this subspace (which corresponds to the remaining $n - m$ qubits). A sequence of gates is represented by an element of $U(2^n)$ that is the product of the "elementary" unitary matrices corresponding to these gates. Consequently, a physical system can operate as a *universal quantum computer* if it supports a set of "elementary" unitary matrices that spans $U(2^n)$ densely.

There are several known universal sets of quantum gates. For example, one can prove [389] that any unitary operation can be approximated with arbitrary precision using Hadamard, $\pi/8$, and CNOT gates. We emphasize that this universality should *not* be understood as implying the possibility of generating an arbitrary unitary operation on n qubits using only polynomially many quantum gates. One can prove that, in fact, there are unitary operations requiring a number of steps that depends exponentially on n, i.e., there are unitary operations that are hard to compute using single-qubit and CNOT gates. Moreover, one can demonstrate that *nearly all* unitary operations are hard to compute, although constructing *explicitly* a class of hard to compute unitary operations on n qubits remains an open problem. Finding unitary transformations that can be performed efficiently – i.e., identifying efficient quantum algorithms – is one of the main goals in quantum computation.

The construction of the universal set of quantum gates consisting of Hadamard, $\pi/8$, and CNOT gates involves the three main steps briefly described below. Formal proofs can be found in Nielsen and Chuang [389].

1. An arbitrary unitary operator can be expressed *exactly* as a product of *two-level unitary operators* [429], i.e., operators acting nontrivially only on a subspace spanned by two computational basis states. More specifically, consider an arbitrary unitary matrix U acting on a d-dimensional space. One can prove [389] that the matrix U may always be written as a product of two-level unitary matrices V_i,

$$U = V_1 V_2 \ldots V_k, \qquad (11.2)$$

where $k \leq d(d-1)/2$. For n-qubits, an arbitrary unitary transformation can be expressed as a product of at most $2^{n-1}(2^n - 1) \simeq \mathcal{O}(4^n)$ two-level unitary matrices. Note that this may require an exponential number of matrices.

2. An arbitrary unitary operation on n qubits can be implemented *exactly* using single qubit and controlled gates (e.g., CNOT gates) [146]. This result can be obtained by combining step 1 with the property that an arbitrary two-level unitary operation on the state space of n qubits can be implemented using $\mathcal{O}(n^2)$ single qubit and CNOT gates [389]. Note that, although the CNOT plus single qubit unitaries form a universal set for quantum computation, this set has an infinite number of elements (i.e., single qubit gates) and would be difficult (if not impossible) to implement.

3. An arbitrary unitary operation on n qubits can be *approximated* with arbitrary precision using the Hadamard gate, phase gate, $\pi/8$ gate, and CNOT gate, i.e., these gates form a *discrete* set of universal operations for quantum computation [72]. We will call this set the *standard set* of universal quantum gates. One can show [389] that these gates can be done in a fault-tolerant manner (for a definition of fault-tolerance, see note on page 366). In fact, the phase gate, which can be constructed using two $\pi/8$ gates, is typically included in the standard set because of its natural role in the fault-tolerant construction. Replacing the $\pi/8$ gates with Toffoli gates (see below) generates another discrete set of universal quantum gates.

It is important to emphasize that approximating quantum circuits that contain single qubit and controlled gates (step 2) using a discrete set of gates (step 3) can be done *efficiently*. Specifically, the *Solovay–Kitaev theorem* [297, 389] states that approximating a circuit that contains $m = n^2 4^n$ CNOT gates and single qubit unitaries to an accuracy ϵ requires $\mathcal{O}(m \log^c(m/\epsilon))$ gates from the discrete set, where $c \approx 2$ is a constant; i.e., the overhead required to perform the simulation is polynomial in $\log(m/\epsilon)$. Nonetheless, there might be an initial exponential cost for representing a given unitary operator U using two-level unitary operations (step 1). The difficulty of identifying interesting families of unitary operators that can be efficiently represented using two-level unitaries translates into the difficulty of finding efficient quantum algorithms that outperform their classical counterparts.

As a final note, we point out that any classical circuit can be replaced by an equivalent circuit containing only reversible *Toffoli* gates. The (classical) Toffoli gate acts on three input bits and has three output bits: two *control bits* (a and b) are unaffected by the action of the gate and the third bit, the *target bit* (c), is flipped when both control bits are set to 1, $c \to c \oplus a \cdot b$. Being a reversible gate, the Toffoli gate has a quantum analog (also known as the controlled-controlled-NOT gate) that acts on three qubits: two control qubits, $|c_1\rangle$ and $|c_2\rangle$, and one target qubit $|t\rangle$. When both control qubits are in state $|1\rangle$, the amplitudes of the target qubit are flipped, i.e., $|110\rangle \to |111\rangle$ and $|111\rangle \to |110\rangle$. The implementation of the (quantum) Toffoli gate using Hadamard, phase, CNOT, and $\pi/8$ gates is illustrated in Figure 11.3. We conclude that any classical circuit (i.e., classical algorithm) that solves a given computational problem can be constructed using Toffoli gates and then "translated" *efficiently* into a quantum circuit. Hence, quantum computers are capable of performing any computation that can be done classically. Moreover, any *efficient* classical circuit can be translated into an *efficient* quantum analog.

11.4   QUANTUM ALGORITHMS

A quantum algorithm is a step-by-step procedure for solving a computational problem on a quantum computer. Classical algorithms can, of course, be performed on a quantum machine, but the term "quantum" usually applies to those algorithms that exploit some intrinsically quantum feature, such as superposition and entanglement. To date, there are only a handful of known quantum algorithms that have the potential to outperform their classical counterparts, most notably the factoring algorithm by Peter Shor [475, 478] and the searching algorithm by Lov Grover [222, 223]. Their key elements are briefly discussed below. In addition, we mention Deutsch's algorithm [138], the first example of a true quantum algorithm, and discuss a few basic aspects regarding the simulation of quantum systems, a likely "killer app" for future "first generation" quantum computers.

11.4.1   Deutsch's algorithm

Deutsch's algorithm solves a rather artificial problem formulated by David Deutsch with the goal of showing that, in principle, there exist computational tasks that quantum computers could solve more easily than classical computers. Some key ideas behind this first quantum algorithm inspired the development of subsequent algorithms, such as Shor's factoring algorithm.

The problem proposed by Deutsch may be described as the following game. Alice selects a number $x \in \{0, 1, \ldots, 2^n - 1\}$ and sends it to Bob. Upon receiving the number, Bob calculates a certain function $f(x) = y$ with $y \in \{0, 1\}$ that is either *constant* (i.e., y is the same for all values of x) or *balanced* (i.e., $y = 0$ for half of the possible values of x and $y = 1$ for the other half) and sends the result back to Alice. Calculating $f(x)$ is assumed to be extremely difficult. Alice may select another number $x' \in \{0, 1, \ldots, 2^n - 1\}$ and repeat the procedure. Her goal is to determine *with certainty* whether the function f calculated by Bob is constant or balanced. How fast can she succeed?

Here, we are not concerned about the algorithm used for calculating $f(x)$, which can be regarded as a "black box function." Alice's problem is just to minimize the number of times this black box needs to be used for evaluating $f(x)$ (since the evaluations are very expensive). We note that, given a classical algorithm for computing f, one can show that there exists a quantum circuit of *comparable efficiency* for evaluating f on a quantum computer with $n + 1$ qubits. The first n qubits (representing the so-called *data register*) encode the input number x in binary form, $x = x_1 x_2 \ldots x_n$, with $x_i \in \{0, 1\}$, while the last qubit (the *target register*) will contain the result of the calculation. The quantum circuit generates a unitary transformation U_f on the corresponding 2^{n+1} dimensional Hilbert space. The transformation is defined by $U_f |x_1, x_2, \ldots, x_n, z\rangle = |x_1, x_2, \ldots, x_n, z \oplus f(x)\rangle$, where $\oplus$ indicates addition modulo 2 and $|z\rangle$ is the initial state of qubit $n + 1$.

Returning to Alice's problem, we note that classically she needs $2^{n-1} + 1$ evaluations of f to establish *with certainty* that the function is constant. Also, proving that f is balanced requires at least two evaluations (if she is lucky and gets a 1 and a 0 after the first two evaluations) and at most $2^{n-1} + 1$ evaluations, since in the worst case scenario Alice may get 1 (or 0) 2^{n-1} times, before getting a 0 (or a 1). However, using a quantum algorithm it is possible to determine whether f is constant or balanced using only one query to the black box. The corresponding algorithm – referred to as the *Deutsch-Jozsa algorithm* – was proposed by David Deutsch and Richard Jozsa in 1992 [139] and improved by Cleve et al. in 1998 [114]. Here, we will not discuss the generic case but focus instead on the simple case corresponding to $n = 1$: how many queries to the black box are required in order to determine whether the function $f : \{0, 1\} \to \{0, 1\}$ is constant or balanced? Classically, the answer is two. Deutsch's quantum algorithm [138] solves this problem using only one evaluation of f. An improved version of the original algorithm is presented below.

The quantum circuit contains two qubits. A black box U_f performs the unitary transformation $|x\rangle|z\rangle \longrightarrow |x\rangle|z \oplus f(x)\rangle$, where $f(x)$ is either *constant* or *balanced*. The specific steps of the algorithm are illustrated in Figure 11.4.

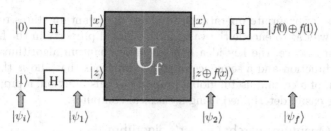

FIGURE 11.4 Quantum circuit implementing Deutsch's algorithm. The black box U_f performs the unitary transformation $|x\rangle|z\rangle \longrightarrow |x\rangle|z \oplus f(x)\rangle$.

1. The qubits are initialized in the state $|\psi_i\rangle = |0\rangle|1\rangle$.
2. Using Hadamard gates, the qubits are prepared as the superposition

$$|\psi_1\rangle = \left[\frac{|0\rangle + |1\rangle}{\sqrt{2}}\right]\left[\frac{|0\rangle - |1\rangle}{\sqrt{2}}\right]. \tag{11.3}$$

3. The unitary transformation U_f is applied to $|\psi_1\rangle$. Since $U_f|x\rangle(|0\rangle - |1\rangle)/\sqrt{2} = (-1)^{f(x)}|x\rangle(|0\rangle - |1\rangle)/\sqrt{2}$, applying U_f generates the state

$$|\psi_2\rangle = \begin{cases} \pm\left[\frac{|0\rangle + |1\rangle}{\sqrt{2}}\right]\left[\frac{|0\rangle - |1\rangle}{\sqrt{2}}\right] & \text{if } f(0) = f(1) \\ \pm\left[\frac{|0\rangle - |1\rangle}{\sqrt{2}}\right]\left[\frac{|0\rangle - |1\rangle}{\sqrt{2}}\right] & \text{if } f(0) \neq f(1) \end{cases} \tag{11.4}$$

4. A Hadamard gate applied to the first qubit gives the final state

$$|\psi_f\rangle = \pm|f(0) \oplus f(1)\rangle\left[\frac{|0\rangle - |1\rangle}{\sqrt{2}}\right], \tag{11.5}$$

where $f(0) \oplus f(1) = 0$ if $f(0) = f(1)$ and 1 otherwise.
5. By measuring the first qubit we determine $f(0) \oplus f(1)$, i.e., whether the function f is constant or balanced.

Remarkably, this quantum algorithm allows us to determine a *global* property of $f(x)$ using only *one* query to the black box. What makes this possible? Most commonly, the answer makes reference to the so-called *quantum parallelism* understood as the ability of a quantum computer to perform multiple evaluations of a function simultaneously. More specifically, the idea is to prepare the data register into a superposition of both possible input states (step 2), then, using the black box once, to encode information about both $f(0)$ and $f(1)$ in the phase of the quantum state (step 3). Indeed, we could say that $|\psi_2\rangle$ "contains information about all possible values of $f(x)$." However, one has to keep in mind that, in fact, the accessible information corresponding to the data register is one bit, so it is impossible to recover all these values. The crucial idea is to identify a *global* property of f that can be specified using one bit and, by performing appropriate operations, to cause the quantum states to interfere in such a way as to encode the value of this global property in the final state. This is, in essence, the key idea behind many quantum algorithms: a clever choice of function and a sequence of transformations that allows the efficient evaluation of a certain useful global property of this function, a property that cannot be easily determined using a classical computer.

11.4.2 Quantum search: Grover's algorithm

The algorithm presented by Lov Grover in 1996 [222, 223] addresses the following problem: given the unstructured list of items $\{e_0, e_1, \ldots, e_{N-1}\}$, find the element e^* that satisfies a certain condition. For convenience we assume $N = 2^n$. Note that, in general, a search problem can have M solutions, with $1 \leq M \leq N$. We will focus on the particular case $M = 1$. Classical algorithms cannot do better than searching through the list, one element at a time, which requires on average $N/2$ steps (i.e., scanning half of the entries). Grover's algorithm, on the other hand, requires only $\mathcal{O}(\sqrt{N})$ operations, generating a "quantum speedup" of order $\sqrt{N}/2$. Note that this speedup does not change the complexity class of the computational task (i.e., the problem remains computationally hard), but may become important when searching through extremely large sets.[8] Furthermore, Bennett *et al.* have demonstrated [47] that Grover's algorithm is optimal, i.e., no quantum search algorithm can run faster than $\mathcal{O}(\sqrt{N})$.

Grover's algorithm can be applied far beyond the search of unstructured classical databases. In fact, its most promising potential application is to speedup search-based classical algorithms for computationally hard problems. As discussed in Section 11.2.2, problems in complexity class NP (outside class P) are hard to solve but, given a *witness*, the solution can be easily verified. For example, factoring a large number m is hard, but checking whether or not $q < m$ is a factor of m is easy. Consequently, one possible approach to solving problems in NP is to search through all possible witnesses. Consider, for example, the *Hamiltonian cycle* problem: given a graph with n vertices, determine whether or not it has a simple cycle that visits every vertex of the graph (called a Hamiltonian cycle). The problem is NP-complete, hence believed to be intractable on a classical computer. One can address it by searching through all $n^n = 2^{n \log n}$ possible orderings of the vertices and checking for the Hamiltonian cycle property. More generally, assuming that an NP problem (of size n) has witnesses that can be specified using $p(n)$ bits, where $p(n)$ is a polynomial in n, the solution of the problem (if one exists) can be found by searching through all $N = 2^{p(n)}$ possible witnesses. Using Grover's quantum algorithm to perform the search provides a speedup of order $\sqrt{N}/2$.

The oracle. Suppose we have an unstructured list $\{e_0, e_1, \ldots, e_{N-1}\}$ (with $N = 2^n$) and wish to find the element $e^* = e_{x^*}$ that satisfies a certain

---

[8]For example, code-breaking problems may involve sets with $N \simeq 10^{16}$ [76].

condition. It is convenient to represent the elements of the list using their index $x = 0, 1, \ldots, 2^n - 1$ and the condition satisfied by the solution e^* using a function $f : \{0, 1, \ldots, N - 1\} \to \{0, 1\}$ with the property $f(x^*) = 1$ and $f(x) = 0$ for $x \neq x^*$. We should keep in mind that *verifying* the condition (e.g., verifying whether or not a certain ordering of vertices represents a Hamiltonian cycle) is computationally easy, i.e., there exists an efficient classical algorithm for calculating $f(x)$. Of course, the algorithm depends on the particular problem we address and we will not consider any specific example. What is important here is that $f(x)$ can be calculated using polynomial resources. Furthermore, the (typically irreversible) classical circuit for calculating f can be modified into a reversible classical circuit (possibly, with a small overhead). The reversible circuit has an n-bit input register (initially set to x) and a one-bit output register (initially set to q) and takes (x, q) to $(x, q \oplus f(x))$. In turn, this reversible circuit can be translated into a quantum circuit that takes $|x\rangle|q\rangle$ to $|x\rangle|q \oplus f(x)\rangle$, where $|x\rangle$ represents the state on an n-qubit index register (x being the index, written in base 2, of element e_x) and $|q\rangle$ is the state of a single-qubit register (the oracle qubit). In other words, for every search problem we can construct a quantum circuit that allows us to efficiently verify whether or not x is the index of the desired solution.

The recipe described above leads to a quantum circuit that generates a unitary operator O – the *oracle* – defined by the following action on the computational basis

$$|x\rangle|q\rangle \xrightarrow{\;\;O\;\;} |x\rangle|q \oplus f(x)\rangle, \tag{11.6}$$

where $\oplus$ is addition modulo 2 and x is a string of 0s and 1s corresponding to the numerical value of the index written in base 2. Since the details of the circuit are problem-dependent, we will treat it as a black box. To verify whether x is the solution of our search, we prepare the state $|x\rangle|0\rangle$ and apply the oracle; if the oracle qubit flips to $|1\rangle$ (i.e., $f(x) = 1$), we found the solution (i.e., $x = x^*$), otherwise $x \neq x^*$.

Next, inspired by Deutsch's algorithm, we put the information about $f(x)$ into the phase of the index register by initializing the oracle qubit in the state $(|0\rangle - |1\rangle)/\sqrt{2}$. Specifically, the action of the oracle becomes

$$|x\rangle \left[\frac{|0\rangle - |1\rangle}{\sqrt{2}} \right] \xrightarrow{\;\;O\;\;} (-1)^{f(x)}|x\rangle \left[\frac{|0\rangle - |1\rangle}{\sqrt{2}} \right]. \tag{11.7}$$

Note that the state of the oracle qubit does not change (and it will remain the same throughout the entire algorithm). We can view the action of the oracle as "marking" the solution to the search problem by changing the phase of the corresponding state $|x^*\rangle$ by π, the other states being unaffected.

The procedure. So far, nothing is gained by replacing the classical algorithm for calculating f with a quantum oracle; we still need (on average) $N/2$ queries to the oracle to solve the search problem. The idea is to make use of quantum

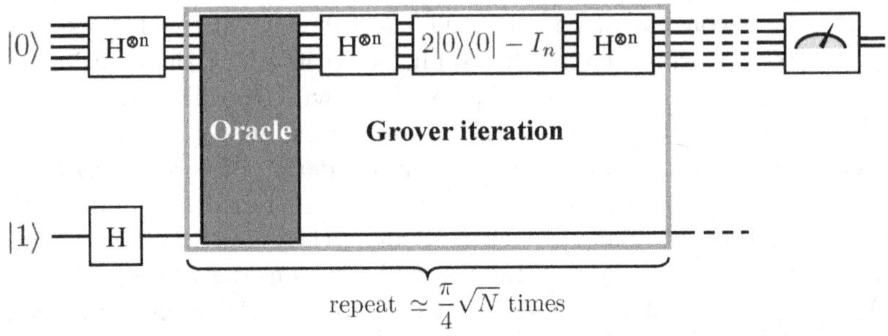

FIGURE 11.5 Schematic circuit for implementing Grover's quantum search algorithm. The Grover iteration is applied $R \simeq \frac{\pi}{4}\sqrt{N}$ times.

superposition by putting the index register into an equal superposition of all possible states, which can be done by applying a Hadamard gate to every qubit in the register (see Figure 11.5). Explicitly, we have

$$|\psi\rangle = H^{\otimes n}|0\rangle^{\otimes n} = \frac{1}{\sqrt{N}} \sum_{x=0}^{N-1} |x\rangle. \tag{11.8}$$

Having prepared the index register in state $|\psi\rangle$ and the oracle qubit in state $(|0\rangle - |1\rangle)/\sqrt{2}$, the search algorithm consists in repeatedly applying a quantum subroutine referred to as the *Grover iteration* (or *Grover operator*). After r iterations, the quantum state of the first n qubits becomes

$$|\psi\rangle_r = \sin(\theta_0 + r\Delta\theta)|x^*\rangle + \frac{\cos(\theta_0 + r\Delta\theta)}{\sqrt{N-1}} \sum_{x \neq x^*} |x\rangle, \tag{11.9}$$

where $\sin\theta_0 = 1/\sqrt{N}$ and $\sin\Delta\theta = 2\sqrt{N-1}/N$. Applying the Grover iteration R times, where $R \simeq \frac{\pi}{4}\sqrt{N}$ [71], results in a rotation $|\psi\rangle \rightarrow |\psi\rangle_R \simeq |x^*\rangle$ of the quantum state of the index register in the two-dimensional subspace spanned by the starting vector $|\psi\rangle$ and the solution $|x^*\rangle$ to the search problem. The final step is to extract x^* (with error probability $\mathcal{O}(1/N)$) by measuring the first n qubits. We note that performing the Grover iteration more than R times increases the error probability, i.e., the state vector is over-rotated.

The quantum circuit for the Grover iteration, which is illustrated schematically in Figure 11.5, implements the following four steps:

1. Apply the oracle O. The sign of the $|x^*\rangle$ component of $|\psi\rangle_r$ is reversed.
2. Apply the Hadamard transform $H^{\otimes n}$ to the index register.
3. Perform a conditional phase shift that reverses the signs of all components except $|0\rangle \equiv |0\rangle^{\otimes n}$. This is represented by the unitary operator $2|0\rangle\langle 0| - I$, which can be implemented using $\mathcal{O}(n)$ quantum gates [389].
4. Apply the Hadamard transform $H^{\otimes n}$ to the index register.

Steps 2-4, the so-called *diffusion transform*, are represented by the following unitary operator

$$H^{\otimes n}(2|0\rangle\langle 0| - I)H^{\otimes n} = 2|\psi\rangle\langle\psi| - I, \qquad (11.10)$$

where $|\psi\rangle$ is the equal superposition given by (11.8) and I is the identity operator. The diffusion transform (11.10) can be viewed as the *inversion about mean* operation. Indeed, one can easily check that applying this operation to a generic state $\sum_k \alpha_k|k\rangle$ produces $\sum_k[-\alpha_k + 2\langle\alpha\rangle]|k\rangle$, where $\langle\alpha\rangle = \sum_k \alpha_k/N$ is the mean value of the coefficients. The entire Grover iteration corresponds to $G = (2|\psi\rangle\langle\psi| - I)O$. Note that the diffusion transform requires $\mathcal{O}(n)$ gates and that the oracle can be efficiently implemented on a quantum computer, although its specific cost depends upon the particular application. Since the full algorithm requires $R \simeq \frac{\pi}{4}\sqrt{N}$ Grover iterations, it provides a quadratic improvement over the $\mathcal{O}(N)$ oracle calls required classically.

11.4.3 Quantum Fourier transform: Shor's algorithm

The quantum algorithms for solving the factoring and discrete logarithm problems, which are based on Shor's *quantum Fourier transform* [475, 478], provide an impressive exponential speedup over the best known classical algorithms and, probably, the clearest evidence so far that quantum computers could address certain problems that are classically intractable. This spectacular discovery has been a significant driving force behind the developments in quantum computing over the past decade. Below, we briefly sketch the main ideas that underlie this class of quantum algorithms focusing on the factoring problem.

Shor's algorithm for integer factorization involves two basic steps: i) reduce the *factoring problem* to the so-called *order-finding* problem and ii) solve the order-finding problem using a quantum algorithm. The first part can be done efficiently on a classical computer. The second component, which represents a particular case of a more general procedure called *phase estimation*, has the quantum Fourier transform (QFT) as key ingredient. It is this application of the QFT that provides the exponential speedup over the (known) classical algorithms. We note that, in addition to integer factorization, phase estimation represents the central component of a few other quantum algorithms.

The original problem is the following: given an integer N, find its prime factors. Note that determining whether N is prime or not can be done efficiently using classical algorithms [123]. Therefore, the original problem reduces to the following task: given an integer N that is not prime, find a nontrivial factor of N. In turn, this task can be reduced to the following order-finding[9] problem: given the function $f_a(x) = a^x \pmod{N}$, where x is integer and a is an element[10] of the multiplicative group of integers modulo N, find its period,

---

[9]The *order* (or *period*) of the element a of a group is the smallest positive integer r such that $a^r = e$, where e is the identity element and a^r denotes the product of r copies of a.

[10]As an element of the the multiplicative group of integers modulo N, a is relatively

i.e., find the smallest integer r for which $f(x+r) = f(x)$. The reduction of factoring to order-finding can be done using the following algorithm:

1. Randomly choose an integer $a < N$; compute the greatest common divisor $\gcd(a, N)$; if $\gcd(a, N) \neq 1$, return it and stop (a nontrivial factor of N has been found); otherwise, continue.

2. Apply the order-finding subroutine to find the period r of the function $f_a(x) = a^x \pmod{N}$.

3. If r is odd or $a^{r/2} = -1 \pmod{N}$, go back to step 1, otherwise continue.

4. At least one of the integers $p = \gcd(a^{r/2}+1, N)$ and $q = \gcd(a^{r/2}-1, N)$ is a nontrivial factor of N; test which one and return it; stop.

The proof of the theorem behind step 4 can be found in, for example, Nielsen and Chuang [389]. The key component of this factoring algorithm is the order-finding subroutine (step 2). The corresponding quantum order-finding circuit is shown schematically in Figure 11.6. For concreteness, we assume that the number N we want to factorize is an L-bit number. The quantum order-finding circuit has two registers [389]: one containing q qubits, with $q \simeq 2L + 1 + \log(2 + \frac{1}{2\epsilon})$, initialized to $|0\rangle = |00\ldots 0\rangle$ and the other containing L qubits initialized to $|1\rangle = |00\ldots 01\rangle$. The algorithm proceeds as follows:

1. Apply Hadamard gates to the first register to create the superposition of $Q = 2^q$ states

$$\frac{1}{\sqrt{Q}} \sum_{j=0}^{Q-1} |j\rangle|1\rangle. \tag{11.11}$$

2. Apply a sequence of controlled-$U_a^{2^k}$ operations, with $0 \leq k < q$, that performs the transformation $|j\rangle|1\rangle \rightarrow |j\rangle|a^j \pmod{N}\rangle$. The controlled-$U_a$ black box performs the unitary transformation $|u\rangle \rightarrow |u\,a \pmod{N}\rangle$, if the control qubit is in the state $|1\rangle$. The controlled-$U_a^{2^k}$ operations can be efficiently implemented using a procedure called *modular exponentiation* and the entire sequence requires $\mathcal{O}(L^3)$ gates. The resulting state is

$$\frac{1}{\sqrt{Q}} \sum_{j=0}^{Q-1} |j\rangle|a^j \pmod{N}\rangle \approx \frac{1}{\sqrt{rQ}} \sum_{s=0}^{r-1}\sum_{j=0}^{Q-1} e^{2\pi i s j/r}|j\rangle|u_s\rangle, \tag{11.12}$$

where r is the order of a modulo N and

$$|u_s\rangle \equiv \frac{1}{\sqrt{r}} \sum_{k=0}^{r-1} \exp\left[\frac{-2\pi i s k}{r}\right] |a^k \pmod{N}\rangle \tag{11.13}$$

---

prime to N, i.e., $\gcd(a, N) = 1$. Note that finding the greatest common divisor (gcd) of two integers can be done efficiently using Euclid's algorithm [490].

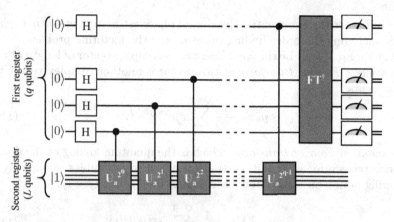

FIGURE 11.6 Schematic circuit for implementing the quantum *order-finding* algorithm. Replacing $U_a \to U$ (unitary operator) and $|1\rangle \to |\psi\rangle$ (eigenstate of U) corresponds to a generic *phase-estimation* algorithm that outputs θ accurate to $q - \log(2 + \frac{1}{2\epsilon})$ bits with probability of success at least $1 - \epsilon$.

is an eigenstate of U_a corresponding to the eigenvalue $\exp\left[\frac{2\pi i s}{r}\right]$. Note that, more generally, the spectrum of a unitary operator U consists of phases, $U|\psi\rangle = e^{i\theta}|\psi\rangle$. Given an eigenvector $|\psi\rangle$, the corresponding θ can be efficiently determined using the quantum *phase estimation* algorithm [389]. Order-finding corresponds to the particular case $U = U_a$.

3. Apply the inverse quantum Fourier transform $FT^\dagger$ to the first register. The resulting quantum states is [389]

$$\frac{1}{\sqrt{r}} \sum_{s=0}^{r-1} |\widetilde{s/r}\rangle \tag{11.14}$$

where $\widetilde{s/r}$ is a good estimator for s/r.

4. Measure the first register. A good estimate (accurate to $2L + 1$ bits) of the phase angle $\theta/2\pi = s/r$ is obtained with probability of success at least $1 - \epsilon$.

5. Extract the period r using the *continued fractions algorithm* [389]. Starting with $\widetilde{s/r}$, the algorithm produces numbers s' and r' with no common factor, such that $s'/r' = s/r$. Check whether r' is the order of a modulo N. If yes, stop. Otherwise, repeat the algorithm.

The order-finding algorithm uses $\mathcal{O}(L^3)$ gates, with the major cost coming from the modular exponentiation. We note that the black box transformation U_a can be implemented using $\mathcal{O}(L^2)$ quantum gates, similar to the cost of the classical algorithm for modular multiplication, and the entire $U_a^{2^k}$ sequence requires $\mathcal{O}(L^3)$ gates. The core of the order-finding subroutine is the inverse quantum Fourier transform (step 3), which requires $\mathcal{O}(L^2)$ gates [389] and is responsible for the exponential speedup of phase-estimation and other related tasks, including the order-finding problem and the factoring problem.

Let $x_0, x_1, \ldots, x_Q$ be the components of a complex vector of length $Q = 2^q$. The *discrete Fourier transform* of this vector is another complex vector with components

$$y_k = \frac{1}{\sqrt{Q}} \sum_{j=0}^{Q-1} x_j e^{2\pi ijk/Q}. \tag{11.15}$$

The *quantum Fourier transform*, which is the quantum analog of the discrete Fourier transform, is a unitary transformation on q qubits defined by the following action on the q-qubit basis states $|0\rangle, |1\rangle, \ldots, |Q-1\rangle$:

$$|j\rangle \xrightarrow{\text{FT}} \frac{1}{\sqrt{Q}} \sum_{k=0}^{Q-1} e^{2\pi ijk/Q} |k\rangle. \tag{11.16}$$

For an arbitrary state, the action of this transformation takes the form $\sum_j x_j |j\rangle \xrightarrow{\text{FT}} \sum_k y_k |k\rangle$, where the amplitudes y_k are given by Eq. (11.15). Being unitary, the FT transformation can be implemented as a quantum circuit. Moreover, this implementation can be done efficiently [389] using $\mathcal{O}(q^2)$ quantum gates. By contrast, the best classical algorithms for computing the discrete Fourier transform, e.g., the fast Fourier transform (FFT), use $\mathcal{O}(q 2^q)$ gates. Note, however, that there is no known procedure to use the quantum Fourier transform to efficiently compute the Fourier transformed amplitudes y_k. This is because, on the one hand, the amplitudes y_k cannot be directly accessed by measurement and, on the other hand, there is no recipe for efficiently preparing a generic input state $\sum_j x_j |j\rangle$.

In Eq. (11.16), j is an integer ($0 \le j < Q$) written in binary representation $j \Rightarrow j_1 j_2 \ldots j_q$, with $j_k \in \{0,1\}$ and $j = j_1 2^{q-1} + j_2 2^{q-2} + \cdots + j_q 2^0$. We introduce the following notation for binary fractions: $j_l 2^{-1} + j_{l+1} 2^{-2} + \cdots + j_m 2^{l-m-1} = 0.j_l j_{l+1} \ldots j_m$. Using this notation, the right hand side of the quantum Fourier transform (11.16) can be written [389] in the form

$$\frac{1}{\sqrt{Q}} \left(|0\rangle + e^{2\pi i\, 0.j_q} |1\rangle \right) \left(|0\rangle + e^{2\pi i\, 0.j_{q-1}j_q} |1\rangle \right) \ldots \left(|0\rangle + e^{2\pi i\, 0.j_1 j_2 \ldots j_q} |1\rangle \right).$$
$$\tag{11.17}$$

This *product representation* is the basis for the construction of an efficient circuit for the QFT. The key idea is to use $\mathcal{O}(q)$ gates to evolve each qubit into the corresponding state of the product (11.17). Consider, as an example, the first qubit, $|j_1\rangle$. Applying a Hadamard gate will generate the state $|0\rangle + (-1)^{j_1} |1\rangle$, where, for simplicity, we omit the overall factor $1/\sqrt{2}$. Since $e^{2\pi i\, 0.j_1} = e^{j_1 \pi i} = (-1)^{j_1}$, we can write the state of the first qubit after the application of the Hadamard gate as $|0\rangle + e^{2\pi i\, 0.j_1} |1\rangle$. Next, we apply a sequence of controlled-R_k gates, where R_k denotes the unitary transformation

$$R_k = \begin{bmatrix} 1 & 0 \\ 0 & e^{2\pi i/2^k} \end{bmatrix}. \tag{11.18}$$

The gate R_k (with $2 \leq k \leq q$) is applied to the first qubit if the qubit k is set, i.e., if $j_k = 1$. Applying the controlled-R_2 gate adds an extra phase $e^{2\pi i j_2/4}$ in front of $|1\rangle$ and the state of the first qubit becomes $|0\rangle + e^{2\pi i\, 0.j_1 j_2}|1\rangle$. After applying the whole sequence, the state of qubit one becomes $|0\rangle + e^{2\pi i\, 0.j_1 j_2 \cdots j_q}|1\rangle$, i.e., the last factor in Eq. (11.17).

How can we understand the role of the inverse QFT in step 3 of the quantum order-finding algorithm? For simplicity, let us assume that $\theta = s/r$ can be expressed exactly in q bits as $\theta = 0.\theta_1\theta_2 \ldots \theta_q$, where $\theta_k \in \{0, 1\}$. In this case, the product $j(\theta) \equiv \theta Q = s/r2^q$ is an integer with the binary representation $j(\theta) \Rightarrow \theta_1\theta_2 \ldots \theta_q$. With these observations, the quantum state (11.12) of the first register, before applying the QFT, can be written as

$$\frac{1}{\sqrt{Q}} \sum_{k=0}^{Q-1} e^{2\pi i\, (0.\theta_1\theta_2 \ldots \theta_q)\, k}|k\rangle = \frac{1}{\sqrt{Q}} \sum_{k=0}^{Q-1} e^{2\pi i j(\theta)k/Q}|k\rangle. \qquad (11.19)$$

A simple comparison between the right hand side of this equation and Eq. (11.16) shows that the output of the inverse Fourier transform is nothing but $|\theta_1\theta_2 \ldots \theta_q\rangle = |j(s/r)\rangle$. Measuring this state in the computational basis provides the values of the bits $\theta_1, \theta_2, \ldots, \theta_q$, hence it represents a direct measurement of s/r. If s/r cannot be expressed exactly in q bits, one can show [389] that, with high probability, the output of the inverse QFT is $|j(\widetilde{s/r})\rangle$, where $\widetilde{s/r}$ is a good estimator for s/r. Note that in Eq. (11.14) we have used the (somehow ambiguous) notation $|j(\widetilde{s/r})\rangle \equiv |\widetilde{s/r}\rangle$. Finally, we note that these basic ideas are behind the use of the quantum Fourier transform in other quantum algorithm based on phase-estimation [389].

11.4.4   Simulation of quantum systems

If there is an application of quantum computation that could be realized in the foreseeable future, it is the simulation of quantum systems [166, 200]. The key challenge in simulating quantum systems using classical computers stems from i) the huge space required to store information about the state vector (which is an element of an exponentially large Hilbert space) and the evolution operator and ii) the inefficiency of calculating the evolution of the system. Let $N = \dim(\mathcal{H})$ be the dimension of the Hilbert space; then, $\log N \simeq n$ is proportional to the number of degrees of freedom of the quantum system. Simulating the evolution of the system on a classical computer requires the manipulation of matrices containing 2^{2n} elements, which involves an exponentially large number of elementary operations. Of course, there are situations when insightful approximations can make the classical simulation feasible. However, there are many important classes of quantum systems (e.g., strongly correlated materials, including the notorious high-temperature cuprate superconductors) for which classical simulation is an intractable problem. We need not go farther that the innocuously looking Hubbard model,

$$H = \sum_{\langle i,j \rangle, \sigma} t_{ij} c_{i\sigma}^\dagger c_{j\sigma} + V_0 \sum_i n_{i\uparrow} n_{i\downarrow}, \tag{11.20}$$

where $\langle i, j \rangle$ are nearest neighbor sites on a d-dimensional lattice. Being able to simulate this type of models could lead to huge advances in understanding the physics of strongly interacting many-body systems, including the properties of correlated materials and interacting topological matter.

In essence, a quantum simulator can be understood as a *controllable* quantum system capable of *emulating* other quantum systems [88]. More specifically, consider a quantum system S that evolves from the initial state $|\psi(0)\rangle \equiv |\psi_0\rangle$ to $|\psi(t)\rangle \equiv |\psi_t\rangle$ via the unitary transformation $U = \exp(-iHt)$, where H is the Hamiltonian of the system. We say that the system S can be simulated if there exists another (controllable) quantum system S' (the simulator) that satisfies the following conditions: i) there is a mapping between the system and the simulator so that $|\psi_0\rangle \leftrightarrow |\phi_0\rangle$ and $|\psi_t\rangle \leftrightarrow |\phi_t\rangle$, where $|\phi_0\rangle$ and $|\phi_t\rangle$ are the initial and final states of the simulator, respectively; ii) the initial state $|\phi_0\rangle$ can be prepared; iii) the unitary transformation $U' = \exp(-iH't)$ describing the evolution from $|\phi_0\rangle$ to $|\phi_t\rangle$ can be engineered (here H' represents the controllable Hamiltonian of the simulator); iv) the final state $|\phi_t\rangle$ (or a relevant property of this state) can be measured. Of course, a (universal) quantum computer satisfies these conditions (plus additional ones). The key question concerns the efficiency of the simulation, more specifically whether or not one can prepare the initial state, engineer the unitary transformation, and measure the final state using resources that depend polynomially on the size (i.e., number of degrees of freedom) of the simulated system.

What is the advantage of using a quantum simulator? First, it solves the information storage problem. Indeed, instead of dealing with 2^n complex numbers, a quantum computer requires just n qubits to store the information about the state vector. Note, however, that this information may not be accessible. Second, to simulate evolution, the quantum computer has to implement unitary operations acting on the 2^n-dimensional Hilbert space. As discussed in Section 11.3, there exist unitary operations that cannot be efficiently approximated, so, in principle, there are Hamiltonian evolutions that cannot be efficiently simulated on a quantum computer. Whether or not such evolutions are realized in nature is an interesting open question. If such processes existed, they could be used to perform information processing beyond the quantum circuit model.

An important class of quantum systems that can be efficiently simulated on a quantum computer is characterized by Hamiltonians of the form

$$H = \sum_{k=1}^{p(n)} H_k, \tag{11.21}$$

where n is the number of particles in the system, $p(n)$ is a polynomial function of n, and H_k are terms that include local interactions, each term acting only on a few particles. The Hubbard model (11.20) is an example of Hamiltonian in this class. The key point is that, unlike the evolution operator e^{-iHt}, the unitary operators $e^{-iH_k t}$ can be easily approximated using quantum circuits because they act on small subspaces of the Hilbert space. However, in general, the terms of the Hamiltonian (11.21) do not commute, hence $e^{-iHt} \neq \prod_k e^{-iH_k t}$. To deal with this complication, the key idea is to divide the time evolution interval $[0, t]$ into n small segments $\Delta t = t/n$ and exploit the following asymptotic approximation theorem: given the Hermitian operators A and B, for any real t we have

$$\lim_{n \to \infty} \left(e^{iAt/n} e^{iBt/n} \right)^n = e^{i(A+B)t}. \tag{11.22}$$

Equation (11.22), also called the Trotter formula, holds even if A and B do not commute. For quantum simulations one can use convenient approximations obtained by modifying the Trotter formula. For example, one can divide the evolution of the system in elementary steps described approximately by the unitary operator $U_{\Delta t} = e^{-iH\Delta t} + \mathcal{O}(\Delta t^3)$, with

$$U_{\Delta t} = \left[e^{-iH_1\Delta t/2} \dots e^{-iH_{p(n)}\Delta t/2} \right] \left[e^{-iH_{p(n)}\Delta t/2} \dots e^{-iH_1\Delta t/2} \right], \tag{11.23}$$

where H_k are the Hamiltonian terms from the right hand side of Eq. (11.21).

Consider now the evolution of a quantum system described by the Hamiltonian (11.21), which acts on an N-dimensional Hilbert space, from some initial state $|\psi_0\rangle$ at $t = 0$ to a final state $|\psi_t\rangle$ at $t > 0$. The evolution of the quantum simulator will be described by a Hamiltonian H' having the same structure as the Hamiltonian in Eq. (11.21). The key steps of a quantum simulation algorithm for this evolution can be summarized as follows [200, 389]

1. *Initial-state preparation.* The physical state $|\psi_0\rangle$ is *represented* by the state $|\phi_0\rangle$ of n qubits, where n is a polynomial function of $\log N$. The quantum simulation problem cannot be solved efficiently unless the quantum register can be *initialized* to the state $|\phi_0\rangle$ using only *polynomial* resources. Examples of efficient state-preparation algorithms include the proposals by Ward et al. [529] (realistic many-particle quantum states on a lattice), Kassal et al. [280] (commonly used chemical wave functions), and Wang et al. [527] (arbitrary energy eigenstates).

2. *Unitary evolution.* The time evolution of the system is *represented* by the unitary evolution of the simulator, $U \leftrightarrow U'$. This evolution is broken into a large number of small time steps Δt and the correspondent "elementary" evolution operator $U'_{\Delta t}$, which is approximated using Eq. (11.23) or similar approximations, is applied $t/\Delta t$ times.

3. *Measurement.* The relevant information is extracted from the simulator by measuring the final state $|\psi_t\rangle$. The state could be uniquely identified using *quantum state tomography* [126], which involves performing

different measurements repeatedly and requires resources that grow exponentially with the size of the system. It is more convenient to identify certain relevant physical quantities, such as correlation functions or spectra of operators, that can be extracted directly [397, 483]. This requires repeating the algorithm only at most a polynomial number of times.

As a final note, we emphasize that even a small-scale quantum simulator (a few hundreds of qubits) can provide access to interesting physics that cannot be simulated using classical computers. This is in sharp contrast with other quantum algorithms (e.g. Shor's), which would require thousands of ideal qubits (many more with quantum error correction) to outperform their classical counterparts. One can reasonably hope that in the foreseeable future quantum simulation will have a profound impact on physics research.

11.5 QUANTUM ERROR CORRECTION

There is no such thing as an isolated quantum system, nor are there perfect quantum gates. Hence, the natural question: can one stabilize a quantum computer against the effects of random noise? The answer is far from obvious. One can start from our experience with classical computers, which are insensitive to noise. Their stability is based on a combination of *amplification* (if the signal it too weak, one amplifies it) and *dissipation* (if a physical bit ends up slightly away from its "0" (or "1") state, a *damped* restoring force will bring it back to the desired state). Dissipation transforms an intrinsically analog system (the physical computer, which is characterized by a continuum of possible states) into a digital device. The price is energy consumption, but the computational task gets done. However, for quantum computation this is not an option. Amplification of unknown quantum states is ruled out by the no-cloning theorem, while dissipation (i.e., irreversibility) is inconsistent with unitary transformations. This may suggest that the intrinsic fragility of coherent quantum states may represent an insurmountable obstacle to building a quantum computer. We note that the root of the problem is not necessarily the size of the system (i.e., number of particles), but the size of the (relevant) Hilbert space. One can easily maintain quantum coherence in a superconductor (which contains many electrons, but has a unique, gapped ground state, with no information content), but it is a totally different task to do it in a quantum computer with, say, a 2^{10^4}-dimensional Hilbert space; this would practically correspond to stabilizing Schrödinger's cat (well, maybe "Schrödinger's virus") into a superposition of dead and alive states (whatever that means for a virus).

Going back to the lessons of classical information processing, we know that error correction is possible and that the key idea is redundant information encoding (see Section 10.2). With quantum information, one faces a few difficulties that have no classical correspondent: (1) Replicating unknown states to generate repetition codes is prohibited by the no-cloning theorem.

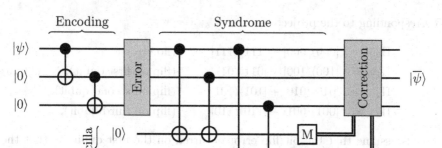

FIGURE 11.7 Circuit that realizes the encoding from Eq. (11.25) and corrects for one single bit-flip error. Ancilla qubits coupled to the data block are measured to check the parity between data qubits. A bit-flip error is corrected using the syndrome information by applying classically-controlled X gates.

The quantum correspondent of the classical encoding $0 \mapsto 000$, $1 \mapsto 111$ *cannot* be $|\psi\rangle \mapsto |\psi\psi\psi\rangle \equiv |\psi\rangle \otimes |\psi\rangle \otimes |\psi\rangle$. (2) Possible errors form a continuum. For example, a generic phase error corresponds to the transformation

$$R_{\frac{\theta}{2}} = \begin{pmatrix} 1 & 0 \\ 0 & e^{i\theta} \end{pmatrix} = e^{\frac{i\theta}{2}} \left[\cos \frac{\theta}{2} I - i \sin \frac{\theta}{2} Z \right], \qquad (11.24)$$

with θ arbitrary. Exact determination of an error would require infinite resources. (3) Measurements destroy quantum information. One cannot simply measure an unknown quantum state to check for possible errors.

Despite these difficulties, quantum error correction is possible [93, 140, 491]. To illustrate the main idea, we consider the simple example of a 3-qubit code. Note that this is not a full quantum code, since it cannot simultaneously correct for both bit (i.e., $|0\rangle \leftrightarrow |1\rangle$) and phase (i.e., $+|x\rangle \leftrightarrow -|x\rangle$) flips, but it contains all the key ingredients.

Encoding. Consider the quantum state $|\psi\rangle = \alpha|0\rangle + \beta|1\rangle$. We encode the state using three physical qubits as $|\psi\rangle \mapsto |\overline{\psi}\rangle$, with

$$|\overline{\psi}\rangle = \alpha|000\rangle + \beta|111\rangle \equiv \alpha|\overline{0}\rangle + \beta|\overline{1}\rangle. \qquad (11.25)$$

Note that $|\overline{\psi}\rangle \neq |\psi\psi\psi\rangle$, so this encoding is not prohibited by the no-cloning theorem. A quantum circuit containing two CNOT gates that realizes the encoding is shown in Figure 11.7.

Error detection. Assume that a bit-flip error has corrupted the state $|\overline{\psi}\rangle$. To correct it, one first needs to detect the error by performing a so-called *syndrome*. In essence, the syndrome involves acquiring information about possible bit-flips on the three qubits. More specifically, there are four error syndromes corresponding to the projectors

$$
\begin{array}{rcll}
\Pi_0 & = & |000\rangle\langle000| + |111\rangle\langle111| & \text{(no error)}, \\
\Pi_1 & = & |100\rangle\langle100| + |011\rangle\langle011| & \text{(flip on first qubit)}, \quad (11.26) \\
\Pi_2 & = & |010\rangle\langle010| + |101\rangle\langle101| & \text{(flip on second qubit)}, \\
\Pi_3 & = & |001\rangle\langle001| + |110\rangle\langle110| & \text{(flip on third qubit)}.
\end{array}
$$

Let us assume that the bit flip error occurred on the first qubit, so that the corrupted state is $|\overline{\psi}_E\rangle = \alpha|100\rangle + \beta|011\rangle$. Measurements corresponding to the projectors (11.26) give $\langle\overline{\psi}_E|\Pi_k|\overline{\psi}_E\rangle = 1$ for $k = 1$ and $\langle\overline{\psi}_E|\Pi_k|\overline{\psi}_E\rangle = 0$ if $k \neq 1$, indicating where the error occurred. Most importantly, these measurements *do not* destroy the qubit superposition. On the other hand, they provide minimal information about the state $|\overline{\psi}_E\rangle$ and *no information* about the encoded state $|\psi\rangle$ (i.e., about the coefficients α and β).

In practice, the encoding qubits are not measured directly, but, instead, we introduce two additional ancilla qubits to extract the syndrome information, as illustrated in Figure 11.7. If no error occurs, the ancilla qubits will remain in the state in which they are prepared, i.e., $|00\rangle$. Assuming that a bit flip error occurred on the first qubit, the quantum state of the five qubit system prior to the syndrome measurement is $\alpha|100\rangle\otimes|11\rangle + \beta|011\rangle\otimes|11\rangle$. Similarly, if an error occurs on the second or third encoding qubit, the state of the ancilla qubits changes to $|10\rangle$ or $|01\rangle$, respectively. Hence, the syndrome information can be extracted by measuring the state of the ancilla qubits.

Error correction. Once a bit flip error is identified, one can correct it using X gates with classical controls (see Figure 11.7). For example, if the measurement corresponds to the ancilla qubits being in the state $|11\rangle$, we know that the "collapsed" three-qubit state is $|\overline{\psi}_E\rangle = \alpha|100\rangle + \beta|011\rangle$, i.e., the desired code state with one bit flip error on the first qubit. The correct code state $|\overline{\psi}\rangle$ can be recovered by applying an X gate to the first qubit. Similarly, to correct a bit flip error detected on the second or third qubit one applies the X gate to the corresponding qubit.

The simple code described above only works if a maximum of one bit flip error occurs. If, for example, there are two errors on qubits two and three, the syndrome will incorrectly indicate a single error on qubit one. We note that one can imagine a similar scheme that performs phase flip error correction. The key observation is that a phase flip in the computational basis, i.e., $\alpha|0\rangle + \beta|1\rangle \rightarrow \alpha|0\rangle - \beta|1\rangle$, acts as a bit flip in the Hadamard basis $|\pm\rangle = (|0\rangle \pm |1\rangle)/\sqrt{2}$. Consequently, we can encode a generic quantum state as $|\psi\rangle \mapsto |\overline{\psi}\rangle = \alpha| + + +\rangle + \beta| - - -\rangle$ and adapt the above procedure to correct for a phase flip error occurring on one of the three qubits. Moreover, one can combine the two codes into a 9-qubit code that corrects for both (single qubit) bit flip errors *and* (single qubit) phase flip errors. The 9-qubit error correcting code, which was first developed by Shor in 1995 [476], has the following basis states

$$|\bar{0}\rangle = \frac{1}{\sqrt{2^3}}(|000\rangle + |111\rangle) \otimes (|000\rangle + |111\rangle) \otimes (|000\rangle + |111\rangle), \qquad (11.27\text{a})$$

$$|\bar{1}\rangle = \frac{1}{\sqrt{2^3}}(|000\rangle - |111\rangle) \otimes (|000\rangle - |111\rangle) \otimes (|000\rangle - |111\rangle). \qquad (11.27\text{b})$$

A generic quantum state is encoded as $|\bar{\psi}\rangle = \alpha|\bar{0}\rangle + \beta|\bar{1}\rangle$. The inner layer of the code corrects for bit flip errors by implementing the majority rule as discussed above, while the outer layer corrects for phase flip errors through a similar implementation of the majority *sign* rule.

Further insight into the basic principles of quantum error correction can be gained by considering the syndrome measurement from a slightly different perspective. Instead of measuring projectors, like those in Eq. (11.26), let us consider the observables $Z_i Z_{i+1} = 1 \otimes \cdots \otimes Z_i \otimes Z_{i+1} \otimes \cdots \otimes 1$ corresponding to the bit string parity of a neighboring pair of qubits (from a three-qubit block). For the 9-qubit code, we consider the following six observables

$$Z_1 Z_2, \quad Z_2 Z_3, \quad Z_4 Z_5, \quad Z_5 Z_6, \quad Z_7 Z_8, \quad Z_8 Z_9. \qquad (11.28)$$

Measuring these observables provides the necessary information about possible bit flip errors without collapsing the quantum state. Assume, for example, a bit flip error on the fifth qubit. In this case, the output of the measurement will be $(+1, +1, -1, -1, +1, +1)$, clearly indicating that the error happened on the fifth qubit. The original state can be recovered by applying the Pauli operator X_5 (i.e., an X gate on the fifth qubit). Similarly, the phase flip sindromes can be obtained by measuring the observables

$$X_1 X_2 X_3 X_4 X_5 X_6, \qquad X_4 X_5 X_6 X_7 X_8 X_9. \qquad (11.29)$$

If a phase flip error occurs on, say, the second qubit, the output of these measurement will be $(-1, +1)$. This indicates that the phase flip happened on one of the first three qubits. The original state can be recovered by applying a phase flip gate to each qubit of the first block, i.e., $Z_1 Z_2 Z_3$. Note that the code works even in the presence of one bit-flip error and one phase flip error (on any two of the nine qubits).

In the actual code the syndromes are implemented using additional ancilla qubits that are initialized to $|0\rangle$ and record what type of error occurred and where, similar to the simple scheme illustrated in Figure 11.7. We emphasize that this scheme corrects generic phase errors $R_{\frac{\theta}{2}}$ given by Eq. (11.24). Indeed, assume that such an error occurs on qubit k. The state of the 9-qubit block plus the relevant ancilla qubits takes the form $\cos\frac{\theta}{2}(\alpha|\bar{0}\rangle + \beta|\bar{1}\rangle) \otimes |0\ldots0\rangle - i\sin\frac{\theta}{2} Z_k(\alpha|\bar{0}\rangle + \beta|\bar{1}\rangle) \otimes |Z_k\rangle$, where $|Z_k\rangle$ is the ancilla state that signals the corresponding error. After measuring the ancilla, we get $|0\ldots0\rangle$ (i.e., no error) with probability $\cos^2\frac{\theta}{2}$, in which case the 9-qubit state collapses to the original state $|\bar{\psi}\rangle$, or $|Z_k\rangle$ (i.e., phase flip in a certain block) with probability

$\sin^2 \frac{\theta}{2}$, in which case the state collapses to $Z_k|\overline{\psi}\rangle$ and can be restored to $|\overline{\psi}\rangle$ by inverting the error indicated by the ancilla.

The operators in Eqs. (11.28) and (11.29) generate an Abelian qroup called the *stabilizer* of the 9-qubit code. The stabilizer formalism [216] describes error correction in terms of operators that *stabilize* certain states. We do not address this important and rather complex topic; instead, we only provide some basic definitions. Consider the so-called *Pauli group* $\mathcal{P}_n$ generated under multiplication by the Pauli matrices X, Y, Z and the identity I (multiplied by ± 1 or $\pm i$) acting on n qubits. The operators (11.28) and (11.29) are examples of elements of $\mathcal{P}_9$. Now let S (the *stabilizer*) be a subgroup of $\mathcal{P}_n$ with the property that i) all its elements commute and ii) $-I$ is not an element of S. Then, the *encoding states* $|\psi\rangle \in \mathcal{C}(S)$ of the stabilizer code consist of all the states with the property $K_i|\psi\rangle = |\psi\rangle$ for all $K_i \in S$. For example, the 3-qubit subspace spanned by $\{|000\rangle, |111\rangle\}$ is stabilized by the Abelian subgroup $S = \{I, Z_1 Z_2, Z_2 Z_3, Z_1 Z_3\}$, which can be obtained from $k = 2$ independent generators (e.g., $Z_1 Z_2$ and $Z_2 Z_3$). In general, if S has k independent generators, the code space $\mathcal{C}(S)$ has dimension 2^{n-k}, i.e., it can encode $n - k$ qubits. Every error $E \in \mathcal{P}_n$ that *anticommutes* with the elements of S takes $|\psi\rangle \in \mathcal{C}(S)$ to an orthogonal subspace; however, such errors can be detected (via syndrome measurements) and corrected. If, on the other hand, E commutes with all $K_i \in S$, then either $E \in S$, in which case it does not affect the state at all, or $E \notin S$, in which case it corrupts the state.

CHAPTER 12

Anyons and Topological Quantum Computation

TOPOLOGICAL QUANTUM COMPUTATION HAS EMERGED AS A RAPIDLY growing research field at the confluence of physics, mathematics, and computer science, bringing together several major themes that fruitfully inform each other. Physics problems, ranging from fundamental questions concerning the foundations of quantum mechanics to problems regarding the classification, realization, and manipulation of topological phases of matter, and mathematical challenges, ranging from identifying the formal tools for the complete classification of topological phases to open problems in computational complexity, converge at this intellectual junction in a combined effort to realize fault-tolerant quantum computation. The convergence of these themes has a significance that goes beyond the simple realization of a certain application – even one as remarkable as a quantum computer – and generates a new paradigm in which the cross fertilization of ideas takes center stage. Thinking about topological quantum matter from the perspective of quantum computation is more creative and fruitful than simply asking the standard questions of many-body condensed matter physics. This is one of the main messages of this book, while the themes involved in this synergy represent its main subject. Therefore, it is fitting to conclude our journey with a chapter on the most explicit manifestation of this new paradigm: topological quantum computation.

12.1 QUANTUM COMPUTATION WITH ANYONS

Topological quantum computation relies on employing *anyons* for encoding and manipulating information in a manner that is resilient against environmental perturbations. Anyons are quasiparticle excitations of two-dimensional topological states of matter having exchange properties that are different from those of both fermions and bosons. Particularly relevant for quantum

DOI: 10.1201/9781003226048-12

computation are the so-called *non-Abelian* anyons, which are quasiparticles that obey non-Abelian braiding statistics. In a topological quantum computer information is encoded nonlocally using multi-quasiparticle states, which makes it immune to errors caused by local perturbations and generates fault-tolerance.[1] In this section, we briefly review the main properties of anyons, highlight their role in quantum computation, and discuss a few examples. For a more detailed discussion of the basic concepts and ideas in topological quantum computation, including the properties of anyons, the reader is referred to the review article by Nayak et al. [384] and the book by Jiannis Pachos [399].

12.1.1 Abelian and non-Abelian anyons

Exchange statistics, a central element of the quantum mechanical description of the world, arises from the indistinguishability of identical quantum particles. The exchange of two identical particles in a many-body quantum system represents a symmetry operation and, consequently, cannot affect the physical properties of the quantum state. However, it may result in a transformation of the wave function that is consistent with this symmetry. In three spatial dimensions (3D), there are only two types of particles: bosons, which are characterized by wave functions that are completely symmetric under particle exchange, and fermions, which are described by completely antisymmetric wave functions. To understand this limitation to only two possibilities, we first notice that the process of adiabatically interchanging a pair of particles *twice* is equivalent to adiabatically taking one particle around the other along a certain closed path C that does not intersect the position r_2 of the second particle. Continuously deforming the path cannot affect the wave function. In 3D any path $C \in \mathbb{R}^3 - \{r_2\}$ can be continuously deformed to the trivial path $C_0(r_1)$ that corresponds to keeping the first particle at its original position. Since, the double exchange process does not modify the wave function, a single interchange can either leave the wave function unchanged (which corresponds to a system of identical bosons) or change its overall sign (for fermions).

The exchange properties of one- and two-dimensional quantum systems are fundamentally different from those of 3D systems. In 1D exchange statistics is not well defined because particle interchange is not possible without the two particles passing through one another. Two dimensions, on the other hand, are a true heaven for quantum statistics. A rich diversity of behaviors is possible in

---

[1] Fault-tolerant quantum computation involves quantum circuit designs and error correction procedures that do not cause errors to cascade. A circuit element is said to be *fault-tolerant* if a single error (generated by some component) causes *at most* one error in the output for each logical qubit block. This definition can be relaxed if the quantum code is able to correct multiple errors (i.e., if r errors can be corrected, then $n_e \leq r$ errors during an operation should not result in more than r errors in the output for each logical qubit). A remarkable consequence of fault-tolerant circuit design is the so-called *threshold theorem* [216], which, basically, states that an arbitrarily large quantum circuit can be successfully implemented to arbitrary accuracy if the physical error rate is below a certain threshold. Gate operations based on braiding of anyons are "naturally" endowed with fault tolerance.

2D [493], essentially because a particle loop C that encircles another particle (at position r_2) cannot be continuously deformed to a trivial path $C_0(r_1)$ without going through the second particle. Hence, in 2D it is possible to have a nontrivial winding of one particle around another, which implies that the initial and final states of the system are not necessarily identical. This sharp difference between 3D and 2D, first realized by Leinaas and Myrheim [324] and by Wilczek [542], is topological in nature and can be expressed mathematically as the difference between the fundamental groups $\pi_1(\mathbb{R}^3-\{0\}) = 1$ (in 3D all loops can be deformed to a point without passing through the origin) and $\pi_1(\mathbb{R}^2-\{0\}) = \mathbb{Z}$ (in 2D there are nontrivial windings around the origin).

Consider now the (counterclockwise) exchange of two identical particles in 2D. As a result of this process, the wave function can change by a phase factor,

$$\psi(r_1, r_2) \longrightarrow e^{i\theta}\psi(r_1, r_2). \tag{12.1}$$

Performing a second counterclockwise exchange does not necessarily lead back to the initial state, i.e., the phase $e^{2i\theta}$ can be arbitrary, which implies that, in general, the phase angle θ can be different from 0 (which corresponds to bosons) and π (for fermions). Particles with $\theta \neq 0, \pi$ are called *anyons* [544] with statistics θ.

For a 2D system of N identical particles, the paths associated with particle exchange belong to topological classes that are in one-to-one correspondence with the elements of the so-called *braid group* (see below, Section 12.1.2). The evolution of the system is given by the action of the braid group on the multi-particle quantum states. The simplest case corresponds to a straight-forward generalization of Eq. (12.1) in which an element of the braid group is represented by a phase factor $e^{im\theta}$, where m is the number of two-particle counterclockwise exchanges minus the number of clockwise processes. Note that $e^{im\theta}$ is a one-dimensional representation[2] of the braid group and that m does not depend on the specific order of the braiding operations, i.e., the representation is Abelian. The quasiparticles described by a wave function that evolves under particle exchange according to a one-dimensional representation of the braid group, $\psi(r_1, \ldots, r_N) \to e^{im\theta}\psi(r_1, \ldots, r_N)$, are called *Abelian anyons*.

In addition to Abelian anyons, in 2D there are other types of quasiparticles that give rise to exchange evolutions corresponding to higher-dimensional representations of the braid group. Assume that a quantum system having N quasiparticles at certain fixed positions is g times degenerate and let ψ_α, with $\alpha = 1, 2, \ldots, g$, be an orthonormal basis for the subspace of degenerate quantum states. An element σ of the braid group will be represented by a $g \times g$ unitary matrix $\rho(\sigma)$ acting on the subspace of degenerate states,

$$\psi_\alpha \longrightarrow [\rho(\sigma)]_{\alpha\beta}\psi_\beta. \tag{12.2}$$

---

[2]A representation assigns a matrix $\rho(g)$ to every element g of a group so that the group operation can be represented by matrix multiplication: $g = g_1 \star g_2 \longrightarrow \rho(g) = \rho(g_1)\rho(g_2)$.

Consequently, exchanging two particles corresponds to a rotation within the degenerate N-quasiparticle Hilbert space. Since, in general, two unitary matrices $\rho(\sigma)$ and $\rho(\sigma')$ corresponding to two different braiding operations do not commute, the evolution of the system depends on the specific order of these operations. The quasiparticles characterized by this type of statistical evolution are called *non-Abelian anyons*. Remarkably, within the degenerate subspace of N non-Abelian anyons all *local* perturbations have vanishing matrix elements. Consequently, the only way to perform unitary transformations within this subspace is by braiding quasiparticles. This property, together with the richness of the braid group, are the key features that enable fault-tolerant quantum computation with non-Abelian anyons.

12.1.2 Braiding

Consider a system of N particles with initial positions $\boldsymbol{R}_1$, $\boldsymbol{R}_2$, ..., $\boldsymbol{R}_N$ at time t_i and final positions $\boldsymbol{R}_1$, $\boldsymbol{R}_2$, ..., $\boldsymbol{R}_N$ at time t_f. The trajectories of the particles correspond to N world lines (strands) in $(d+1)$-dimensional space-time originating at points $(\boldsymbol{R}_n, t_i)$ and terminating at $(\boldsymbol{R}_m, t_f)$. Let us consider first the 3D case (i.e., $d = 3$). If the particles are distinguishable, all strands $(\boldsymbol{R}_n, t_i) \rightarrow (\boldsymbol{R}_n, t_f)$ can be continuously deformed into straight lines parallel to the time direction corresponding to particles sitting at fixed positions. If the particles are indistinguishable, their trajectories may have different initial and final points, $\boldsymbol{R}_n \neq \boldsymbol{R}_m$, and the corresponding sets of N world lines fall into different topological classes corresponding to the elements P of the *permutation group* S_N. Note that S_N has two one-dimensional irreducible representations, the *trivial representation*, $P \rightarrow 1$, and the *sign representation*, $P \rightarrow \text{sign}(P)$, which describe the evolution of a system of N identical bosons or fermions, respectively.

In 2D, the strands associated with N indistinguishable particles are in one-to-one correspondence with the elements of the *braid group* $\mathcal{B}_N$. An element of the braid group can be represented graphically as a set of world lines originating at initial positions and terminating at final positions, which correspond to N distinct points on the horizontal axis, time being represented by the vertical axis (see Figure 12.1). It is critical to distinguish between clockwise and counterclockwise exchanges by specifying whether a given strand passes over or under another strand. As shown in Figure 12.1, panels (c) and (g), this key information does not only involve the real space trajectories of the particles, but also their time evolution along these trajectories.

An elementary braid operation corresponds to the counterclockwise exchange of particles i and $i+1$, as shown in Figure 12.1(d) and (e). Algebraically, these elementary operations can be represented in terms of generators σ_i, with $1 \leq i \leq N-1$. A clockwise exchange of particles i and $i+1$ corresponds to the inverse σ_i^{-1}. Note that $\sigma_i^{-1} \neq \sigma_i$, i.e., $\sigma_i^2 \neq 1$. An arbitrary element of the braid group can be written as a product of generators, which corresponds graphically (up to continuous deformations) to the vertical stacking of

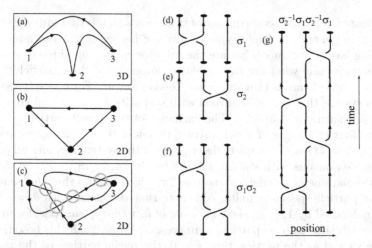

FIGURE 12.1 Graphical representation of real-space trajectories (a-c) and quasiparticle world lines (d-g) for a system of $N = 3$ identical particles. The trajectories in panels (a-c) correspond to the same element of the permutation group S_3, $(1, 2, 3) \rightarrow (3, 1, 2)$, which uniquely determines the evolution of the system in 3D (a), but not in 2D (b-c). The evolution in (b) corresponds to the element of the braid group shown in panel (f), while the evolution in (c) is represented by the element in panel (g). These elements can be expressed in terms of the elementary operations shown in panels (d) and (e). Note that in the sequence $\sigma_2^{-1} \sigma_1 \sigma_2^{-1} \sigma_1$ the first elementary operation is the rightmost one, i.e., σ_1, followed by σ_2^{-1}, etc.

drawings representing elementary operations (see Figure 12.1). The generators of the braid group satisfy the following basic relations

$$\sigma_i \sigma_j \neq \sigma_j \sigma_i \quad \text{for } |i - j| = 1, \qquad \sigma_i \sigma_j = \sigma_j \sigma_i \quad \text{for } |i - j| \geq 2,$$
$$\sigma_i \sigma_{i+1} \sigma_i = \sigma_{i+1} \sigma_i \sigma_{i+1}. \tag{12.3}$$

Note that the elements P_i of the permutation group S_N representing elementary permutations, $i \rightleftharpoons i + 1$ satisfy similar relations, with the notable difference that $P_i^2 = 1$, while $\sigma_i^2 \neq 1$. As a consequence, the permutation group is finite (specifically, S_N has $N!$ elements), while the braid group is infinite. As we will see later in this chapter, the richness of the braid group plays a key role in enabling topological quantum computation.

12.1.3 Particle types, fusion rules, and exchange properties

In a system of N identical fermions, bringing together two particles results in a composite object that behaves like a boson. Moreover, two spin-1/2 fermions can be combined into either a spin-0 or a spin-1 boson. This type of properties are also present in anyonic systems, but generally the situation is more complicated. The main consequence of this observation is that anyonic systems contain, in general, multiple types of anyons; hence, one has to address the following basic questions: what are the rules for combining different species of quasiparticles and what are the exchange properties of these particles?

The simplest case is that involving Abelian anyons. For concreteness, consider a system that contains anyons with statistics $\theta = \pi/m$ and composites made of k such quasiparticles. The exchange of two $k = 2$ particles is equivalent to four exchanges of $k = 1$ anyons, therefore the $k = 2$ composites have statistics $\theta = 4\pi/m$. A complete description of the system includes all particle species, i.e., anyons with statistics $\theta = \pi/m, 4\pi/m, 9\pi/m, \ldots$. Note that the statistics parameter θ is defined modulo 2π, which implies that the number of distinct particle species is finite. Also note that the boson with $\theta = 0$, which is often denoted by **1**, (e.g., the composite of m π/m-type anyons, for m even) behaves trivially, as if no particle were present. We will call this boson, which can be viewed as the particle type $k = 0$, the *trivial* particle or the *vacuum*. Finally, since θ is only defined up to 2π, we can ascribe the topological number $-k$ to the particle species $m - k$ (for m even) or $2m - k$ (for m odd), since these particles have statistics $k^2\pi/m$. The anyon $-k$ is the antiparticle of the k anyon, as combining them generates the vacuum. Note that certain situations (e.g., if $2k = m$) the antiparticle can coincide with the particle.

The rules for combining two anyons into a larger composite are called *fusion rules*. For Abelian anyons, the fusion rule is simple: $\phi_n \times \phi_k = \phi_{n+k}$, where ϕ_k denotes the k-type particle and $\phi_n \times \phi_k$ means that ϕ_n is fused with ϕ_k. The situation becomes more complicated for non-Abelian anyons because different *fusion channels* are possible, similarly to combining two spin-1/2 particles into either a spin-0 boson or a spin-1 boson. In general, the fusion rules for a set of particles $\phi_0 \equiv \mathbf{1}, \phi_a, \phi_b, \ldots$ can be written as

$$\phi_a \times \phi_b = \sum_c N_{ab}^c \phi_c, \tag{12.4}$$

which expresses the fact that fusing particles ϕ_a and ϕ_b can result in any of the particles ϕ_c for which $N_{ab}^c \neq 0$. For Abelian anyons we have $N_{nk}^q = \delta_{n+k,q}$, i.e., one always gets a single type of composite particle. By contrast, in the case of non-Abelian anyons there is at least one pair (ϕ_a, ϕ_b) that has multiple fusion channels, $\sum_c N_{ab}^c > 1$. Note that there are theories in which ϕ_a and ϕ_b can fuse to form ϕ_c in several distinct ways, i.e., $N_{ab}^c > 1$. Here, we will only consider examples when N_{ab}^c is either 0 or 1.

Next, let us assume that we want to fuse more than two anyons. There is a freedom to choose the ordering in which pairwise fusion processes take place. Consider, for example, the fusion of three anyons ϕ_a, ϕ_b, and ϕ_c into a particle ϕ_d. One can either fuse ϕ_a and ϕ_b into an intermediate quasiparticle ϕ_i, followed by the fusion of ϕ_i with ϕ_c, or fuse ϕ_b and ϕ_c into ϕ_j, followed by the fusion with ϕ_a. Since the total fusion channel ϕ_d is fixed, the final outcome is the same, but the intermediate two-anyon states may be different. Nonetheless, these intermediate states span the same *anyonic Hilbert space*,

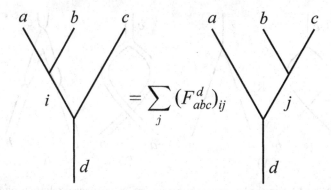

FIGURE 12.2 Correspondence between three-anyon fusion processes with fixed total fusion channel (d) but different fusion orderings. The F matrix can be viewed as a rotation in the intermediate fusion space.

so that the two choices described above correspond to different choices of basis for this space. The relation between these bases is given by the so-called *fusion* or F matrix, which can be viewed as a rotation in the intermediate fusion space. Using the F matrix, any two fusion orderings can be mapped into each other, as illustrated in Figure 12.2.

Consider now two particles a and b that fuse in a particular channel c. Braiding the two particles cannot change the fusion channel, since the total topological charge of the pair is not a local property – it could be measured, for example, along a large loop enclosing the particles well before they become a "local composite" – and does not depend on the evolution of the pair, as long as braiding with a third particle is not considered. Consequently, a counterclockwise exchange of particles a and b could also be viewed as a half twist of the c particle and corresponds to multiplying the wave function of the system by a phase factor R^c_{ab}. This behavior is illustrated in Figure 12.3. We emphasize that in order to fully specify the braiding statistics of a system of anyons it is required to specify i) the particle species, ii) the fusion rules N^c_{ab}, iii) the F matrix, and iv) the R matrix.

We close this section with a remark on the relation between spin and statistics. In the familiar cases involving bosons and fermions, rotating a particle around itself generates a phase factor $e^{i2\pi s}$, where the spin s is an integer for bosons and a half-integer for fermions. For anyons, the situation is similar, except that in this case spin can take any value. To get a better intuition about the spin-statistics relation, we use "worldribbons" (instead of worldlines) to describe the evolution of the system, as we already mentioned in Chapter 2 (Section 2.4). Let us now consider k exchanges of the particles a and b with given fusion channel c. The spins ascribed to these particles are s_a, s_b, and s_c, respectively, and the phase generated by a counterclockwise rotation of a spin s particle by and angle ϕ is $e^{i\phi s}$. Assuming that the evolution of the

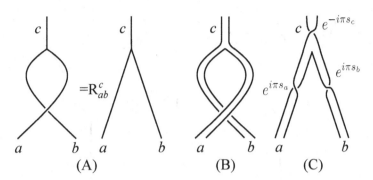

FIGURE 12.3 (A) Symbolic representation of a counterclockwise exchange of anyons a and b which fuse into a particle of type c. The exchange process generates a phase R_{ab}^c. (B) Representation of the exchange process in (A) using worldribbons. (C) The continuous deformation of the ribbon produces a topologically equivalent process containing half-twists that correspond to rotations of the particles around themselves.

system remains invariant under continuous deformations of the worldribbons, the total exchange process has to be equivalent to performing k π-twists of the ribbons (counterclockwise for particles a and b and clockwise for c), as illustrated in Figure 12.3. This equivalence can be expressed as the following spin-statistics theorem [34]

$$(R_{ab}^c)^k = e^{i\pi k s_a} e^{i\pi k s_b} e^{-i\pi k s_c}. \tag{12.5}$$

Note that exchanging particle a and its antiparticle $\bar{a}$, both having spin s and the fusion channel being the vacuum ($c = 1$, $s_1 = 0$), results in the familiar spin-statistics relation $R_{a\bar{a}}^1 = e^{i2\pi s}$, where s can take arbitrary values.

12.1.4 Fault-tolerance from non-Abelian anyons

The standard model of quantum computation involves three basic steps: initialization, unitary evolution, and measurement. The first step has the role to encode the classical information representing the input of the computation as a state vector in a large Hilbert space $\mathcal{H}$. This initial state $|\psi_i\rangle$ is then evolved unitarily (e.g., by applying a set of quantum gates) into a final state $|\psi_f\rangle = U|\psi_i\rangle$, from which the (classical) output of the computation is extracted by performing a measurement.

In the topological quantum computation scheme, the Hilbert space $\mathcal{H}$ is a subspace of the total Hilbert space of the system that consists of the degenerate ground states with a fixed number of non-Abelian anyons at fixed positions. A possible way to initialize the system is by creating particle-antiparticle pairs from the vacuum. The dimension of the Hilbert space $\mathcal{H}$, which corresponds to the number of all possible in-between fusion outcomes (the total fusion channel being in this case the vacuum), increases exponentially with the number of anyonic pairs. Note that the pairs have conjugate quantum numbers, since the total quantum number of a pair (before braiding with other particles) is the same as that of the vacuum, and provide a known state to start the computation process. The initialization is then completed by performing a controlled unitary operation U_{ψ_i} to put the quasiparticles in the desired initial state $|\psi_i\rangle$.

The controlled unitary operations U_{ψ_i} and U are performed by braiding the anyons, i.e., dragging the quasiparticles around one another in some specified way.[3] Assume that we start with n anyonic pairs, each having a specified fusion channel (e.g., the vacuum). When particles belonging to different pairs braid, the fusion channels of the pairs change, which corresponds to a rotation in the Hilbert space $\mathcal{H}$. After completing the unitary evolution, the output of the computation is retrieved by fusing the anyons in some specified way and measuring the anyonic charge of the outcomes, ϕ_1, ϕ_2, ..., which encode the result of the computation. Note that the braiding algorithm depends on the computation that is performed, but also on the initial state and the fusion procedure. Also note that a non-Abelian anyonic system is suitable for quantum computation if the set of transformations generated by the braiding of quasiparticles contains all operations needed for computation.

In general, errors represent a huge obstacle to building a functional quantum computer. Errors may occur when quantum information is either stored or processed. The main source of errors affecting a qubit that stores quantum information is interaction with the environment, which causes decoherence. As a result of this interaction, the quantum information is not "contained" by the qubit alone, but by the total system qubit + environment, hence, it is practically lost. Even if we eliminate the coupling to the environment, there could still be systematic errors stemming from the imperfection of the applied quantum gates. For example, a (desired) rotation of the qubit state by an angle θ may actually be a rotation by $\theta + \delta\theta$. Even when $\delta\theta \ll \theta$, applying the gate many times may significantly affect the total unitary transformation U and, consequently, the result of the computation. The possibility of error correction [93, 217, 417, 476, 491] is a cornerstone of quantum computation. However, error correction is only possible below a certain threshold error rate of about one error in $10^4 - 10^6$ operations. This is an extremely stringent constraint that demands quantum computation schemes with intrinsic low error rates. Topological quantum computation is the most promising known candidate.

How robust is the topological quantum computation scheme? First, we note that topological quantum memories are protected against errors due to local interactions with the environment. For example, the fusion channel of two anyons can only be changed by braiding them with other anyons and is immune to any local perturbation applied to the system. This is in sharp contrast with quantum information being stored using local qubits (e.g., the spin degree of freedom in a quantum dot); coupling to the environment (e.g., nuclear spins)

---

[3]In practice, it may be convenient to implement braiding using a measurement-only approach, without physically moving anyons [66]; see below, Section 12.3.

results in the rapid loss of this information. Topological quantum computers are also protected against systematic errors due to imprecise gate operations. In this scheme, the unitary operations depend only on the topology of the braid and not on the details of the quasiparticle paths. Note that the exact positions of the particles after performing the braid is not a real concern; one can always imagine an initialization scheme based on the creation of particle-antiparticle pairs and a measurement involving the fusion of anyons, so that the computation corresponds to a set of links, rather than open braids, and represents a completely topological process.

Finally, we note that probabilistic errors due to, for example, finite temperature do affect topological quantum computation. These errors are generated by the presence of additional, "unwanted" quasiparticles that braid with the "working" anyons and change their fusion channel. The main sources for these additional quasiparticles are disorder (e.g., impurities and defects) and thermal fluctuations. Note, however, that even in the presence of "unwanted" quasiparticles, certain types of processes do not cause an error, e.g., when a thermally created pair encircles a "working" anyon and then reannihilates, or when one of the particles from the pair annihilates an existing anyon. Generally, the system is protected against the generation of "unwanted" quasiparticles by a finite energy gap. Working at temperatures much lower than this gap suppresses exponentially the errors generated by thermally excited quasiparticle pairs. Also, disorder becomes an issue when its strength is comparable with (or larger than) the energy gap. In this case, one can view disorder as a strong (local) perturbation that (locally) closes the energy gap.

We conclude that the fault-tolerant nature of the anyonic system stems from i) the nonlocal encoding of quantum information in the fusion states of the anyons and ii) the presence of a finite energy gap. Local perturbations to the Hamiltonian due to interaction with the environment cannot alter the fusion states [80], as long as they are weak compared to the energy gap. In addition, the unitary operations depend only on the topology of the braid providing topological quantum computation with immunity to unitary errors due to imperfect gate operations.

12.1.5   Ising anyons

The most commonly discussed type of non-Abelian anyon is the so-called *Ising anyon*. The importance of this model stems from its relevance in the context of several promising routes for the physical realization of topological quantum states that support non-Abelian anyons. For many years, the main focus of research was on the fractional quantum Hall state with filling factor $\nu = 5/2$ [549]. Moore and Read suggested that the quasiparticle excitations of this state are non-Abelian anyons [369]. The statistical properties of these quasiparticles were shown [385] to be essentially the same as those of Ising anyons, up to an additional Abelian component to their statistics. Currently, intense research efforts are dedicated to the realization of zero-energy bound states in p-wave superconductors (see Chapter 8). The statistical properties of these quasiparticles are also described by the Ising anyon model, as we discuss below (Section 12.3).

The Ising anyon model contains three types of particles: the vacuum, 1, the non-Abelian anyon, σ, and the fermion, ψ. The vacuum fuses trivially with the other particles, $\sigma \times 1 = \sigma$ and $\psi \times 1 = \psi$; the nontrivial fusion rules are

$$\sigma \times \sigma = 1 + \psi, \qquad \sigma \times \psi = \sigma, \qquad \psi \times \psi = 1. \tag{12.6}$$

Note the existence of two fusion channels for a pair of σ anyons, the vacuum and the fermion ψ. Also note that two fermions condense to the vacuum, hence the total fermion number is not well defined, but its parity is and, consequently, represents a measurable quantity. For example, in the case of Majorana zero modes (MZMs) realized in a p-wave topological superconductor the fermions ψ are unpaired electrons; when paired (i.e., fused), they are "absorbed" into the vacuum (the superconducting condensate). The fusion rule for the Majorana zero modes (i.e., the σ particles) is often expressed in terms of a pair of MZMs being "empty" (fusion channel 1) or "occupied" by an electron (fusion channel ψ).

Consider the fusion of three anyons with fixed total fusion channel. The only case involving more than one intermediate quasiparticle corresponds to the fusion of three σ anyons, which can only have σ as their total fusion channel. Since $\sigma \times \sigma = 1 + \psi$, the intermediate fusion space is two-dimensional and we can choose the fusion states $|\sigma, \sigma \to 1\rangle$ and $|\sigma, \sigma \to \psi\rangle$ as a basis. Note that a generic fusion state $|\phi_a, \phi_b \to \phi_c\rangle$ represents a state containing two anyons, ϕ_a and ϕ_b that, if fused, generate the anyon ϕ_c. Using the basis specified above, the F matrix is given by

$$F^{\sigma}_{\sigma\sigma\sigma} = \frac{1}{\sqrt{2}} \begin{pmatrix} 1 & 1 \\ 1 & -1 \end{pmatrix}. \tag{12.7}$$

The meaning of this matrix can be easily understood from the diagram in Figure 12.2, with $a, b, c, d = \sigma$ and the intermediate quasiparticles (i and j) being either the vacuum 1 or the fermion ψ. For a derivation of this equation using the so-called *pentagon identity*, a simple consistency equation satisfied by the F matrix [512], see, for example, the book by Jiannis Pachos [399]. Equation (12.7) implies that the intermediate fusion outcomes can be nontrivially transformed by changing the fusion ordering.

The R matrix that gives the phase corresponding to taking one anyon around another (when their fusion channel is fixed) can be obtained [399] using the so-called *hexagon identity* [512]. Explicitly, we have

$$R^{1}_{\sigma\sigma} = e^{-\pi i/8}, \qquad R^{\psi}_{\sigma\sigma} = e^{3\pi i/8}, \qquad R^{1}_{\psi\psi} = -1, \qquad R^{\sigma}_{\sigma\psi} = i. \tag{12.8}$$

Note that exchanging two ψ fermions changes the sign of the state, as expected. On the other hand, exchanging two σ anyons generates a phase that depends on the fusion channel of the pair.

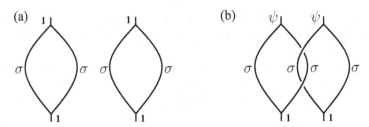

FIGURE 12.4 Worldlines representing two pairs of Ising anyons, σ, generated from the vacuum. (a) The anyons are fused back to the vacuum. (b) Nontrivial braiding-induced evolution of the fusion states. After one anyon from the first pair is circulated around an anyon from the second pair, the fusion of σ anyons generates two ψ fermions.

How can one exploit Ising anyons for topological quantum computation? In essence, since a pair of σ quasiparticles has two fusion channels, we can view it as a qubit with states $|0\rangle$ and $|1\rangle$ corresponding to the fusion outcome being 1 and ψ, respectively. However, having a pair prepared in a given channel (say 1), one can only change the fusion channel by braiding it with other anyons. Therefore, for a single qubit we need, in fact, two pairs of σ anyons. Indeed, assume that the pairs are generated from vacuum (see Figure 12.4), so that initially the fusion state of each pair is 1. Next, we braid two anyons that belong to different pairs, as shown in Figure 12.4. Note that the two anyons (say, anyons 2 and 3) do not have a direct fusion channel (i.e., if fused, they will generate a superposition of 1 and ψ). To determine the effect of the braiding operation, we first change the fusion order using the F matrix so that anyons 2 and 3 fuse to particle α, where α is either 1 or ψ; then we account for the 2π rotation by applying $R^\alpha_{\sigma\sigma}$ twice, and, finally, we restore the original fusion order using F^{-1}. Explicitly, the fusion state of the first pair changes after braiding with the third particle according to

$$|\sigma, \sigma \to 1\rangle \Longrightarrow \sum_\beta |\sigma, \sigma \to \beta\rangle B_{\beta 1} = e^{-i\pi/4}|\sigma, \sigma \to \psi\rangle, \qquad (12.9)$$

with a B matrix (or *braid matrix*) given by

$$B_{\beta\gamma} = \sum_\alpha [(F^{-1})^\sigma_{\sigma\sigma\sigma}]_{\beta\alpha} (R^\alpha_{\sigma\sigma})^2 [F^\sigma_{\sigma\sigma\sigma}]_{\alpha\gamma} = e^{-i\pi/4} \begin{pmatrix} 0 & 1 \\ 1 & 0 \end{pmatrix}_{\beta\gamma}, \qquad (12.10)$$

where we have used the explicit form of the F and R matrices given by equations (12.7) and (12.8). We conclude that braiding with the third particle changes the fusion channel of the first pair from 1 to ψ (up to an overall phase factor). A similar conclusion holds for the second pair. Hence, the four σ anyons with the total fusion channel 1, have a two-dimensional intermediate fusion space $\mathcal{H}$ spanned by the states $||11\rangle\rangle$ and $||\psi\psi\rangle\rangle$, where

$||ab\rangle\rangle = |\sigma, \sigma \to a\rangle \otimes |\sigma, \sigma \to b\rangle$ is a shorthand notation for the fusion states of the two pairs. The two-dimensional space $\mathcal{H}$, which corresponds to one qubit, can be used to store quantum information nonlocally.

We note that similar considerations apply if one pair of anyons is created from the vacuum and the other one from a ψ fermion, i.e., when the total fusion channel is ψ. The corresponding basis states of the intermediate fusion space are $||1\psi\rangle\rangle$ and $||\psi 1\rangle\rangle$ and, as before, braiding two anyons from different pairs will transform one state into the other. For three pairs of σ particles with total fusion channel $\mathbf{1}$, the intermediate fusion states $||111\rangle\rangle$, $||1\psi\psi\rangle\rangle$, $||\psi 1\psi\rangle\rangle$, and $||\psi\psi 1\rangle\rangle$ span a four-dimensional space $\mathcal{H}$ that corresponds to two qubits. Braiding the anyons induces rotations within this fusion space. A similar space can be obtained using three pairs of anyons with total fusion channel ψ. In general, n pairs of σ anyons with fixed total fusion channel (i.e., either $\mathbf{1}$ or ψ) will generate a 2^{n-1}-dimensional Hilbert space, which can be viewed as the Hilbert space of $n-1$ qubits. Logical operations can be performed by braiding the quasiparticles. However, for Ising anyons the operations generated by braiding are not sufficient to implement all possible unitary transformations [178, 179]. In other words, Ising anyons are not capable of *universal* topological quantum computation. Nonetheless, in order to construct arbitrary unitary operations, it is only necessary to supplement braiding with a single-qubit $\pi/8$ phase gate and a two-qubit measurement, which can be done using non-topological operations [79]. Of course, these operations are not topologically protected. However, it was shown [79] that error correction protocols can be implemented, provided the $\pi/8$ phase gate and the two-qubit measurement are accurate to within 14% and 38%, respectively, which is not a serious constraint.

12.1.6 Fibonacci anyons

The *Fibonacci anyon* model or the *golden theory*, combines simplicity and richness to provide one of the simplest models of non-Abelian statistics that supports universal quantum computation. The model contains two types of particles: the vacuum, $\mathbf{1}$, and the non-Abelian quasiparticle, τ. The only non-trivial fusion rule is

$$\tau \times \tau = \mathbf{1} + \tau. \tag{12.11}$$

When fusing three anyons that have a fixed total fusion channel, the only process involving more than one intermediate particle consists of fusing three τ anyons into a τ particle. The correspondence between the intermediate fusion states associated with different orderings is given by the following F matrix

$$F^{\tau}_{\tau\tau\tau} = \begin{pmatrix} \phi^{-1} & \phi^{-\frac{1}{2}} \\ \phi^{-\frac{1}{2}} & -\phi^{-1} \end{pmatrix}, \tag{12.12}$$

where $\phi = (1+\sqrt{5})/2$ is the *golden ratio*. Finally, the phase acquired during the (counterclockwise) exchange of two Fibonacci anyons is described by the R matrix

$$R^{1}_{\tau\tau} = e^{4\pi i/5}, \qquad R^{\tau}_{\tau\tau} = -e^{2\pi i/5} \tag{12.13}$$

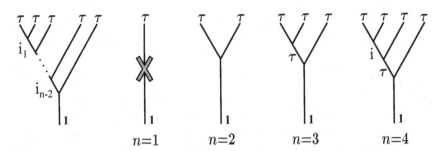

FIGURE 12.5 Fusion processes for n Fibonacci anyons with total fusion channel **1**. The symbols i and i_k designates intermediate fusion outcomes (i.e., particles **1** or τ). For $n = 1$, the "disappearance" of a τ anyon is an impossible process. Note that $i_{n-2} = \tau$ (for $n \geq 3$), which implies that the dimension of the Hilbert space for n particles fusing to **1** is the same as that of $n - 1$ anyons fusing to τ.

Using the F and R matrices, one can determine the unitary operation resulting from any braid performed on any number of particles $n \geq 2$.

To understand the structure of the fusion space of a system of Fibonacci anyons, let us consider the fusion of n non-Abelian particles with total fusion channel **1**, as shown in Figure 12.5. The particle generated by the fusion of the first $k + 1$ anyons is i_k, which can be either **1** or τ. Note that fusing a τ anyon with the vacuum generates a τ particle, $\tau \times \mathbf{1} = \tau$. Hence, if $i_k = 1$ we necessarily have $i_{k+1} = \tau$. Also, to obtain the vacuum one has to fuse two τ particles, which implies $i_{n-2} = \tau$. Consequently, the dimension of the fusion space $\mathcal{H}$ for n anyons, which represents the number of independent ways in which the particles could fuse to give total charge **1**, will be given by the number N_n of independent sequences $(i_1, i_2, \ldots, i_{n-2})$ that are consistent with the constraints. As shown in Figure 12.5, for $n = 1$ we have the forbidden process $\tau \rightarrow \mathbf{1}$ (i.e., $N_1 = 0$), while for $n = 2, 3$ the fusion tree is unique (i.e., $N_2 = N_3 = 1$). On the other hand, for $n = 4$ we have a two-dimensional fusion space corresponding to two possible intermediate outcomes: $i = 1$ and $i = \tau$. In general, the dimension of the fusion space can be obtained using a simple recursion relation. Let i_1 be the outcome of fusing the first two anyons. If $i_1 = 1$, the remaining $n - 2$ anyons can fuse in N_{n-2} distinct ways, while if $i_1 = \tau$, there are N_{n-1} ways to fuse this anyon with the other $n - 2$ τ particles. Consequently,

$$N_n = N_{n-2} + N_{n-1}, \qquad \text{with } N_1 = 0 \text{ and } N_2 = 1. \qquad (12.14)$$

The solution to this recursion relation is the famous *Fibonacci sequence*: $0, 1, 1, 2, 3, 5, 8, 13, \ldots$. Note that the ratio of two successive numbers in this series is a (rapidly converging) approximation of the golden ratio, $N_n/N_{n-1} \approx \phi$. Hence, the dimension of the fusion space $\mathcal{H}$ grows exponentially with the number of anyons at a rate $N_n \propto \phi^n$.

To construct a qubit, we can start with two pairs of non-Abelian anyons (τ_1, τ_2) and (τ_3, τ_4) created from the vacuum. The dimension of the corresponding Hilbert space $\mathcal{H}$ is $N_4 = 2$. We note that a two-dimensional space can also be obtained using three anyons with total fusion channel τ, as evident from the rightmost diagram in Figure 12.5. In fact, alternative schemes for encoding qubits based on three anyons have been proposed [178]. Returning to the four-anyon scheme, we note that a convenient basis, given, for example, by the intermediate fusion states $\|ab\rangle\rangle = |\tau_1, \tau_2 \to a\rangle \otimes |\tau_3, \tau_4 \to b\rangle$, provides the qubit states for encoding quantum information, $|0\rangle = \|11\rangle\rangle$ and $|1\rangle = \|\tau\tau\rangle\rangle$.

To perform unitary operations on the logical states of the qubit we braid the quasiparticles. Consider, for example, exchanging the first two particles. This elementary braid operation corresponds to the generator σ_1 of the braid group (see Figure 12.1) and is represented by the matrix

$$\rho(\sigma_1) \equiv R_{\tau\tau} = \begin{pmatrix} R_{\tau\tau}^1 & 0 \\ 0 & R_{\tau\tau}^\tau \end{pmatrix}, \tag{12.15}$$

where the R matrix is given by Eq. (12.13). A representation of the generator σ_2 (which corresponds to exchanging the second and third anyons) in terms of R and F matrices can be obtained following the main arguments leading to Eq. (12.10),

$$\rho(\sigma_2) = (F_{\tau\tau\tau}^\tau)^{-1} R_{\tau\tau} F_{\tau\tau\tau}^\tau. \tag{12.16}$$

Remarkably, one can show [417] that the unitary operations $\rho(\sigma_1)$ and $\rho(\sigma_2)$ are dense in SU(2), i.e., one can generate an arbitrary element of SU(2) with accuracy ϵ using a number of operations that scales polynomially with $\log(1/\epsilon)$. This means that any single gate operation can be realized by braiding alone. Furthermore, one can show that two-qubit controlled gates can also be implemented with arbitrary accuracy using braiding alone [67, 178, 250]. We conclude that the Fibonacci anyon model allows for universal topological quantum computation solely through braiding.

12.2 ANYONS AND TOPOLOGICAL QUANTUM PHASES

The realization of a topological quantum computer rests, ultimately, on the existence of topological phases of matter that support non-Abelian anyons and on our ability to identify and manipulate such states. The task at hand is twofold and involves both the theoretical understanding of the physics governing the topological phases of matter, as well as the actual experimental realization and study of these phases. At the time of this writing, the most promising real systems that could support non-Abelian anyons are quantum Hall systems in two-dimensional electron gases and p-wave superconductors in solid state hybrid structures. Their perceived potential is based on theoretical predictions and partial indirect experimental evidence.

The quantum Hall fluids[4] are phases of a two-dimensional electron gas in the presence of a strong magnetic field. The filling factor $\nu = 5/2$ [549], which is believed to realize the so-called *Pfaffian state* proposed by Moore and Green [369], has been the focus of research for many years. The excitations of the Pfaffian state are non-Abelian anyons that are similar to the Ising anyons (but have an additional Abelian component to their statistics [384]). Several schemes to realize universal quantum computation in the $\nu = 5/2$ fractional quantum Hall state have been proposed [79, 81, 130, 180]. If the filling fraction $\nu = 12/5$ [556] is in the universality class of the so-called $\mathbb{Z}_3$ Read–Rezayi state [427], it is expected to support non-Abelian quasiparticles that are similar to the Fibonacci anyons, i.e., are capable of universal topological quantum computation through braiding. Presently, however, it is not entirely clear how to actually transport the quasiparticles in a quantum Hall fluid, i.e., how to perform braiding operation. A comprehensive discussion of various aspects concerning the realization of topological quantum computation using quantum Hall states can be found in the review article by Nayak et al. [384].

The most transparent route toward the realization of non-Abelian anyons is based on p-wave superconductors, which were shown to support vortices with anyonic statistics [261, 426]. The chiral p-wave superconductor (see Section 6.2.3) is equivalent to Kitaev's honeycomb model [291], a lattice model of anisotropically interacting spins that can be mapped onto a system of Majorana fermions, and, essentially, represents another embodiment of the Pfaffian state. In Chapters 8 and 9, we have discussed various aspects regarding the realization of p-wave superconducting states in solid-state heterostructures and cold atom systems. Below, in Section 12.3, we will address a few specific points related to the use of Majorana bound states as a platform for topological quantum computation. But before we address these issues, it is instructive to briefly comment on some generic features of topological models that are directly relevant to quantum computation. We focus on Chern–Simons field theories and particularly on how the properties of Abelian and non-Abelian anyons can be captured within these theories.

It is worth noting, however, that lattice theories also play a key role in understanding the connection between topological quantum computation and the physics of many-body systems capable of supporting topological phases. The very birth of topological quantum computation was triggered by Alexei Kitaev's idea to convert a quantum error-correcting code – so-called *toric code* – into a Hamiltonian describing an interacting many-body system [296, 292]. Hence, the toric code, which is an example of a *stabilizer code*, can also be viewed as the Hamiltonian of a two-dimensional interacting spin system. The model is the simplest example of a so-called *quantum double model*, $D(G)$. Here G is a finite group that labels a spin-like Hilbert space with basis $\{|g\rangle : g \in G\}$ defined on each link of a two-dimensional lattice. The toric code corresponds to $G = \mathbb{Z}_2$. In these models, the ground state of the Hamiltonian represents the

---

[4]For a review of the quantum Hall effect, see, for example, [415] and [129].

stabilizer space of the code, while the quasiparticle excitations are anyons. The properties of the quasiparticles can be determined by studying the *irreducible representations* of $D(G)$. For an elementary introduction to quantum double models see, for example, Ref. [399].

12.2.1   Abelian Chern–Simons field theories

Quantum field theory is a basic theoretical tool for describing the low-energy physics of condensed matter systems. A topological quantum field theory (TQFT) is characterized by observables (e.g., correlation functions) that are invariant under smooth coordinate transformations (i.e., diffeomorphisms). In other words, the theory is insensitive to the metric of space-time (e.g., to changes in the shape of the space-time manifold on which the system is defined) and the correlation functions are topological invariants. Note that these are precisely the defining features of a topological phase of matter: a system is in a topological phase if all its low-energy, long-wavelength observable properties are invariant under smooth deformations of space-time. In other words, a system is in a topological phase if its low-energy effective theory is a TQFT. Some basic features of TQFTs and a few examples were discussed in Chapters 5 and 6. Here, we are interested in the relation between TQFTs and the properties of Abelian and non-Abelian anyons.

The simplest example of a TQFT is the Abelian Chern–Simons theory, which is defined by the action

$$S_{CS}[a] = \frac{k}{4\pi} \int d\boldsymbol{r}^2 dt\, \epsilon^{\mu\nu\rho} a_\mu \partial_\nu a_\rho, \tag{12.17}$$

where ϵ is the antisymmetric tensor, a is a $U(1)$ gauge field, and the indices μ, ν, and ρ take the values 0 (for the time direction), 1, and 2. Note that the action (12.17) is independent of any metric and is invariant under gauge transformations $a_\mu \to a_\mu + \partial_\mu \omega$. When k is an odd integer, Eq. (12.17) corresponds to the low-energy effective theory of the Laughlin quantum Hall state at filling fraction $\nu = 1/k$.

The topological phase is protected by a finite energy gap, which is not captured by the Chern–Simons theory. Hence, the natural question: does the low-energy effective theory have anything to say about the properties of the quasiparticles, which are excitations of the system above the energy gap? It turns out that the quasiparticle properties are in fact part of the universal low-energy physics of the system and emerge naturally in a Chern–Simons theory. Indeed, consider a conserved current $j^\mu \equiv (\rho, \boldsymbol{j})$ that couples to the gauge field. For now, the quasiparticles are not dynamical, but move along classical trajectories consistent with the current $\boldsymbol{j}$. The Lagrangian density becomes $\mathcal{L} = \frac{k}{4\pi} \epsilon^{\mu\nu\rho} a_\mu \partial_\nu a_\rho - a_\mu j^\mu$ and we have the following classical Euler–Lagrange equations

$$j^\mu = \frac{k}{2\pi} \epsilon^{\mu\nu\rho} \partial_\nu a_\rho. \tag{12.18}$$

Note that Eq. (12.18) makes current conservation explicit, $\partial_\mu j^\mu = 0$. Consider now the equation corresponding to $\mu = 0$, i.e., $j_0 \equiv \rho = \frac{k}{2\pi} \nabla \times a$, which tells us that the quasiparticle density is locally proportional to the "magnetic field" associated with the gauge field a_μ. Hence, we can view the Chern–Simons coupling as a pure constraint that rigidly attaches a flux $q\, 2\pi/k$ to each particle that carries a_μ-charge $q = \int dr^2 j_0$. Upon moving one particle with $q = 1$ around another (i.e., performing two quasiparticle exchanges), the Chern–Simons flux gives rise to an Aharonov–Bohm phase [537] $2\theta = 2\pi/k$. Consequently, the quasiparticles of the Chern–Simons theory (12.17) have Abelian braiding statistics with $\theta = \pi/k$, i.e., they are Abelian anyons.

Further progress can be made by addressing the following question: what are the fundamental observables of a topological field theory? Clearly, these observables cannot be local, since the topological phase is insensitive to any local perturbation. On the other hand, we know that finding the topological charge of a quasiparticle requires taking another (test) particle around it. The test charge could be obtained by creating a pair of anyons from the vacuum, evolving them along some worldlines, then fusing them back to the vacuum. The corresponding operator is the so-called *Wilson loop*

$$W_\gamma = \exp\left(iq \oint_\gamma dx^\mu a_\mu(x)\right),\qquad (12.19)$$

where q is the charge of the particle and γ a given looping trajectory. Note that the operator W_γ is gauge-invariant and independent of any metric. Consider now the ground state expectation value of a product of r Wilson loops, $\langle W(\Gamma)\rangle = \langle\psi_0|\prod_i W_{\gamma_i}|\psi_0\rangle$, where $1 \leq i \leq r$ indexes the (closed) trajectory γ_i of a particle with charge q_i, Γ denotes the entire set of loops, and $|\psi_0\rangle$ is the vacuum state. The expectation value can be evaluated in the path integral formalism, $\langle W(\Gamma)\rangle = \int \mathcal{D}a\, W(\Gamma)e^{iS_{CS}[a]}$, and we obtain [399, 413]

$$\langle W(\Gamma)\rangle = \exp\left(\frac{i\pi}{k}\sum_{i,j} q_i q_j \Phi(\gamma_i, \gamma_j)\right),\qquad (12.20)$$

where i and j take all possible values and $\Phi(\gamma_i, \gamma_j)$ is the Gauss integral

$$\Phi(\gamma_i, \gamma_j) = \frac{1}{4\pi}\oint_{\gamma_i} dx^\mu \oint_{\gamma_j} dy^\nu \epsilon_{\mu\nu\rho} \frac{(x-y)^\rho}{|x-y|^3}.\qquad (12.21)$$

When the loops γ_i and γ_j do not intersect[5], $\Phi(\gamma_i, \gamma_j)$ is an integer [413] representing the linking between the two loops, i.e., the number of times γ_i winds around γ_j. Since the Chern–Simons action and the Wilson loops are independent of the metric, $\Phi(\gamma_i, \gamma_j)$ and $\langle W(\Gamma)\rangle$ depend only on the topology of the

---

[5] To eliminate ambiguities in the case $i = j$, we follow the so-called *framing* prescription [553]: the loop γ_i is slightly displaced in a given direction so that it generates a distinguishable loop γ_i'. The two loops define the edges of a ribbon. The integral $\Phi(\gamma_i, \gamma_i)$ is then substituted by the well-defined quantity $\Phi(\gamma_i, \gamma_i')$.

loops, i.e., on how the particles that evolve along the corresponding worldlines are braided. For example, consider four quasiparticles created pairwise from the vacuum which are fused back to the vacuum after braiding one particle from the first pair with a particle from the second pair. Assuming, for simplicity, $q_i = 1$, the expectation value of the Wilson loops gives the phase factor $\exp(2\pi i/k)$, which corresponds to the double exchange of two Abelian anyons with $\theta = \pi/k$. We conclude that the phase change due to braiding of Abelian quasiparticles is efficiently captured by the expectation value of the Wilson loop operators.

12.2.2 Non-Abelian Chern–Simons field theories

The above considerations can be generalized to non-Abelian Chern–Simons theories. In essence, similar to the Abelian case, the anyonic properties of the quasiparticles, including their fusion rules and braiding statistics [227, 553], are part of the universal low-energy physics of the system and can be determined within the low-energy effective field theory that describes the topological phase. Below, without delving too much into technical aspects, we briefly comment on how the existence of different types of quasiparticles and their fusion properties emerge naturally in Chern–Simons theories. The braiding properties of these quasiparticles can be obtained by calculating expectation values of Wilson loops, following a procedure that generalizes the Abelian scheme sketched above.

Consider the non-Abelian Chern–Simons field theory defined by the action

$$S_{CS}[a] = \frac{k}{4\pi} \int d^3x\, \epsilon^{\mu\nu\rho} \mathrm{Tr}\left(a_\mu \partial_\nu a_\rho + i\frac{2}{3} a_\mu a_\nu a_\rho \right). \tag{12.22}$$

Under a generic gauge transformation $U(x) \in G$, the action (12.22) transforms as $S_{CS}[a] \to S_{CS}[a] + 2\pi k m$, where the integer m is the *winding number* of the gauge transformation, i.e., the number of times U spans the whole group (also see Chapter 6). Since we require the expectation values of all observables to be gauge-invariant, the quantity $\exp(iS_{CS}[a])$ has to be gauge-invariant, which means that the coupling k (called the *level* of the theory) is necessarily an integer. For example, when $G = SU(2)$ and $k = 2$, we talk about an $SU(2)$ level 2 Chern–Simons theory and we use the notation $SU(2)_2$ to designate it.

Focusing on the group $SU(2)$, we define the non-Abelian Wilson loop associated with a curve γ as

$$W_\gamma = \mathrm{Tr}\left[\boldsymbol{P} \exp\left(iq_j \oint_\gamma T^a a_\mu^a\, dx^\mu \right) \right], \tag{12.23}$$

where $\boldsymbol{P}$ is the path-ordering operator, $q_j = 2j$ is an integer, $T^a = \rho_j(J^a)$ is a $(2j+1)$-dimensional representation[6] of the $SU(2)$ algebra, and $a_\mu = T^a a_\mu^a$.

---

[6]See footnote on page 205.

One can show that considering $j > k/2$, where k is the level of the Chern–Simons theory, will generate the same expectation value of the Wilson loop as a certain representation with $0 \leq j \leq k/2$ [155]. In other words, there are only $k+1$ distinct representations: $0, 1/2, \ldots, k/2$. With this observation, we are ready for the key point: each of the nonequivalent representations of the $SU(2)$ algebra corresponds to a distinct quasiparticle type. We conclude that the $SU(2)_k$ theory has $k+1$ types of anyons, each corresponding to a different representation ρ_j of the $SU(2)$ group.

What about the fusion properties of these quasiparticles? Consider two particles corresponding to the j_1 and j_2 representations of $SU(2)$. Each of the particles/representations has an associated Hilbert space spanned by the states $|j_1, m_{j_1}\rangle$ and $|j_2, m_{j_2}\rangle$, respectively. The tensor product $|j_1, m_{j_1}\rangle \otimes |j_2, m_{j_2}\rangle$ provides a basis for the composite Hilbert space associated with the pair of particles. Another possible basis is given by the states $|j, m_j\rangle$, where $|j_1 - j_2| \leq j \leq j_1 + j_2$, as dictated by the composition rules for angular momentum. The corresponding representations ρ_{j_1} and ρ_{j_2} can be combined in a similar manner into a $(2j_1 + 1) \times (2j_2 + 1)$-dimensional representation, which, in turn, can be written as a direct sum of ρ_j representations. Note, however, that not all representations are distinct. As a result, one can show [399] that the upper value for j is $j_{\max}(j_1, j_2) = \min[j_1 + j_2, k - (j_1 + j_2)]$. The composition of (nonequivalent) representation can be expressed symbolically as

$$\rho_{j_1} \times \rho_{j_2} = \sum_{j \leq |j_1 - j_2|}^{j_{\max}(j_1, j_2)} \rho_j, \tag{12.24}$$

where the sum should be understood as a direct sum. Equation (12.24) gives the fusion rules for the $SU(2)$ level-k theory. Considering, for example, the $SU(2)_2$ theory, we have three types of particles corresponding to the representations ρ_0, $\rho_{\frac{1}{2}}$, and ρ_1 and satisfying the nontrivial fusion rules $\rho_{\frac{1}{2}} \times \rho_{\frac{1}{2}} = \rho_0 + \rho_1$, $\rho_{\frac{1}{2}} \times \rho_1 = \rho_{\frac{1}{2}}$, and $\rho_1 \times \rho_1 = \rho_0$. Note that these are exactly the fusion rules satisfied by Ising anyons, with ρ_0, $\rho_{\frac{1}{2}}$, and ρ_1 corresponding to 1, σ, and ψ, respectively.

We conclude that the $SU(2)$ level-k Chern–Simons theory has $k+1$ distinct particles that correspond to the nonequivalent representations ρ_j of the $SU(2)$ group, where $j = 0, 1/2, 1, \ldots, k/2$. The fusion properties of these quasiparticles correspond to the decomposition rules (12.24) for tensor products of irreducible $SU(2)$ representations. The braiding properties of the anyonic system can be determined by calculating expectation values of Wilson loops. In general, the Wilson loop operators can be viewed as creation or annihilation operators for the degrees of freedom in a topological phase, which are gauge- and metric-independent. A many-body interacting system will support a topological phase if its effective low-energy theory is a topological quantum field theory, i.e., if its low-energy degrees of freedom can be mapped into loops or, alternatively, into string nets [329].

12.3 TOPOLOGICAL QUANTUM COMPUTATION WITH MAJO-RANA ZERO MODES

The possibility of topologically protected storage and processing of quantum information using Majorana zero modes (MZMs) localized at vortices or domain walls is based on the non-Abelian nature of these objects. We will first illustrate the non-Abelian properties of the MZMs, then we will show how these properties can be exploited to implement quantum computation.

12.3.1 Non-Abelian statistics

In a two-dimensional (2D) spinless chiral superconductor MZMs are carried by vortices (see Figure 6.7 on page 202). The non-Abelian statistics is associated with the exchange of these vortices. On the other hand, in 1D systems (e.g., Majorana wires) exchange statistics is not well defined. However, one can meaningfully discuss exchange properties in a network of 1D wires [16], where Majorana-carrying domain walls can be manipulated without overlapping the MZMs (hence, without splitting the ground state degeneracy). Here, we focus on the 2D case [261]. We emphasize that the objects whose exchange statistics we are studying are *Majorana-carrying vortices*, although we may simply call them vortices or MZMs.

Consider a 2D spinless chiral superconductor (SC) containing a collection of $2n$ MZMs localized at vortices that are well separated from each other. The positions of the vortices are given by the position vectors r_i ($1 \leq i \leq 2n$) and magnetic flux through each vortex is $h/2e$. The presence of a vortex changes the phase of the superconducting order parameter, so that cycling the vortex once corresponds to a 2π phase change [261]. In general, a phase shift by ϕ of the SC order parameter is equivalent to rotating the phase of the electronic annihilation operator by $\phi/2$, $\hat{\psi}(r) \to e^{i\phi/2}\hat{\psi}(r)$. Consequently, the Majorana operator corresponding to the ith MZM has the form

$$\gamma_i = \int d^2r \left[u_i(r)e^{i\frac{\phi_i}{2}}\hat{\psi}(r) + u_i^*(r)e^{-i\frac{\phi_i}{2}}\hat{\psi}^\dagger(r) \right], \qquad (12.25)$$

where $u_i(r)$ is the MZM wave function and the phase angle ϕ_i depends on the relative angles between r_i and the position vectors of the other vortices. Taking (adiabatically) the vortex i (and the corresponding MZM) around vortex j results in a 2π phase shift, $\phi_i \to \phi_i + 2\pi$, which corresponds to a sign change of the Majorana operator, $\gamma_i \to -\gamma_i$. To account for the 2π phase shifts associated with going around a vortex we introduce branch cuts, as shown in Figure 12.6. Upon exchange processes, whenever the MZM bounded by vortex i passes through the branch cut, it picks up an additional minus sign.

Consider now the counterclockwise exchange of vortices i and j. As a result of this process, γ_i and γ_j swap positions and γ_i acquires an additional minus sign, as it crosses a branch cut (see Figure 12.6). Specifically, we have $\gamma_i \to -\gamma_j$ and $\gamma_j \to \gamma_i$. This transformation is implemented by the unitary

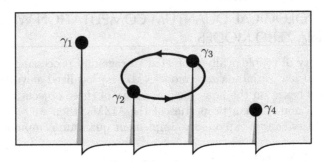

FIGURE 12.6 Four Majorana-carrying vortices in a 2D topological superconductor. The branch cuts correspond to a 2π phase shift of the superconducting
order parameter. Upon exchanging the central vortices, γ_2 crosses a branch
cut and acquires an extra minus sign. As a result, $\gamma_2 \to -\gamma_3$ and $\gamma_3 \to \gamma_2$.

operator

$$U_{i,j} = \frac{1}{\sqrt{2}}(1 + \gamma_i\gamma_j) = \exp\left(\frac{\pi}{4}\gamma_i\gamma_j\right). \tag{12.26}$$

Indeed, one can easily check that the transformed γ_i operator is
$U_{i,j}\gamma_i U_{i,j}^\dagger = -\gamma_j$, while $U_{i,j}\gamma_j U_{i,j}^\dagger = \gamma_i$. We note that $U_{i,i+1}$ is a representation of the braid group generators σ_i introduced in Section 12.1.2.

To show that this is a non-Abelian representation, consider the particular
case involving four vortices (i.e., $n = 2$). The MZMs can be paired into two
complex fermions,

$$f_1 = \frac{1}{2}(\gamma_1 + i\gamma_2), \qquad\qquad f_2 = \frac{1}{2}(\gamma_3 + i\gamma_4). \tag{12.27}$$

The system has four degenerate ground states and a natural basis for the
ground state subspace is given by the eigenvectors $|n_1 n_2\rangle = |n_1\rangle \otimes |n_2\rangle$, where
$|n_i\rangle$ is an eigenstate of the fermion occupation operator $\hat{n}_i = f_i^\dagger f_i$. Assume
that we start in a particular basis state $|\Psi_i\rangle = |n_1 n_2\rangle$ and perform a counter-
clockwise exchange of γ_2 and γ_3. The final state can be found by expressing
$U_{2,3}$ in terms of fermion operators. Explicitly, we have

$$|\Psi_f\rangle = U_{2,3}|n_1 n_2\rangle = \frac{1}{\sqrt{2}}\left[|n_1 n_2\rangle + i(-1)^{n_1}|1-n_1, 1-n_2\rangle\right]. \tag{12.28}$$

Note that $|\Psi_i\rangle$ and $|\Psi_f\rangle$ have the same total fermion parity and that the
exchange of γ_2 and γ_3 corresponds to a rotation of the state vector within
the ground state manifold. This nontrivial transformation reveals the non-
Abelian statistics of the Majorana-carrying vortices (since exchanging Abelian
anyons can only generate a phase factor, i.e., $|\Psi_f\rangle = e^{i\theta}|\Psi_i\rangle$). Furthermore,
performing m sequential exchanges, $U_{i_m,j_m}\ldots U_{i_1,j_1}$, results in a final state
that depends not only on what exchanges are performed, but also on the
order in which they are carried out (i.e., on the element of the braid group

represented by $U_{i_m,j_m} \ldots U_{i_1,j_1}$). The nontrivial quantum evolutions generated by braiding operations are key elements of topological quantum computation with MZMs.

12.3.2 Fusion of Majorana zero modes

Fusion rules provide another unique signature of non-Abelian anyonic behavior. As discussed in Section 12.1.3, non-Abelian anyons (unlike Abelian particles) have multiple fusion channels. For the purpose of demonstrating experimentally the non-Abelian nature of MZMs, determining their fusion rules may be technically less challenging than determining their exchange statistics. A charge-sensing fusion-rule experiment for Majorana wires has been proposed in Ref. [3]. The basic idea is illustrated in Figure 12.7(a). A semiconductor (SM) wire partially coated with a mesoscopic superconducting island with charging energy E_c is coupled to a bulk superconductor (SC). The island charge and the Josephson energy that characterizes the coupling to the bulk SC are controlled by gate potentials. When $E_J \gg E_c$, the wire hosts two Majorana modes at its ends and is characterized by degenerate ground states $|\Psi_e\rangle$ and $|\Psi_o\rangle$ with even and odd fermion parity, respectively. Increasing the barrier potential ($E_J \ll E_c$) restores charging energy and converts the degenerate parity states into nondegenerate eigenstates with charges Q_o and Q_e.

The protocol for detecting the fusion rules for MZMs [3] involves two SC islands and three basic steps, as illustrated in Figure 12.7(b). (I) Starting from a unique ground state with total charge Q_{tot} create two MZMs (out of the vacuum) by lowering the outer barriers. We assume that lowering the barriers is an adiabatic process. Let $f_{14} = (\gamma_1 + i\gamma_4)/2$ be the fermion operator associated with the two Majoranas. As a result of lowering the gates, the system evolves into the state $|0_{14}\rangle$, which is the eigenstate with eigenvalue 0 of the fermion number operator $\hat{n}_{14} = f_{14}^\dagger f_{14}$. (II) Create two additional MZMs, γ_2 and γ_3, out of the vacuum by raising the central barrier. The system evolves into the quantum state $|0_{14}0_{23}\rangle$, which is an eigenstate of $\hat{n}_{14}$ and $\hat{n}_{23}$. One can redefine the fermion basis in terms of occupation numbers for the left and right Majorana pairs $\hat{n}_{12}$ and $\hat{n}_{34}$. In the new basis we have

$$|0_{14}0_{23}\rangle = \frac{1}{\sqrt{2}} \left(|0_{12}0_{34}\rangle + |1_{12}1_{34}\rangle \right). \tag{12.29}$$

(III) Restore charging energy by raising the outer barriers. This fuses the anyons corresponding to γ_1 and γ_2, as well as the pair γ_3, γ_4. Let $|Q_L, Q_R\rangle$ be the unique ground state of the double-island system, where Q_L and Q_R (with $Q_L + Q_R = Q_{\text{tot}}$) designate the charges on the left and right islands, respectively. Also, assume that the lowest excited state with opposite (relative to the ground state) fermion parities on each island is $|Q_L - 1, Q_R + 1\rangle$. Upon restoring charging energy, the state $|0_{12}0_{34}\rangle$ evolves into the ground state $|Q_L, Q_R\rangle$, while $|1_{12}1_{34}\rangle$ evolves into the excited state $|Q_L - 1, Q_R + 1\rangle$.

The (idealized) protocol described above corresponds to the evolution

$$|Q_{\text{tot}}\rangle \to \frac{1}{\sqrt{2}} \left(|Q_L, Q_R\rangle + e^{i\alpha}|Q_L - 1, Q_R + 1\rangle \right) \tag{12.30}$$

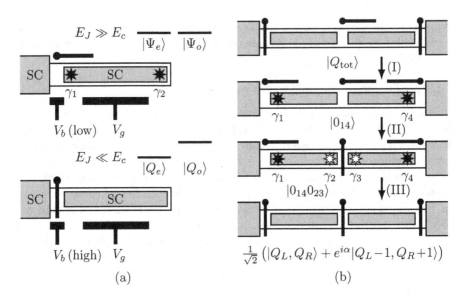

FIGURE 12.7 (a) Gate-controlled "parity-to-charge" converter. A SM wire is covered by a SC island (with charging energy E_c) and coupled to a bulk SC. The back gate voltage V_g tunes the charge on the island. The Josephson energy E_J is controlled via the barrier potential V_b. Low (high) barrier potential is symbolized by an open (closed) "valve." When $E_J \gg E_c$, the wire supports MZMs (symbolized by stars) localized at its ends and has degenerate ground states with even and odd fermion parity. (b) Protocol for detecting the fusion rules for MZMs using charge sensing (see main text).

generated by creating two MZM pairs, γ_1, γ_4 and γ_2, γ_3, out of the vacuum and fusing the anyons in a different order. The superposition in Eq. (12.30) is a consequence of the anyons having two fusion channels, the vacuum and the fermion, i.e., being non-Abelian. The experimental signature is a probabilistic charge measurement of a given island characterized by even and odd electron numbers occurring with equal probability.

12.3.3 Quantum information processing

Storing quantum information in a system of MZMs can be done following the basic ideas discussed earlier in the context of Ising anyons. Using the so-called "dense" encoding approach, one can store $n-1$ qubits in $2n$ MZMs with fixed total fermion parity. The basis states of the encoding space are labeled by the eigenvalues ± 1 of the parity operators $i\gamma_{2j-1}\gamma_{2j}$, where $1 \leq j \leq n-1$. Note that, given the eigenvalues of the first $n-1$ operators, the eigenvalue of $i\gamma_{2n-1}\gamma_{2n}$ is fixed by the total fermion parity. Hence, the dimension of the encoding space is 2^{n-1}, which allows the storage of $n-1$ qubits. In this scheme, one can easily entangle qubits by braiding the MZMs.

However, in a highly entangled state an error that occurs in one qubit can propagate to many other qubits corrupting the quantum information stored in the system. An alternative approach – the so-called "sparse" encoding – uses $4n$ MZMs to store n qubits by imposing the condition $\gamma_{4j-3}\gamma_{4j-2}\gamma_{4j-1}\gamma_{4j} = -1$. This means that the fermion parity for each set $(4j - 3, 4j - 3, 4j - 1, 4j)$ of four MZMs is even. Since there are $2n$ parity operators, $i\gamma_{2j-1}\gamma_{2j}$ with $1 \leq j \leq 2n$, and n parity conditions, the dimension of the encoding space is 2^n. The basis states for qubit j correspond to the eigenvalues of the parity operator $i\gamma_{4j-3}\gamma_{4j-2} = i\gamma_{4j-1}\gamma_{4j} = \pm 1$. In this scheme it is easier to limit the propagation of errors, but entangling qubits requires performing measurements.

Quantum computation requires the implementation of a universal set of quantum gates. One possible universal set consists of the Hadamard gate (H), the $\pi/8$ gate (T), and the controlled-Z gate. Unfortunately, MZMs are Ising-type anyons, hence, they are not capable of implementing universal quantum computation by braiding alone. More specifically, the T gate cannot be realized through braiding and has to be implemented in a "conventional," topologically unprotected way. For example, one can bring two MZMs close together for a short time τ, then separate them again. When the distance between the MZMs is short (so that they overlap), the corresponding states are split in energy by a quantity ΔE, which generates a dynamical phase factor. The evolution of the quantum state described by the unitary operator

$$U = \begin{pmatrix} 1 & 0 \\ 0 & e^{i\Delta E\tau} \end{pmatrix}. \qquad (12.31)$$

Of course, getting exactly $\Delta E\tau = \pi/4$ would require perfect control over this process, which is practically impossible. Error correction is, therefore, necessary. In this case, it is the error correction process (rather than the gate itself) that takes advantage of the topologically protected operations. For example, one can use so-called *magic state distillation* [79, 81] to enhance the fidelity of the T gate. Relying on topologically protected operations to achieve the required fidelity means fewer qubits and gate operations, as compared to conventional quantum computation.

The other gates from the universal set can be implemented via MZM braiding and (for sparse encoding) additional measurements. For example, in the sparse encoding scheme, applying a Hadamard gate on the jth qubit can be done by performing an exchange of the MZMs γ_{4j-2} and γ_{4j-1}. However, entangling two qubits cannot be done directly. To apply a controlled-Z gate on two qubits, which are encoded in 8 MZMs, one has to first perform a parity measurement (on two MZMs) in order to switch to dense encoding with 6 MZMs. Then, the controlled-Z operation is done by performing a braid and, finally, one returns to the sparse encoding by introducing an ancillary MZM pair and performing another measurement [131]. As a final remark, we mention that recent algebraic number theory studies [302, 464] have shown that the number of (topologically unprotected) T gates required to implement a given quantum algorithm can be significantly reduced. This number can be further reduced by taking advantage of certain techniques based on using ancilla qubits and measurements [63, 400]. These results demonstrate that quantum algorithms could be implemented quite efficiently using Majorana zero modes, should these anyons be physically available.

We should point out that, to realize braiding operations, the adiabatic transport of MZMs (and, more generally, of non-Abelian anyons) around each other is not a requirement. Instead, one can implement the desired gate operations through adaptive series of topological charge measurements [66]. This measurement-only approach does not involve the consumption of entanglement resources – in particular it does not require a highly entangled initial state – thus allowing for computations of indefinite length. In addition, the scheme can be optimized with respect to various constraints imposed by specific physical architectures, e.g., layout and the relative difficulty of implementing different types of measurements [507]. Examples of measurements that could be used in this type of approach include interferometric measurements [384] and electric charge sensing (using e.g., quantum dots [3]).

We conclude this section with a brief summary of topological quantum computation, linking together some of the main themes of this book. Topological quantum systems that host non-Abelian anyons are characterized by entangled ground states. A pair of anyons – e.g., a pair quasiparticle excitations emerging from the vacuum – should be viewed as the result of some sort of "twist" in the entanglement structure of the vacuum. Braiding the anyons has nontrivial effects because moving an anyon changes the "twist" pattern, which, in turn, affects the other anyons. In some sense, they "talk" to each other through the nontrivial vacuum, which essentially determines their properties. Therefore, it is not surprising that the topological quantum field theories capture the properties of the quasiparticles, as discussed above. In topological system that hosts anyons, quantum information is not encoded locally, in the anyons themselves, but nonlocally, in the "twists" of the vacuum associated with their presence. This information can be (partially) accessed by bringing the anyons together, i.e., fusing them. The computational Hilbert space is given by the fusion states, not by some "local states" of the anyons, which would be similar to the states of a "standard" qubit realized, for example, using a spin-1/2 system. Measurement-only braiding schemes, which can be implemented using ancilla anyons (in the spirit of the quantum teleportation protocol), do not directly access the quantum information encoded in the fusion states of the computational anyons (hence, they do not "destroy" it), but modify the associated "twist pattern" similar to braiding operations, thus implementing the desired gates. The key to the successful implementation of topological quantum computation is a robust topological system, a system in which the "twist pattern" that encodes the information is not (easily) affected by disorder, thermal fluctuations, and other perturbations generated by the coupling to the environment.

12.4 OUTLOOK: QUANTUM COMPUTATION AND TOPOLOGICAL QUANTUM MATTER

Are we on the verge of a technological revolution driven by new ways of processing information based on the ability to control and manipulate non-Abelian topological quantum systems? Possibly. However, foreseeing when this will happen and to what result is not for the scientists, but for the planners and the soothsayers. Science, on the other hand, is about understanding. For a physicist, this means to be able to a) make predictions about certain behaviors of physical systems and b) verify those predictions by performing experiments. From this perspective, the topological quantum world awaits us full of marvels and possible surprises.

Taking the search for Majorana bound states as a paradigmatic example, we have learned that when a system expected to host a certain quantum state cannot be obtained by just picking it out of the ground, the way to go is to built it "from scratch." Materials, heterostructure, and device engineering are major themes that are here to stay. Their relevance and implications go way beyond topological quantum systems, but the lesson emerging in this field is that raising this type of engineering to a level that provides access to new physics is within reach.

This "new physics" has intrinsic ties with quantum information and quantum computation. Essentially, being able to control and manipulate individual quantum states, particularly if they are many-body quantum states "living" in a large Hilbert space, could provide access to "nature's way" of doing computations. Is this fundamentally different from our well-known classical framework? Answering this question will ultimately involve finding solutions to a multitude of problems ranging from engineering robust qubits and identifying new methods to perform quantum measurements quickly and accurately, to designing and optimizing quantum algorithms and solving outstanding problems in computational complexity.

As always, the best solution (or even a solution) might not be in sight. Perhaps, zero-energy Majorana bound states are not the way to practically perform topological quantum computation. Nonetheless, what is done in the search for Majorana and its possible implementation as a platform for quantum computation (i.e., looking for new materials, optimizing heterostructure and device engineering, identifying techniques for efficiently modeling complex quantum devices, finding better quantum algorithms and error correction methods, etc.) is exactly what needs to be done. Maybe some Fibonacci-type anyon will be the answer. Still, to get there we need mastery over materials, nanostructures, and measurement techniques. Aiming for a specific goal (e.g., the Majorana) serves this need very well, even if the outcome will not be the expected one.

What about an "epic failure"? What if quantum computation is not possible *in principle*? This would perhaps be a prize even bigger than quantum computation itself, since we would have discovered a *principle* that makes quantum computation impossible. This brings us to the most fundamental level of this interrogation of the topological quantum world, an inquiry that touches upon the foundations of quantum mechanics and the principles of elementary particle physics. Is there a fundamental information-theoretic principle that constrains all possible physical theories? Can we learn something about the nature of elementary particles by studying emergent physics in our "universe-in-a-box" – the condensed matter systems in our labs? Can topological order survive nonzero temperature? Does this have anything to do with the classical and the quantum worlds being different? Is $P \neq NP$? And, after all, what is the relevance of all these problems in the Age of AI?

Bibliography

[1] S. Aaronson. The limits of quantum computers. *Scientific American*, (298):62, 2008.

[2] S. Aaronson. *Quantum Computing since Democritus*. Cambridge University Press, 2013.

[3] David Aasen, Michael Hell, Ryan V. Mishmash, Andrew Higginbotham, Jeroen Danon, Martin Leijnse, Thomas S. Jespersen, Joshua A. Folk, Charles M. Marcus, Karsten Flensberg, and Jason Alicea. Milestones toward majorana-based quantum computing. *Phys. Rev. X*, (6):031016, 2016.

[4] A. A. Abrikosov, L. P. Gor'kov, and I. E. Dzyaloshinski. *Methods of Quantum Field Theory in Statistical Physics*. Dover Publications, Inc., New York, 1963.

[5] A. Acke. *Gravitation Explained by Gravitoelectromagnetism*. LAP LAMBERT Academic Publishing, 2018.

[6] S. L. Adler. Axial-vector vertex in spinor electrodynamics. *Phys. Rev.*, (177):2426, 1969.

[7] I. Affleck. Quantum spin chains and the Haldane gap. *J. Phys.: Cond. Matt.*, (1):3047, 1989.

[8] I. Affleck, T. Kennedy, E. H. Lieb, and H. Tasaki. Rigorous results on valence-bond ground states in antiferromagnets. *Phys. Rev. Lett.*, (59):799, 1987.

[9] Y. Aharonov and J. Anandan. Phase change during a cyclic quantum evolution. *Phys. Rev. Lett.*, (58):1593, 1987.

[10] Y. Aharonov and D. Bohm. Significance of electromagnetic potentials in the quantum theory. *Phys. Rev.*, (115):485, 1959.

[11] A. R. Akhmerov, J. Nilsson, and C. W. J. Beenakker. Electrically detected interferometry of Majorana fermions in a topological insulator. *Phys. Rev. Lett.*, (102):216404, 2009.

[12] E. Alba, X. Fernandez-Gonzalvo, J. Mur-Petit, J. K. Pachos, and J. J. Garcia-Ripoll. Seeing topological order in time-of-flight measurements. *Phys. Rev. Lett.*, (107):235301, 2011.

[13] A. Alexandradinata, Zhijun Wang, and B. Andrei Bernevig. Topological insulators from group cohomology. *Phys. Rev. X*, (6):021008, 2016.

[14] J. Alicea. Majorana fermions in a tunable semiconductor device. *Phys. Rev. B*, (81):125318, 2010.

[15] J. Alicea. New directions in the pursuit of Majorana fermions in solid state systems. *Reports on Progress in Physics*, (75):076501, 2012.

[16] J. Alicea, Y. Oreg, G. Refael, F. von Oppen, and M. P. A. Fisher. Non-Abelian statistics and topological quantum information processing in 1D wire networks. *Nature Physics*, (7):412, 2011.

[17] A. Altland and B. D. Simons. *Condensed Matter Field Theory*. Cambridge University Press, 2010.

[18] A. Altland and M. R. Zirnbauer. Nonstandard symmetry classes in mesoscopic normal-superconducting hybrid structures. *Phys. Rev. B*, (55):1142, 1997.

[19] L. Alvarez-Gaumé and E. Witten. Gravitational anomalies. *Nuclear Physics B*, (234):269, 1984.

[20] M. H. Anderson, J. R. Ensher, M. R. Matthews, C. E. Wieman, and E. A. Cornell. Observation of Bose-Einstein condensation in a dilute atomic vapor. *Science*, (269):198, 1995.

[21] P. W. Anderson. Absence of diffusion in certain random lattices. *Phys. Rev.*, (109):1492, 1958.

[22] P. W. Anderson. Resonating valence bonds: A new kind of insulator? *Materials Research Bulletin*, (8):153, 1973.

[23] Y. Ando. Topological insulator materials. *Journal of the Physical Society of Japan*, (82):102001, 2013.

[24] Y. Ando and L. Fu. Topological crystalline insulators and topological superconductors: From concepts to materials. *Annual Review of Cond. Matt. Phys.*, (6):361, 2015.

[25] N. P. Armitage, E. J. Mele, and A. Vishwanath. Weyl and Dirac semimetals in three-dimensional solids. *Rev. Mod. Phys.*, (90):015001, 2018.

[26] N. P. Armitage and Liang Wu. On the matter of topological insulators as magnetoelectrics. *SciPost Phys.*, (6):046, 2019.

[27] S. Arora and B. Barak. *Computational Complexity: A Modern Approach*. Cambridge University Press, 2009.

[28] D. Arovas, J. R. Schrieffer, and F. Wilczek. Fractional statistics and the quantum Hall effect. *Phys. Rev. Lett.*, (53):722, 1984.

[29] A. Aspect, E. Arimondo, R. Kaiser, N. Vansteenkiste, and C. Cohen-Tannoudji. Laser cooling below the one-photon recoil energy by velocity-selective coherent population trapping. *Phys. Rev. Lett.*, (61):826, 1988.

[30] A. Aspect, J. Dalibard, and G. Roger. Experimental test of Bell's inequalities using time-varying analyzers. *Phys. Rev. Lett.*, (49):1804, 1982.

[31] M. Atala, M. Aidelsburger, J. T. Barreiro, D. Abanin, T. Kitagawa, E. Demler, and I. Bloch. Direct measurement of the Zak phase in topological Bloch bands. *Nature Physics*, (9):795, 2013.

[32] F. T. Avignone, S. R. Elliott, and J. Engel. Double beta decay, Majorana neutrinos, and neutrino mass. *Rev. Mod. Phys.*, (80):481, 2008.

[33] A. Bérut, A. Arakelyan, A. Petrosyan, S. Ciliberto, R. Dillenschneider, and E. Lutz. Experimental verification of Landauer's principle linking information and thermodynamics. *Nature*, (483):187, 2012.

[34] F. A. Bais, P. van Driel, and M. de Wild Propitius. Quantum symmetries in discrete gauge theories. *Phys. Lett. B*, (280):63, 1992.

[35] A. Bansil, H. Lin, and T. Das. Colloquium: Topological band theory. *Rev. Mod. Phys.*, (88):021004, 2016.

[36] J. Bardeen, L. N. Cooper, and J. R. Schrieffer. Microscopic theory of superconductivity. *Phys. Rev.*, (106):162, 1957.

[37] C.-E. Bardyn, M. A. Baranov, C. V. Kraus, E. Rico, A. İmamoglu, P. Zoller, and S. Diehl. Topology by dissipation. *New Journal of Physics*, (15):085001, Aug 2013.

[38] J. G. Bednorz and K. A. Müller. Possible high T_c superconductivity in the Ba-La-Cu-O system,. *Zeit. Phys. B*, (64):189, 1986.

[39] C. W. J. Beenakker. Annihilation of colliding Bogoliubov quasiparticles reveals their Majorana nature. *Phys. Rev. Lett.*, (112):070604, 2014.

[40] C. W. J. Beenakker. Search for Majorana fermions in superconductors. *Annual Review of Condensed Matter Physics*, (4):113, 2013.

[41] J. S. Bell. On the Einstein podolsky Rosen paradox. *Physics*, (1):195, 1964.

[42] J. S. Bell. *Speakable and Unspeakable in Quantum Mechanics: Collected Papers on Quantum Philosophy*. Cambridge University Press, 1987.

[43] J. S. Bell and R. Jackiw. A PCAC puzzle: $\pi^0 \rightarrow \gamma\gamma$ in the σ-model. *Nuovo Cimento A*, (60):47, 1969.

[44] W. A. Benalcazar, B. A. Bernevig, and T. L. Hughes. Electric multipole moments, topological multipole moment pumping, and chiral hinge states in crystalline insulators. *Phys. Rev. B*, (96):245115, 2017.

[45] P. Benioff. The computer as a physical system: A microscopic quantum mechanical Hamiltonian model of computers as represented by Turing machines. *J. Stat. Phys.*, (22):563, 1980.

[46] C. H. Bennett. The thermodynamics of computation – A review. *Int. J. Theor. Phys.*, (21):905, 1982.

[47] C. H. Bennett, E. Bernstein, G. Brassard, and U. Vazirani. Strengths and weaknesses of quantum computing. *SIAM Journal on Computing*, (26):1510, 1997.

[48] C. H. Bennett and G. Brassard. Quantum cryptography: Public key distribution and coin tossing. *Proc. of IEEE International Conference on Computers, Systems and Signal Processing, IEEE, New York,*, page 175, 1984.

[49] C. H. Bennett, G. Brassard, C. Crepeau, R. Jozsa, A. Peres, and W. K. Wootters. Teleporting an unknown quantum state via dual classical and Einstein-Podolsky-Rosen channels. *Phys. Rev. Lett.*, (70):1895, 1993.

[50] C. H. Bennett, G. Brassard, and N. D. Mermin. Quantum cryptography without Bell's theorem. *Phys. Rev. Lett.*, (68):557, 1992.

[51] C. H. Bennett and S. J. Wiesner. Communication via one- and two-particle operators on Einstein-Podolsky-Rosen states. *Phys. Rev. Lett.*, (69):2881, 1992.

[52] V. L. Berezinskii. Destruction of long-range order in one-dimensional and two-dimensional systems having a continuous symmetry group I. Classical systems. *Sov. Phys. JETP*, (32):493, 1971.

[53] B. Beri and N. R. Cooper. $\mathbb{Z}_2$ Topological insulators in ultracold atomic gases. *Phys. Rev. Lett.*, (107):145301, 2011.

[54] B. A. Bernevig. *Topological Insulators and Topological Superconductors*. Princeton University Press, New Jersey, 2013.

[55] B. A. Bernevig, T. L. Hughes, and S.-C. Zhang. Quantum spin Hall effect and topological phase transition in HgTe quantum wells. *Science*, (314):1757, 2006.

[56] C. Bernhardt. *Quantum Computing for Everyone*. The MIT Press, 2019.

[57] M. V. Berry. Quantal phase factors accompanying adiabatic changes. *Proc. Roy. Soc. London*, (A392):45, 1984.

[58] G. Bihlmayer, O. Rader, and R. Winkler. Focus on the Rashba effect. *New Journal of Physics*, (17):050202, 2015.

[59] I. Bloch. Ultracold quantum gases in optical lattices. *Nature Physics*, (1):23, 2005.

[60] I. Bloch, J. Dalibard, and W. Zwerger. Many-body physics with ultracold gases. *Rev. Mod. Phys.*, (80):885, 2008.

[61] B. Blok and X.-G. Wen. Effective theories of the fractional quantum Hall effect: Hierarchy construction. *Phys. Rev. B*, (42):8145, 1990.

[62] G. E. Blonder, M. Tinkham, and T. M. Klapwijk. Transition from metallic to tunneling regimes in superconducting microconstrictions: Excess current, charge imbalance, and supercurrent conversion. *Phys. Rev. B*, (25):4515, 1982.

[63] A. Bocharov, M. Roetteler, and K. M. Svore. Efficient Synthesis of Universal Repeat-Until-Success Quantum Circuits. *Phys. Rev. Lett.*, (114):080502, 2015.

[64] A. Bohm, A. Mostafazadeh, H. Koizumi, Q. Niu, and J. Zwanziger. *The Geometric Phase in Quantum Systems: Foundations, Mathematical concepts, and Applications in Molecular and Condensed Matter Physics*. Springer-Verlag, Berlin, 2003.

[65] C. J. Bolech and E. Demler. Observing Majorana bound states in p-wave superconductors using noise measurements in tunneling experiments. *Phys. Rev. Lett.*, (98):237002, 2007.

[66] P. Bonderson, M. Freedman, and C. Nayak. Measurement-only topological quantum computation. *Phys. Rev. Lett.*, (101):010501, 2008.

[67] N. E. Bonesteel, L. Hormozi, G. Zikos, and S. H. Simon. Braid topologies for quantum computation. *Phys. Rev. Lett.*, (95):140503, 2005.

[68] M. Born and V. Fock. Beweis des Adiabatensatzes. *Z. Phys.*, (51):165, 1928.

[69] R. Bott. The periodicity theorem for the classical groups and some of its applications. *Advances in Mathematics*, (4):353, 1970.

[70] A. Bouhon, A. M. Black-Schaffer, and R.-J. Slager. Wilson loop approach to fragile topology of split elementary band representations and topological crystalline insulators with time-reversal symmetry. *Phys. Rev. B*, (100):195135, 2019.

[71] Michel Boyer, Gilles Brassard, Peter Hyer, and Alain Tapp. Tight bounds on quantum searching. *Fortschritte der Physik*, (46):493, 1998.

[72] P. O. Boykin, T. Mor, M. Pulver, V. Roychowdhury, and F. Vatan. On universal and fault-tolerant quantum computing. *Information Processing Letters*, (75):101, 2000.

[73] C. C. Bradley, C. A. Sackett, J. J. Tollett, and R. G. Hulet. Evidence of Bose-Einstein condensation in an atomic gas with attractive interactions. *Phys. Rev. Lett.*, (75):1687, 1995.

[74] B. Bradlyn, J. Cano, Z. Wang, M. G. Vergniory, C. Felser, R. J. Cava, and B. Andrei Bernevig. Beyond Dirac and Weyl fermions: Unconventional quasiparticles in conventional crystals. *Science*, (353):aaf5037, 2016.

[75] B. Bradlyn, L. Elcoro, J. Cano, M. G. Vergniory, Z. Wang, C. Felser, M. I. Aroyo, and B. A. Bernevig. Topological quantum chemistry. *Nature*, (547):298, 2017.

[76] G. Brassard. Searching a quantum phone book. *Science*, (275):627, 1997.

[77] G. Brassard, C. Crepeau, D. Mayers, and L. Salvail. A brief review on the impossibility of quantum bit commitment. *arXiv e-print quant-ph/9712023*, 1997.

[78] B. Braunecker and P. Simon. Interplay between classical magnetic moments and superconductivity in quantum one-dimensional conductors: Toward a self-sustained topological Majorana phase. *Phys. Rev. Lett.*, (111):147202, 2013.

[79] S. Bravyi. Universal quantum computation with the $\nu = 5/2$ fractional quantum Hall state. *Phys. Rev. A*, (73):042313, 2006.

[80] S. Bravyi, M. B. Hastings, and S. Michalakis. Topological quantum order: Stability under local perturbations. *J. Math. Phys.*, (51), 2010.

[81] S. Bravyi and A. Kitaev. Universal quantum computation with ideal Clifford gates and noisy ancillas. *Phys. Rev. A*, (71):022316, 2005.

[82] S. B. Bravyi and A. Y. Kitaev. Quantum codes on a lattice with boundary. *arXiv e-print quant-ph/9811052*, 1998.

[83] H.-P. Breuer and F. Petruccione. *The Theory of Open Quantum Systems*. Oxford Univ. Press, 2007.

[84] C. Brouder, G. Panati, M. Calandra, C. Mourougane, and N. Marzari. Exponential localization of wannier functions in insulators. *Phys. Rev. Lett.*, (98):046402, 2007.

[85] P. W. Brouwer and V. Dwivedi. Homotopic classification of band structures: Stable, fragile, delicate, and stable representation-protected topology. *e-print arXiv:2306.13713*, 2023.

[86] D. Bruss. Characterizing entanglement. *J. Math. Phys.*, (43):4237, 2002.

[87] P. M. R. Brydon, S. Das Sarma, Hoi-Yin Hui, and Jay D. Sau. Topological Yu-Shiba-Rusinov chain from spin-orbit coupling. *Phys. Rev. B*, (91):064505, 2015.

[88] I. Buluta and F. Nori. Quantum simulators. *Science*, (326):108, 2009.

[89] A. A. Burkov, M. D. Hook, and L. Balents. Topological nodal semimetals. *Phys. Rev. B*, (84):235126, 2011.

[90] A.A. Burkov. Weyl metals. *Annual Review of Cond. Matt. Phys.*, (9):359, 2018.

[91] P. Busch, M. Grabowski, and P. J. Lahti. *Operational Quantum Physics*. Springer-Verlag, Berlin Heidelberg, 1995.

[92] A. R. Calderbank, E. M. Rains, P. W. Shor, and N. J. A. Sloane. Quantum error correction via codes over GF(4). *IEEE Trans. Inform. Theory*, (44):1369, 1998.

[93] A. R. Calderbank and Peter W. Shor. Good quantum error-correcting codes exist. *Phys. Rev. A*, (54):1098, 1996.

[94] A. H. Castro Neto, F. Guinea, N. M. R. Peres, K. S. Novoselov, and A. K. Geim. The electronic properties of graphene. *Rev. Mod. Phys.*, (81):109, 2009.

[95] C. Chamon, R. Jackiw, Y. Nishida, S.-Y. Pi, and L. Santos. Quantizing Majorana fermions in a superconductor. *Phys. Rev. B*, (81):224515, 2010.

[96] C.-Z. Chang, J. Zhang, X. Feng, J. Shen, Z. Zhang, M. Guo, et al. Experimental observation of the quantum anomalous Hall effect in a magnetic topological insulator. *Science*, (340):167, 2013.

[97] M.-C. Chang and Q. Niu. Berry curvature, orbital moment, and effective quantum theory of electrons in electromagnetic fields. *J. Phys.: Cond. Matt.*, (20):193202, 2008.

[98] X. Chen, Z.-C. Gu, Z.-X. Liu, and X.-G. Wen. Symmetry protected topological orders and the group cohomology of their symmetry group. *Phys. Rev. B*, (87):155114, 2013.

[99] X. Chen, Z.-C. Gu, and X.-G. Wen. Local unitary transformation, long-range quantum entanglement, wave function renormalization, and topological order. *Phys. Rev. B*, (82):155138, 2010.

[100] X. Chen, Z.-C. Gu, and X.-G. Wen. Classification of gapped symmetric phases in one-dimensional spin systems. *Phys. Rev. B*, (83):035107, 2011.

[101] S.-S. Chern and J. Simons. Characteristic forms and geometric invariants. *Ann. Math.*, (1):48, 1974.

[102] C. Chin, R. Grimm, P. Julienne, and E. Tiesinga. Feshbach resonances in ultracold gases. *Rev. Mod. Phys.*, (82):1225, 2010.

[103] C.-K. Chiu and A. P. Schnyder. Classification of reflection-symmetry-protected topological semimetals and nodal superconductors. *Phys. Rev. B*, page 205136, 2014.

[104] C.-K. Chiu, J. C. Y. Teo, A. P. Schnyder, and S. Ryu. Classification of topological quantum matter with symmetries. *Rev. Mod. Phys.*, (88):035005, 2016.

[105] C.-K. Chiu, H. Yao, and S. Ryu. Classification of topological insulators and superconductors in the presence of reflection symmetry. *Phys. Rev. B*, (88):075142, 2013.

[106] G. Y. Cho and J.l E. Moore. Topological BF field theory description of topological insulators. *Annals of Physics*, (326):1515, 2011.

[107] T.-P. Choy, J. M. Edge, A. R. Akhmerov, and C. W. J. Beenakker. Majorana fermions emerging from magnetic nanoparticles on a superconductor without spin-orbit coupling. *Phys. Rev. B*, (84):195442, 2011.

[108] S. Chu. Nobel lecture: The manipulation of neutral particles. *Rev. Mod. Phys.*, (70):685, 1998.

[109] S. Chu, L. Hollberg, J. E. Bjorkholm, A. Cable, and A. Ashkin. Three-dimensional viscous confinement and cooling of atoms by resonance radiation pressure. *Phys. Rev. Lett.*, (55):48, 1985.

[110] S. B. Chung, H.-J. Zhang, X.-L. Qi, and S.-C. Zhang. Topological superconducting phase and Majorana fermions in half-metal/superconductor heterostructures. *Phys. Rev. B*, (84):060510, 2011.

[111] A. Church. An unsolvable problem of elementary number theory. *Am. J. Math.*, (58):345, 1936.

[112] H. O. H. Churchill, V. Fatemi, K. Grove-Rasmussen, M. T. Deng, P. Caroff, H. Q. Xu, and C. M. Marcus. Superconductor-nanowire devices from tunneling to the multichannel regime: Zero-bias oscillations and magnetoconductance crossover. *Phys. Rev. B*, (87):241401, 2013.

[113] J. I. Cirac and P. Zoller. Quantum computations with cold trapped ions. *Phys. Rev. Lett.*, (74):4091, 1995.

[114] R. Cleve, A. Ekert, C. Macchiavello, and M. Mosca. Quantum algorithms revisited. *Proc. R. Soc. London A*, (454):339, 1998.

[115] W. K. Clifford. *Mathematical Papers.* (ed. R. Tucker), London: Macmillan, 1882.

[116] C. Cohen-Tannoudji. Nobel lecture: Manipulating atoms with photons. *Rev. Mod. Phys.*, (70):707, 1998.

[117] C. Cohen-Tannoudji. *Atoms in Electromagnetic Fields*. World Scientific, 2004.

[118] C. Cohen-Tannoudji and D. Guery-Odelin. *Advances in Atomic Physics: An Overview*. World Scientific, 2011.

[119] S. A. Cook. The Complexity of Theorem-Proving Procedures. In *Proceedings of the Third Annual ACM Symposium on Theory of Computing*, STOC '71, page 151. ACM Press, New York, 1971.

[120] N. R. Cooper. Optical flux lattices for ultracold atomic gases. *Phys. Rev. Lett.*, (106):175301, 2011.

[121] E. Cornfeld and A. Chapman. Classification of crystalline topological insulators and superconductors with point group symmetries. *Phys. Rev. B*, (99):075105, 2019.

[122] Ph. Courteille, R. S. Freeland, D. J. Heinzen, F. A. van Abeelen, and B. J. Verhaar. Observation of a Feshbach resonance in cold atom scattering. *Phys. Rev. Lett.*, (81):69, 1998.

[123] R. Crandall and C. Pomerance. *Prime Numbers: A Computational Perspective*. Springer, New York, 2005.

[124] M. O. T. Cunha, J. A. Dunningham, and V. Vedral. Entanglement in single-particle systems. *Proceedings of the Royal Society of London A: Mathematical, Physical and Engineering Sciences*, (463):2277, 2007.

[125] J. Dalibard, F. Gerbier, G. Juzeliunas, and P. Öhberg. Colloquium: Artificial gauge potentials for neutral atoms. *Rev. Mod. Phys.*, (83):1523, 2011.

[126] G. M. D'Ariano, M. G. A. Paris, and M. F. Sacchi. Quantum tomography. *Adv. Imaging Electron Phys.*, (128):205, 2003.

[127] A. Das and B. K. Chakrabarti. Colloquium: Quantum annealing and analog quantum computation. *Rev. Mod. Phys.*, (80):1061, 2008.

[128] A. Das, Y. Ronen, Y. Most, Y. Oreg, M. Heiblum, and H. Shtrikman. Zero-bias peaks and splitting in an Al-InAs nanowire topological superconductor as a signature of Majorana fermions. *Nature Physics*, (8):887, 2012.

[129] S. Das Sarma and A. Pinczuk (editors). *Perspectives in quantum Hall effects: Novel quantum liquids in low-dimensional semiconductor structures*. Wiley, New York, 1997.

[130] S. Das Sarma, M. Freedman, and C. Nayak. Topologically protected qubits from a possible non-Abelian fractional quantum Hall state. *Phys. Rev. Lett.*, (94):166802, 2005.

[131] S. Das Sarma, M. Freedman, and C. Nayak. Majorana zero modes and topological quantum computation. *Npj Quantum Information*, (1):15001, 2015.

[132] S. Das Sarma, C. Nayak, and S. Tewari. Proposal to stabilize and detect half-quantum vortices in strontium ruthenate thin films: Non-Abelian braiding statistics of vortices in a $p_x + ip_y$ superconductor. *Phys. Rev. B*, (73):220502, 2006.

[133] S. Das Sarma and H. Pan. Disorder-induced zero-bias peaks in majorana nanowires. *Phys. Rev. B*, (103):195158, 2021.

[134] S. Das Sarma, J. D. Sau, and T. D. Stanescu. Splitting of the zero-bias conductance peak as smoking gun evidence for the existence of the Majorana mode in a superconductor-semiconductor nanowire. *Phys. Rev. B*, (86):220506, 2012.

[135] P. C. W. Davies. The implications of a cosmological information bound for complexity, quantum information and the nature of physical law. *Fluctuation and Noise Letters*, (07):C37, 2007.

[136] K. B. Davis, M. O. Mewes, M. R. Andrews, N. J. van Druten, D. S. Durfee, D. M. Kurn, and W. Ketterle. Bose-Einstein condensation in a gas of sodium atoms. *Phys. Rev. Lett.*, (75):3969, 1995.

[137] M. T. Deng, C. L. Yu, G. Y. Huang, M. Larsson, P. Caroff, and H. Q. Xu. Anomalous zero-bias conductance peak in a nb–InSb nanowire–Nb hybrid device. *Nano Letters*, (12):6414, 2012.

[138] D. Deutsch. Quantum theory, the Church-Turing principle and the universal quantum computer. *Proc. R. Soc. Lond. A*, (400):97, 1985.

[139] D. Deutsch and R. Jozsa. Rapid solution of problems by quantum computation. *Proceedings of the Royal Society of London A: Mathematical, Physical and Engineering Sciences*, (439):553, 1992.

[140] S. J. Devitt, K. Nemoto, and W. J. Munro. Quantum error correction for beginners. *Rep. Prog. Phys.*, (76):076001, 2013.

[141] D. Dieks. Communication by EPR devices. *Phys. Lett. A*, (92):271, 1982.

[142] M. Diez, D. I. Pikulin, I. C. Fulga, and J. Tworzydło. Extended topological group structure due to average reflection symmetry. *New Journal of Physics*, (17):043014, 2015.

[143] W. Diffie and M. Hellman. New directions in cryptography. *IEEE Trans. Inf. Theory*, (IT-22):644, 1976.

[144] P. A. M. Dirac. The quantum theory of the electron. *Proceedings of the Royal Society of London A: Mathematical, Physical and Engineering Sciences*, (117):610, 1928.

[145] P. A. M. Dirac. *The Principles of Quantum Mechanics*. Oxford University Press Inc., New York, 1958.

[146] D. P. DiVincenzo. Two-bit gates are universal for quantum computation. *Phys. Rev. A*, (51):1015, 1995.

[147] L.-M. Duan, E. Demler, and M. D. Lukin. Controlling spin exchange interactions of ultracold atoms in optical lattices. *Phys. Rev. Lett.*, (91):090402, 2003.

[148] M. I. Dyakonov. *Will We Ever Have a Quantum Computer?* Springer, Cham, 2020.

[149] F. J. Dyson. The threefold way. Algebraic structure of symmetry groups and ensembles in quantum mechanics. *J. Math. Phys.*, (3):1199, 1962.

[150] P. Dziawa, B. J. Kowalski, K. Dybko, R. Buczko, A. Szczerbakow, M. Szot, E. Lusakowska, T. Balasubramanian, B. M. Wojek, M. H. Berntsen, O. Tjernberg, and T. Story. Topological crystalline insulator states in $Pb_{1-x}Sn_xSe$. *Nature Materials*, (11):1023, 2012.

[151] K. Efetov. *Supersymmetry in Disorder and Chaos*. Cambridge University Press, 1997.

[152] A. Einstein, B. Podolsky, and N. Rosen. Can quantum-mechanical description of physical reality be considered complete? *Phys. Rev.*, (47):777, 1935.

[153] J. Eisert and D. Gross. Multi-partite entanglement. In D. Bruss and G. Leuchs, editors, *Lectures on quantum information*. Wiley-VCH, Weinheim, 2006.

[154] A. K. Ekert. Quantum cryptography based on Bell's theorem. *Phys. Rev. Lett.*, (67):661, 1991.

[155] S. Elitzur, G. Moore, A. Schwimmer, and N. Seiberg. Remarks on the canonical quantization of the Chern-Simons-Witten theory. *Nucl. Phys. B*, (326):108, 1989.

[156] S. R. Elliott and M. Franz. Colloquium: Majorana fermions in nuclear, particle, and solid-state physics. *Rev. Mod. Phys.*, (87):137, 2015.

[157] D. V. Else, B. Bauer, and C. Nayak. Prethermal phases of matter protected by time-translation symmetry. *Phys. Rev. X*, (7):011026, 2017.

[158] A. M. Essin and M. Hermele. Classifying fractionalization: Symmetry classification of gapped $\mathbb{Z}_2$ spin liquids in two dimensions. *Phys. Rev. B*, (87):104406, 2013.

[159] A. M. Essin, J. E. Moore, and D. Vanderbilt. Magnetoelectric polarizability and axion electrodynamics in crystalline insulators. *Phys. Rev. Lett.*, (102):146805, 2009.

[160] Z. F. Ezawa. *Quantum Hall Effects: Recent Theoretical and Experimental Developments (3rd Edition)*. World Scientific, 2013.

[161] M. Fannes, B. Nachtergaele, and R. F. Werner. Exact antiferromagnetic ground states of quantum spin chains. *EPL*, (10):633, 1989.

[162] U. Fano. Effects of configuration interaction on intensities and phase shifts. *Phys. Rev.*, (124):1866, 1961.

[163] B. E. Feldman, M. T. Randeria, J. Li, S. Jeon, Y. Xie, Z. Wang, I. K. Drozdov, B. A. Bernevig, and A. Yazdani. High-resolution studies of the Majorana atomic chain platform. *Nature Physics*, (13):286, 2017.

[164] P. Fendley. Critical Points in Two-Dimensional Replica Sigma Models. In A. M. Tsvelik, editor, *New Theoretical Approaches to Strongly Correlated Systems*. Kluwer Academic Publishers, The Netherlands, 2006.

[165] H. Feshbach. Unified theory of nuclear reactions. *Annals of Physics*, (5):357, 1958.

[166] R. P. Feynman. Simulating physics with computers. *Int. J. Theor. Phys.*, (21):467, 1982.

[167] R. P. Feynman and A. R. Hibbs. *Quantum Mechanics and Path Integrals (Emended edition)*. Dover, 2005.

[168] L. Fidkowski, X. Chen, and A. Vishwanath. Non-Abelian topological order on the surface of a 3D topological superconductor from an exactly solved model. *Phys. Rev. X*, (3):041016, 2013.

[169] L. Fidkowski and A. Kitaev. Effects of interactions on the topological classification of free fermion systems. *Phys. Rev. B*, (81):134509, 2010.

[170] L. Fidkowski and A. Kitaev. Topological phases of fermions in one dimension. *Phys. Rev. B*, (83):075103, 2011.

[171] A. D. K. Finck, D. J. Van Harlingen, P. K. Mohseni, K. Jung, and X. Li. Anomalous modulation of a zero-bias peak in a hybrid nanowire-superconductor device. *Phys. Rev. Lett.*, (110):126406, 2013.

[172] G. Floquet. Sur les équations différentielles linéaires à coefficients périodiques. *Ann. l'École Norm. Supér.*, (12):47, 1883.

[173] V. Fock. Über die Beziehung zwischen den Integralen der quantenmechanischen Bewegungsgleichungen und der Schrödingerschen Wellengleichung. *Z. Phys.*, (49):323, 1928.

[174] E. Fradkin. *Field Theories of Condensed Matter Physics*. Cambridge University Press, 2013.

[175] M. Franz and L. Molenkamp (Editors). *Topological Insulators*. Elsevier, 2013.

[176] E. Fredkin and T. Toffoli. Conservative logic. *Int. J. Theor. Phys.*, (21):219, 1982.

[177] D. S. Freed and G. W. Moore. Twisted equivariant matter. *Annales Henri Poincare*, (14):1927, 2013.

[178] H. M. Freedman, M.l Larsen, and Z. Wang. A modular functor which is universal for quantum computation. *Commun. Math. Phys.*, (227):605, 2002.

[179] H. M. Freedman, M.l Larsen, and Z. Wang. The two-eigenvalue problem and density of Jones representation of braid groups. *Commun. Math. Phys.*, (228):177, 2002.

[180] M. Freedman, C. Nayak, and K. Walker. Towards universal topological quantum computation in the $\nu = \frac{5}{2}$ fractional quantum Hall state. *Phys. Rev. B*, (73):245307, 2006.

[181] M. H. Freedman, A. Kitaev, M. J. Larsen, and Z. Wang. Topological quantum computation. *Bull. Amer. Math. Soc.*, (40):31, 2003.

[182] S. J. Freedman and J. F. Clauser. Experimental test of local hidden-variable theories. *Phys. Rev. Lett.*, (28):938, 1972.

[183] S. Frolov. Quantum computing's reproducibility crisis: Majorana fermions. *Nature*, (592):350, 2021.

[184] L. Fu. Electron teleportation via Majorana bound states in a Mesoscopic superconductor. *Phys. Rev. Lett.*, (104):056402, 2010.

[185] L. Fu and C. L. Kane. Time reversal polarization and a Z_2 adiabatic spin pump. *Phys. Rev. B*, (74):195312, 2006.

[186] L. Fu and C. L. Kane. Topological insulators with inversion symmetry. *Phys. Rev. B*, (76):045302, 2007.

[187] L. Fu and C. L. Kane. Superconducting proximity effect and Majorana fermions at the surface of a topological insulator. *Phys. Rev. Lett.*, (100):096407, 2008.

[188] L. Fu and C. L. Kane. Josephson current and noise at a superconductor/quantum-spin-Hall-insulator/superconductor junction. *Phys. Rev. B*, (79):161408, 2009.

[189] L. Fu, C. L. Kane, and E. J. Mele. Topological insulators in three dimensions. *Phys. Rev. Lett.*, (98):106803, 2007.

[190] Liang Fu. Topological crystalline insulators. *Phys. Rev. Lett.*, (106):106802, 2011.

[191] Y. Fuji, F. Pollmann, and M. Oshikawa. Distinct trivial phases protected by a point-group symmetry in quantum spin chains. *Phys. Rev. Lett.*, (114):177204, 2015.

[192] K. Fujikawa and H. Suzuki. *Path Integrals and Quantum Anomalies.* Clarendon Press, Oxford, 2004.

[193] V. Galitski and I. B. Spielman. Spin–orbit coupling in quantum gases. *Nature*, (494):49, 2013.

[194] J. J. Garcja-Ripoll, M. A. Martin-Delgado, and J. I. Cirac. Implementation of Spin Hamiltonians in Optical Lattices. *Phys. Rev. Lett.*, (93):250405, 2004.

[195] K. Gödel. Über formal unentscheidbare Sätze der Principia Mathematica und verwandter Systeme I. *Monatshefte für Mathematik und Physik*, (38):173, 1931.

[196] M. Geier, L. Trifunovic, M. Hoskam, and P. W. Brouwer. Second-order topological insulators and superconductors with an order-two crystalline symmetry. *Phys. Rev. B*, (97):205135, 2018.

[197] A. K. Geim and K. S. Novoselov. The rise of graphene. *Nature Materials*, (6):183, 2007.

[198] I. M. Gelfand and M. A. Neumark. On the embedding of normed rings into the ring of operators in Hilbert space. *Rec. Math. [Mat. Sbornik] N.S.*, (12):197, 1943.

[199] M. Gell-Mann and M. Lévy. The axial vector current in beta decay. *Il Nuovo Cimento (1955-1965)*, (16):705, 1960.

[200] I. M. Georgescu, S. Ashhab, and F. Nori. Quantum simulation. *Rev. Mod. Phys.*, (86):153, 2014.

[201] W. Gerlach and O. Stern. Das magnetische Moment des Silberatoms. *Zeitschrift für Physik*, (9):353, 1922.

[202] Manuel Gessner, Claude Fabre, and Nicolas Treps. Superresolution limits from measurement crosstalk. *Phys. Rev. Lett.*, (125):100501, 2020.

[203] V. L. Ginzburg and L. D. Landau. On the theory of superconductivity. *Zh. Eksp. Teor. Fiz.*, (20):1064, 1950.

[204] N. Gisin, G. Ribordy, W. Tittel, and H. Zbinden. Quantum cryptography. *Rev. Mod. Phys.*, (74):145, 2002.

[205] R. G. Glauber. Coherent and incoherent states of the radiation field. *Phys. Rev.*, (131):2766, 1963.

[206] R. G. Glauber. The quantum theory of optical coherence. *Phys. Rev.*, (130):2529, 1963.

[207] N. Goldenfeld. *Lectures on Phase Transitions and the Renormalization Group.* Perseus Publishing, 1992.

[208] D. Goldin and P. Wegner. The interactive nature of computing: Refuting the strong Church–Turing thesis. *Minds and Machines*, (18):17, 2008.

[209] N. Goldman, E. Anisimovas, F. Gerbier, P. Öhberg, I. B. Spielman, and G. Juzeliūnas. Measuring topology in a laser-coupled honeycomb lattice: From Chern insulators to topological semi-metals. *New Journal of Physics*, (15):013025, 2013.

[210] N. Goldman, G. Juzeliunas, P. Ohberg, and I. B. Spielman. Light-induced gauge fields for ultracold atoms. *Rep. Prog. Phys.*, (77):126401, 2014.

[211] N. Goldman, I. Satija, P. Nikolic, A. Bermudez, M. A. Martin-Delgado, M. Lewenstein, and I. B. Spielman. Realistic time-reversal invariant topological insulators with neutral atoms. *Phys. Rev. Lett.*, (105):255302, 2010.

[212] J. Goldstone, A. Salam, and S. Weinberg. Broken symmetries. *Phys. Rev.*, (127):965, 1962.

[213] J. P. Gordon. Noise at optical frequencies; information theory. In *P. A. Miles (ed.), Quantum Electronics and Coherent Light; Proceedings of the International School of Physics Enrico Fermi, Course XXXI*, pages 156–181. Academic Press, New York, 1964.

[214] J. P. Gordon and A. Ashkin. Motion of atoms in a radiation trap. *Phys. Rev. A*, (21):1606, 1980.

[215] D. Gottesman. Class of quantum error-correcting codes saturating the quantum Hamming bound. *Phys. Rev. A*, (54):1862, 1996.

[216] D. Gottesman. *Quantum Error Correction. Ph.D. thesis.* California Institute of Technology, Pasadena, CA, 1997.

[217] D. Gottesman. Theory of fault-tolerant quantum computation. *Phys. Rev. A*, (57):127, 1998.

[218] M. Greiner, O. Mandel, T. Esslinger, T. W. Hansch, and I. Bloch. Quantum phase transition from a superfluid to a Mott insulator in a gas of ultracold atoms. *Nature*, (415):39, 2002.

[219] E. Grosfeld, N. R. Cooper, A. Stern, and R. Ilan. Predicted signatures of p-wave superfluid phases and Majorana zero modes of fermionic atoms in RF absorption. *Phys. Rev. B*, (76):104516, 2007.

[220] E. Grosfeld and A. Stern. Observing Majorana bound states of Josephson vortices in topological superconductors. *Proc. Natl. Acad. of Sci.*, (108):11810, 2011.

[221] E. P. Gross. Structure of a quantized vortex in boson systems. *Il Nuovo Cimento (1955-1965)*, (20):454, 1961.

[222] L. K. Grover. A Fast Quantum Mechanical Algorithm for Database Search. In *Proceedings of the Twenty-eighth Annual ACM Symposium on Theory of Computing*, STOC '96, page 212. ACM Press, New York, 1996.

[223] L. K. Grover. Quantum mechanics helps in searching for a needle in a Haystack. *Phys. Rev. Lett.*, (79):325, 1997.

[224] G. Grynberg and C. Robilliard. Cold atoms in dissipative optical lattices. *Physics Reports*, (355):335, 2001.

[225] Z.-C. Gu and X.-G. Wen. Tensor-entanglement-filtering renormalization approach and symmetry-protected topological order. *Phys. Rev. B*, (80):155131, 2009.

[226] Z.-C. Gu and X.-G. Wen. Symmetry-protected topological orders for interacting fermions: Fermionic topological nonlinear σ models and a special group supercohomology theory. *Phys. Rev. B*, (90):115141, 2014.

[227] E. Guadagnini, M. Martellini, and M. Mintchev. Wilson lines in Chern-Simons theory and link invariants. *Nucl. Phys. B*, (330):575, 1990.

[228] F. D. M. Haldane. Spontaneous dimerization in the s=1/2 Heisenberg antiferromagnetic chain with competing interactions. *Phys. Rev. B*, (25):4925, 1982.

[229] F. D. M. Haldane. Fractional quantization of the Hall effect: A hierarchy of incompressible quantum fluid states. *Phys. Rev. Lett.*, (51):605, 1983.

[230] F. D. M. Haldane. Nonlinear field theory of large-spin Heisenberg antiferromagnets: Semiclassically quantized solitons of the one-dimensional easy-axis Neel state. *Phys. Rev. Lett.*, (50):1153, 1983.

[231] F. D. M. Haldane. Model for a quantum Hall effect without landau levels: Condensed-matter realization of the "Parity Anomaly". *Phys. Rev. Lett.*, (61):2015, 1988.

[232] B. I. Halperin. Statistics of quasiparticles and the hierarchy of fractional quantized Hall states. *Phys. Rev. Lett.*, (52):1583, 1984.

[233] J. H. Hannay. Angle variable holonomy in adiabatic excursion of an integrable Hamiltonian. *J. Phys. A Math. Gen.*, (18):221, 1985.

[234] T. H. Hansson, V. Oganesyan, and S. L. Sondhi. Superconductors are topologically ordered. *Ann. Phys.*, (313):497, 2004.

[235] M. Haque, O. S. Zozulya, and K. Schoutens. Entanglement between particle partitions in itinerant many-particle states. *Journal of Physics A: Mathematical and Theoretical*, (42):504012, 2009.

[236] S. Haroche, J. C. Gay, and G. Grynberg (Editors). *Atomic Physics 11: Proceedings of the Eleventh International Conference on Atomic Physics*. World Scientific, 1989.

[237] F. Harper, R. Roy, M. S. Rudner, and S. L. Sondhi. Topology and Broken Symmetry in Floquet Systems. *Annual Review of Cond. Matt. Phys.*, (11):345, 2020.

[238] M. Z. Hasan, G. Chang, I. Belopolski, G. Bian, S.-Y. Xu, and J.-X. Yin. Weyl, Dirac and high-fold chiral fermions in topological quantum matter. *Nature Reviews Materials*, (6):784, 2021.

[239] M. Z. Hasan and C. L. Kane. Colloquium: Topological insulators. *Rev. Mod. Phys.*, (82):3045, 2010.

[240] M. Z. Hasan and J. E. Moore. Three-dimensional topological insulators. *Ann. Review.Condensed Matter Physics*, (2):55, 2011.

[241] T. W. Hänsch and A. L. Schawlow. Cooling of gases by laser radiation. *Optics Communications*, (13):68, 1975.

[242] J. J. He, T. K. Ng, P. A. Lee, and K. T. Law. Selective equal-spin Andreev reflections induced by Majorana fermions. *Phys. Rev. Lett.*, (112):037001, 2014.

[243] O. Heinonen (Editor). *Composite Fermions: A Unified View of the Quantum Hall Regime.* World Scientific, 1998.

[244] S. Helgason. *Differential Geometry, Lie Groups, and Symmetric Spaces.* Academic Press, 1978.

[245] G. Herzberg and H. C. Longuet-Higgins. Intersection of potential energy surfaces in polyatomic molecules. *Discuss. Faraday Soc.*, (35):77, 1963.

[246] J. D. Hidary. *Quantum Computing: An Applied Approach.* Springer, Cham, 2021.

[247] D. Hilbert and W. Ackermann. *Principles of Mathematical Logic.* Springer-Verlag, 1928.

[248] A. S. Holevo. Bounds for the quantity of information transmitted by a quantum communication channel. *Problems of Information Transmission*, (9):177, 1973.

[249] J. P. Home, D. Hanneke, J. D. Jost, J. M. Amini, D. Leibfried, and D. J. Wineland. Complete methods set for scalable ion trap quantum information processing. *Science*, (325):1227, 2009.

[250] L. Hormozi, G. Zikos, N. E. Bonesteel, and S. H. Simon. Topological quantum compiling. *Phys. Rev. B*, (75):165310, 2007.

[251] M. Horodecki, J. Oppenheim, and A. Winter. Partial quantum information. *Nature*, (436):673, 2005.

[252] Paweł Horodecki, Łukasz Rudnicki, and Karol Życzkowski. Five open problems in quantum information theory. *PRX Quantum*, (3):010101, 2022.

[253] R. Horodecki, P. Horodecki, M. Horodecki, and K. Horodecki. Quantum entanglement. *Rev. Mod. Phys.*, (81):865, 2009.

[254] D. Hsieh, D. Qian, L. Wray, Y. Xia, Y. S. Hor, R. J. Cava, and M. Z. Hasan. A topological Dirac insulator in a quantum spin Hall phase. *Nature*, (452):970, 2008.

[255] T. H. Hsieh, H. Lin, J. Liu, W. Duan, A. Bansil, and L. Fu. Topological crystalline insulators in the SnTe material class. *Nature Communication*, (3):982, 2012.

[256] S.-M. Huang, S.-Y. Xu, I. Belopolski, et al. New type of Weyl semimetal with quadratic double Weyl fermions. *Proc. Natl. Acad. Sci.*, (113):1180, 2016.

[257] S. Inouye, M. R. Andrews, J. Stenger, H.-J. Miesner, D. M. Stamper-Kurn, and W. Ketterle. Observation of Feshbach resonances in a Bose-Einstein condensate. *Nature*, (392):151, 1998.

[258] I. Garcja Irastorza. An introduction to axions and their detection. *SciPost Phys. Lect. Notes*, page 45, 2022.

[259] C. Isham. *Modern Differential Geometry for Physicists.* World Scientific, Singapore, 1989.

[260] E. Ising. Beitrag zur Theorie des Ferromagnetismus. *Z. Physik*, (31):253, 1925.

[261] D. A. Ivanov. Non-Abelian statistics of half-quantum vortices in p-wave superconductors. *Phys. Rev. Lett.*, (86):268, 2001.

[262] J. K. Jain. Composite-fermion approach for the fractional quantum Hall effect. *Phys. Rev. Lett.*, (63):199, 1989.

[263] D. Jaksch and P. Zoller. Creation of effective magnetic fields in optical lattices: the Hofstadter butterfly for cold neutral atoms. *New Journal of Physics*, (5):56, 2003.

[264] D. Jaksch and P. Zoller. The cold atom Hubbard toolbox. *Annals of Physics*, (315):52, 2005. Special Issue.

[265] P. S. Jessen and I. H. Deutsch. Optical lattices. *Adv. Atm. Mol. Opt. Phys.*, (37):95, 1996.

[266] L. Jiang, T. Kitagawa, J. Alicea, A. R. Akhmerov, D. Pekker, G. Refael, J. I. Cirac, E. Demler, M. D. Lukin, and P. Zoller. Majorana fermions in equilibrium and in driven cold-atom quantum wires. *Phys. Rev. Lett.*, (106):220402, 2011.

[267] Vaughan F. R. Jones. A polynomial invariant for knots via von Neumann algebras. *Bull. Amer. Math. Soc.*, (12):103, 1985.

[268] G. Jotzu, M. Messer, R. Desbuquois, M. Lebrat, T. Uehlinger, D. Greif, and T. Esslinger. Experimental realization of the topological Haldane model with ultracold fermions. *Nature*, (515):237, 2014.

[269] R. Jozsa and B. Schumacher. A new proof of the quantum noiseless coding theorem. *J. Mod. Opt.*, (41):2343, 1994.

[270] T. Jungwirth, Q. Niu, and A. H. MacDonald. Anomalous Hall effect in ferromagnetic semiconductors. *Phys. Rev. Lett.*, (88):207208, 2002.

[271] G. Juzeliunas, J. Ruseckas, and J. Dalibard. Generalized Rashba-Dresselhaus spin-orbit coupling for cold atoms. *Phys. Rev. A*, (81):053403, 2010.

[272] Juzeliunas, G. and Ruseckas, J. and Öhberg, P. and Fleischhauer, M. Light-induced effective magnetic fields for ultracold atoms in planar geometries. *Phys. Rev. A*, (73):025602, 2006.

[273] V. Kalmeyer and R. B. Laughlin. Equivalence of the resonating-valence-bond and fractional quantum Hall states. *Phys. Rev. Lett.*, (59):2095, 1987.

[274] C. L. Kane and E. J. Mele. Quantum spin Hall effect in graphene. *Phys. Rev. Lett.*, (95):226801, 2005.

[275] C. L. Kane and E. J. Mele. Z_2 topological order and the quantum spin Hall effect. *Phys. Rev. Lett.*, (95):146802, 2005.

[276] A. Kapustin. Symmetry protected topological phases, anomalies, and cobordisms: Beyond group cohomology. *e-print arXiv:1403.1467*, 2014.

[277] R. M. Karp. Reducibility Among Combinatorial Problems. In R. E. Miller and J. W. Thatcher, editors, *Complexity of Computer Computations*, page 85. Plenum, New York, 1972.

[278] M. Kasevich and S. Chu. Laser cooling below a photon recoil with three-level atoms. *Phys. Rev. Lett.*, (69):1741, 1992.

[279] V. Kasirajan. *Fundamentals of Quantum Computing: Theory and Practice.* Springer, 2021.

[280] I. Kassal, S. P. Jordan, P. J. Love, M. Mohseni, and A. Aspuru-Guzik. Polynomial-time quantum algorithm for the simulation of chemical dynamics. *Proc. Natl. Acad. Sci. U.S.A.*, (105):18681, 2008.

[281] T. Kato. On the adiabatic theorem of quantum mechanics. *J. Phys. Soc. Jap.*, (5):435–439, 1950.

[282] M. König, S. Wiedmann, C. Brüne, A. Roth, H. Buhmann, L. W. Molenkamp, X.-L. Qi, and S.-C. Zhang. Quantum spin Hall insulator state in HgTe quantum wells. *Science*, (318):766, 2007.

[283] B. Keimer, S. A. Kivelson, M. R. Norman, S. Uchida, and J. Zaanen. From quantum matter to high-temperature superconductivity in copper oxides. *Nature*, (518):179, 2015.

[284] G. Kells, D. Meidan, and P. W. Brouwer. Near-zero-energy end states in topologically trivial spin-orbit coupled superconducting nanowires with a smooth confinement. *Phys. Rev. B*, (86):100503, 2012.

[285] R. Kennedy and C. Guggenheim. Homotopy theory of strong and weak topological insulators. *Phys. Rev. B*, (91):245148, 2015.

[286] W. Ketterle and N.J. Van Druten. *Evaporative Cooling of Trapped Atoms. Number 37 in Advances in Atomic, Molecular, and Optical Physics*, page 181. Academic Press, 1996.

[287] E. Khalaf, H. C. Po, A. Vishwanath, and H. Watanabe. Symmetry indicators and anomalous surface states of topological crystalline insulators. *Phys. Rev. X*, (8):031070, 2018.

[288] D. E. Kharzeev. The chiral magnetic effect and anomaly-induced transport. *Progress in Particle and Nuclear Physics*, (75):133, 2014.

[289] H. Kim, A. Palacio-Morales, T. Posske, L. Rozsa, K. Palotas, L. Szunyogh, M. Thorwart, and R. Wiesendanger. Toward tailoring Majorana bound states in artificially constructed magnetic atom chains on elemental superconductors. *Science Advances*, (4):eaar5251, 2018.

[290] R. D. King-Smith and David Vanderbilt. Theory of polarization of crystalline solids. *Phys. Rev. B*, (47):1651, 1993.

[291] A. Kitaev. Fault-tolerant quantum computation by anyons. *Annals Phys.*, (303):2, 2003.

[292] A. Kitaev. Anyons in an exactly solved model and beyond. *Annals of Physics*, (321):2, 2006.

[293] A. Kitaev. Periodic table for topological insulators and superconductors. *AIP Conference Proceedings*, (1134):22, 2009.

[294] A. Kitaev and C. Laumann. Topological phases and quantum computation. *e-print arXiv:0904.2771*, 2009.

[295] A. Kitaev and J. Preskill. Topological entanglement entropy. *Phys. Rev. Lett.*, (96):110404, 2006.

[296] A. Y. Kitaev. *Proceedings of the 3rd International Conference of Quantum Communication and Measurement*. Editors O. Hirota, A. S. Holevo, and C. M. Caves, New York, Plenum, 1997.

[297] A. Y. Kitaev. Quantum computations: Algorithms and error correction. *Russ. Math. Surv.*, (52):1191, 1997.

[298] A. Y. Kitaev. Unpaired Majorana fermions in quantum wires. *Physics-Uspekhi*, (44):131, 2001.

[299] T. Kitagawa, E. Berg, M. Rudner, and E. Demler. Topological characterization of periodically driven quantum systems. *Phys. Rev. B*, (82):235114, 2010.

[300] M. Kjaergaard, K. Wölms, and K. Flensberg. Majorana fermions in superconducting nanowires without spin-orbit coupling. *Phys. Rev. B*, (85):020503, 2012.

[301] K. v. Klitzing, G. Dorda, and M. Pepper. New method for high-accuracy determination of the fine-structure constant based on quantized Hall resistance. *Phys. Rev. Lett.*, (45):494, 1980.

[302] V. Kliuchnikov, D. Maslov, and M. Mosca. Fast and efficient exact synthesis of single qubit unitaries generated by clifford and t gates. *Quantum Inf. Comput.*, (13):607, 2013.

[303] D. E. Knuth. *The Art of Computer Programming, Volumes 1-4A*. Addison-Wesley, 2011.

[304] S. Kobayashi and K. Nomizu. *Foundations of Differential Geometry*. Wiley Classics Library, 2009.

[305] M. Kohmoto, B. I. Halperin, and Y.-S. Wu. Diophantine equation for the three-dimensional quantum Hall effect. *Phys. Rev. B*, (45):13488, 1992.

[306] W. Kohn. Periodyc thermodynamics. *J. Stat. Phys.*, (103):417, 2001.

[307] S. S. Kondov, W. R. McGehee, W. Xu, and B. DeMarco. Disorder-induced localization in a strongly correlated atomic hubbard gas. *Phys. Rev. Lett.*, (114):083002, 2015.

[308] D. J. Kosterlitz and J. M. Thouless. Ordering, metastability and phase transitions in two-dimensional systems. *Journal of Physics C: Solid State Physics*, (6):1181, 1973.

[309] C. V. Kraus, S. Diehl, P. Zoller, and M. A. Baranov. Preparing and probing atomic Majorana fermions and topological order in optical lattices. *New Journal of Physics*, (14):113036, 2012.

[310] K. Kraus. *States, Effects, and Operations: Lecture Notes in Physics, 190*. Springer-Verlag, Berlin Heidelberg, 1983.

[311] J. Kruthoff, J. de Boer, J. van Wezel, C. L. Kane, and R.-J. Slager. Topological classification of crystalline insulators through band structure combinatorics. *Phys. Rev. X*, (7):041069, 2017.

[312] L. D. Landau. Oscillations in a Fermi-liquid. *Sov. Phys. JETP*, (5):101, 1957.

[313] L. D. Landau. Theory of Fermi-liquids. *Sov. Phys. JETP*, (3):920, 1957.

[314] L. D. Landau and E. M. Lifshitz. *Statistical Physics Part 1, 3rd Ed. (Course of Theoretical Physics Vol. 5)*. Pergamon Press, Oxford, 1994.

[315] R. Landauer. Irreversibility and heat generation in the computing process. *IBM J. Res. Dev.*, (5):183, 1961.

[316] J. Langbehn, Y. Peng, L. Trifunovic, F. von Oppen, and P. W. Brouwer. Reflection-symmetric second-order topological insulators and superconductors. *Phys. Rev. Lett.*, (119):246401, 2017.

[317] R. LaPierre. *Introduction to Quantum Computing*. Springer, 2021.

[318] A. Lau, J. van den Brink, and C. Ortix. Topological mirror insulators in one dimension. *Phys. Rev. B*, (94):165164, 2016.

[319] R. B. Laughlin. Anomalous quantum Hall effect: An incompressible quantum fluid with fractionally charged excitations. *Phys. Rev. Lett.*, (50):1395, 1983.

[320] K. T. Law, Patrick A. Lee, and T. K. Ng. Majorana fermion induced resonant Andreev reflection. *Phys. Rev. Lett.*, (103):237001, 2009.

[321] P. A. Lee and T. V. Ramakrishnan. Disordered electronic systems. *Rev. Mod. Phys.*, (57):287, 1985.

[322] A. J. Leggett. A theoretical description of the new phases of liquid ^{3}He. *Rev. Mod. Phys.*, (47):331, 1975.

[323] M. Leijnse and K. Flensberg. Introduction to topological superconductivity and Majorana fermions. *Semiconductor Science and Technology*, (27):124003, 2012.

[324] J. M. Leinaas and J. Myrheim. On the theory of identical particles. *Nuovo Cimento*, (37B):1, 1977.

[325] P. D. Lett, R. N. Watts, Ch. I. Westbrook, W. D. Phillips, P. L. Gould, and H. J. Metcalf. Observation of atoms laser cooled below the Doppler Limit. *Phys. Rev. Lett.*, (61):169, 1988.

[326] L. Levin. Universal search problems (Russian, 1973); translated by B. A. Trakhtenbrot. A survey of Russian approaches to perebor (brute-force searches) algorithms. *Annals of the History of Computing*, (6):384, 1984.

[327] M. Levin and A. Stern. Classification and analysis of two-dimensional Abelian fractional topological insulators. *Phys. Rev. B*, (86):115131, 2012.

[328] M. Levin and X.-G. Wen. Detecting topological order in a ground state wave function. *Phys. Rev. Lett.*, (96):110405, 2006.

[329] M. A. Levin and X.-G. Wen. String-net condensation: A physical mechanism for topological phases. *Phys. Rev. B*, (71):045110, 2005.

[330] L. B. Levitin. On the quantum measure of information. In *Proceedings of the Fourth All-Union Conference on Information and Coding Theory*. Sec. II, Tashkent, 1969.

[331] M. Lewenstein, A. Sanpera, and V. Ahufinger. *Ultracold atoms in optical lattices: Simulating quantum many-body systems*. Oxford University Press, 2012.

[332] M. Lewenstein, A. Sanpera, V. Ahufinger, B. Damski, A. Sen(De), and U. Sen. Ultracold atomic gases in optical lattices: mimicking condensed matter physics and beyond. *Advances in Physics*, (56):243, 2007.

[333] J. Li, H. Chen, I. K. Drozdov, A. Yazdani, B. A. Bernevig, and A. H. MacDonald. Topological superconductivity induced by ferromagnetic metal chains. *Phys. Rev. B*, (90):235433, 2014.

[334] R. Li, J. Wang, X. Qi, and S.-C. Zhang. Dynamical axion field in topological magnetic insulators. *Nature Physics*, (6):284, 2010.

[335] T. Li, L. Duca, M. Reitter, F. Grusdt, E. Demler, M. Endres, M. Schleier-Smith, I. Bloch, and U. Schneider. Bloch state tomography using wilson lines. *Science*, (352):1094, 2016.

[336] D. A. Lidar (Editor) and T. A. Brun (Editor). *Quantum Error Correction*. Cambridge University Press, 2013.

[337] E. M. Lifshitz. Anomalies of electron characteristics of a metal in the high pressure region. *Sov. Phys. JETP*, (11):1130, 1960.

[338] E. M. Lifshitz and L. P. Pitaevskii. *Statistical Physics: Theory of the Condensed State (Course of Theoretical Physics Vol. 9)*. Pergamon Press, Oxford, 1980.

[339] N. H. Lindner, G. Refael, and V. Galitski. Floquet topological insulator in semiconductor quantum wells. *Nature Physics*, (7):490, 2011.

[340] J. Liu, A. C. Potter, K. T. Law, and P. A. Lee. Zero-bias peaks in the tunneling conductance of spin-orbit-coupled superconducting wires with and without Majorana end-states. *Phys. Rev. Lett.*, (109):267002, 2012.

[341] A. M. Lobos, R. M. Lutchyn, and S. Das Sarma. Interplay of disorder and interaction in Majorana quantum wires. *Phys. Rev. Lett.*, (109):146403, 2012.

[342] H. C. Longuet-Higgens, U. Öpik, M. H. L. Pryce, and R. A. Sack. Studies of the John-Teller effect. *Proc. Roy. Soc. London*, (A244):1, 1958.

[343] Y.-M. Lu and A. Vishwanath. Theory and classification of interacting integer topological phases in two dimensions: A Chern-Simons approach. *Phys. Rev. B*, (86):125119, 2012.

[344] A. W. W. Ludwig. Topological phases: Classification of topological insulators and superconductors of non-interacting fermions, and beyond. *Physica Scripta*, (2016):014001, 2016.

[345] R. M. Lutchyn, J. D. Sau, and S. Das Sarma. Majorana fermions and a topological phase transition in semiconductor-superconductor heterostructures. *Phys. Rev. Lett.*, (105):077001, 2010.

[346] B. Q. Lv, H. M. Weng, B. B. Fu, X. P. Wang, H. Miao, J. Ma, P. Richard, X. C. Huang, L. X. Zhao, G. F. Chen, Z. Fang, X. Dai, T. Qian, and H. Ding. Experimental discovery of Weyl semimetal TaAs. *Phys. Rev. X*, (5):031013, 2015.

[347] J. Maciejko, T. L. Hughes, and S. C. Zhang. The quantum spin Hall effect. *Annual Review of Condensed Matter Physics*, (2):31, 2011.

[348] J. Maciejko, X.-L. Qi, H. D. Drew, and S.-C. Zhang. Topological quantization in units of the fine structure constant. *Phys. Rev. Lett.*, (105):166803, 2010.

[349] A. P. Mackenzie and Y. Maeno. The superconductivity of Sr_2RuO_4 and the physics of spin-triplet pairing. *Rev. Mod. Phys.*, (75):657, 2003.

[350] R. MacKenzie. Path integral methods and applications. *e-print arXiv:quant-ph/0004090*, 2000.

[351] Y. Maeno, S. Kittaka, T. Nomura, S. Yonezawa, and K.i Ishida. Evaluation of spin-triplet superconductivity in Sr_2RuO_4. *J. Phys. Soc. Jap.*, (81):011009, 2012.

[352] E. Majorana. Teoria simmetrica dell'elettrone e del positrone. *Nuovo Cimento*, (5):171, 1937.

[353] D. J. E. Marsh, K. C. Fong, E. W. Lentz, L. Šmejkal, and M. N. Ali. Proposal to detect dark matter using axionic topological antiferromagnets. *Phys. Rev. Lett.*, (123):121601, 2019.

[354] B. R. Martin and G Shaw. *Particle Physics (2nd Edition)*. Manchester Physics, John Wiley & Sons, 2008.

[355] I. Martin and A. F. Morpurgo. Majorana fermions in superconducting helical magnets. *Phys. Rev. B*, (85):144505, 2012.

[356] J. M. Martyn, Z. M. Rossi, A. K. Tan, and I. L. Chuang. Grand unification of quantum algorithms. *PRX Quantum*, (2):040203, 2021.

[357] V. Mastropietro and D. C. Mattis. *Luttinger Model: The First 50 Years and Some New Directions*. World Scientific, 2013.

[358] S. Matsuura, P.-Y. Chang, A. P. Schnyder, and S. Ryu. Protected boundary states in gapless topological phases. *New Journal of Physics*, (15):065001, 2013.

[359] C.A. Mead and D. G. Truhlar. On the determination of Born-Oppenheimer nuclear motion wave functions including complications due to conical intersections and identical nuclei. *J. Chem. Phys.*, (70):2284, 1979.

[360] R. Merkle. Secure communications over insecure channels. *Comm. of the ACM*, (21):294, 1978.

[361] A. Messiah. *Quantum Mechanics*. Dover, 2014.

[362] M. A. Metlitski, C. L. Kane, and M. P. A. Fisher. Bosonic topological insulator in three dimensions and the statistical Witten effect. *Phys. Rev. B*, (88):035131, 2013.

[363] H.-J. Mikeska and A. K. Kolezhuk. *Quantum Magnetism*, chapter One-dimensional magnetism, page 1. Springer Berlin Heidelberg, Berlin, Heidelberg, 2004.

[364] G. P. Mikitik and Yu. V. Sharlai. Manifestation of Berry's phase in metal physics. *Phys. Rev. Lett.*, (82):2147, 1999.

[365] G. P. Mikitik and Yu. V. Sharlai. Berry phase and de Haas–van Alphen effect in $LaRhIn_5$. *Phys. Rev. Lett.*, (93):106403, 2004.

[366] T. Mizushima, Y. Tsutsumi, T. Kawakami, M. Sato, M. Ichioka, and K. Machida. Symmetry-protected topological superfluids and superconductors: From the basics to ^{3}He. *J. Phys. Soc. Jap.*, (85):022001, 2016.

[367] A. J. Moerdijk, B. J. Verhaar, and A. Axelsson. Resonances in ultracold collisions of ^{6}Li, ^{7}Li, and ^{23}Na. *Phys. Rev. A*, (51):4852, 1995.

[368] R. S. K. Mong, A. M. Essin, and J. E. Moore. Antiferromagnetic topological insulators. *Phys. Rev. B*, (81):245209, 2010.

[369] G. Moore and N. Read. Nonabelions in the fractional quantum Hall effect. *Nuclear Physics B*, (360):362, 1991.

[370] J. E. Moore and L. Balents. Topological invariants of time-reversal-invariant band structures. *Phys. Rev. B*, (75):121306, 2007.

[371] T. Morimoto and A. Furusaki. Topological classification with additional symmetries from clifford algebras. *Phys. Rev. B*, (88):125129, 2013.

[372] T. Morimoto, A. Furusaki, and C. Mudry. Anderson localization and the topology of classifying spaces. *Phys. Rev. B*, (91):235111, 2015.

[373] O. Motrunich, K. Damle, and D. A. Huse. Griffiths effects and quantum critical points in dirty superconductors without spin-rotation invariance: One-dimensional examples. *Phys. Rev. B*, (63):224204, 2001.

[374] N. F. Mott and R. Peierls. Discussion of the paper by de Boer and Verwey. *Proceedings of the Physical Society*, (49):72, 1937.

[375] V. Mourik, K. Zuo, S. M. Frolov, S. R. Plissard, E. P. A. M. Bakkers, and L. P. Kouwenhoven. Signatures of Majorana fermions in hybrid superconductor-semiconductor nanowire devices. *Science*, (336):1003, 2012.

[376] S. Murakami and S. Kuga. Universal phase diagrams for the quantum spin Hall systems. *Phys. Rev. B*, (78):165313, 2008.

[377] S. Nadj-Perge, I. K. Drozdov, B. A. Bernevig, and Ali Yazdani. Proposal for realizing Majorana fermions in chains of magnetic atoms on a superconductor. *Phys. Rev. B*, (88):020407, 2013.

[378] S. Nadj-Perge, I. K. Drozdov, J. Li, H. Chen, S. Jeon, J. Seo, A. H. MacDonald, B. A. Bernevig, and A. Yazdani. Observation of Majorana fermions in ferromagnetic atomic chains on a superconductor. *Science*, (346):602, 2014.

[379] N. Nagaosa, J. Sinova, S. Onoda, A. H. MacDonald, and N. P. Ong. anomalous Hall effect. *Rev. Mod. Phys.*, (82):1539, 2010.

[380] M. Nakahara. *Geometry, Topology and Physics*. Adam Hilger, Bristol, 1990.

[381] Y. Nambu. Quasi-particles and gauge invariance in the theory of superconductivity. *Phys. Rev.*, (117):648, 1960.

[382] R. Nandkishore and D. A. Huse. Many-body localization and thermalization in quantum statistical mechanics. *Annual Review of Cond. Matt. Phys.*, (6):15, 2015.

[383] C. Nash and S. Sen. *Topology and Geometry for Physicists*. Academic Press, London, 1983.

[384] C. Nayak, S. H. Simon, A. Stern, M. Freedman, and S. Das Sarma. Non-Abelian anyons and topological quantum computation. *Rev. Mod. Phys.*, (80):1083, 2008.

[385] C. Nayak and F. Wilczek. $2n$-Quasihole states realize 2^{n-1}-dimensional spinor braiding statistics in paired quantum Hall states. *Nuclear Physics B*, (479):529, 1996.

[386] D.M. Nenno, C.A.C. Garcia, J. Gooth, C. Felser, and P. Narang. Axion physics in condensed-matter systems. *Nat. Rev. Phys.*, (2):682, 2020.

[387] H.B. Nielsen and M. Ninomiya. A no-go theorem for regularizing chiral fermions. *Physics Letters B*, (105):219, 1981.

[388] H.B. Nielsen and M. Ninomiya. The Adler-Bell-Jackiw anomaly and Weyl fermions in a crystal. *Physics Letters B*, (130):389, 1983.

[389] M. A. Nielsen and I. L. Chuang. *Quantum Computation and Quantum Information, 10th Anniversary Edition*. Cambridge University Press, 2010.

[390] J. Nilsson, A. R. Akhmerov, and C. W. J. Beenakker. Splitting of a cooper pair by a pair of Majorana bound states. *Phys. Rev. Lett.*, (101):120403, 2008.

[391] P. Nimbe, B.A. Weyori, and A.F. Adekoya. Models in quantum computing: A systematic review. *Quantum Inf Process*, (20):80, 2021.

[392] Q. Niu, X. Wang, L. Kleinman, W.-M. Liu, D. M. C. Nicholson, and G. M. Stocks. Adiabatic dynamics of local spin moments in itinerant magnets. *Phys. Rev. Lett.*, (83):207, 1999.

[393] K. Nomura, S. Ryu, A. Furusaki, and N. Nagaosa. Cross-correlated responses of topological superconductors and superfluids. *Phys. Rev. Lett.*, (108):026802, 2012.

[394] K. S. Novoselov, A. K. Geim, S. V. Morozov, D. Jiang, M. I. Katsnelson, I. V. Grigorieva, S. V. Dubonos, and A. A. Firsov. Two-dimensional gas of massless Dirac fermions in graphene. *Nature*, (438):197, 2005.

[395] T. Oka and S. Kitamura. Floquet engineering of quantum materials. *Annual Review of Cond. Matt. Phys.*, (10):387, 2019.

[396] Y. Oreg, G. Refael, and F. von Oppen. Helical liquids and Majorana bound states in quantum wires. *Phys. Rev. Lett.*, (105):177002, 2010.

[397] G. Ortiz, J. E. Gubernatis, E. Knill, and R. Laflamme. Quantum algorithms for fermionic simulations. *Phys. Rev. A*, (64):022319, 2001.

[398] G. Ortiz and R. M. Martin. Macroscopic polarization as a geometric quantum phase: Many-body formulation. *Phys. Rev. B*, (49):14202, 1994.

[399] J. K. Pachos. *Introduction to Topological Quantum Computation*. Cambridge University Press, 2012.

[400] A. Paetznick and K. Svore. Repeat-until-success: Non-deterministic decomposition of single-qubit unitaries. *Quantum Inf. Comput.*, (14):1277, 2014.

[401] S. Pancharatnam. Generalized theory of interference, and its applications. *Proc. Indian Acad. Sci.*, (A44):247, 1956.

[402] C. M. Papadimitriou. *Computational Complexity.* Addison-Wesley, 1994.

[403] Y. Peng, F. Pientka, L. I. Glazman, and F. von Oppen. Strong localization of Majorana end states in chains of magnetic adatoms. *Phys. Rev. Lett.,* (114):106801, 2015.

[404] A. Peres. *Quantum Theory: Concepts and Methods.* Kluwer Academic Publishers, 1995.

[405] D. S. Petrov, C. Salomon, and G. V. Shlyapnikov. Scattering properties of weakly bound dimers of fermionic atoms. *Phys. Rev. A,* (71):012708, 2005.

[406] W. D. Phillips. Nobel Lecture: Laser cooling and trapping of neutral atoms. *Rev. Mod. Phys.,* (70):721, 1998.

[407] F. Pientka, L. I. Glazman, and F. von Oppen. Topological superconducting phase in helical Shiba chains. *Phys. Rev. B,* (88):155420, 2013.

[408] L. P. Pitaevskii. Vortex lines in an imperfect Bose gas. *Soviet Physics JETP,* (13):451, 1961.

[409] M. B. Plbnio and S. Virmani. An introduction to entanglement measures. *Quantum Info. Comput.,* (7):1, 2007.

[410] H. C. Po, A. Vishwanath, and H. Watanabe. Symmetry-based indicators of band topology in the 230 space groups. *Nat. Commun.,* (8):50, 2017.

[411] H. C. Po, H. Watanabe, and A. Vishwanath. Fragile topology and wannier obstructions. *Phys. Rev. Lett.,* (121):126402, 2018.

[412] F. Pollmann, A. M. Turner, E. Berg, and M. Oshikawa. Entanglement spectrum of a topological phase in one dimension. *Phys. Rev. B,* (81):064439, 2010.

[413] A.M. Polyakov. Fermi-Bose transmutations induced by gauge fields. *Mod. Phys. Lett. A,* (3):325, 1988.

[414] A. C. Potter and Patrick A. Lee. Engineering a $p + ip$ superconductor: Comparison of topological insulator and Rashba spin-orbit-coupled materials. *Phys. Rev. B,* (83):184520, 2011.

[415] R. Prange and S. M. Girvin (editors). *The Quantum Hall effect.* Springer-Verlag, New York, 1990.

[416] J. Preskill. *Fault Tolerant Quantum Computation.* in *Introduction to Quantum Computation,* ed. H.K. Lo, S. Popescu and T.P. Spiller, World Scientific, 1998.

[417] J. Preskill. *Lecture Notes for Physics 219: Quantum Computation.* available at http://www.theory.caltech.edu/people/preskill/ph229/# lecture, 2004.

[418] Niu Q. and D. J. Thouless. Quantised adiabatic charge transport in the presence of substrate disorder and many-body interaction. *J. Phys. A: Math. Gen.,* (17):2453, 1984.

[419] X.-L. Qi, T. L. Hughes, and S.-C. Zhang. Topological field theory of time-reversal invariant insulators. *Phys. Rev. B,* (78):195424, 2008.

[420] X.-L. Qi, E. Witten, and S.-C. Zhang. Axion topological field theory of topological superconductors. *Phys. Rev. B*, (87):134519, 2013.

[421] X.-L. Qi and S.-C. Zhang. Topological insulators and superconductors. *Rev. Mod. Phys.*, (83):1057, 2011.

[422] K. De Raedt, K. Michielsen, H. De Raedt, B. Trieu, G. Arnold, M. Richter, Th. Lippert, H. Watanabe, and N. Ito. Massively parallel quantum computer simulator. *Comput. Phys. Commun.*, (176):121, 2007.

[423] Y. Ran, Y. Zhang, and A. Vishwanath. One-dimensional topologically protected modes in topological insulators with lattice dislocations. *Nature Physics*, (5):298, 2009.

[424] E. I. Rashba and Sheka. V. I. Symmetry of energy bands in crystals of wurtzite type: II. Symmetry of bands including spin-orbit interaction. *Fiz. Tverd. Tela: Collected Papers*, (2):162, 1959.

[425] N. Read. Excitation structure of the hierarchy scheme in the fractional quantum Hall effect. *Phys. Rev. Lett.*, (65):1502, 1990.

[426] N. Read and D. Green. Paired states of fermions in two dimensions with breaking of parity and time-reversal symmetries and the fractional quantum Hall effect. *Phys. Rev. B*, (61):10267, 2000.

[427] N. Read and E. Rezayi. Beyond paired quantum Hall states: Parafermions and incompressible states in the first excited Landau level. *Phys. Rev. B*, (59):8084, 1999.

[428] N. Read and S. Sachdev. Large-N expansion for frustrated quantum antiferromagnets. *Phys. Rev. Lett.*, (66):1773, 1991.

[429] M. Reck, A. Zeilinger, H. J. Bernstein, and P. Bertani. Experimental realization of any discrete unitary operator. *Phys. Rev. Lett.*, (73):58, 1994.

[430] R. Resta. Macroscopic polarization in crystalline dielectrics: The geometric phase approach. *Rev. Mod. Phys.*, (66):899, 1994.

[431] R. Resta and D. Vanderbilt. Theory of Polarization: A Modern Approach. In *Physics of Ferroelectrics: A Modern Perspective*. ed. K.M. Rabe, C. H. Ahns, and J. M. Triscone. Springer Verlag, 2007.

[432] Y. Rinott, T. Shoham, and G Kalai. Statistical aspects of the quantum supremacy demonstration. *e-print arXiv:2008.05177*, 2020.

[433] R. L. Rivest, A. Shamir, and L. M. Adleman. A method of obtaining digital signatures and public-key cryptosystems. *Comm. ACM*, (21):120, 1978.

[434] R. Roy. Topological phases and the quantum spin Hall effect in three dimensions. *Phys. Rev. B*, (79):195322, 2009.

[435] R. Roy and F. Harper. Periodic table for Floquet topological insulators. *Phys. Rev. B*, (96):155118, 2017.

[436] M. S. Rudner and N. H. Lindner. Band structure engineering and non-equilibrium dynamics in Floquet topological insulators. *Nature Reviews Physics*, (2):229, 2020.

[437] M. S. Rudner, N. H. Lindner, E. Berg, and M. Levin. Anomalous edge states and the bulk-edge correspondence for periodically driven two-dimensional systems. *Phys. Rev. X*, (3):031005, 2013.

[438] J. Ruseckas, G. Juzeliunas, P. Öhberg, and M. Fleischhauer. Non-Abelian gauge potentials for ultracold atoms with degenerate dark states. *Phys. Rev. Lett.*, (95):010404, 2005.

[439] A. I. Rusinov. Superconcductivity near a paramagnetic impurity. *JETP Lett.*, (9):85, 1969.

[440] S. Ryu. Interacting topological phases and quantum anomalies. *Physica Scripta*, (2015):014009, 2015.

[441] S. Ryu, J. E. Moore, and A. W. W. Ludwig. Electromagnetic and gravitational responses and anomalies in topological insulators and superconductors. *Phys. Rev. B*, (85):045104, 2012.

[442] S. Ryu, A. P. Schnyder, A. Furusaki, and A. W. W. Ludwig. Topological insulators and superconductors: Tenfold way and dimensional hierarchy. *New Journal of Physics*, (12):065010, 2010.

[443] J. J. Sakurai and J. J. Napolitano. *Modern Quantum Mechanics*. Pearson, 2014.

[444] J. Samuel and R. Bhandari. General setting for Berry's phase. *Phys. Rev. Lett.*, (60):2339, 1988.

[445] M. Sato, Y. Takahashi, and S. Fujimoto. Non-Abelian topological order in s-wave superfluids of ultracold fermionic atoms. *Phys. Rev. Lett.*, (103):020401, 2009.

[446] J. D. Sau, R. M. Lutchyn, S. Tewari, and S. Das Sarma. Generic new platform for topological quantum computation using semiconductor heterostructures. *Phys. Rev. Lett.*, (104):040502, 2010.

[447] J. D. Sau, B. Swingle, and S. Tewari. Proposal to probe quantum non-locality of Majorana fermions in tunneling experiments. *Phys. Rev. B*, (92):020511, 2015.

[448] L. Savary and L. Balents. Quantum spin liquids. *e-print arXiv:1601.03742*, 2016.

[449] F. Schindler, A. M. Cook, M. G. Vergniory, Z. Wang, S. S. P. Parkin, B. A. Bernevig, and T. Neupert. Higher-order topological insulators. *Science Advances*, (4):eaat0346, 2018.

[450] F. Schmidt-Kaler, H. Häffner, M. Riebe, S. Gulde, G. P. T. Lancaster, T. Deuschle, C. Becher, C. F. Roos, J. Eschner, and R. Blatt. Realization of the Cirac–Zoller controlled-NOT quantum gate. *Nature*, (422):408, 2003.

[451] Ch. Schneider, D. Porras, and T. Schaetz. Experimental quantum simulations of many-body physics with trapped ions. *Rep. Prog. Phys.*, (75):024401, 2012.

[452] P. Schneider, L. abd Beck, J. Neuhaus-Steinmetz, L. Rozsa, Th. Posske, J. Wiebe, and Wiesendanger R. Precursors of Majorana modes and their length-dependent energy oscillations probed at both ends of atomic Shiba chains. *Nat. Nanotechnol.*, (17):384, 2022.

[453] A. P. Schnyder and P. M. R. Brydon. Topological surface states in nodal superconductors. *Journal of Physics: Condensed Matter*, (27):243201, 2015.

[454] A. P. Schnyder, S. Ryu, A. Furusaki, and A. W. W. Ludwig. Classification of topological insulators and superconductors in three spatial dimensions. *Phys. Rev. B*, (78):195125, 2008.

[455] M. Schreiber, S. S. Hodgman, P. Bordia, H. P. Lüschen, M. H. Fischer, R. Vosk, E. Altman, U. Schneider, and I. Bloch. Observation of many-body localization of interacting fermions in a quasirandom optical lattice. *Science*, (349):842, 2015.

[456] N. Schuch, D. Perez-Garcia, and I. Cirac. Classifying quantum phases using matrix product states and projected entangled pair states. *Phys. Rev. B*, (84):165139, 2011.

[457] B. Schumacher. Quantum coding. *Phys. Rev. A*, (51):2738, 1995.

[458] B. Schumacher and M. Westmoreland. *Quantum Processes Systems, and Information*. Cambridge University Press, New York, 2010.

[459] B. Schutz. *Geometrical Methods of Mathematical Physics*. Cambridge University Press, Cambridge, 1980.

[460] A. Schwarz. Topological quantum field theories. *arXiv e-print hep-th/0011260*, 2000.

[461] K. I. Seetharam, C.-E. Bardyn, N. H. Lindner, M. S. Rudner, and G. Refael. Controlled population of Floquet-Bloch states via coupling to bose and Fermi baths. *Phys. Rev. X*, (5):041050, 2015.

[462] A. Sekine and K. Nomura. Chiral magnetic effect and anomalous Hall effect in antiferromagnetic insulators with spin-orbit coupling. *Phys. Rev. Lett.*, (116):096401, 2016.

[463] A. Sekine and K. Nomura. Axion electrodynamics in topological materials. *J. Appl. Phys.*, (129):141101, 2021.

[464] P. Selinger. Efficient Clifford+T approximation of single-qubit operators. *e-print arXiv:1212.6253*, 2012.

[465] T. Senthil. Symmetry-Protected Topological Phases of Quantum Matter. *Ann. Rev. Cond. Mat. Phys.*, (6):299, 2015.

[466] C. E. Shannon. A mathematical theory of communication. *Bell System Tech. J.*, (27):379–423, 623–656, 1948.

[467] L. B. Shao, S.-L. Zhu, L. Sheng, D. Y. Xing, and Z. D. Wang. Realizing and detecting the quantum Hall effect without Landau levels by using ultracold atoms. *Phys. Rev. Lett.*, (101):246810, 2008.

[468] A. Shapere and F. Wilczek. *Geometric Phases in Physics*. World Scientific, Singapore, 1989.

[469] S.-Q. Shen. *Topological Insulators: Dirac Equation in Condensed Matters*. Springer Series in Solid-State Sciences, Springer, 2013.

[470] Y. Shevy, D. S. Weiss, P. J. Ungar, and S. Chu. Bimodal speed distributions in laser-cooled atoms. *Phys. Rev. Lett.*, (62):1118, 1989.

[471] H. Shiba. Classical Spins in superconductors. *Prog. Theor. Phys.*, (40):435, 1968.

420 ■ Bibliography

[472] K. Shiozaki and S. Fujimoto. Dynamical axion in topological superconductors and superfluids. *Phys. Rev. B*, (89):054506, 2014.

[473] K. Shiozaki and M. Sato. Topology of crystalline insulators and superconductors. *Phys. Rev. B*, (90):165114, 2014.

[474] K. Shiozaki, M. Sato, and K. Gomi. Atiyah-hirzebruch spectral sequence
in band topology: General formalism and topological invariants for 230
space groups. *Phys. Rev. B*, (106):165103, 2022.

[475] P. W. Shor. Algorithms for quantum computation: Discrete logarithms
and factoring. *Proc. 35nd Annual Symposium on Foundations of Computer Science*, IEEE Computer Society Press, page 124, 1994.

[476] P. W. Shor. Scheme for reducing decoherence in quantum computer
memory. *Phys. Rev. A*, (52):R2493, 1995.

[477] P. W. Shor. Fault-tolerant quantum computation. *Proc. 37nd Annual
Symposium on Foundations of Computer Science*, IEEE Computer Society Press, page 56, 1996.

[478] P. W. Shor. Polynomial-time algorithms for prime factorization and
discrete logarithms on a quantum computer. *SIAM J. Computing*,
(26):1484, 1997.

[479] B. Simon. Holonomy, the quantum adiabatic theorem, and Berry's
phase. *Phys. Rev. Lett.*, (51):2167, 1983.

[480] M. Sitte, A. Rosch, E. Altman, and L. Fritz. Topological insulators in
magnetic fields: Quantum Hall effect and edge channels with a nonquantized θ term. *Phys. Rev. Lett.*, (108):126807, 2012.

[481] Graeme Smith, Joseph M. Renes, and John A. Smolin. Structured codes
improve the bennett-brassard-84 quantum key rate. *Phys. Rev. Lett.*,
(100):170502, 2008.

[482] A. A. Soluyanov, D. Gresch, Z. Wang, Q. Wu, M. Troyer, X. Dai, and
B. A. Bernevig. Type-II Weyl semimetals. *Nature*, (527):495, 2015.

[483] R. Somma, G. Ortiz, J. E. Gubernatis, E. Knill, and R. Laflamme.
Simulating physical phenomena by quantum networks. *Phys. Rev. A*,
(65):042323, 2002.

[484] Z. Song, Z. Fang, and C. Fang. (d–2)-Dimensional edge states of rotation
symmetry protected topological states. *Phys. Rev. Lett.*, (119):246402,
2017.

[485] Z. Song, T. Zhang, Z. Fang, and C. Fang. Quantitative mappings between symmetry and topology in solids. *Nat Commun*, (9):3530, 2018.

[486] T. D. Stanescu, B. Anderson, and V. Galitski. Spin-orbit coupled Bose-
Einstein condensates. *Phys. Rev. A*, (78):023616, 2008.

[487] T. D. Stanescu and S. Das Sarma. Proximity-induced low-energy renormalization in hybrid semiconductor-superconductor Majorana structures. *Phys. Rev. B*, (96):014510, 2017.

[488] T. D. Stanescu, V. Galitski, and S. Das Sarma. Topological states in
two-dimensional optical lattices. *Phys. Rev. A*, (82):013608, 2010.

[489] T. D. Stanescu and S. Tewari. Majorana fermions in semiconductor nanowires: Fundamentals, modeling, and experiment. *J. Phys.: Cond. Matt.*, (25):233201, 2013.

[490] H. M. Stark. *An Introduction to Number Theory.* MIT Press, Cambridge, 1978.

[491] A. M. Steane. Error correcting codes in quantum theory. *Phys. Rev. Lett.*, (77):793, 1996.

[492] J. Stenger, S. Inouye, M. R. Andrews, H.-J. Miesner, D. M. Stamper-Kurn, and W. Ketterle. Strongly enhanced inelastic collisions in a Bose-Einstein condensate near Feshbach resonances. *Phys. Rev. Lett.*, (82):2422, 1999.

[493] A. Stern. Anyons and the quantum Hall effect—A pedagogical review. *Annals of Physics*, (323):204, 2008.

[494] P. Streda. Theory of quantised Hall conductivity in two dimensions. *Journal of Physics C: Solid State Physics*, (15):L717, 1982.

[495] W. P. Su, J. R. Schrieffer, and A. J. Heeger. Solitons in polyacetylene. *Phys. Rev. Lett.*, (42):1698, 1979.

[496] R. S. Sutor. *Dancing with Qubits: How Quantum Computing Works and How it can Change the World.* Pakt Publishing, Birmingham, 2019.

[497] L. Szilard. Uber die entropieverminderung in einen thermodynamischen system bei eingriffen intelligenter wesen. *Z. Phys.*, (53):840, 1929.

[498] Y. Tanaka, Z. Ren, T. Sato, K. Nakayama, S. Souma, T. Takahashi, K. Segawa, and Y. Ando. Experimental realization of a topological crystalline insulator in SnTe. *Nat. Phys.*, (8):800, 2012.

[499] J. C. Y. Teo and C. L. Kane. Topological defects and gapless modes in insulators and superconductors. *Phys. Rev. B*, (82):115120, 2010.

[500] S. Tewari, S. Das Sarma, C. Nayak, C. Zhang, and P. Zoller. Quantum computation using vortices and Majorana zero modes of a $p_x + ip_y$ superfluid of fermionic cold atoms. *Phys. Rev. Lett.*, (98):010506, 2007.

[501] S. Tewari, C. Zhang, S. Das Sarma, C. Nayak, and D.-H. Lee. Testable signatures of quantum nonlocality in a two-dimensional Chiral p-wave superconductor. *Phys. Rev. Lett.*, (100):027001, 2008.

[502] D. J. Thouless. Quantization of particle transport. *Phys. Rev. B*, (27):6083, 1983.

[503] D. J. Thouless, M. Kohmoto, M. P. Nightingale, and M. den Nijs. Quantized Hall conductance in a two-dimensional periodic potential. *Phys. Rev. Lett.*, (49):405, 1982.

[504] E. Tiesinga, B. J. Verhaar, and H. T. C. Stoof. Threshold and resonance phenomena in ultracold ground-state collisions. *Phys. Rev. A*, (47):4114, 1993.

[505] C. G. Timpson. *Quantum Information Theory and the Foundations of Quantum Mechanics.* Oxford University Press, 2013.

[506] M. Tinkham. *Introduction to Superconductivity.* Dover, New York, 2004.

[507] A. Tran, A. Bocharov, B. Bauer, and P. Bonderson. Optimizing Clifford gate generation for measurement-only topological quantum computation with Majorana zero modes. *SciPost Phys.*, (8):091, 2020.

[508] L. Trifunovic and P. W. Brouwer. Higher-order bulk-boundary correspondence for topological crystalline phases. *Phys. Rev. X*, (9):011012, 2019.

[509] L. Trifunovic and P. W. Brouwer. Higher-order topological band structures. *Physica Status Solidi (b)*, (258):2000090, 2021.

[510] Mankei Tsang, Ranjith Nair, and Xiao-Ming Lu. Quantum theory of superresolution for two incoherent optical point sources. *Phys. Rev. X*, (6):031033, 2016.

[511] D. C. Tsui, H. L. Stormer, and A. C. Gossard. Two-dimensional magnetotransport in the extreme quantum limit. *Phys. Rev. Lett.*, (48):1559, 1982.

[512] V. G. Turaev. *Quantum Invariants of Knots and 3-Manifolds*. Walter de Gruyter, Berlin, New York, 1994.

[513] A. M. Turing. On computable numbers, with an application to the Entscheidungsproblem. *Proc. Lond. Math. Soc. 2*, (42):230, 1936.

[514] A. M. Turner and A. Vishwanath. Beyond band insulators: Topology of semi-metals and interacting phases. *e-print arXiv:1301.0330*, 2013.

[515] A. Uhlmann. Parallel transport and "quantum holonomy" along density operators. *Rep. Math. Phys.*, (24):229, 1986.

[516] W. G. Unruh. Maintaining coherence in quantum computers. *Phys. Rev. A*, (51):992, 1995.

[517] M. M. Vazifeh and M. Franz. Self-organized topological state with Majorana fermions. *Phys. Rev. Lett.*, (111):206802, 2013.

[518] F. Verstraete and J. I. Cirac. Matrix product states represent ground states faithfully. *Phys. Rev. B*, (73):094423, 2006.

[519] F. Verstraete, J. I. Cirac, J. I. Latorre, E. Rico, and M. M. Wolf. Renormalization-group transformations on quantum states. *Phys. Rev. Lett.*, (94):140601, 2005.

[520] A. Vishwanath and T. Senthil. Physics of three-dimensional bosonic topological insulators: Surface-deconfined criticality and quantized magnetoelectric effect. *Phys. Rev. X*, (3):011016, 2013.

[521] D. Vollhardt and P. Wölfle. *The Superfluid Phases of Helium 3*. Taylor and Francis, London, 1990.

[522] G. E. Volovik. Fermion zero modes on vortices in chiral superconductors. *JETP Letters*, (70):609, 1999.

[523] G. E. Volovik. *The Universe in a Helium Droplet*. Clarendon, Oxford, 2003.

[524] C. Wang and T. Senthil. Interacting fermionic topological insulators/superconductors in three dimensions. *Phys. Rev. B*, (89):195124, 2014.

[525] D.-S. Wang. A comparative study of universal quantum computing models: Toward a physical unification. *Quantum Engineering*, (3):e85, 2021.

[526] F. Wang and A. Vishwanath. Spin-liquid states on the triangular and Kagome lattices: A projective-symmetry-group analysis of Schwinger boson states. *Phys. Rev. B*, (74):174423, 2006.

[527] H. Wang, S. Ashhab, and F. Nori. Quantum algorithm for simulating the dynamics of an open quantum system. *Phys. Rev. A*, (83):062317, 2011.

[528] Z. Wang, X.-L. Qi, and S.-C. Zhang. Topological order parameters for interacting topological insulators. *Phys. Rev. Lett.*, (105):256803, 2010.

[529] N. J. Ward, I. Kassal, and A. Aspuru-Guzik. Preparation of many-body states for quantum simulation. *J. Chem. Phys.*, (130):194105, 2009.

[530] J. Watrous. *The Theory of Quantum Information*. Cambridge University Press, Cambridge, U.K., 2018.

[531] A. Wehrl. General properties of entropy. *Rev. Mod. Phys.*, (50):221, 1978.

[532] X.-G. Wen. Topological orders in rigid states. *Int. J. Mod. Phys.*, (B4):239, 1990.

[533] X.-G. Wen. Quantum orders and symmetric spin liquids. *Phys. Rev. B*, (65):165113, 2002.

[534] X.-G. Wen. *Quantum Field Theory of Many-Body Systems*. Oxford University Press, 2004.

[535] X.-G. Wen and Q. Niu. Ground-state degeneracy of the fractional quantum Hall states in the presence of a random potential and on high-genus Riemann surfaces. *Phys. Rev. B*, (41):9377, 1990.

[536] X.-G. Wen, F. Wilczek, and A. Zee. Chiral spin states and superconductivity. *Phys. Rev. B*, (39):11413, 1989.

[537] X.-G. Wen and A. Zee. On the possibility of a statistics-changing phase transition. *Journal de Physique*, (50):1623, 1989.

[538] X.-G. Wen and A. Zee. Classification of Abelian quantum Hall states and matrix formulation of topological fluids. *Phys. Rev. B*, (46):2290, 1992.

[539] Keola Wierschem and P. Sengupta. Characterizing the Haldane phase in quasi-one-dimensional spin-1 Heisenberg antiferromagnets. *Mod. Phys. Lett. B*, (28):1430017, 2014.

[540] U.-J. Wiese. Ultracold quantum gases and lattice systems: Quantum simulation of lattice gauge theories. *Annalen der Physik*, (525):777, 2013.

[541] S. Wiesner. Conjugate coding. *SIGACT News*, (15):78, 1983.

[542] F. Wilczek. Quantum mechanics of fractional-spin particles. *Phys. Rev. Lett.*, (49):957, 1982.

[543] F. Wilczek. Two applications of axion electrodynamics. *Phys. Rev. Lett.*, (58):1799, 1987.

[544] F. Wilczek. *Fractional Statistics and Anyon Superconductivity.* World Scientific, Singapore, 1990.

[545] F. Wilczek. The birth of axions. *Current Content,* (16):8, 1991.

[546] F. Wilczek. Majorana returns. *Nature Physics,* (5):614, 2009.

[547] F. Wilczek and A. Zee. Appearance of gauge structure in simple dynamical systems. *Phys. Rev. Lett.,* (52):2111, 1984.

[548] M. M. Wilde. *Quantum Information Theory.* Cambridge University Press, New York, 2013.

[549] R. Willett, J. P. Eisenstein, H. L. Störmer, D. C. Tsui, A. C. Gossard, and J. H. English. Observation of an even-denominator quantum number in the fractional quantum Hall effect. *Phys. Rev. Lett.,* (59):1776, 1987.

[550] D. Wineland and H. Dehmelt. Proposed $10^{14}\delta\nu < \nu$ laser fluorescence spectroscopy on Tl^+ mono-ion oscillator III (side band cooling). *Bull. Am. Phys. Soc.,* (20):637, 1975.

[551] D. J. Wineland and W. M. Itano. Laser cooling of atoms. *Phys. Rev. A,* (20):1521, 1979.

[552] E. Witten. Topological quantum field theory. *Commun. Math. Phys.,* (117):353, 1988.

[553] E. Witten. Quantum field theory and the Jones polynomial. *Commun. Math. Phys.,* (121):351, 1989.

[554] T. G. Wong. *Introduction to Classical and Quantum Computing.* Rooted Grove, Omaha, 2022.

[555] W. K. Wootters and W. H. Zurek. A single quantum cannot be cloned. *Nature,* (229):802, 1982.

[556] J. S. Xia, W. Pan, C. L. Vicente, E. D. Adams, N. S. Sullivan, H. L. Stormer, D. C. Tsui, L. N. Pfeiffer, K. W. Baldwin, and K. W. West. Electron correlation in the second Landau level: A competition between many nearly degenerate quantum phases. *Phys. Rev. Lett.,* (93):176809, 2004.

[557] Y. Xia, D. Qian, D. Hsieh, L. Wray, A. Pal, H. Lin, A. Bansil, D. Grauer, Y. S. Hor, R. J. Cava, and M. Z. Hasan. Observation of a large-gap topological-insulator class with a single Dirac cone on the surface. *Nature Physics,* (5):398, 2009.

[558] D. Xiao, M.-C. Chang, and Q. Niu. Berry phase effects on electronic properties. *Rev. Mod. Phys.,* (82):1959, 2010.

[559] S.-Y. Xu, I. Belopolski, N. Alidoust, M. Neupane, et al. Discovery of a Weyl fermion semimetal and topological Fermi arcs. *Science,* (349):613, 2015.

[560] B. Yan and C. Felser. Topological materials: Weyl semimetals. *Annual Review of Cond. Matt. Phys.,* (8):337, 2017.

[561] S. A. Yang, H. Pan, and F. Zhang. Dirac and Weyl Superconductors in three dimensions. *Phys. Rev. Lett.,* (113):046401, 2014.

[562] A. C. Yao. Quantum circuit complexity. *Proc. of the 34th Ann. IEEE Symp. on Foundations of Computer Science,* page 352, 1993.

[563] D Yoshioka. *The Quantum Hall Effect.* Springer, 2002.

[564] L. Yu. Bound state in superconductors with paramagnetic impurities. *Acta Phys. Sin.*, (21):75, 1965.

[565] J. Zak. Berry's phase for energy bands in solids. *Phys. Rev. Lett.*, (62):2747, 1989.

[566] B. Zeng, X. Chen, D.-L. Zhou, and X.-G. Wen. Quantum information meets quantum matter – From quantum entanglement to topological phase in many-body systems. *e-print arXiv:1508.02595*, 2015.

[567] H. Zhai. Degenerate quantum gases with spin–orbit coupling: A review. *Rep. Prog. Phys.*, (78):026001, 2015.

[568] C. Zhang, S. Tewari, R. M. Lutchyn, and S. Das Sarma. $p_x + ip_y$ Superfluid from s-wave interactions of fermionic cold atoms. *Phys. Rev. Lett.*, (101):160401, 2008.

[569] R.-X. Zhang, Y.-T. Hsu, and S. Das Sarma. Higher-order topological Dirac superconductors. *Phys. Rev. B*, (102):094503, 2020.

[570] Y. Zhang, Y.-W. Tan, H. L. Stormer, and P. Kim. Experimental observation of the quantum Hall effect and Berry's phase in graphene. *Nature*, (438):201, 2005.

[571] L. Zhou and D.-J. Zhang. Non-Hermitian Floquet topological matter – A review. *Entropy*, (25), 2023.

[572] S.-L. Zhu, H. Fu, C.-J. Wu, S.-C. Zhang, and L.-M. Duan. Spin Hall effects for cold atoms in a light-induced gauge potential. *Phys. Rev. Lett.*, (97):240401, 2006.

[573] M. R. Zirnbauer. Riemannian symmetric superspaces and their origin in random-matrix theory. *J. Math. Phys.*, (37):4986, 1996.

[574] E. Zohar, J. I. Cirac, and B. Reznik. Quantum simulations of lattice gauge theories using ultracold atoms in optical lattices. *Rep. Prog. Phys.*, (79):014401, 2016.

[575] A. A. Zyuzin and A. A. Burkov. Topological response in Weyl semimetals and the chiral anomaly. *Phys. Rev. B*, (86):115133, 2012.

Index

Printed in the United States
by Baker & Taylor Publisher Services